Mastering ArcGIS

Fourth Edition

Maribeth Price
South Dakota School of Mines and Technology

 Higher Education

Boston   Burr Ridge, IL   Dubuque, IA   New York   San Francisco   St. Louis
Bangkok   Bogotá   Caracas   Kuala Lumpur   Lisbon   London   Madrid   Mexico City
Milan   Montreal   New Delhi   Santiago   Seoul   Singapore   Sydney   Taipei   Toronto

McGraw Hill **Higher Education**

MASTERING ARCGIS, FOURTH EDITION

Published by McGraw-Hill, a business unit of The McGraw-Hill Companies, Inc., 1221 Avenue of the Americas, New York, NY 10020. Copyright © 2010 by The McGraw-Hill Companies, Inc. All rights reserved. Previous editions © 2008, 2006, and 2004. No part of this publication may be reproduced or distributed in any form or by any means, or stored in a database or retrieval system, without the prior written consent of The McGraw-Hill Companies, Inc., including, but not limited to, in any network or other electronic storage or transmission, or broadcast for distance learning.

Some ancillaries, including electronic and print components, may not be available to customers outside the United States.

♲ This book is printed on recycled, acid-free paper containing 10% postconsumer waste.

1 2 3 4 5 6 7 8 9 0 QPD/QPD 0 9

ISBN 978—0—07—352284—5
MHID 0—07—352284—8

Publisher: *Thomas D. Timp*
Executive Editor: *Margaret J. Kemp*
Director of Development: *Kristine Tibbetts*
Senior Marketing Manager: *Lisa Nicks*
Project Manager: *Melissa M. Leick*
Senior Production Supervisor: *Kara Kudronowicz*
Lead Media Project Manager: *Judi David*
Associate Design Coordinator: *Brenda A. Rolwes*
Cover Designer: *Studio Montage, St. Louis, Missouri*
(USE) Cover Image: © *Getty Images*
Typeface: *10/12 Times Roman*
Printer: *Quebecor World Dubuque, IA*

Library of Congress Cataloging-in-Publication Data

Price, Maribeth Hughett, 1963-
  Mastering ArcGIS / Maribeth Price. -- 4th ed.
    p. cm.
  Includes index.
  ISBN 978—0—07—352284—5 — ISBN 0—07—352284—8 (hard copy : alk. paper) 1. ArcGIS. 2.
Geographic information systems. I. Title.
  G70.212.P74 2010
  910.285--dc22
                        2008050704

www.mhhe.com

Table of Contents

PREFACE..................................IX

INTRODUCTION 1

What Is GIS? 1
  Concepts....................................1
    What is GIS?..........................1
    A history of GIS 3
    What can a GIS do? 4
    New trends and directions in GIS 4
    What do GIS professionals do? 6
    GIS project management 6
    Project case study: A thorny issue 8
    Types of GIS projects 11
    Planning a GIS project...........11

Example of a GIS Project Proposal 14

PART I. GIS DATA AND MAPS

CHAPTER 1. GIS DATA 17

Mastering the Concepts 17
  GIS Concepts 17
    Representing real-world objects 17
    Coordinate systems.................21
    Modeling feature behavior with topology 21
    Data quality..........................22
    Citing GIS data sources 24
  About ArcGIS 25
    ArcGIS overview....................25
    Data files in ArcGIS 26
    About ArcToolbox.................31
    The Help System 31
  Summary.................................33
  Chapter Review Questions...........34

Mastering the Skills......................35
  Teaching Tutorial........................35
  Exercises 51
  Skills Reference 52

CHAPTER 2. MAPPING GIS DATA 63

Mastering the Concepts 63
  GIS Concepts 63
    Map scale concepts................63
    Types of maps........................64
    Classifying numeric data 67
    Displaying rasters 70
  About ArcGIS 72

    Map documents........................72
    Data frame coordinate systems 73
  Summary 74
  Chapter Review Questions 75

Mastering the Skills......................77
  Teaching Tutorial 77
  Exercises 93
  Skills Reference 95

CHAPTER 3. PRESENTING GIS DATA 107

Mastering the Concepts...................107
  GIS Concepts 107
    Basic elements of map design...............107
    Choosing symbols.................110
    Choosing coordinate systems.................112
  About ArcGIS 115
    Maps and reports in ArcGIS 115
    Assigning map scales.............115
    Setting up scale bars 116
    Labeling and annotation..........117
  Summary 119
  Chapter Review Questions 120

Mastering the Skills......................121
  Teaching Tutorial 121
  Exercises 135
  Skills Reference 136

CHAPTER 4. ATTRIBUTE DATA 153

Mastering the Concepts...................153
  GIS Concepts 153
    Overview of tables.................153
    Database Management Systems...........154
    Queries on tables...................154
    Joining and relating tables 155
    Summarizing tables................157
    Field types...........................158
  About ArcGIS 160
    Tables in ArcGIS 160
    Table formats 162
    Editing and calculating fields................162
  Summary 163
  Chapter Review Questions 164

Mastering the Skills......................165
  Teaching Tutorial 165
  Exercises 177
  Skills Reference 179

PART II. GIS ANALYSIS

CHAPTER 5. QUERIES 191

Mastering the Concepts 191
GIS Concepts 191
About queries..191
Attribute queries 192
Spatial queries 194
About ArcGIS 196
Selection states 196
Interactive selection 198
Selecting by attributes.............................198
Selecting by location................................199
Choosing the selection method 199
Creating layers from selected features ...200
Summary 201
Chapter Review Questions......................202

Mastering the Skills..............................203
Teaching Tutorial....................................203
Exercises 213
Skills Reference 214

CHAPTER 6. SPATIAL JOINS... 221

Mastering the Concepts 221
GIS Concepts 221
What is a spatial join?..............................221
Cardinality 223
Types of spatial joins 224
Feature geometry and spatial joins 225
Coordinate systems and distance joins ..227
About ArcGIS 228
Choosing the join type 228
Setting up a spatial join...........................229
Summary 235
Chapter Review Questions......................236

Mastering the Skills..............................237
Teaching Tutorial....................................237
Exercises 248
Skills Reference 249

CHAPTER 7. GEOPROCESSING 251

Mastering the Concepts 251
GIS Concepts 251
Map overlay 251
Other spatial analysis functions 255
About ArcGIS 258
About geoprocessing 258
Using ArcToolbox 260
Summary 262

Chapter Review Questions 263

Mastering the Skills..............................264
Teaching Tutorial 264
Exercises 279
Skills Reference 280

CHAPTER 8. RASTER ANALYSIS..285

Mastering the Concepts...............................285
GIS Concepts 285
Raster data 285
Raster analysis 286
About ArcGIS 293
Storing rasters 293
Using Spatial Analyst 294
Summary 299
Chapter Review Questions 300

Mastering the Skills..............................301
Teaching Tutorial 301
Exercises 315
Skills Reference 317

CHAPTER 9. NETWORK
ANALYSIS 327

Mastering the Concepts...............................327
GIS Concepts 327
About networks...327
Types of networks....................................328
Network analysis......................................329
Generic trace solvers...............................330
Utility trace solvers 332
About ArcGIS 333
The Utility Network Analyst toolbar 333
Building networks 335
Summary 336
Chapter Review Questions 337

Mastering the Skills..............................338
Teaching Tutorial 338
Exercises 348
Skills Reference 349

CHAPTER 10. GEOCODING......359

Mastering the Concepts...............................359
GIS Concepts 359
What is geocoding?...................................359
How does geocoding work?.....................360
Available geocoding styles 363
The reference data....................................365
Adding *x-y* coordinates 365
About ArcGIS 366

The geocoding process 366
Setting up an address locator 367
Summary 369
Chapter Review Questions 370

Mastering the Skills **371**
Teaching Tutorial 371
Exercises 382
Skills Reference 382

PART III. DATA MANAGEMENT

CHAPTER 11. COORDINATE SYSTEMS 387

Mastering the Concepts **387**
GIS Concepts 387
About coordinate systems 387
Geographic coordinate systems 388
Spheroids and datums 389
Map projections 390
Raster coordinate systems 392
Common projection systems 395
Choosing projections 397
About ArcGIS 397
Labeling coordinate systems 397
On-the-fly reprojection 399
Projecting data 400
Troubleshooting coordinate systems 400
Summary 403
Chapter Review Questions 404

Mastering the Skills **405**
Teaching Tutorial 405
Exercises 420
Skills Reference 421

CHAPTER 12. BASIC EDITING 425

Mastering the Concepts **425**
GIS Concepts 425
Snapping features 425
Creating adjacent polygons 427
About ArcGIS 427
The Editor toolbar 427
About editing 428
How editing works 429
Summary 431
Chapter Review Questions 432

Mastering the Skills **433**
Teaching Tutorial 433
Exercises 447
Skills Reference 448

CHAPTER 13. MORE EDITING TECHNIQUES 459

Mastering the Concepts **459**
GIS Concepts 459
Topology 459
Combining features 461
Buffering features 462
Complex polygon topology 463
Stream digitizing 463
About ArcGIS 464
Using different sketching tools 464
Changing existing features 465
Topological editing 466
Editing annotation 467
Summary 467
Chapter Review Questions 468

Mastering the Skills **469**
Teaching Tutorial 469
Exercises 484
Skills Reference 485

CHAPTER 14. GEODATABASES 501

Mastering the Concepts **501**
About ArcGIS 501
About geodatabases 501
Creating geodatabases 503
Creating feature datasets 503
Using default values 504
Setting up domains 504
Split and merge policies 506
About subtypes 508
Summary 509
Chapter Review Questions 510

Mastering the Skills **511**
Teaching Tutorial 511
Exercises 526
Skills Reference 527

CHAPTER 15. METADATA 533

Mastering the Concepts **533**
GIS Concepts 533
The metadata standards 533
Data quality issues 534
The metadata format 538
Acquiring metadata 539
Metadata development 540
About ArcGIS 541
The Metadata Editor 541
Metadata templates 542
Summary 544

Chapter Review Questions.........................545

Mastering the Skills......................................**546**
Teaching Tutorial......................................546
Exercises 553
Skills Reference 554

GLOSSARY................................ 559

SELECTED ANSWERS 573

INDEX.. 585

ALPHABETICAL SKILLS INDEX...601

CONVERSION TABLE............... 610

Skills Reference List

Chapter 1 GIS Data

Starting ArcMap or ArcCatalog 52
Starting ArcToolbox..................................... 52
Connecting and disconnecting from folders .. 53
Connecting to an Internet service 53
Setting options 53
Viewing the contents of a folder 54
Viewing and setting layer properties 54
Examining the coordinate system 54
Previewing layer geography 55
Creating and viewing thumbnails 55
Identifying features 55
Previewing a table 56
Viewing metadata 57
Setting symbols for a layer 57
Creating new symbols 58
Labeling features using dynamic labels 59
Managing ArcCatalog files 60
Searching for data....................................... 60

Chapter 2 Mapping GIS Data

Starting ArcMap... 95
Managing toolbars 96
Adding data 96
Dragging and dropping files to ArcMap 96
Saving an ArcMap document 97
Opening an ArcMap document 97
Controlling display of layers 97
Setting Map Tips 98
Grouping layers for display 98
Using the Zoom/Pan tools and bookmarks..... 99
Measuring features 100
Finding features by searching attributes 100
Setting data frame and layer properties 101
Clipping to a layer 101
Examining the coordinate system................. 101
Setting the data frame coordinate system 102
Creating a map based on an attribute............ 103
Classifying attributes for a map 104
Modifying the appearance of the legend 105
Creating a map layer or map group layer 106

Chapter 3 Presenting GIS Data

Composing the data frames 136
Setting up the map page 137
Using the Layout toolbar.............................. 138
Setting the scale or extent............................ 139
Labeling features interactively 139
Setting the reference scale 141
Converting dynamic labels to annotation 142
Adding a north arrow.................................. 143
Adding titles and text.................................. 143
Adding graphics to layouts 143
Adding a legend.. 144
Changing the appearance of a legend 144
Adding a scale bar 147
Adding neatlines, backgrounds, and shadows 148
Adding pictures 148
Creating a map from a template.................... 149
Printing a map.. 150
Exporting a map as a picture file 150
Creating a simple graph 151

Chapter 4 Attribute Data

Adding or removing a table 179
Opening a table 179
Modifying a table's appearance.................... 180
Selecting records using attributes 182
Getting statistics for a field 183
Summarizing on a field 184
Exporting a table 184
Adding or deleting fields 185
Calculating fields 187
Editing fields in a table............................... 187
Creating joins and relates 188
Creating a new table 189
Deleting a table.. 189
Opening Excel Data..................................... 190

Chapter 5 Queries

Setting the selectable layers........................ 214
Selecting features interactively..................... 215
Selecting features using shapes 215
Clearing a selection 216
Using the Selection tab................................ 216
Selecting by attributes 217
Selecting by location 218
Creating a layer from selected features......... 219
Changing the selection options.................... 219

Chapter 6 Spatial Joins

Performing a spatial join.............................. 249

Chapter 7 Geoprocessing

Performing an intersection........................... 280
Creating buffers 281
Clipping a layer 282
Appending layers.. 282
Dissolving.. 283
Using tools in ArcToolbox 284

Chapter 8 Raster Analysis

Turning on Spatial Analyst........................... 317
Managing grids... 317
Setting the analysis environment.................. 318
Converting between grids and features......... 319
Reclassifying a grid..................................... 320
Calculating cell statistics 321
Calculating neighborhood statistics.............. 321
Calculating zonal statistics 322
Using surface functions 323
Calculating a viewshed................................ 324
Creating a density map 324
Interpolating between points 325
Finding a least-cost path.............................. 326

Chapter 9 Network Analysis

Examining network properties...................... 349
Performing a trace 350
Setting general tracing options 350
Changing the results format......................... 351
Using weights when tracing 351
Viewing flow directions 352
Adding and clearing flags and barriers......... 353
Finding a path... 354
Finding the shortest path 354
Finding connections and loops 355
Tracing upstream/downstream 355
Tracing with accumulation 355
Building a simple network............................ 356

Chapter 10 Geocoding

Creating an address locator 383
Geocoding addresses 384
Rematching a geocoding result..................... 384
Interactive matching.................................... 385
Adding x-y data 386

Chapter 11 Coordinate Systems

Defining coordinate systems of data sets 421
Projecting feature classes 422
Projecting rasters 423

Chapter 12 Basic Editing

Beginning an editing session 448
Setting snapping options.............................. 449
Selecting features for editing........................ 449
Moving features... 450
Rotating features... 450
Deleting features 451
Creating new features 451
Using sketch context menus 452
Creating multipart features........................... 455

Creating features with square corners........... 455
Creating adjacent polygons.......................... 456
Using the Attributes window 457
Saving changes during an edit session.......... 458
Stopping an edit session 458
Switching to another folder or geodatabase.. 458

Chapter 13 More Editing

Using the sketching tools............................. 485
Modifying a feature 489
Reshaping a feature 490
Flipping a line... 490
Mirroring features....................................... 491
Dividing a line 491
Splitting lines exactly 492
Moving a feature an exact distance............... 492
Using Copy Parallel..................................... 492
Trimming lines 493
Extending lines 493
Merging features 493
Union of features 494
Intersection of features 494
Clipping features 494
Creating donut polygons.............................. 495
Buffering features 496
Editing shared features 496
Editing annotation....................................... 498

Chapter 14 Geodatabases

Creating a geodatabase 527
Deleting a geodatabase 527
Deleting feature classes 527
Deleting feature datasets.............................. 527
Creating a feature dataset............................. 528
Creating a feature class 529
Importing feature classes 530
Creating attribute domains........................... 531
Setting default values or domains................. 532

Chapter 15 Metadata

Editing metadata... 554
Creating a metadata template....................... 555
Importing or exporting metadata 555
A metadata reference map 556

Preface

Welcome to *Mastering ArcGIS*, a detailed primer on learning the ArcGIS™ software by ESRI®, Inc. This book is designed to offer everything you need to master the basic elements of GIS.

> **Notice:** ArcGIS™, ArcView™, ArcMap™, ArcCatalog™, ArcInfo™, ArcInfo Workstation™, ArcEditor™, and the other program names used in this text are registered trademarks of ESRI, Inc. The software names and the screen shots used in the text are reproduced by permission. For ease of reading only, the ™ symbol has been omitted from the names; however, no infringement or denial of the rights of ESRI® is thereby intended or condoned by the author.

What's new in the fourth edition?

Some major changes have been implemented for this edition of the book. First, the material has been updated to Version 9.3 of ArcGIS. Either version 9.2 or version 9.3 should work with this edition, although minor adjustments to tutorials will be needed for version 9.2 users.

Second, the order of the chapters has changed substantially. Three major sections now make up the book. The first section covers the basics of GIS data, mapping, and tables. The second section focuses on GIS analysis. The third section focuses on data input and management. I have moved the complete discussion of coordinate systems to Chapter 11 although earlier chapters introduce some fundamental ideas. The goal was to repeat key fundamentals early and save the full discussion until the student has more experience and is working on data management skills. A new chapter on metadata will please many instructors, I hope.

The chapter formats have also changed. Each Concepts section has been subdivided into a section which discusses fundamental GIS concepts and a section which focuses on the ArcGIS software. I hope I have succeeded in placing greater emphasis on background knowledge and in clarifying what is general and what is specific to the ESRI products. As part of this process, I have increased coverage of the general concepts in many chapters, including greater emphasis on data quality issues and eliminating some less-frequently-used aspects of the software.

I would like to thank the many people who have used and commented on this book, and I hope that it continues to serve their needs in the rapidly evolving world of GIS.

Previous experience

This book assumes that the reader is comfortable using Windows™ to carry out basic tasks such as copying files, moving directories, opening documents, exploring folders, and editing text and word processing documents. Previous experience with maps and map data is also helpful. No previous GIS experience or training is necessary to use this book.

Elements of the package

This learning system includes a textbook with a CD-ROM, including:

> ➤ Fifteen chapters on the most important capabilities of ArcGIS.
> ➤ Comprehensive tutorials in every chapter to learn the skills, with each step demonstrated in a video clip.
> ➤ A set of exercises, map documents, and data for practicing skills independently.
> ➤ Reference sections on skills with video clips demonstrating each one.

Philosophy

This text reflects the author's personal philosophies and prejudices developed from 14 years of teaching GIS at an engineering school. The main goal is not to train geographers, but to provide students in any field with GIS skills and knowledge. It is assumed that most students using this book already have a background of discipline-specific knowledge and skills upon which to draw and are seeking to apply geospatial techniques within their own knowledge domains.

> ➤ **GIS is best learned by doing it, not by studying it.** The laboratory is THE critical component of the course, and theory is introduced sparingly and integrated with experience. Hence this book is heavy on experience and light on theory.

> ➤ **Independent work and projects are critical to learning GIS.** This book includes a wealth of exercises in which the student must find solutions independently without a cookbook recipe of steps. A wise instructor will also require students to develop an independent project.

> ➤ **GIS analysis should be taught before GIS data development.** Analysis is more interesting and provides the student with a greater incentive to tackle the challenging tasks of creating data. Also, data development guidelines and rules make more sense to someone who has used the data first.

This book assumes that the student has access to ArcView 9.2 or ArcGIS Desktop 9.2 or higher. Some GIS functions require the more expensive ArcEditor or ArcInfo licenses, but they are not covered in this text. The Spatial Analyst extension is required to do the activities in Chapter 8.

Chapter sequence

The book contains an introduction and 15 chapters. Each chapter includes roughly one week's work for a three-credit semester course, although Chapter 8 is best covered over two weeks. This book intentionally contains more material than the average GIS class can cover during a single semester, and instructors may choose which parts to cover in a course.

An introductory chapter describes GIS and gives some examples of how it is used. It also provides an overview of GIS project management and how to develop a project and a proposal. The remaining chapters can be divided into three sections. Sections 2 and 3 could be done in either order depending on instructor preferences.

> **I. GIS Data and Maps:** Chapters 1–4 introduce basic skills for working with GIS data, creating maps and layouts, and working with tables.

II. GIS Analysis: Chapters 5–10 introduce different analysis techniques, including queries, spatial joins, overlay, raster analysis, geocoding, and network analysis

III. Data Entry and Management: Chapters 11–15 cover coordinate systems, editing, geodatabases, and metadata.

Chapter layout

Each chapter is organized into the following sections:

> **Concepts:** provides basic background material for understanding geographic concepts and how they are specifically implemented within ArcGIS. A set of review questions follows the concepts section.

> **Tutorial:** contains a step-by-step tutorial demonstrating the concepts and skills. The tutorials begin with detailed instructions, which gradually become more general as mastery is built. Every step in the tutorial is demonstrated by accompanying video clips.

> **Exercises:** presents a series of problems to build skill in identifying the appropriate techniques and applying them without step-by-step help. Through these exercises, the student builds an independent mastery of GIS processes. Answers or solution methods are included for selected exercises. A full answer and methods document is available for instructors at the McGraw-Hill Instructor Web site.

> **Skills Reference:** provides step-by-step instructions for carrying out the most frequent and important tasks being learned. This material is also demonstrated with video clips.

The CD-ROM contains all the data needed to follow the tutorials and complete the exercises.

Instructors should use judgment in assigning exercises, as the typical class would be stretched to complete all the exercises in every chapter. Very good students can complete the entire chapter in 3–5 hours, most students would need 6–8 hours, and a few students would require 10 or more hours. Students with extensive computer experience generally find the material easier than others.

Using this text

In working through this book, the following sequence of steps is suggested:

> READ through the Concepts sections to get familiar with the principles and techniques.
> SKIM the Skills Reference section to see what techniques will be learned. You may wish to view the videos at this time or simply use them later for reference as needed.
> ANSWER the Chapter Review Questions to test your comprehension of the material.
> WORK the Tutorial section for a step-by-step tutorial and explanation of key techniques.
> REREAD the Concepts section to reinforce the ideas.
> PRACTICE by doing the Exercises.

Using the tutorial

The tutorial provides step-by-step practice and introduces details on how to perform specific steps. Students should be encouraged to think about the steps as they are performed, and not just race to get to the end.

It is important to follow the directions carefully. Skipping a step or doing it incorrectly may result in a later step not working properly. Saving often will make it easier to go back and correct a mistake in order to continue. Occasionally a step will not work due to differences between computer systems or software versions. Having an experienced user nearby to identify the problem can help. If one isn't available, however, just skip the step and move on without it.

Using the videos

The CD distributed with the book contains two types of videos. The Tutorial Videos demonstrate each step of the tutorial. They are numbered in the text for easy reference. The Skills Reference Videos show how to perform generic tasks, such as deleting a file. The videos are intended as an alternate learning strategy. It would be tedious to watch all of them. Instead, use them in the following situations:

> ➤ When the student does not understand the written instructions or cannot find the correct menu or button.
> ➤ When a step cannot be made to work properly.
> ➤ When a reminder is needed to do a previously learned skill in order to complete a step.
> ➤ When a student finds that watching the videos enhances learning.

The CD contains an index file with hyperlinks to each clip. As you work through a chapter, keep the index on the screen, and click the appropriate link to see a video. The tutorial clips are distinguished by numbered steps and the Skills Reference Videos are distinguished by their headings.

Installing the CD-ROM components

The CD-ROM includes HTML index documents, a folder of video clips, and a self-installing archive of the tutorial data. See the README file on the CD for any last-minute changes to the installation instructions below. A splash screen with a use agreement and hyperlinks to the data and tutorials should appear when the CD is inserted in the drive. It can be started manually from the CD, if necessary, by double-clicking the Start_Here.exe file.

To view the videos:

The tutorial video links appear in blue in a table, and the Skills Reference videos appear below. Click on a blue video hyperlink to start a video. For best resolution, size the video window as large as possible. See the detailed instructions at the start of the Chapter 1 tutorial.

To install the training data:

The mgisdata.exe archive contains a folder with the documents and data needed to do the tutorials and exercises. The student must copy this folder to his own hard drive. If more than one person on the computer is using this book, then each person should make her own copy of the data in a separate folder. The data on the CD is a self-extracting zip file that requires approximately 100 MB of disk space for installation. Follow these instructions to install the data:

➢ Place the book CD in your computer's CD-ROM drive and wait for the CD splash screen to appear. Indicate your acceptance of the license agreement.

➢ Click on the link to Install Data. If a dialog window appears asking whether to Run or Save the data, choose Run.

➢ When a dialog box appears asking whether to Open or Save the data, choose Open. Don't choose Save because it will only copy the data archive instead of extracting it.

➢ Click the button with ellipses to set the folder to which to extract the data [see (a) in the figure]. The data will be placed in a folder called mgisdata in whatever location you choose. In other words, if you select C:\temp as the target folder, then the data will be placed in C:\temp\mgisdata.

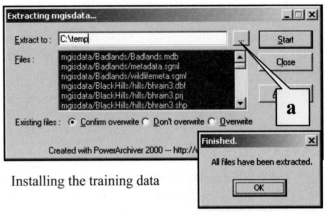

Installing the training data

➢ Click Start to begin installing the data. It may take several minutes. Wait until you see the Finished window announcing that all the files have been extracted and then click OK.

System requirements

To use the tutorials and do the exercises in this book, the student must have access to a computer with the following characteristics:

Minimum hardware:

    PC-Intel™-platform computer with 1.6 GHz processor or better and 1 GB RAM

    Suitable sound/graphics card (nearly all systems have one)

Software:

    Windows Vista™ (Ultimate, Enterprise, Business, Home Premium), Windows 2000™, or Windows XP™ (Home Edition and Professional)

    A Web browser, such as Netscape or Internet Explorer, or Microsoft® Word

    Microsoft® MediaPlayer™ 10.0 or higher (can be downloaded free from Microsoft®)

    ArcView™ 9.2 or higher, or ArcGIS™ Desktop 9.2 or higher

For assistance in acquiring or installing these components, contact your system administrator, hardware/software provider, or local computer store.

Acknowledgments

I would like to thank many people who made this book possible. Governor Janklow of South Dakota funded a three-month summer project in 2000 that got the book started, as part of his Teaching with Technology program. Many students in my GIS classes between 2000 and 2008 tested the text and exercises and helped immensely in making sure the tutorials were clear and worked correctly. Reviewers of previous editions, including Richard Aspinall, Joe Grengs, Tom Carlson, Susan K. Langley, Henrietta Loustsen, Xun Shi, Richard Lisichenko, John Harmon, Michael Emch, Jim Sloan, Sharolyn Anderson, Talbot Brooks, Qihao Weng, Jeanne Halls, Mark Leipnik, Michael Harrison, Ralph Hitz, Olga Medvedkov, James W. Merchant, Raymond L. Sanders, Jr., Yifei Sun, Fahui Wang, John Harmon, Michael Haas, Jason Kennedy, Dafna Kohn, Henrietta Laustsen, Jessica Moy, James C. Pivirotto, Peter Price, and Judy Sneller, provided detailed and helpful comments, and the book is better than it would have been without their efforts. I also thank the reviewers who provided valuable advice for the fourth edition.

Reviewers for the Fourth Edition:

Dave Verbyla, *University of Alaska, Fairbanks*
Birgit Mühlenhaus, *Macalester College*
Jason Duke, *Tennessee Technological University*
Darla Munroe, *Ohio State University*
Wei-Ning Xiang, *University of North Carolina at Charlotte*
L. Joe Morgan, *University of North Carolina at Greensboro*

Marketing Review

Sarah Harris, *Denison University*

ESRI, Inc. was prompt and generous in its granting of permission to use the screenshots, data, and other materials throughout the text. They also provided a prerelease version of ArcGIS 9.3 so that I could include those changes in time for the fourth edition. I thank George Sielstad, Eddie Childers, Mark Rumble, Tom Junti, and Patsy Horton for their generous donations of data. I am grateful to Dale Nickels and Steve Bauer for their long-term computer lab administration without which I could not have taught GIS courses or developed this book. I thank Linda Heindel for organizing student feedback and assisting with the initial round of edits on the first draft. I thank editors Tom Lyon and Lisa Bruflodt of McGraw-Hill for their unfailing encouragement and enthusiasm about the book as it took shape, as well as for their excellent feedback. I thank the McGraw-Hill team working on the fourth edition, especially Margaret Kemp and Robin Reed. I continue to be grateful, as well, to the third edition staff, Debra Henricks and Joyce Watters. I am grateful to Daryl Pope, who first started me in GIS, and to John Suppe, who encouraged me to return to graduate school and continue doing GIS on a fascinating study of Venus. I thank my partner David Stolarz, who provided unfailing encouragement when it seemed as though the editing on this edition would never end. Last, and certainly not least, I thank Curtis Price and my daughters Ginny and Maddi for their understanding and support during the many, many hours I spent creating this book.

Introduction

What Is GIS?

Objectives

> ➤ Developing a basic understanding of what GIS is, its operations, and its uses

> ➤ Getting familiar with GIS project management

> ➤ Learning to plan a GIS project and write a project proposal

Concepts

What is GIS?

GIS stands for Geographic Information System. In practical terms, a GIS is a set of computer tools that allows people to work with data that are tied to a particular location on the earth. Although many people think of a GIS as a computer mapping system, its functions are broader and more sophisticated than that. A GIS is a database that is designed to work with map data.

For example, consider the accounting department of the local telephone company. They maintain a large computer database of their customers, in which they store the name, address, phone number, type of service, and billing information for each customer. This information is only incidentally tied to where customers live; they can carry out most of the important functions (billing, for example) without needing to know where each house is. Of course, they need to have addresses for mailing bills, but it is the post office that worries about where the houses actually ARE. This type of information is called **aspatial data**, meaning that it is not tied, or is only incidentally tied, to a location on the earth's surface.

In the service department, however, they need to work with **spatial data** to provide the telephone services. When hundreds of people call in after a power outage, the service department must analyze the distribution of the calls and isolate the location where the outage occurred. When a construction company starts work on a street, workers must be informed of the precise location of buried telephone cables. If a developer builds a new neighborhood, the company must be able to determine the best place to tie into the existing network so that the services are efficiently distributed from the main trunk lines. When technicians prepare lists of house calls for the day, they need to plan the order of visits to minimize the amount of driving time. In these tasks, location is a critical aspect of the job, and the information is spatial.

In this example, two types of software are used. The accounting department uses special software called a database management system, or DBMS, which is optimized to work with large volumes of aspatial data. The service department needs access to a database that is optimized for working with spatial data, a Geographic Information System. Because these two types of software are related, they often work together, and they may access the same information. However, they do different things with the data.

A GIS is built from a collection of hardware and software components.

> **A computer hardware platform.** Due to the intensive nature of spatial data storage and processing, a GIS was once limited primarily to large mainframe computers or expensive workstations. Today it can run on a typical desktop personal computer.

> **GIS software.** The software varies widely in cost, ease of use, and level of functionality but should offer at least some minimal set of functions as described in the next few paragraphs. In this book, we study one particular package that is very powerful and widely used, but others are available and may be just as suitable for certain applications.

> **Data storage.** Some projects use only the hard drive of the GIS computer. Other projects may require more sophisticated solutions if very large volumes of data are being stored or multiple users need access to the same data sets. Today many data sets are stored in digital warehouses and accessed by many users over the Internet. Backup capability is an issue when expensive and irreplaceable data are being used. Compact disc writers and/or tape systems are highly useful for backing up and sharing data.

> **Data input hardware.** Many GIS projects require sophisticated data entry tools. Digitizer tablets enable the shapes on a paper map to be entered as features in a GIS data file. Scanners create digital images of paper maps. An Internet connection provides easy access to large volumes of GIS data. High-speed connections are preferred, as GIS data sets may constitute tens or hundreds of megabytes or more.

> **Information output hardware.** A quality color printer capable of letter-size prints provides the minimum desirable output capability for a GIS system. Printers that can handle map-size output (36 in. × 48 in.) will be required for many projects.

> **GIS data.** Data comes from a variety of sources and in a plethora of formats. Gathering data, assessing its accuracy, and maintaining it usually constitute the longest and most expensive part of a GIS project.

> **GIS personnel**. A system of computer and hardware is useless without trained and knowledgeable people to run it. The contribution of professional training to successful implementation of a GIS is often overlooked.

GIS software varies widely in functionality, but any system claiming to be a GIS should provide the following functions at a minimum:

> Data entry from a variety of sources, including digitizing, scanning, text files, and the most common spatial data formats. Ways to export information to other programs should also be provided.

> Data management tools, including tools for building data sets, editing spatial features and their attributes, and managing coordinate systems and projections.

> Thematic mapping (displaying data in map form), including symbolizing map features in different ways and combining map layers for display.

> Data analysis functions for exploring spatial relationships in and between map layers.

> ➤ Map layout functions for creating soft and hard copy maps with titles, scale bars, north arrows, and other map elements.

Geographic Information Systems are put to many uses, but providing the means to collect, manage, and analyze data to produce information for better decision making is the common goal and the strength of each GIS. This book is a practical guide to understanding and using a particular Geographic Information System called ArcGIS. Using this book, you can learn what a GIS is, what it does, and how to apply its capabilities to solve real-world problems.

A history of GIS

Geographic Information Systems have grown from a long history of cartography begun in the lost mists of time by early tribesmen who made sketches on hides or formed crude models of clay as aids to hunting for food or making war. Ptolemy, an astronomer and geographer from the second century B.C., created one of the earliest known atlases, a collection of world, regional, and local maps and advice on how to draw them, which remained essentially unknown to Europeans until the 15th century. Translated into Latin, it became the core of Western geography, influencing cartographic giants, such as Gerhard Mercator, who published his famous world map in 1569. The 17th and 18th centuries saw many important developments in cartography, including the measurement of a degree of longitude by Jean Picard in 1669, the discovery that the earth flattens toward the poles, and the adoption of the Prime Meridian that passes through Greenwich, England. In 1859, French photographer and balloonist Gaspard Felix Tournachon founded the art of remote sensing by carrying large-format cameras into the sky. In an oft-cited early example of spatial analysis, Dr. John Snow mapped cholera deaths in central London in September 1854 and was able to locate the source of the outbreak—a contaminated well. However, until the 20th century, cartography remained an art and a science carried out by laborious calculation and hand drawing. In 1950, Jacqueline Tyrwhitt made the first explicit reference to map overlay techniques in an English textbook on town and country planning, and Ian McHarg was one of the early implementers of the technique for highway planning.

As with many other endeavors, the development of computers inspired cartographers to see what these new machines might do. The early systems developed by these groups, crude and slow by today's standards, nevertheless laid the groundwork for modern Geographic Information Systems. Dr. Roger Tomlinson, head of an Ottawa group of consulting cartographers, has been called the "Father of GIS" for his promotion of the idea to use computers for mapping and for his vision and effort in developing the Canada Geographic Information System (CGIS) in the mid-1960s. Another pioneering group, the Harvard Laboratory for Computer Graphics and Spatial Analysis, was founded in the mid-1960s by Howard Fisher. He and his colleagues developed a number of early programs between 1966 and 1975, including SYMAP, CALFORM, SYMVU, GRID, POLYVRT, and ODYSSEY. Other notable developers included professors Nystuen, Tobler, Bunge, and Berry from the Department of Geography at Washington University during 1958–1961. In 1970, the U.S. Bureau of the Census produced its first geocoded census and developed the early DIME data format based on the CGIS and POLYVRT data representations. DIME files were widely distributed and were later refined into the TIGER format. These efforts had a pronounced effect on the development of data models for storing and distributing geographic information.

In 1969, Laura and Jack Dangermond founded the Environmental Systems Research Institute (ESRI), which pioneered the powerful idea of linking spatial representation of features with attributes in a table, the core idea that revolutionized the industry and launched the development of Arc/Info, a program whose descendents have captured about 90% of today's GIS market.

Other vendors are still active in developing GIS systems, which include packages MAPINFO, MGA from Intergraph, and IDRISI from Clark University.

What can a GIS do?

A GIS works with many different applications: land-use planning, environmental management, sociological analysis, business marketing, and more. Any endeavor that uses spatial data can benefit from a GIS. For example, researchers at the U.S. Department of Agriculture Rocky Mountain Research Station in Rapid City are conducting a study of elk habitat in the Black Hills of South Dakota and Wyoming. They have placed radio transmitter collars on about 70 elk bulls and cows (Fig. I.1c). Using the collars and a handheld antenna, they track the animals and obtain their locations. Several thousand locations have been collected (Fig. I.1a) for the scientists to study the characteristics of the habitat where elk spend time.

The elk locations are entered into a GIS system for record keeping and analysis. Each location becomes a point with attached information, including the animal ID number and the date and time of the sighting. Information about vegetation, slope, aspect, elevation, water availability, and other site factors can be derived by overlaying the points on other data layers, allowing the biologists to compare the characteristics of sites utilized by the elk. Figure I.1b shows a map of distances to major roads in the central part of the study area. The elk locations tend to cluster in the darker roadless areas, and statistical analysis demonstrates this observation empirically. The next section describes a GIS model to evaluate elk habitat.

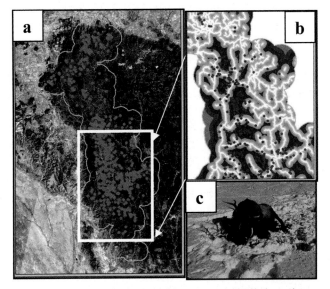

Fig. I.1. Analyzing elk habitat use. (a) Elk locations and study area. (b) Locations on a map of distance to nearest road. (c) Collaring an elk.

There are many applications of GIS in almost every human endeavor, including business, defense and intelligence, engineering and construction, government, health and human services, conservation, natural resources, public safety, education, transportation, utilities, and communication. In June 2008, the Industries section of the ESRI Web site at www.esri.com listed 50 different application areas in these categories, each one with examples, maps, and case studies. Instead of reading through a list here, go to the site and see the latest applications.

New trends and directions in GIS

The GIS industry has grown exponentially since its inception. Starting with the mainframe and then the desktop computer, GIS began as a relatively private endeavor focused on small clusters of specialized workers who spent years developing expertise with the software and data. Since then, the development of the Internet and the rapidly advancing field of computer hardware are driving some significant changes in the industry.

Proliferation of options for data sharing

Instead of storing data sets locally on individual computers or intraorganization network drives, more people are serving large volumes of data over the Internet to remote locations within an organization, across organizations, and to the general public. In the past, large collections of data were hosted by various organizations through clearinghouses sponsored by the National Spatial Data Infrastructure (NSDI) organization. The data usually existed as GIS data files in various formats and required significant expertise and the right software to download and use. Today, *Internet Map Servers* are designed to bring GIS data to the general public. The newer services utilize a point-and-click approach that users can learn in a few minutes and that automatically manage many of the data issues that previously frustrated users. Another class of data providers that give wide access to spatial data, although not specifically GIS applications, include Web sites, such as GoogleEarth, MapQuest, and Microsoft Virtual Earth.

Proliferation of options for working with GIS data

In the early days, people who wanted to do GIS had to purchase a large expensive program and learn to use it. Now, a wide variety of scaled applications permits different levels of use for different levels of need. Not every user must have the full program. There are map servers for people who just need to view and print maps, free download software for viewing interactive map publications, and scaled-down versions of the full program with fewer options.

Many organizations are turning to Server GIS as a less expensive alternative to purchasing large numbers of GIS licenses. Many workers need GIS, but they only use a small subset of GIS functions on a daily basis. A Server GIS can provide both data and a customizable set of viewing and analysis functions to users without a GIS license. All they need is a Web browser. Because of more simple and low-cost ways of accessing GIS data and functions, the user base has expanded dramatically.

Expansion of GIS into wireless technology

More people are collecting and sharing data using handheld wireless devices, such as handheld computers and Global Positioning System (GPS) units. These units can now access Internet data and map servers directly so that users in the field can download background data layers, collect new data, and transmit it directly back to the large servers. Even cell phones are moving into GPS and location services, and someday it may be possible to have a GIS unit on your phone.

Emphasis on open source solutions

Instead of relying on proprietary, specialized software, more GIS functions are now implemented within open source software and hardware. GIS data are now more often stored using engines from commercial database platforms and utilize the same development environments as the rest of the computer industry. This trend makes it easier to have the GIS communicate with other programs and computers and enhances interoperability between systems and parts of systems.

Customization

With emphasis on open source solutions, new opportunities have developed in creating customized applications based on a fundamental set of GIS tools, such as a hydrology tool or a wildlife management tool. These custom applications gather the commonly used functions of GIS into a smaller interface, introduce new knowledge, and formalize best practices into an easy-to-use interface. Customization requires a high level of expertise in object-oriented computer programming.

Enterprise GIS

Enterprise GIS integrates a server with multiple ways to access the same data, including traditional GIS software programs, Web browser applications, and wireless mobile devices. The goal is to meet the data needs of many different levels of users and to provide access to non-traditional users of GIS. The Enterprise GIS is the culmination of the other trends and capabilities already mentioned. The costs and challenges in developing and maintaining an Enterprise GIS are significant, but the rewards and cost savings can also be substantial.

What do GIS professionals do?

These days, GIS professionals play a variety of roles. A few broad categories can be defined.

Primary Data Providers create the base data that form the backbone of many GIS installations. Surveyors and land-planning professionals contribute precise measurement of boundaries. Photogrammetrists develop elevation and other data from aerial photography or the newer laser altimetry (LIDAR) systems. Remote sensing professionals extract all kinds of manmade and natural information from a variety of satellite and airborne measurement systems. Experts in global positioning systems (GPS), which provide base data, are also important.

Applications GIS usually involves professionals trained in other fields, such as geography, hydrology, land-use planning, business, utilities, and so on, who utilize GIS as part of their work. Specialists in mathematics and statistics develop new ways to analyze and interpret spatial data. For these folks, GIS is an added skill and a tool to make their work more efficient, productive, and valuable.

Development GIS involves skilled software and hardware engineers who build and maintain the GIS software itself, as well as the hardware components upon which it relies—not only computers and hard drives and plotters but also GPS units, wireless devices, scanners, digitizers, and other systems that GIS could not function without. This group also includes an important class of GIS Developers who create customized applications from the basic building blocks of existing GIS software.

Distributed Database GIS involves computer science professionals with a background in networking, Internet protocols, and/or database management systems. These specialists set up and maintain the complex server and network systems that allow data services, Server GIS, and Enterprise GIS to operate.

GIS project management

A GIS project may be a small effort spanning a couple of days by a single person, or it can be an ongoing concern of a large organization with hundreds of people participating. Large or small, however, projects often follow the generalized model shown in Figure I.2, and most new users learn GIS through a project approach. A project usually begins with an assessment of needs. What specific issues must be studied? What kind of information is needed to support decision making? What functions must the GIS perform? How long will the project last? Who will be using the data? What funding is available for start-up and long-term support?

Without a realistic idea of what the system must accomplish, it is impossible to design it efficiently. Users may find that some critical data are absent or that resources have been wasted acquiring data that no one ever uses. In a short-term project the needs are generally clear-cut. A long-term organizational system will find that its needs evolve over time, requiring periodic reassessment. A well-designed system will adapt easily to future modification. The creator of a haphazard system may be constantly redoing previous work when changes arise.

In studies seeking specific answers to scientific or managerial questions, a methodology or **model** must be chosen. Models convert the raw data of the project into useful information using a well-defined series of steps and assumptions. Creating a landslides hazards map provides a simple example. One might define a model such that, if an area has a steep slope and consists of a shale rock unit, then it should be rated as hazardous. The raw data layers of geology and slope can thus be used to create the hazard map. A more complex model might also take into account the dip (bedding angle) of the shale units. Models can be very simple, as in this example, or more complex, like the elk model presented earlier.

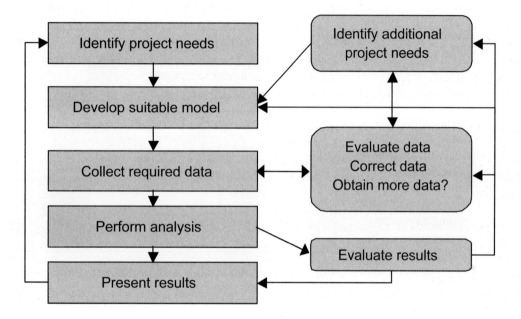

Fig. I.2. Generalized flow diagram representing steps in a GIS project

Once needs are known and the appropriate models have been designed, data collection can begin. GIS data are stored as layers, with each layer representing one type of information, such as roads or soil types. The needs dictate which data layers are required and how accurate they must be. A source for each data layer must be found. In some cases data can be obtained free from other organizations. General base layers, such as elevation, roads, streams, political boundaries, and demographics, are freely available from government sources (although the accuracy and level of detail are not always what one might wish). More specific data must often be developed in-house. For example, a utility company would need to develop its own layers showing electric lines and substations—no one else would be likely to have it.

The spatial detail and accuracy of the data must be evaluated to ensure that it is able to meet the needs of the project. For example, an engineering firm creating the site plan for a shopping mall could download elevation contours of the site for free. However, standard 10-foot contours cannot provide the detailed surface information needed by the engineers. Instead, the firm might contract with a surveying firm to measure contours at half-foot intervals.

After the data are assembled, the analysis can begin. During the analysis phase, it is not unusual to encounter problems that might require making changes to the model and/or data. Thus, the steps of model development, data collection, and analysis often become iterative, as experience gained is used to refine the process. The final result must be checked carefully against reality in order to recognize any shortcomings and to provide guidelines for improving future work.

Finally, no project is complete until the results have been presented, perhaps informally to a supervisor, perhaps published in a scientific journal, or perhaps as the focus of a heated public meeting. The form of the presentation influences the format and style of the presentation, which can take the form of maps, reports, presentation slides, and so on.

Project case study: A thorny issue

A project recently commissioned by Badlands National Park (Fig. I.3) can serve as an example of a GIS project in action. Canada thistle (*Cirsium arvense*) is a highly invasive, noxious weed that is widespread throughout the park. Proposed control measures included spraying with Tordon, a broad-based herbicide. However, the herbicide also kills native plants, and it has a long residence time in soils, leading to potential groundwater contamination. The park approached the author with a proposal to use GIS analysis to better locate areas of infestation, thereby minimizing the amount of pesticide used.

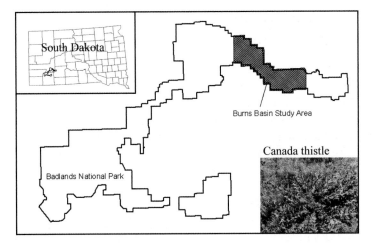

Fig. I.3. Burns Basin study area in Badlands National Park

Step 1: Identify project needs

Work began by identifying the project objectives in order to write the proposal. The objectives centered on a specific research question: Can the occurrence of Canada thistle be predicted from knowledge of environmental factors at known thistle localities? The objectives included:

> ➤ Correlating known thistle occurrence to environmental parameters such as soil type
> ➤ Developing a model to predict the occurrence of thistle
> ➤ Testing the model predictions using additional field surveys
> ➤ Finishing predictions/testing prior to the anticipated field spraying operation in June

Step 2: Develop model

For this project a **multivariate regression** model was chosen. The familiar **linear regression** model calculates a best-fit line through a cloud of points in an *x-y* graph (Fig. I.4), and the equation of the line $y = mx + b$ can then be used to predict values of y (dependent

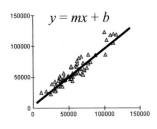

Fig. I.4. Linear regression

variable) for any proposed value of *x* (independent variable). A multivariate regression applies this concept to multiple independent variables at once. In GIS, each independent variable is represented by a map layer. Since thistle presence is a true/false variable, a variation of the multivariate regression model, **logistic regression**, must be used.

Analysis of the literature on Canada thistle revealed that the presence of water and anthropogenic disturbance are significant factors in thistle infestations. Likely other factors included soil types, water availability, vegetation communities, the prior existence of thistle, and prevailing winds.

The author and the park prepared a joint research proposal and received funding from the National Park Service Geographic Information Systems Funding Program (Midwest Regional Office). The project had a $10,000 budget and a six-month duration. The project proposal is included at the end of this chapter. The short time frame and small budget dictated that the project focus on data that were easily available.

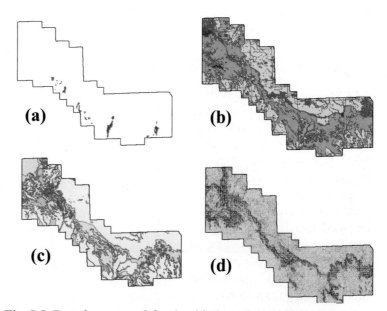

Fig. I.5. Data layers used for the thistle project: (a) Known thistle localities, (b) Soils, (c) Geology, and (d) Slope.

Step 3: Collect data

The park provided most of the data layers, including geology, soil types, vegetation communities, slope, aspect, water bodies, and roads. The author also included vegetation and soil moisture indices derived from Landsat satellite imagery. The park also had a layer of known thistle localities plotted during a horseback survey in 1998. Figure I.5 shows some of the data layers included in the project.

Reliable layers for hiking trails, wind direction, and wetlands would have been welcome additions but were dismissed for reasons of time and economy. The scope and budget of a project often require decisions of this sort, at the risk of leaving out critical information. Some projects have more clearly defined layers that are critical to the project. In this one, the important factors were not entirely known, so the approach adopted was to include everything easily available that might have an influence.

Data collection for the project included locating all of the layers and bringing them together. Additional data management tasks included merging separate files for each quadrangle into single map layers and removing information outside the study area (clipping). Analysis functions were used to create slope and aspect maps from the original digital elevation models and to calculate indices from the satellite images.

Step 4: Perform analysis

As work began, a need to revise the proposed model quickly surfaced due to two difficulties. First, the logistic regression requires input locations representing the absence of thistles as well as their presence. A no-thistle data layer was constructed by assuming that the survey found all the thistle areas and by randomly choosing other points outside those areas that were assumed to be thistle free. Second, logistic regression works for numerical data only. Geology, soil type, and vegetation are nonnumeric quantities and could not be included in the regression. We brought in an additional model based on a binomial statistical analysis of the abundance of thistles relative to the area fraction of each variable. The two models would be run independently, and the results would be combined to produce the final result.

This part of the project demonstrates that the initial planning often cannot foresee all eventualities and that additional data or a revised model may be needed to complete a project. Thus, the procedure shown in Figure I.2 may not necessarily follow a linear progression, but it may include cycles of reevaluating the data or model, finding additional data, or making corrections to the procedures.

The logistic regression calculated correlations between the numeric variables and the thistle locations/nonlocations and produced a map showing the probability of finding thistle at any location on the map. The results are shown in Figure I.6a. The white areas represent a low probability of finding thistle, and the brown areas represent a high probability.

In the binomial model, the number of thistle locations falling inside a particular map unit (e.g., a soil type) was compared with the number of locations that would be expected if the locations were random. (If random, the expected number is proportional to the area fraction of each map unit.) The comparisons were converted to Z-scores in which a positive value greater than 2 indicates an affinity of the site to thistles, and a negative score less than -2 indicated an avoidance of the unit by thistles. Values between -2 and 2 were neutral with respect to thistles. The analysis was applied separately to the soil, geology, and vegetation layers. The results of the three layers were added together to determine a combined Z-score. Figure I.6b shows the model predictions. The blue areas indicate a low probability of thistles; the orange and brown areas predict a high probability of thistles.

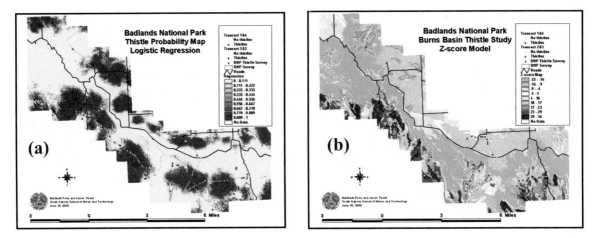

Fig. I.6. Thistle study results: (a) Probability map from logistic regression, showing likelihood of infestation, (b) Combined Z-score map showing areas thistles avoid (blue) and prefer (browns).

Step 5: Evaluate results

Once the analysis is complete, the results must be evaluated for validity. In this case, we chose to hike transects across the park, stopping at regular intervals to record the presence or absence of thistles (systematic sampling). The GIS helped to define the transects, using criteria that each transect should cross both high and low thistle probabilities, should begin and end near a road and should not cross any impassibly steep areas.

Figure I.7 shows portions of two transects. The red points indicate the presence of thistles, and the black points were thistle free. Note that virtually all of the red points occur in high-probability areas, yet many black points occur there as well. After statistical evaluation of the results, the study concluded that the model could locate areas where thistles were able to grow, but it could not well identify where thistles were actually growing at any given time.

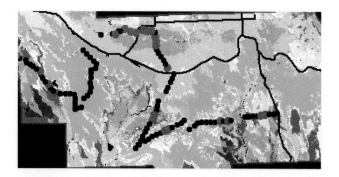

Fig. I.7. Portions of field survey plots on the model prediction map. Black points are thistle free; red points have thistles.

The final phase of the study included identifying future areas for improvement. Using current infestations and prevailing winds and hiking trails to identify areas at risk of future infestation could be very helpful. Also, the application of the model to the entire park could have provided useful information. Had more funding been available, the study could have improved the model and initiated another cycle as shown in the project flow diagram (Fig. I.2).

Step 6: Present results

The results were presented at several conferences, and a final report was submitted to the park in time to plan for the spraying (which unfortunately was cancelled for other reasons).

Types of GIS projects

The Badlands thistle study represents a **project-oriented GIS** in which the scale and time frame of the project are well defined, and the project life is finite. At the other end of the spectrum lies the **enterprise GIS,** designed for long-term support of critical functions within an organization. A city planning department, for example, is continually collecting and updating information as new parcels are developed, zoning laws change, owners and tax rates fluctuate, and regulations require more and more documentation. An enterprise GIS undergoes many project cycles in which data gathering and analysis are ongoing concerns, not finite tasks. They also can encounter special issues, including sharing of data by multiple users, keeping track of updates, protecting privacy, and managing large volumes of data. However, the tasks of defining needs, collecting and evaluating data, performing analysis, and presenting results are just as applicable to the large system, even though they proceed in a more circular and ongoing fashion.

Planning a GIS project

Learning GIS by doing exercises from a book can take you only so far. The best way to learn is to develop and carry out a GIS project, ideally under the supervision of a more experienced user who can help solve the inevitable difficulties encountered in the real world with real data. Preparing a project proposal is an excellent way to develop a project. In many cases a proposal is required to obtain funding or approval for work to begin. Even if not required, however, the exercise of producing a proposal helps focus the work and provides immediate feedback on the feasibility of the project. Steps toward building a proposal include the following (in practice, the first four steps often occur in overlapping cycles until a suitable project is found):

➢ Decide on a general topic of interest.
➢ Formulate a specific question to be answered or a product to be made.
➢ Decide on an approach or model to use.
➢ Determine the data needed and evaluate its availability and quality.
➢ Write the proposal according to established guidelines.

Step 1: Find a general topic

Deciding on a topic can be difficult, as it involves matching your interests with suitable data. If you work for an organization, then a topic or a problem may suggest itself immediately, and most of the data may already be present in-house. If not, choose an interesting topic and begin checking whether appropriate data are available. For a first project, it is wise to choose a problem for which most of the data needed are already available in GIS format.

Many projects sound terrific until it becomes clear that the needed data are impossible to obtain or are not of sufficient quality for the intended analysis. Since it is common to develop a topic only to find that the data resources are inadequate, working backwards from the data is often helpful. Examine the data sources available, and use them as a springboard to the imagination. Peruse your organization's data. Search the Internet, especially government sites such as the U.S. Geological Survey or the Environmental Protection Agency. Talk to colleagues. Visit the local GIS expert and ask for ideas and data holdings.

Step 2: Formulate a specific question

Once a topic is identified, you must formulate a specific research question. This step ensures that the project will yield a useful result. Too often, projects are conducted by gathering a lot of information and then analyzing it to see if anything interesting turns up. Often nothing significant emerges, and the supervisor or instructor is less than impressed. For example, one might propose to "study thistles in Badlands National Park." In the first place, the topic is not even a question. Second, it is too broad, giving no guidance as to the data layers required or the analysis to be performed.

A good research question is focused, limited, and answerable. Look again at the research question for the thistle case study: "Can the occurrence of Canada thistle be predicted from knowledge of environmental factors at known thistle localities?" The question has an answer: yes, no, or maybe. It narrows down the data requirements to thistle locations and environmental factors, enabling one to start tracking down layers to see what is available. It is unlikely to go off on irrelevant tangents such as the palatability of thistles to wildlife.

Projects may alternatively be structured around a specific product rather than a research question, and some fall more naturally into this category. For example, an agency could attempt to develop a 1:250,000 earthquake hazards map for southern California.

Step 3: Develop the methodology

Once the question or product is known, a methodology must be defined, with the major steps clearly articulated. In the thistle project, a formal model was implemented (logistic regression). In other projects, a sequence of analysis steps constitutes the methodology. Using the earthquake hazards map example, the model would include defining the important independent variables (fault proximity, fault activity, population density, soil acceleration potential), which would determine the dependent variable (earthquake hazard rating). The goal is to develop a sequence of steps which clearly indicates the data layers needed and defines how they will be manipulated to give the final result.

Step 4: Find and evaluate the data

Before the project can be considered viable, you must ensure that all of the data layers required by the model are readily available, are adequate to the proposed task, and are affordable based on the resource limitations of the project (both money and time). Failure to ascertain these conditions can derail the project later on, sometimes fatally, if a critical data set turns out to be unobtainable. Find all the data sources before committing to the project.

Step 5: Write the proposal

Once steps 1–4 have been done, the proposal nearly writes itself. If the proposal is a formal document being submitted to an agency (or instructor) for funding or approval, be sure to obtain the relevant proposal guidelines and follow them carefully, *especially the page limits*. If you are doing the proposal as an exercise, follow the fairly standard outline below. Use the exact headings shown in boldface to organize the proposal.

Introduction. Give a brief overview of the general problem you are addressing and why it is important. State the overall goals of the project and transition gracefully to a statement of the research question near the end.

Objectives. State the specific objectives to be accomplished or products to be created. A bulleted list is often easiest to read. One to four objectives are ample for most projects.

Methodology. Explain the rationale behind the methodology and list the **major** steps. Standard procedures can be glossed over, but unique aspects (such as how to calculate the carbon sequestration potential from the county soil maps) should be clearly defined.

Data sources. List the data layers required and provide proper citations for each one.

Work plan. List the major tasks and the estimated completion dates.

Budget (if applicable). List the itemized budget items and amounts and total funds and/or in-kind contributions requested.

The length of the proposal depends both on the complexity of the project and the guidelines set forth by a funding agency. As a general rule, a simple class project can be succinctly described in one to two pages. Extended or complex projects require additional description.

Once the project proposal is complete, bring it to a GIS professional to review, if one is available. He or she can identify any problems with methodology or data that might impair the success of the project or suggest alternate approaches not previously considered. When all is in order, start the project and have fun with it.

Example of a GIS Project Proposal

Prediction of thistle-infested areas in Badlands National Park using a GIS model

Dr. Maribeth Price
South Dakota School of Mines and Technology
Rapid City, SD 57701
November 18, 1999

Introduction

Canada thistle (*Cirsium arvense*) is a highly-invasive, noxious weed that is widespread throughout Badlands National Park (BADL). The widespread occurrence of this weed causes potential problems such as displacing native plants, harboring predators in black-footed ferret reintroduction sites, and causing economic damage to adjacent agricultural lands. The Canada thistle population exceeds the park's current chemical, mechanical, and biological control programs. Thus, an intensive herbicide treatment program has been proposed.

Tordon™, a picloramine compound manufactured by Dow Chemical, has been shown to be effective in thistle control. However, the herbicide also kills native plants and has a long residence time in soils, creating a significant potential for groundwater contamination. Limiting the herbicide application to infested areas reduces environmental risk and also provides significant cost savings. Unfortunately, the rugged nature of Badlands National Park makes field identification of thistle-infested areas both cost and time prohibitive.

An alternative approach includes developing a GIS model to predict the occurrence of thistle in the park based on environmental factors of known infestations, such as soil type, slope, and water availability. Such a model might assist BADL in targeting the herbicide applications to areas of highest need, as well as providing information on thistle habitat, which could help in locating additional infestations throughout the park.

Objectives

This project will ascertain whether a GIS model based on environmental variables at known thistle localities can be used to predict the occurrence of thistle throughout the park. The specific objectives include:

> ➤ Correlating known thistle occurrence to environmental parameters such as soil type
> ➤ Developing a model to predict the occurrence of thistle
> ➤ Testing the model predictions using additional field surveys
> ➤ Finishing predictions/testing prior to an anticipated field spraying operation in June

Methodology

The GIS analysis will use standard multivariate regression techniques to identify potential spatial correlations between approximately 50 field-mapped thistle patches and various terrain and vegetation parameters and to develop a predictive model for thistle infestation. In addition to site factors, studies of Canada thistle suggest that infestation is affected by water availability and anthropogenic disturbances. Parameters tested will include soil type, geologic unit, vegetation community, slope, aspect, distance to water, distance to roads, and Landsat-derived Normalized Difference Vegetation Index (NDVI) and Normalized Difference Moisture Index (NDMI). The final product will include a map showing the predicted thistle density and distribution for the 2000 growing season.

After the start of the growing season, and prior to spraying, the map will be field checked to develop an accuracy estimate for the regression model. Systematic sampling sites along several transects will be located using the BADL Trimble GPS unit, and the presence/absence of thistle will be measured and compared to the model predictions. The transects will be chosen such that they adequately represent all levels of thistle probability developed by the model, they start and end near roads, and they do not cross areas of impassibly steep slopes.

Data sources

The South Dakota School of Mines and the Badlands National Park possess all of the data, software, and equipment needed to conduct the analysis.

1998 Horseback Survey of thistle infestations (shapefile)
1:24,000 vegetation communities map (coverage)
1:24,000 USGS Geologic Map and 1:24,000 SSURGO soil maps (coverages)
1:24,000 USGS Digital Elevation Model quadrangles (grids)
1:24,000 USGS Digital Line Graph roads, streams, and water bodies (coverages)
May 03, 1998 and November 17, 1998 Landsat TM images

Work plan

Jan-Feb 2000	Compile data from SDSMT and BADL sources.
	Merge quads into single layers.
	Clip layers to study area boundary in Burns Basin.
	Calculate NDVI and NDMI from Landsat images.
Mar-Apr 2000	Perform multivariate regression analysis to produce predictive map.
	Create metadata for final layers.
	Choose field transects and sampling interval.
May 2000	Field check predictive map.
June 2000	Analyze results and write final report for BADL.

Budget

Salary and Benefits	
Faculty, 3 weeks	$3,900
Graduate student, 5 weeks	$2,600
Undergraduate, 1 week for field work	$350
Travel	
6 trips to BADL by car, ~700 miles @ $0.27	$200
Ropes, stakes, misc. field equipment	$50
Overhead	
43% of salaries	<u>$2,945</u>
Total requested	$10,045

Chapter 1. GIS Data

Objectives

> Understanding how real-world features are represented by GIS data

> Knowing the differences between the raster and vector data models

> Getting familiar with the basic elements of data quality and metadata

> Learning the different types of GIS files used by ArcGIS

> Learning to use ArcCatalog to view and manage GIS data

> Learning about layers and their properties

Mastering the Concepts

GIS Concepts

Representing real-world objects as a map

To work with maps on a computer requires developing methods to store different types of map data and the information associated with it. Objects in the real world, such as cities, roads, soils, rivers, and topography, must first be portrayed as map objects, such as those on a paper topographic map. These map objects must then be encoded for storage on a computer.

Many different data formats have been invented to encode data for use with GIS programs; however, most follow one of two basic approaches: the **vector model** or the **raster model**. In either approach, the critical task includes representing the information at a point, or over a region in space, using x and y coordinate values (and sometimes z for height). The x and y coordinates are the spatial data. The information being represented, such as a soil type or a chemical analysis of a well, is called the attribute data. Raster and vector data models both store spatial and attribute data, but they do it in different ways.

Both data systems are **georeferenced**, meaning that the information is tied to a specific location on the earth's surface. One can use a variety of different coordinate systems for georeferencing, as we will see in Chapter 11. As long as the coordinate systems match, we can display any two spatial data sets together and have them appear in the correct spatial relationship to one another.

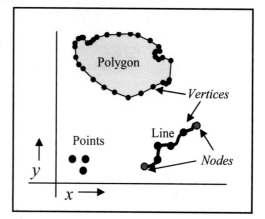

Fig. 1.1. The vector data model uses a series of x-y locations to represent points, lines, and polygon areas.

The vector model

Vector data uses a series of x-y locations to store information (Fig. 1.1). Three basic vector objects

17

exist: points, lines, and polygons. These objects are called **features**. **Point** features are used to represent objects that have no dimensions, such as a well or a sampling locality. **Line** features represent objects in one dimension, such as a road or a utility line. **Polygons** are used to represent two-dimensional areas, such as a parcel or a state. In all cases, the features are represented using one or more *x-y* coordinate locations (Fig. 1.1). A point consists of a single *x-y* coordinate pair. A line includes two or more pairs of coordinates—the endpoints of the line are termed **nodes**, and each of the intermediate points is called a **vertex**. A polygon is a group of **vertices** that define a closed area.

To some extent, the type of object used to represent features depends on the scale of the map. A large river would be represented as a line on a map of the United States because at that scale it is too small for its width to encompass any significant area on the map. If one is viewing a USGS topographic map, however, the river encompasses an area and might be represented as a polygon.

In GIS, like features are grouped into data sets called **feature classes** (Fig. 1.2). Roads and rivers are different types of features and would be stored in separate feature classes. A feature class can only contain one kind of geometry—it can include point features, line features, or polygon features but never a combination. In addition, objects in a feature class have information stored about them, such as their names or populations. This information is called the **attributes**. A river and a highway would not be found in the same feature class because their information would be different—flow measurements for one versus pavement type for the other.

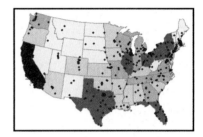

Fig. 1.2. A states feature class and a cities feature class

Vector GIS systems mostly use a **georelational** data model in which the spatial features are linked to attributes in a separate table by means of a unique feature identification code, or **FID** (Fig. 1.3). Each feature corresponds to one and only one line (record) in the table. The attributes for a state might include its name, abbreviation, and population. More recently **object-oriented** data models have gained favor, in which the spatial coordinates and the attribute data are stored together in a single database file. This approach saves overhead in linking the two aspects together and also helps ensure the integrity of features.

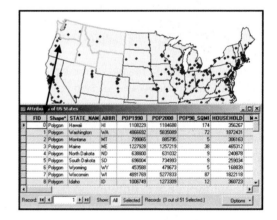

Fig. 1.3. Each state is represented by a spatial feature (polygon), which is linked to the attributes.

Regardless of which model is used, when a state is highlighted on the map, its matching attributes are highlighted in the table, and vice versa. It is this live link between the spatial and attribute information that gives the GIS system its power. It enables us, for example, to create a map in which the states are colored based on their populations (Fig. 1.2). This **thematic mapping** is only one example of how linked attributes can be used to analyze geographic information.

Feature classes can be stored in several different formats. Some data formats, such as shapefiles, only contain one feature class. Others, called **feature datasets**, can contain multiple feature classes that are in some way related to each other. For example, a feature dataset called Transportation might contain the feature classes Roads, Traffic Lights, Railroads, and Canals.

The benefits of the vector data model are many. First, it can store individual features such as roads and parcels with a high degree of precision. Second, the linked attribute table provides great flexibility in the number and type of attributes that can be stored about each feature. Third, the vector model is ideally suited to mapmaking because of the high precision and detail of features that can be obtained. The vector model is also a compact way of storing data, typically requiring a tenth of the space of a raster with similar information. Finally, the vector model is ideally suited to certain types of analysis problems, such as determining perimeters and areas, detecting adjacency of features, and modeling flow through networks.

However, the vector model has some drawbacks. First, it is poorly adapted to storing continuously varying surfaces such as elevation or precipitation. Contour lines (as on topographic maps) have been used for many years to display surfaces, but calculating derived information from contours, such as slope, flow direction, and aspect, is difficult. Finally, some types of analysis are more time-consuming to perform with vectors.

The raster model

The raster model has the benefit of simplicity. A set of spatial data, such as a land-use map, is represented as a series of small squares, called **cells** or **pixels** (Fig. 1.4). Each pixel has a numeric code indicating the land use, and the raster is stored as an array of numbers. To display it, a different color is assigned to each code value.

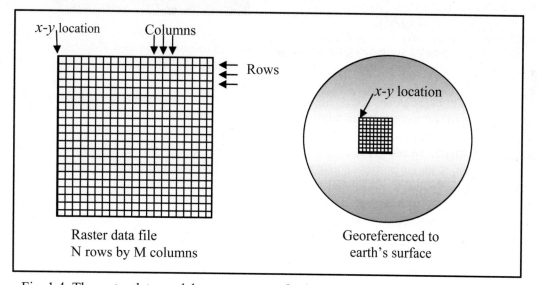

Fig. 1.4. The raster data model uses an array of values to represent a map. The raster is tied to a real-world location using the *x-y* coordinates of the upper-left corner.

A raster data set is laid out as a series of rows and columns. Each pixel has an "address" indicated by its position in the array, such as row = 3 and column = 6. Georeferencing a map in an *x-y* coordinate system requires four numbers: an *x-y* location for one pixel in the raster data set and the size of the pixel in the *x* and *y* directions. Usually the upper-left corner is chosen as the known location, and the *x* and *y* pixel dimensions are the same so that the pixels are square. From these four numbers, it is possible to calculate the coordinates of every other pixel based on its row and column position. In this sense, the georeferencing of the pixels in a raster data set is implicit—one need not store the *x-y* location of every pixel.

The *x* and *y* dimensions of each pixel define the **resolution** of the raster data. The higher the resolution, the more precisely the data can be represented. Consider the 90-meter resolution roads raster in Figure 1.5. The three colors represent three different numeric values indicating primary, secondary, and primitive roads. Since the raster cell dimensions are 90 meters, the roads are represented as much wider than they actually are, and they appear blocky rather than forming smooth curves. A 10-meter resolution raster could represent the roads more accurately; however, the file size would increase by 9×9, or 81 times.

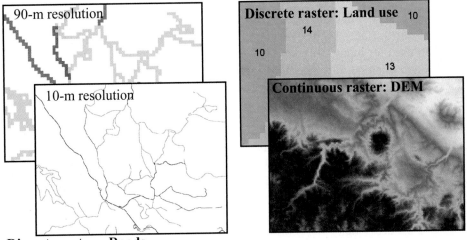

Fig. 1.5. Discrete rasters store categorical data such as land use or road types. Continuous rasters store data which vary smoothly over a surface, such as elevation or rainfall.

Two styles of raster data can be stored (Fig. 1.5). A **discrete** raster represents individual objects such as wells, roads, or parcels. It has relatively few values that tend to repeat themselves in adjacent cells. Categorical data such as land use, which falls into a few named categories, is discrete also. A **digital raster graphic** (DRG) is a scanned image of a topographic map, and each element (contours, roads) is portrayed using a different color. A **continuous** raster data set is one with a large range of numeric values that can range smoothly from one location to another, forming a surface or field. A **digital elevation model** (DEM) is an example of continuous data—cells are unlikely to have the same elevation value as their neighbors. Satellite images and digital air photos are other examples of continuous data.

The raster model mitigates some of the drawbacks of vectors. It is ideally suited to storing continuous and rapidly changing discontinuous information because each cell can have a value completely different from its neighbors. Many analyses are simple and rapid to perform, and an extensive set of analysis tools for rasters far outstrips those available for vectors.

The drawbacks of rasters lie chiefly in two areas. First, they suffer from trade-offs between precision and storage space to a greater extent than vectors do. The second major drawback of rasters concerns their inability to store multiple attributes. A raster file is an array of cells with numeric values, and each cell has only one value. To store both geology and infiltration values for an area requires storing two separate rasters. Vector files, by contrast, can store hundreds of attribute values for each spatial feature and can handle text data more efficiently.

Coordinate systems

Both raster and vector data rely on *x-y* values to locate data to a particular spot on the earth's surface. The *x-y* values of the coordinate pairs can vary, however. The choice of values and units to store a data set is called its **coordinate system**. Consider a standard topographic map, which actually has three different coordinate systems marked on it. The corners are marked with degrees of latitude and longitude. Another set of markings indicates a scale in meters representing the UTM, or Universal Transverse Mercator, coordinate system. A third set of markings shows a scale in feet, corresponding to a State Plane coordinate system. Any location on the map can be represented by three different *x-y* pairs corresponding to one of the three coordinate systems (Fig. 1.6). A global positioning system (GPS) unit also has this flexibility. It can be set to record a location in degrees, UTM meters, State Plane feet, or other coordinate systems as well.

When creating a vector or raster data set, one must choose a coordinate system and units for storing the *x-y* values. It is also important to label the data in such a way that the user knows which coordinate system has been selected and what the units for the *x-y* values are. If someone needed to calculate the distance between two *x-y* locations, knowing whether the units were feet or meters would

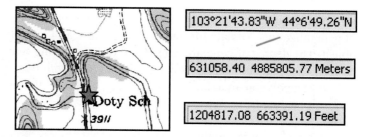

Fig. 1.6. A location can be stored using different coordinate systems and units. The *x-y* location of this school is shown in degrees, UTM meters, and State Plane feet.

be critical to finding the right answer. Thus, every GIS data set must have a label that records the type of coordinate system and units used to store the *x-y* data inside it. You will learn more about coordinate systems in succeeding chapters.

Modeling feature behavior with topology

Two basic vector models exist, **spaghetti models** and **topological models**. A spaghetti model stores features of the file as independent objects, unrelated to each other. Simple and straightforward, this type of model is found in many types of applications that store spatial data. It is also commonly used to transfer vector features from one GIS system to another.

A topological data model stores features, but it also contains information about how the features are spatially related to each other. Many types of spatial relationships might be of interest, for example, whether two parcels share a common boundary (**adjacency**), whether two water lines are attached to each other (**connectivity**), whether a company sprayed pesticide over the same area on two different occasions (**overlap**), or whether a highway connects to a crossroad or has an overpass (**intersection**). Although computer algorithms can determine whether these spatial relationships exist between features in a spaghetti model, storing explicit information about the relationships can save time if the relationships must be used repeatedly.

Another application of topology involves analyzing the **logical consistency** of features. Logical consistency evaluates whether a data model or data set accurately represents the real-world relationships between features. For example, two adjacent states must share a common boundary that is exactly the same (the real-world situation), even though the states are stored in the data model as two separate features with two boundaries that coincide (Fig. 1.7). Lines representing streets should connect if the roads they represent meet. A line or a polygon boundary should not cross over itself.

Finally, topology can be used to better model the real-world behavior of features. In a network topology, for example, the connections between features are explicitly modeled so that flow through the network can be analyzed. Applications of networks include water in streams, traffic along roads, flights in and out of airline hubs, or utilities through pipes or electrical systems.

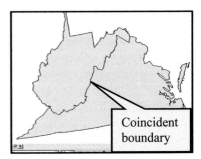

Fig. 1.7. A coincident boundary gets stored twice but is the same for both features.

Data quality

Representing real-world objects as points, lines, polygons, or rasters always involves some degree of generalization. No data file can exactly capture all the spatial or attribute qualities of any object. The degree of generalization often varies with the scale. On a standard topographic map, a river has a width and can be modeled as a polygon with two separate banks. A city would be shown as a polygon area. For a national map, however, the river would simply be shown as a line, and a city would be shown as a point.

Even a detailed representation of an object is not always "true." Rivers and lakes can enlarge in size during a spring flood event or shrink during a drought. The boundary of a city changes over time as the city grows. Users of GIS data must never forget that the data they collect and use will contain flaws both large and small, and that the user has an ethical and legal responsibility to ensure that the data used for a particular purpose are sufficient and appropriate to the task. When evaluating the quality of a data set, geospatial professionals consider the following aspects.

Geometric accuracy refers to the *x-y* values of a feature class or raster. How closely do the locations correspond to the actual location on the earth's surface? Geometric accuracy is usually a function of the original scale at which data are collected and of how they were obtained. Surveying is one of the most accurate ways to position features. GPS units have an accuracy that ranges from centimeters to tens of meters. Maps derived from aerial photography or satellite imagery can vary widely in geometric quality based on factors such as the scale of the image, the resolution of the image, imperfections and distortions in the imaging system, and the types of corrections applied to the image. In Figure 1.8, notice that the vector road in white is offset in places from the road as it appears in the aerial photo. These differences can arise from digitizing errors in the creation of the roads, geometric distortions from the camera or satellite, or other factors.

Fig. 1.8. Aerial photo near Woodenshoe Canyon, Utah. Source: Google Earth and Tele Atlas.

Moreover, not every boundary can be as precisely located as a road. Imagine that you wish to delineate the land-cover types *forest, shrubland, grassland,* and *bare rock* in this photo. Where would you draw the line between *shrub* and *grassland*? At what point does the *shrubland* become *forest*? Six different people given this photo would come up with six different maps. Some boundaries would match closely; others would vary as each person makes a subjective decision about where to place each boundary.

Thematic accuracy refers to the attributes. Some types of data are relatively straightforward to record, such as the name of a city or the number of lanes in a road. Even in this situation, the value of a feature might be incorrectly recorded. Other types of information can never be known exactly. Population data, for example, is collected through a process of surveying and self-reporting that takes many months. It is impossible to include every single person. Moreover, people are born and die during the survey process or are moving in and out of towns. Population data can never be more than an estimate. These difficulties don't mean that it is pointless to collect the data. However, it is important to understand the limitations and potential biases associated with thematic data.

Resolution refers to the sampling interval at which data are acquired. Resolution may be spatial, thematic, or temporal. Spatial resolution indicates at what distance interval measurements are taken or recorded. What is the size of a single pixel of satellite data? If collecting GPS points by driving along a road, at what interval is each point collected? Thematic resolution can be impacted by grouping associated with data collection. If collecting information on the percent crown cover in a forest, is each measurement reported as a continuous value (32%, 78%) or as a classified range (Low, Medium, High)? Temporal resolution indicates how frequently measurements are taken. Census data are collected every ten years. Temperature data taken at a climate station might be recorded every 15 minutes, but it might also be reported as a monthly or yearly average.

Precision refers to either the number of significant digits used to record a measurement or the statistical variation of a repeated single measurement. Many people confuse precision with accuracy, but it is important to understand the distinction. Imagine recording your body temperature with an oral digital thermometer that records to a thousandth of a degree and getting the value of 99.894 degrees Fahrenheit. This measurement would be considered precise. However, imagine that you took the reading immediately after drinking a cup of hot coffee. This action throws off the thermometer reading so that it does not record your true body temperature. Thus, the measurement is precise, but it is not accurate.

Evaluating the quality of a data set can be difficult, especially if the data were created by someone else. Professionals who create data incur an obligation to evaluate the quality of the data, as well as possible, and to provide a report that summarizes the spatial and thematic accuracy so that users can properly determine whether a data set is suited to a particular purpose. Producers should also provide information on other aspects of a data set, such as what geographic area it covers, what coordinate system it uses, what the information in the attribute tables means, how a potential user can access the data, and more. If the original data was created or compiled by others, the producer must also give proper credit to the originators. Such information about a data set is called **metadata** (Fig. 1.9). The content and format of metadata is

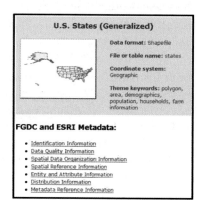

Fig. 1.9. Metadata

established by the Federal Geographic Data Committee, and metadata that follows these standards and has a certain minimal set of items is referred to as FGDC compliant.

Metadata provides a summary of the content and spatial extent of a data set. Organizations assemble collections of metadata to allow potential users to search and evaluate data sets before they are obtained. Much like an entry in an electronic library catalog that allows people to review information about a library book before they order it through interlibrary loan, metadata allows a user to search and locate data sets with a particular theme or geographic extent. Once a candidate data set is identified, the user can explore the full metadata record to determine if the data set is appropriate for the particular application. If so, the metadata itself tells the user where the data are located, how it can be obtained, and what cost might be associated with ordering it.

Metadata also records the access and use constraints on GIS data. GIS data can be copyrighted and its uses restricted to certain people or certain actions. Some GIS data, including most data sets derived from federal agencies, can be freely copied and redistributed with credit given to the originating agency. Other data is developed by companies, and the rights are licensed to specific users. Often the license includes the right to distribute maps or other static copies derived from the data, but not the data itself. Every user is responsible for understanding the applicable use constraints placed on any data set and for abiding by them. Failure to do so can result in civil and criminal penalties against the individual or the organization he or she works for.

Citing GIS data sources

Ethical and professional considerations require that any map, publication, or report that you produce should cite the data source(s) used and give proper attribution and credit to the originators of the data. Since GIS data can come from a variety of sources, several different styles of citation may be employed. The metadata is often a good source of information for these citations, or sometimes the site where the data were obtained is a good source. The best practice is to record the citation when the data are copied, and then you will have it when you need it for a report or a map. The following examples demonstrate the styles for various types of GIS data.

Data retrieved from a local area network:

Database name [type of medium]. Producer location, city/state/province: Producer name, year.

Black Hills National Forest Database [computer file]. Custer, South Dakota: Black Hills National Forest Service, 2004.

Rapid City Parcels Database [computer file]. Rapid City, SD: Rapid City Planning Department, 2006.

Data from a CD-ROM or commercial online database

Database name [type of medium]. (Publication year) Producer location, city/state/province: Producer name. Available: Supplier/Database identifier number or URL if applicable [Access date].

ESRI Data and Maps [DVD]. (2006) Redlands, CA: ESRI, Inc. [October, 2008].

RapidMap 2.2 [online database]. (2008) Rapid City, South Dakota: Pennington County – Rapid City GIS. Available: http://www.rcgov-gis.org/interactive.htm [June 6, 2008].

Data downloaded from a remote FTP server

Database name [type of medium]. (Year) Producer location city/state/province: Producer name. Available FTP: address/path/file [Access date].

Geographic Names Information System [downloaded file]. (2008) Reston, VA: Unites States Geological Survey. Available FTP: http://geonames.usgs.gov/domestic/download_data.htm [May 21, 2008].

1:24,000 Digital Line Graphs (DLG) [downloaded file]. (2000) Sioux Falls, South Dakota: USGS EROS Data Center. Available FTP: http://edc2.usgs.gov/geodata [January 7, 2007].

Data provided upon request by agency or individual

Data name [type of medium]. Producer location city/state/province: Producer name, year.

Badlands National Park GIS Data [CD-ROM]. Interior, South Dakota: National Park Service— Badlands National Park, by request, 2006.

About ArcGIS

ArcGIS overview

ArcGIS is developed and sold by Environmental Systems Research Institute, Inc. (ESRI). It has a long history and has been through many versions and changes. Originally developed for large mainframe computers, in the last 10 years it has metamorphosed from a system based on typed commands to a graphical user interface (GUI), which makes it much easier to use. Because of the size and complexity of the program (actually a suite of programs), and because users have come to depend on certain aspects of the software, much of the code is carried forward and included in the new versions. Knowing this background helps a student of ArcGIS understand the nature of the ArcGIS system and helps explain some of its odd characteristics.

The older core of the ArcGIS system was called Arc/Info and included a basic set of programs—Arc, ArcEdit, and ArcPlot (Fig. 1.10)—which utilized the coverage data model and was built in a database program called **INFO** that appears primitive today. All of the programs were command based, meaning that the user typed commands into a window to make the program work.

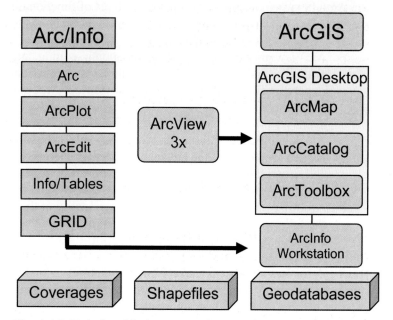

Fig. 1.10. Relationship between ESRI products and data formats

The difficulty of learning Arc/Info prompted ESRI to create ArcView, which was easier to use but not as powerful as Arc/Info. ArcView was designed primarily to view and analyze spatial data rather than create it. ArcView also used a simpler data model, called the shapefile, although it could read coverages and convert them to shapefiles. Beginners in GIS often learned ArcView first and then began learning Arc/Info as their needs and abilities advanced.

ArcGIS, released in 2001, is a synthesis of the powerful Arc/Info system with the easy-to-use interface of ArcView, updated to use the latest advances in desktop computing and database technology. It contains two programs, collectively referred to as ArcGIS Desktop.

> ArcMap provides the means to display, analyze, and edit spatial data and data tables. Similar in appearance to its ArcView predecessor, it nevertheless contains powerful new functionality.

> ArcCatalog is a tool for viewing and managing spatial data files. It resembles Microsoft Windows Explorer, but it is specially designed to work with GIS data. It should **always** be used to delete, copy, rename, or move spatial data files.

In addition, ArcGIS Desktop contains ArcToolbox, a collection of tools and functions for operations in ArcCatalog and ArcMap, such as converting between data formats, managing map projections, and performing analysis. Users may create and add their own tools or scripts for special or often-used tasks. The ESRI Web site at www.esri.com has a large library of scripts and tools that can be downloaded to extend the ArcGIS functionality.

Finally, the original Arc/Info command-line software can still be accessed in the additional module called Workstation ArcInfo, which is still used by organizations that may be tied to the older coverage model for various reasons, such as having a large number of specialized programs written in the older AML programming language.

The ArcGIS system also provides different levels of functionality that all use the same basic interface. Users can save money by buying only the functions they need.

> ArcView provides all of the basic mapping, editing, and analysis functions for shapefiles and geodatabases and is the level of functionality most users will require on a regular basis. It includes ArcMap, ArcCatalog, and a subset of ArcToolbox functions.

> ArcEditor includes all the functions of ArcView but adds editing capabilities needed to work with the advanced aspects of the geodatabase, such as topology and network editing. Additional functions reside in ArcToolbox at this level.

> ArcInfo provides access to the full functionality of the ArcGIS Desktop tools and the full version of ArcToolbox. In addition, it includes the original core Arc/Info software, now called Workstation ArcInfo.

This book focuses almost exclusively on the functions available with an ArcView license, although it mentions some of the additional capabilities as appropriate. Users can read the software documentation to learn more about the advanced topics.

The ESRI system of GIS programs, then, is a fairly complex set of tools with a long history, designed to work with a number of different data formats, also with a long history. We turn now to a discussion of how ArcGIS stores data.

Data files in ArcGIS

ArcGIS can read a variety of different file formats. Many of these come from older versions of the software. Some can come from other programs such as image processing packages and computer-aided design (CAD) systems. Table 1.1 lists many of the data sets than can be used in ArcGIS with the icons showing how they appear in ArcCatalog.

Table 1.1. Types of files and data sources used by ArcGIS

File type	Description
Shapefiles	Shapefiles are vector feature classes developed for the early version of ArcView and have been carried over into ArcGIS.
Coverages	A coverage is the vector data format developed for Arc/Info and is the oldest of the data formats.
Geodatabases	Geodatabases represent an entirely new model for storing spatial information with additional capabilities.
Database connections	Database connections permit users to log in to and utilize data from an RDBMS geodatabase.
Layer files	A layer file references a feature class and stores information about its properties, such as how it should be displayed.
Rasters and grids	Rasters represent thematic maps or images by arrays of numbers stored in binary format (base 2). Grids are a special raster format used with the Spatial Analyst extension.
Tables	Tables can exist as separate data objects that are unassociated with a spatial data set.
Internet servers	Many organizations now make data available over the Internet. Users can connect to these data sources and download information for their work.
TINs	TINs are Triangulated Irregular Networks that store 3D surface information, such as elevation, using a set of nodes and triangles.
CAD drawings	Data sets created by CAD programs can be read by ArcGIS, although they cannot be edited or analyzed unless they are converted to shapefiles or geodatabases.

Shapefiles

Shapefiles are georelational spaghetti data models developed for the early version of ArcView. A shapefile contains one feature class composed of points or lines or polygons but never a mixture. The attributes are stored in a **dBase** file. Shapefiles can, however, store **multipart features (or multifeatures)**, which are single features made of multiple objects. For example, the state of Hawaii requires multiple polygons to represent the different islands, but it can be stored as a multifeature so that it has only one record in the attribute table.

Although a shapefile appears as one icon in ArcCatalog, it is actually composed of multiple data files that can be seen individually in Windows Explorer (Fig. 1.11). The rivers shapefile has eight different files associated with it. The .shp file stores

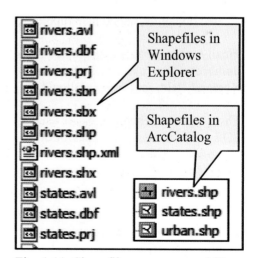

Fig. 1.11. Shapefiles are groups of files but appear as single entries in ArcCatalog.

27

the coordinate data, the .dbf file stores the attribute data, and the .shx file stores a spatial index that speeds drawing and analysis. These first three files are required for every shapefile to function properly. Additional files may also be present: the .prj file stores projection information, the .avl file is a stored legend, and the .xml file contains metadata. Note that to copy a shapefile to a new location, all of these files must be moved together. ArcCatalog takes care of this automatically, but Windows Explorer does not.

In a shapefile attribute table, the first two columns of data are reserved for storing the feature identification code (FID) and the coordinate geometry (Shape) field. These fields are created and maintained by ArcGIS and cannot be modified by the user. All other fields are added by the user.

Geodatabases

A geodatabase is an object-oriented model. It can contain many different objects, including multiple feature classes, geometric networks, tables, rasters, and other objects. Figure 1.12 shows a geodatabase named rapidnets. Feature classes may exist as individual objects in a geodatabase (as do the restaurants or schools), or they may be grouped into feature datasets. A feature dataset contains a collection of related feature classes with the same coordinate system, such as the Utilities feature dataset in Figure 1.12.

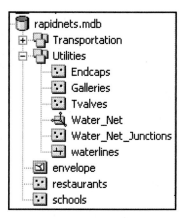

Fig. 1.12. A geodatabase containing two feature data sets and several feature classes.

A feature dataset can also store topological associations between feature classes. The Utilities feature dataset in Figure 1.12 contains a **network topology** constructed from its feature classes. The Water_Net and Water_Net_Junctions are additional feature classes associated with this network. Chapter 9 covers some special analysis functions that can be used with networks.

Feature datasets may also contain **planar topology**, which tracks spatial relationships within or between layers. To create topology, the user specifies topological rules, for example, that counties should not have gaps between them or that lines should always meet. Such errors are commonly introduced during the creation and editing of data; geodatabase topology assists in finding and correcting them. Editing with topology requires an ArcEditor or ArcInfo license. Simple topology created on the fly, called map topology, may be used when editing with an ArcView license.

Finally, geodatabases may contain rules that assist in entering and validating attribute data. Called **domains**, these rules specify which values or range of values may be entered in a particular field; a percent field, for example, should only contain numbers between 0 and 100.

Three types of geodatabases are used by ArcGIS: personal geodatabases, file geodatabases, and SDE geodatabases. The behavior of the three types is similar, but the data storage formats and capabilities differ. They are described in Chapter 14.

Coverages

A **coverage** is the vector data format developed for Arc/Info and is the oldest of the data formats. Like shapefiles, coverages are composed of multiple files on the disk and even data spread among multiple folders. All of the spatial and attribute information for coverages are stored in INFO format data files. A coverage data set includes a folder containing several data files with an .adf (arc data file) extension. In addition, more files are stored in a folder called info that must be in the same directory. A folder containing one or more coverages is called a workspace, and it includes the info folder, as well as folders for each coverage.

Coverages contain multiple feature classes, and some feature classes can be combined to create new feature classes. For example, a polygon feature class requires a point feature class to form polygon labels and a line, or arc, feature class to form the boundaries of the polygons (Fig. 1.13). From these two feature classes, the polygon feature class is created. Coverages store topological information on how the polygons are constructed from arcs (arc-node topology). Each polygon is composed of individual arcs and a label point (Fig. 1.13). Adjacent polygons share the same arc, so it needs to be stored only once. Coverages may store other types of topology, such as arcs that make up a network or multiple polygons combined together to make regions.

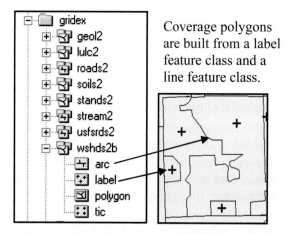

Coverage polygons are built from a label feature class and a line feature class.

Fig. 1.13. Coverages usually contain multiple feature classes.

Attributes for coverages are stored in INFO tables with special names: a polygon attribute table (.pat), arc attribute table (.aat), region attribute table (.rat), and so on. All coverage feature classes have a *cover#* field and a *cover-id* field, where *cover* is the name of the coverage. (For example, a coverage named roads would have fields called *roads#* and *roads-id*.) The *cover#* field is analogous to the FID in a shapefile. The cover-id is a numeric identification code that can be modified by the user.

VERY IMPORTANT TIP: Do not use Windows to copy or delete coverages, shapefiles, and geodatabases. These data sets may span multiple files and folders, and they might not be copied or deleted correctly. Always use ArcCatalog to delete or copy spatial data sets.

Rasters

Rasters in ArcGIS can take a variety of different formats. The native format is called a grid and is required for analysis with the Spatial Analyst extension. Other raster files can only be displayed and include common files formats, such as JPEG, TIF, GEOTIF, BMP, MrSID, and raw binary files (BIP, BIL, BSQ). A list of supported raster formats can be found in the ArcMap Help under the index heading "rasters, formats." Most rasters consist of the data itself plus a header that gives information about the file, such as its number of rows and columns and its coordinate system. This information may be stored in a separate file or as the first part of the binary raster. Rasters can also be stored inside geodatabases.

CAD files

Data sets created by CAD programs can be read by ArcGIS, although they cannot be edited or analyzed unless they are converted to shapefiles or geodatabases. A CAD file may contain multiple feature classes, which correspond to the layers of the drawing, and can be opened separately and viewed just like feature classes in a coverage or a geodatabase. One can also access CAD drawings that portray all the features in the CAD file with preset symbols. In a drawing, the feature classes are not accessible individually.

Not all CAD data sets use real-world coordinate systems such as UTM or State Plane; instead they use a local coordinate system referenced to an arbitrary origin. Before viewing with other GIS data, the coordinate system of the drawing must be transformed from the local system to a real-world system. This process requires knowledge of either the transformation parameters or the real-world coordinates of selected control points in the CAD data set.

Database connections

A user can connect to a database management system (DBMS) on a network through a database connection. This connection can be open, or it can require a login and password, depending on how the system administrator is managing the database security. Once inside, the user can access tabular data according to the permissions set up by the administrator. If the DBMS is also an SDE geodatabase, the user can access spatial data as well as tabular data.

Internet servers

Many organizations now make data available over the Internet, and the incentives and capabilities of sharing GIS data with minimal effort are fostered by organizations such as the OpenGIS Consortium (OGC). OGC is a nonprofit, international, voluntary organization that creates standards and best practices to facilitate data sharing.

Free data services have expanded explosively with players, such as GoogleEarth, MapQuest, and Microsoft Virtual Earth, providing access to huge volumes of image and map data with simple spatial and attribute searches. These services are designed to be used online in a Web interface, and most do not provide data directly to users for use in a GIS such as ArcMap. Although the data quality and documentation procedures are not designed for professional-level work and should be used cautiously, the sheer volume and popularity of these sites introduces many people to GIS techniques and data. Other sites, such as the United States Geological Survey's National Map, does allow users to download some of the data.

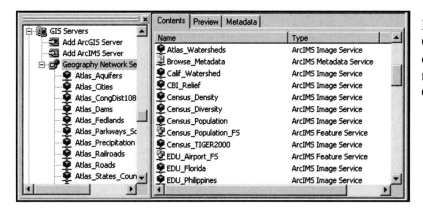

Fig. 1.14. The Geography Network offers many types of map data to use in GIS projects.

Other providers give access to data that can be displayed and queried from ArcMap and ArcCatalog. Users can connect to these data sources and download information for their work. To connect, you need to know the URL of the service, such as www.geographynetwork.com (Fig. 1.14). Two types of services are offered. An **image service** allows people to display the information and print out a map from it but will not allow people to change how it is displayed or make a copy of the data. A **feature service** allows people to download the data, view it, and save the features as a shapefile for later use.

About ArcToolbox

ArcToolbox contains tools for managing and analyzing data. The tools are organized into a hierarchical system of toolboxes containing related tools (Fig. 1.15). The functions available in the toolbox depend on the software license obtained. Users holding only an ArcView license will find fewer tools than users with ArcEditor or ArcInfo. If the user has purchased optional program extensions to ArcGIS, such as Spatial Analyst, that functionality will appear as additional toolboxes. Users may also create their own toolboxes inside ArcToolbox, fill them with frequently used tools, or create new tools. Tools can be run from either ArcMap or ArcCatalog.

Fig. 1.15. ArcToolbox

ArcToolbox is part of the ArcGIS "geoprocessing environment." Geoprocessing means to string together functions and tools that take input data, manipulate it, and produce a desired output. The geoprocessing environment includes an application called ModelBuilder, which allows users to manipulate tools graphically and save tool sequences for use again and again (Fig. 1.16). These models can streamline processing when several analysis steps are always repeated in the same order. Users can also write scripts or programs, that string together analysis steps. Like models, scripts can be used over and over to perform the same series of functions. Advanced users will want to learn more about the geoprocessing environment, models, and scripting because all three help streamline GIS work and can add flexibility and power to the user's repertoire.

The geoprocessing environment utilizes environment settings that control many aspects of how tools work. For example, users can set a default working directory where all outputs are placed or specify that resulting layers are always placed in the same coordinate system. The default settings provide reasonable service for all the exercises in this book. Users interested in advanced geoprocessing will need to learn more about these settings prior to changing the defaults.

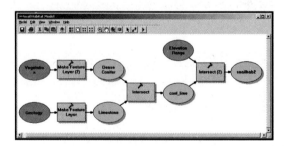

Fig. 1.16. The ModelBuilder application

The Help system

ArcGIS includes extensive help files with important information on how to execute tasks as well as background information on data, models, and functions. Users who make frequent use of Help will find their knowledge and skills growing dramatically. Help is requested from the Help entry on the main menu bar. On the left are three tabs allowing the user to search for information, and on the right is the current entry on display (Fig. 1.17). Four search methods are available. The Contents tab shows an organized outline of material, much like a library of books. The Index tab contains a wealth of frequently used entries. Typing a word in the box on top causes the window to jump through the index to the matching word. The Search tab allows the user to enter a word or phrase and search the entire Help text. Finally, the Favorites tab can be used to save entries that are frequently consulted.

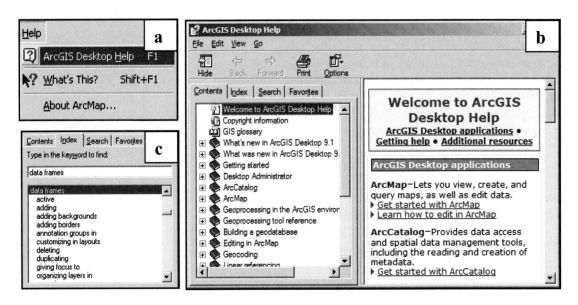

Fig. 1.17. ArcGIS Desktop help: (a) getting help, (b) the Help window showing the Contents tab, (c) the Index tab

Summary

➢ A GIS is designed as a database system that uses both spatial and aspatial data in order to answer questions about where things are and how they are related. It has many functions, including creating data, making maps, and analyzing relationships.

➢ Raster data employ arrays of values representing conditions on the ground within a square called a pixel. The array is georeferenced to a ground location using a single x-y point.

➢ Vector data use sequences of x-y coordinates to store point, line, or polygon features. Every feature is linked to an attribute table containing information about the feature.

➢ Every GIS data set has a coordinate system defined for stored x-y coordinate values. Many different coordinate systems are used, so each data set must be labeled with information about the coordinate system.

➢ Data are stored as simple spaghetti models or as topological models. Topological models can better model feature behavior and aid in locating and correcting geometric errors.

➢ Every GIS user has a responsibility to ensure that data are suitable for the proposed application. Data quality is measured in terms of geometric accuracy, thematic accuracy, resolution, and precision.

➢ Metadata stores information about GIS data layers to help people understand and use them properly. Metadata can be created in ArcCatalog, and the files are automatically copied and updated along with the data sets.

➢ GIS software by ESRI, Inc., has a long history with several major transformations along the way. The current version of ArcGIS Desktop employs a menu-based interface, with optional access to the older command-line functionality of Arc/Info. The Desktop consists of three programs: ArcMap, ArcCatalog, and ArcToolbox.

➢ ArcGIS uses a variety of data formats, old and new, including shapefiles, coverages, geodatabases, grids, images, TINs, and CAD drawings.

➢ ArcCatalog contains many functions for creating data, exploring files, and managing GIS data. It also provides tools for viewing and editing metadata.

➢ ArcToolbox contains functions for processing, managing, and analyzing GIS data. Users may customize it by building models or writing scripts to repeat often-used sequences.

VERY IMPORTANT TIP: Do not use Windows to copy or delete coverages, shapefiles, and geodatabases. These data sets may span multiple files and folders and might not be copied or deleted correctly. Always use ArcCatalog to delete or copy spatial data sets to prevent problems.

IMPORTANT TIP: Although spaces are permitted in the names of files and folders, they can cause problems for some GIS functions. It is recommended NEVER to use spaces when naming files and folders that will contain GIS data or to let spaces appear in any folders above them.

Chapter Review Questions

1. Explain the difference between the terms feature, feature class, and feature dataset.

2. Explain the relationship between Arc/Info and ArcGIS Desktop.

3. John and Mary are collecting GPS data together. John's GPS says their location is at (631058,4885805). Mary's GPS says their location is at (1204817,663391). Explain what is going on. What must be done to make the GPS units agree?

4. If each of the following data were stored as rasters, state which ones would be discrete and which ones would be continuous: rainfall, soil type, voting districts, temperature, slope, and vegetation type.

5. Imagine you are looking at a geodatabase that contains 50 states, 500 cities, and 100 rivers. How many feature classes are there? How many features? How many attribute tables? How many total records in all the attribute tables?

6. Would raster or vector be a better format for storing land ownership parcels? Give at least three reasons for your choice.

7. Scott is walking the boundary of a wetland area to map it. His GPS records locations to the nearest 0.10 meters. Is the boundary he creates accurate? Is it precise? Explain your reasoning.

8. Imagine a feature class of agricultural fields with attributes for the crop and the organic matter content of the soil. What issues might impact the thematic accuracy of each attribute?

9. Which aspects of a geodatabase help the user maintain logical consistency of data?

10. Construct an appropriate citation for the data that comes with this book.

Mastering the Skills

Teaching Tutorial

Preparing to begin

Each step of the tutorials in this book is illustrated by a video clip on the book's CD. You can view these clips whenever you want a demonstration of one of the steps in the tutorial. To view videos, do the following.

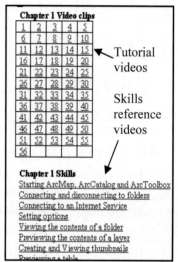

➜ Place the book's CD in the computer's CD-ROM drive. Wait for the splash screen to appear.

➜ Click the button to accept the license agreement. The main window appears (Fig. 1.18).

➜ Size the document window to a narrow strip on the left side of the screen. Put your ArcMap window on the right so that you can see it also.

➜ In the Chapter 1 section, click on the number of the tutorial video you want to see. Windows Media Player will start playing the clip. If asked whether you want to open the clip in Windows Explorer, say NO.

➜ Size the play window as large as possible for best resolution.

➜ When the video finishes, click the Minimize button in the upper-right corner of the Media Player window to get it out of the way.

Fig. 1.18. The Video Index provides links to demonstration videos.

➜ The headings under the Skills section contain links to performing different skills introduced in the chapter. Use these videos as a reference if you have forgotten how to do something.

➜ Before starting the tutorial, make sure that you have installed the mgisdata folder from the CD to the computer hard drive. See the Preface for detailed instructions.

Beginning the tutorial

The following examples provide step-by-step instructions for doing basic tasks and solving basic problems in ArcGIS. The steps you need to do are highlighted with an arrow ➜; follow them carefully. Click on the video number in the Video Index to view a demonstration of the steps.

We will begin with an overview of ArcCatalog just to highlight some of its capabilities as a preview to what GIS is about. You will learn more about these functions in the chapters to come.

TIP: In these tutorials, values you must enter are shown in this font: **type this**.

Adding connections to folders

ArcCatalog (and ArcMap) access data through **connections**, which are links to folders containing GIS data. By default, the main computer hard drive will always show as a connection (C:\).

→ Start ArcCatalog.

1→ Examine the folder tree on the left side and find the default connection, C:\.

1→ Click the plus sign next to it to expand the contents of the drive and see the subfolders.

Although you can navigate through folders to find any data on C:\ from the default connection, you can also establish connections to subfolders, thus creating handy shortcuts to frequently used data. You may already have one shortcut to the mgisdata folder, or it may be absent. If no mgisdata connection exists, then you can add one.

1→ Click the minus sign by the C:\ connection to collapse it again.

1→ Look for a connection to the mgisdata folder as shown by the red oval in Figure 1.19. The first part of the name may be different, depending on where the data were installed (such as C:\student\mgisdata).

1→ If the connection is already there, go to step 3.

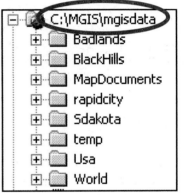

Fig. 1.19. A connection to the mgisdata folder in ArcCatalog

 2→ To add the connection, click the Connect to Folder button.

2→ Navigate to the directory containing your mgisdata and click on the mgisdata *folder* to select it. Do not select any of the subfolders, just the mgisdata folder.

2→ Click OK to add the connection.

Adding connections is critical for accessing data not on C:\, such as on a second disk drive, a network drive, or a CD-ROM drive. Once you add a connection, it will show up in ArcMap also and be present until you delete it. Connections tend to build up over time, so once in a while go through them and delete ones no longer being used. Also, having multiple pathways to the same folder, such as through connections to C:/MGIS/mgisdata and C:/MGIS/mgisdata/World, can cause unstable behavior in ArcGIS. Keep connections uncluttered and simple.

 3→ Locate the connection to the mgisdata folder just added. Click to select it and then click the Disconnect button. It disappears from the list.

3→ Click the Connect button, navigate to the mgisdata folder, and add the connection again.

Before we start exploring data with ArcCatalog, however, we will see how to use the Options in ArcCatalog to control some of the options associated with displaying data.

4→ Click the Contents tab in the ArcCatalog main window.

4→ Adjust the folder tree by clicking the plus signs until you can see the mgisdata\MapDocuments folder and click it to highlight it.

4➔ Click Tools > Options on the main menu bar.

4➔ Click the General tab.

The General tab can control which types of files and services appear in the Catalog. By default, all are shown. You can also choose to hide or show file extensions, such as .shp or .mxd.

4➔ Uncheck the box next to Hide File Extensions.

4➔ Click OK to close the Options menu and apply the changes. Notice that the map documents now appear with an .mxd extension.

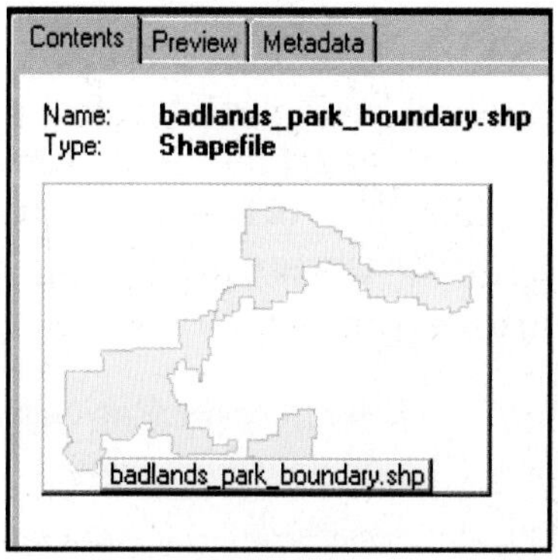

ArcCatalog gives users many ways to get information about data. The left window shows a folder tree. The right window has three tabs: **Contents**, **Preview**, and **Metadata** (Fig. 1.20). The Contents tab shows what is inside a folder or a geodatabase. The Preview tab allows the user to explore the spatial or attribute data of a feature class. The Metadata tab is used to view or edit metadata.

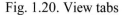

Fig. 1.20. View tabs

The Contents tab

The Contents tab is used to explore the contents of folders and geodatabases and is similar to Windows Explorer. It is the fastest option and shows information about what is currently highlighted in the folder tree. It offers four view styles for content (from left to right in Figure 1.21): Large Icons, List, Details, and Thumbnails.

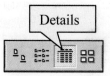

Fig. 1.21. Content view styles

5➔ Make sure that the Contents tab is clicked.

5➔ Click on the button for each view style to examine it, and then end by clicking the Details button (Fig. 1.21).

6➔ Expand the mgisdata folder, if necessary, and click on the Rapidcity folder to highlight it. Examine its contents (Fig. 1.22). (It will not match the figure exactly.)

6➔ Expand the plus sign next to the Rapidcity folder to see the types of spatial and attribute data that it contains.

6➔ Click the plus sign next to the citybnd layer. The expanded list shows each of the feature classes of the coverage. The shapefiles do not expand because they can only have one feature class.

6➔ Expand the TM_24sep98MS raster to see the seven bands in the image. Each band shows a different range of light wavelengths measured by the Landsat satellite.

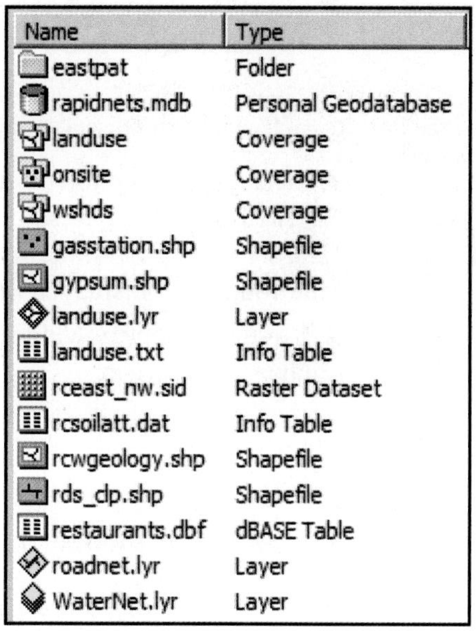

Name	Type
eastpat	Folder
rapidnets.mdb	Personal Geodatabase
landuse	Coverage
onsite	Coverage
wshds	Coverage
gasstation.shp	Shapefile
gypsum.shp	Shapefile
landuse.lyr	Layer
landuse.txt	Info Table
rceast_nw.sid	Raster Dataset
rcsoilatt.dat	Info Table
rcwgeology.shp	Shapefile
rds_clp.shp	Shapefile
restaurants.dbf	dBASE Table
roadnet.lyr	Layer
WaterNet.lyr	Layer

Fig. 1.22. Data sets in ArcCatalog

6➔ Click one of the entries in this folder and notice the information that appears in the main window. A coverage lists all the feature classes. A shapefile has a name and a **thumbnail** picture. Click on each type of data you can find in this folder and watch the Contents window update.

TIP: You can click the edge of a contents column and drag it to change the column width.

---

1. How many geodatabases are there in the folder mgisdata\Rapidcity? _____ How many coverages? _____ How many tables? _____ How many rasters? _____ How many layer files? _____ How many shapefiles? _____ How many feature datasets does the rapidnets geodatabase contain? _____ How many total feature classes does eastpat have? _____

7➔ Close the Rapidcity folder contents by clicking on the minus sign in the box next to it. Click the plus sign to expand the Oregon folder.

7➔ Expand the oregon.mdb personal geodatabase (Fig. 1.23). Expand the Transportation **feature dataset** to see the **feature classes** (including two line feature classes and one point feature class).

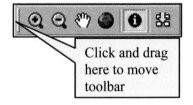

Fig. 1.23. A geodatabase in ArcCatalog

The Preview tab

The Preview tab helps explore the contents of each data set, including both the map data and the attribute data. We will begin by exploring the Tools toolbar (Fig. 1.24).

8➔ Select the counties feature class in the mgisdata\Oregon\oregon geodatabase, and click the Preview tab. The view changes to display the polygons.

9➔ Locate the Identify tool and click on it.

9➔ Place the tool on top of one of the counties and click. The Identify Results box will appear, and the county will flash on the screen. The attributes of the county are displayed in the Identify window.

9➔ Close the Identify window by clicking the X in the upper-right corner.

Fig. 1.24. The Geography toolbar

2. What is the name of the county in the northeast corner of Oregon? _____

Toolbars in ArcMap can be moved and docked at different locations, even outside the program. Hold down the Ctrl key while moving a toolbar to prevent it from docking.

10➔ Locate the toolbar with the Identify button again (called the Geography toolbar). At its top or left, find a faint gray line. This is its handle.

10➔ Click the handle and drag the toolbar out of the ArcCatalog window. Then click it and drag it to a spot with the other menu bars at the top of the window. Leave it wherever you like.

Now we will experiment with the zooming buttons.

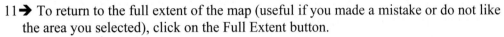

11➔ Click the Zoom In tool. Place the cursor at the upper-left corner of the state and click and hold. Continue holding the mouse button down and drag a box down and right to include a couple of counties. When finished, let go of the mouse button.

11➔ Click once in the lower corner of the map and notice that the view both zooms in and places the point you clicked at the center.

11➔ To return to the full extent of the map (useful if you made a mistake or do not like the area you selected), click on the Full Extent button.

11➔ Click on the Zoom Out tool and click in the upper-left corner of the state again. The view zooms out with the point clicked at the center. Both Zoom In and Zoom Out place the clicked spot at the center of the new field of view.

TIP: You can also draw a box using the Zoom Out tool. If a large box is drawn, the view zooms out a little bit. If a small box is drawn, the view zooms out a large amount.

11➔ Click on the **Pan** tool and then click and drag inside the display window to move the map around.

You can use the Preview tab to create thumbnails of a data set that will appear in the Contents tab when the data set is viewed.

12➔ Click the Contents tab and notice the thumbnail picture. It shows the entire country (the original data source from which the Oregon counties were extracted).

12➔ Click on the Preview tab again. Make sure that the counties layer is highlighted.

12➔ Click the Create Thumbnail button to create a snapshot of the current view. Nothing apparently happens, until . . .

12➔ Click the Contents tab. The updated thumbnail appears.

Not only can you see a data set's features in Preview mode, but you can also look at its attribute table, change the appearance of the table, and even add and delete fields.

13➔ Make sure that the Preview tab is active and then click on the cities feature class in the oregon geodatabase.

13➔ Choose Table from the drop-down menu that currently reads Geography. The cities table appears.

13➔ Scroll to the right through the end of the table, noting all of the fields.

3. How many records (rows) are there in this table? _____

13➔ Hold the cursor over the right edge of the FEATURE field until it turns into a double arrow bar. Click and drag the edge to the left to reduce the width of the column.

13➔ Click the Options button in the table window (if necessary, expand the window to the right to see it better).

Note the menu options. Find searches the table for particular text. Add Field adds a new field to the table, and Export saves a copy of the table as a .dbf file. Other options are also available.

14➔ Click the Options button again, if necessary, and choose Find.

14➔ Type **Portland** in the Find what box and click Find Next.

14➔ The cursor jumps to the record for Portland and highlights the city name.

14➔ Close the Find box.

A context menu gives access to several commands relating to the individual fields.

15➔ Right-click the field name POP_98 to display a context menu. Choose Sort Descending. (Make sure you scroll back to the top of the table to see the largest city, which should be Portland.)

4. Which town has the smallest (non-zero) population? _____

15➔ Right-click the POP_98 field and choose Statistics to see basic statistics and a frequency diagram of the values. Close the Statistics menu when done looking.

15➔ Right-click the NAME field and choose Freeze/Unfreeze. This places the field to the left of the table and keeps it there as you scroll to the right.

TIP: More than one field can be frozen at a time. Unfreezing allows the field to scroll again, although it remains on the left side until ArcCatalog closes.

You can even add or delete fields from the table in Preview mode.

16➔ Click the Options menu and choose Add Field.

16➔ Type **MAYOR** for the field name.

16➔ Use the drop-down box to set the field type to Text.

16➔ Enter a length of **25** for the field. Click OK.

16➔ Scroll to the right to see the new field added to the end of the table. To add data to the fields, you would need to edit the table in ArcMap. We won't try that yet.

17➔ Right-click the new MAYOR field and choose Delete Field. Choose Yes in the warning box.

TIP: Except for the Add/Delete field options, none of these operations changes the data that are stored on the disk. You can sort, freeze, and do statistics without changing the source data.

Feature class properties

Every data set has various properties that can be viewed and set in ArcCatalog. Each different type of data (shapefiles, coverages, rasters, etc.) can have different properties. We will examine some of these properties now, and future chapters will show how to work with the properties.

18➔ Expand the contents of the Rapidcity folder in your mgisdata folder.

18➔ Right-click the gas_stations shapefile and choose Properties from the menu.

18➔ Click the General tab. There is not much to set here.

18➔ Click the Fields tab. You can view, add, and delete fields here, too, and you will learn to do so in Chapter 4.

18➔ Click the Indexes tab.

Feature classes can have two types of indices. An attribute index can be created for individual fields and enhances performance when searching that field. A spatial index decreases the time needed to draw and query the layer.

18➔ Click on the XY Coordinate System tab.

The XY Coordinate System is an important property of a feature class. Every feature class has a coordinate system and should have a label like this one. You will learn more about coordinate systems later, but for now focus on finding two important pieces of information.

5. What is the name of this coordinate system? _____ What linear units are used to store the *x-y* values? _____

Shapefiles, coverages, rasters, and geodatabases all have different properties and some of the tabs will be different.

19➔ Close the Shapefile Properties window.

19➔ Expand the oregon geodatabase in mgisdata\Oregon so the feature classes are visible.

19➔ Right-click the parks feature class and choose Properties. There are many tabs here, some similar to the Shapefile tabs. Close the Properties box when finished.

19➔ Right-click the oregon.mdb geodatabase and examine its properties. Notice that it only has two tabs, General and Domains. Close the window when finished.

19➔ Right-click the Transportation feature dataset and open its properties. Examine each of the tabs and close it when finished.

Recall that a feature dataset can contain multiple feature classes, but all must share a common coordinate system. Thus, the coordinate system is defined for the feature dataset rather than for the feature classes within it.

Although shapefiles and geodatabase feature classes are spaghetti data models, feature datasets can be used to store topology information as a planar topology for finding and fixing errors or as a network topology for analyzing flow. Both topologies are built from existing feature classes.

20➔ Expand the rapidnets geodatabase in the mgisdata\Rapidcity folder.

20➜ Expand the Transportation feature dataset. Notice it has a single feature class, roads, and two feature classes that form the network, Road_Net and Road_Net_Junctions. Preview each of the feature classes.

20➜ Expand the Utilities feature dataset. It has many feature classes that form part of the network, including waterlines, Endcaps, Tvalves, and Galleries. It also has the two feature classes forming the network, Water_Net and Water_Net_Junctions.

21➜ Expand the fivestate geodatabase in the mgisdata\Southdakota folder.

21➜ Expand the stuff feature dataset. It has three feature classes and a planar topology named stuff_Topology.

21➜ Right-click the stuff_Topology feature class and choose Properties.

21➜ Click the General tab. Notice that errors are currently present in this topology, awaiting correction.

21➜ Click the Feature Classes tab. It shows that all three feature classes are participating in the topology.

21➜Click the Rules tab. It shows the rules that have been set up for this topology. Close the Topology Properties window.

22➜ Finally, examine the properties of one of the rasters and one of the tables in the mgisdata\Rapidcity folder.

Layer files

A **layer file** provides a way to set and store properties related to a feature class, such as how it should be displayed. A layer file is based on a feature class, but instead of storing the feature data inside the layer file, it only stores a pointer to where the feature class resides on the hard drive. The layer file can be displayed in ArcCatalog. It can also be added to ArcMap.

23➜ Make sure the Preview tab is clicked, and expand the folder tree to show the feature classes in the usdata geodatabase in the mgisdata\Usa folder.

23➜ Click on the states feature class.

23➜ Use the Zoom In tool to draw a box around the conterminous 48 states.

24➜Right-click the states feature class and choose Create Layer.

24➜ Navigate to the Usa folder, if necessary, and type in StatesLayer as the name of the layer file. Click Save.

24➜ Click on the new StatesLayer.lyr entry in the folder tree and examine the preview it shows of the data.

24➜ Right-click the StatesLayer layer file and choose Properties from the context menu (or double-click it to accomplish the same thing).

You are no longer looking at the feature class properties as you did before. You are looking at the Layer Properties. Notice all the tabs in this window. Each tab sets various properties of the layer and will store the settings in the layer file. Note that we are not changing the underlying data set used to create the layer, the states feature class. The properties in the layer file only act upon that basic data. The same feature class can be referenced by multiple layer files.

First, we will change the **symbol** used to draw the states. The layer is assigned a random symbol when the layer file is created, but it might not be the one we want.

25➔ Click on the Symbology tab.

25➔ Click on the button containing the symbol being used to draw the states.

25➔ The Symbol Selector appears. Click on a different symbol to select it.

25➔ Click OK to finalize your choice and close the Symbol Selector window.

25➔ Click OK to close the Layer Properties window and make the change.

26➔ Open the properties for the **StatesLayer** layer file again.

26➔ Click on the Symbol to open the Symbol Selector. Scroll down to see all of the available symbols.

26➔ Click on the More Symbols button. A list appears with the top two checked, your user name and ESRI. These entries are called styles.

Styles are sets of symbols stored together. The first checked one is your personal style, which is currently empty because you have not created or saved any symbols. The second is the default style, ESRI. The other entries are styles containing more symbols.

26➔ Click on the More Symbols button again to show the list, if necessary, and choose the Geology 24K style. The new symbols are added to the bottom of the current symbol display, so you won't see them yet.

26➔ Scroll down to view the new geology symbols added to the ones available.

26➔ The geology symbols don't look appropriate for this map, so click on More Symbols and choose Geology 24K again to uncheck the entry and remove the symbols from the symbol window.

> **TIP:** Styles can be viewed and managed using the **Style Manager**, which makes it easy to examine styles, copy symbols between styles, and edit symbol properties. See ArcGIS Help.

If the available symbols aren't quite what is wanted, they can be edited. A few simple edits can be made in the Symbol Selector window.

27➔ Choose one of the solid fill colors from the ESRI style again, such as Jade.

27➔ Click on the Fill Color button and choose a light blue color.

27➔ Click on the Outline Color button and choose a medium gray.

27➔ Click OK to close the Symbol Selector window.

27➔ Choose Apply in the Layer Properties window. Move the Layer Properties window a little, if necessary, so part of the states map can be seen.

> **TIP:** As a general rule, choosing Apply enacts the changes you have just made and leaves the window open. Choosing OK enacts the changes and closes the window.

Next we will create a layer file from a point feature class.

28➜ Close the Layer Properties window.

28➜ Right-click the cities layer in the usdata geodatabase and choose Create Layer. Save the layer in the Usa folder and name it **StateCapitals**.

28➜ Open the properties for the StateCapitals layer.

This layer contains all the cities, not just the state capitals. However, we can specify that only a subset of the feature class be included as part of the layer. We do this by establishing a **definition query**. A query is a database term for finding features that meet a stated condition, and a definition query extracts the features that will form the layer. Like all other layer properties, this query does not affect the original file.

29➜ Click on the Definition Query tab.

29➜ Click on the Query Builder button.

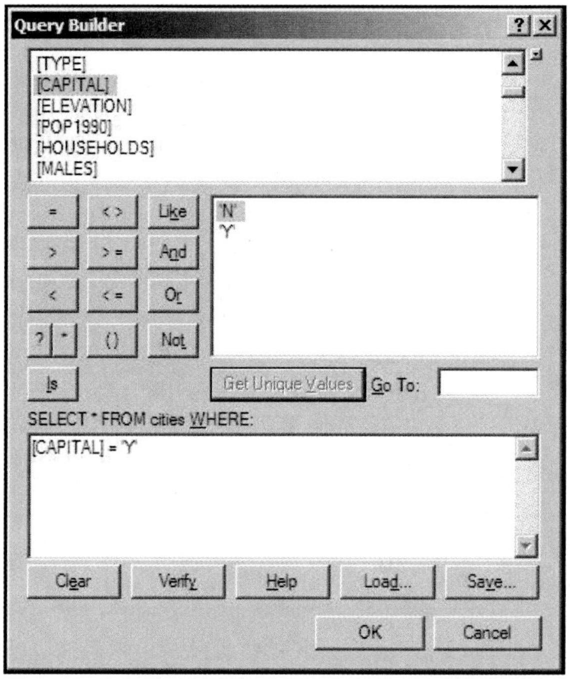

The Query Builder lets us enter the condition defining the capitals (Fig. 1.25). The cities attribute table has a field named [CAPITAL] that contains a 'Y' if the city is a capital and an 'N' if it is not. We will enter a statement describing the query in the lower box. The entire query will look like [CAPITAL] = 'Y'. Follow the directions carefully—it is better to use the buttons rather than trying to type it.

Fig. 1.25. The Query Builder window

29➜ Scroll down the field list to find the [CAPITAL] field and double-click it to enter it in the Query box. Make sure it appears before going on.

29➜ Click on the button with the equals (=) sign.

29➜ Click the Get Unique Values button to see a list of possible values in the field.

29➜ Double-click on 'Y' to enter it in the box.

29➜ To test that your query is properly entered, click the Verify button. If it is successfully verified, click OK. If not, click the Clear button and try again.

TIP: If the query is incorrectly entered, it is likely that no features will show up on the map. Thus, testing the query expression with the Verify button is very important.

29➜ Click OK and OK to establish your changes. Now only 48 cities should appear.

30➜ Open the properties window for StateCapitals again.

30➜ Click on the Symbology tab and click on the symbol to change it.

30➜ Scroll down the symbols window in the Symbol Selector and find the Star 3 symbol. Click on it to select it.

30➜ Change the symbol color to yellow and click OK and OK to view the changes.

31➜ Open the properties window for StateCapitals and click on the Source tab.

31➜ Examine the Data Source and notice that it refers to a feature class called cities and includes the location on the disk. This is the feature class upon which the StateCapitals layer is based. The original feature class still contains all the cities, even though we are only looking at a subset of them in this layer.

31➜ Close the Layer Properties window.

A group layer stores references to multiple feature classes and allows them to be displayed and changed together.

32➜ Right-click the Usa folder and choose New > Group Layer.

32➜ The new group layer appears with the default name, New Group Layer.lyr, highlighted in blue. Type the new name USAGroup and press Enter.

33➜ Open the Properties for the USAGroup layer.

33➜ Click on the Group tab and Choose Add.

33➜ Navigate to the Usa folder and select the StatesLayer layer file. Hold down the Ctrl-key and also select the StateCapitals layer file. Click Add.

33➜ Click Apply to make the change to the group layer. The states and capitals appear in the window. The states appear with labels (in Version 9.3).

You can add either layer files or feature classes to a group layer and set the properties separately for each one.

34➜ Click Add again on the Group tab, and double-click the usdata geodatabase to see the feature classes inside it.

34➜ Click the rivers feature class.

34➜ Hold down the Ctrl-key and click on cities to select it also. Click Add and OK. Both feature classes will be added to the group layer.

Now this group layer is starting to look like a map, but it needs work. The layers are cluttered. The cities dominate the composition and obscure several areas. We can use the **scale range** to set a maximum scale at which the cities appear so that they only show up when the user zooms in.

35➜ Open the USAGroup layer properties and click the Group tab.

35➜ Click on the cities layer to highlight it, and then click the Properties button.

35➜ Click the General tab.

35➜ Change the Layer Name from cities to Cities.

35➜ In the Scale Range section, fill the button to Don't show layer when zoomed.

35➜ Set the Out Beyond limit to 1:5,000,000 (one to 5 million) by typing just the number 5000000 in the box.

35➔ Click OK and OK to make the changes to the group layer.

The cities have disappeared from the map, but they will show up again when you zoom in to a smaller area.

36➔ Click on the Zoom In tool and draw a box around the state of Maine. The cities should show up. Zoom in a little closer, if necessary, until they appear.

36➔ Notice that the cities are drawn on top of the capitals so that each capital star symbol has a city symbol on top.

37➔ Open the USAGroup properties and highlight the StateCapitals layer in the Groups tab. Click the Up arrow to move the capitals layer to the top of the list. (The layers are drawn from bottom to top, so the capital stars will be drawn last.)

37➔ Click OK and notice that the star capital symbols are now clearer.

37➔ Click the Full Extent button and then zoom back in to the conterminous 48 states.

Labeling features

Layer files can also store information for creating labels for feature classes. They are called **dynamic labels** because they are created and updated each time the view of the map changes. An algorithm is used to find overlaps between conflicting labels. If some labels cannot be placed without overlapping others, they will be omitted from the map.

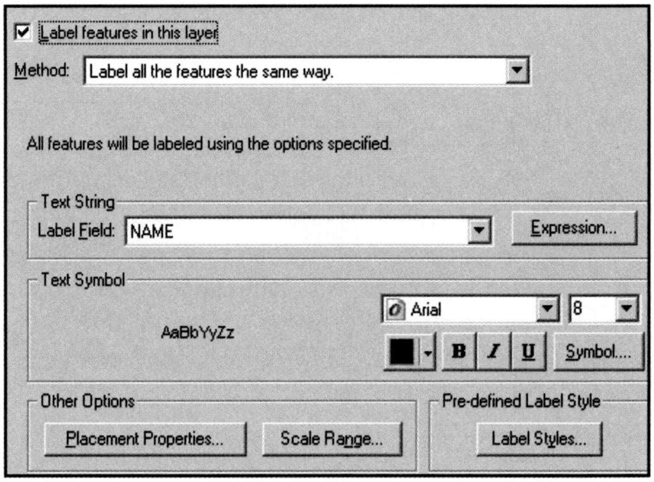

Fig. 1.26. Set the labeling properties of dynamic labels using the Labels tab in the Layer Properties window.

38➔ Open the USAGroup properties, click on the Groups tab, highlight StatesLayer, and choose Properties (Fig. 1.26).

38➔ Click on the General tab and rename the layer States.

38➔ Click on the Labels tab in the Layer Properties window.

38➔ The box to label the features is already checked.

38➔ Change the Label Field to STATE_NAME.

38➔ Make sure the text symbol is 8-point Arial Bold. Click OK and OK.

Notice that the northeast looks cluttered. We can modify the Placement Properties of dynamic labels to help manage this clutter.

39➔ Open the label properties for the States layer in the USAGroup layer again.

39➔ Click the Placement Properties button and click the Placement tab. Examine the settings.

39➔ Check the box to only place labels inside the polygon and to make them always horizontal. Click OK, OK, and OK.

The map now looks neater, although some of the state labels are missing. Unfortunately dynamic labels don't offer much control. For greater control of labels you must use annotation, which is described in Chapter 3. However, if you zoom in closer, the missing labels will appear.

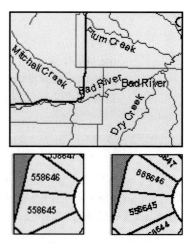

40➔ Zoom in to the New England area, and the missing state names will appear.

40➔ Use the Full Extent button again and zoom in to the conterminous states.

Placement properties vary depending on whether points, lines, or polygons are being labeled (Fig. 1.27). Polygon labels are generally placed horizontal or straight along the longest boundary of the polygon. Line labels can be splined along features.

Fig. 1.27. Label placement options for lines and polygons

41➔ Open the properties for the rivers layer in the USAGroup layer.

41➔ Click on the General tab and change the name to **Rivers**.

41➔ Click on the Symbology tab and choose a dark blue color for the river symbol.

42➔ Click on the Labels tab and choose to label the features.

42➔ Make sure the Label Field is set to NAME.

42➔ Change the Text Symbol to 8-point Italic Arial in a dark blue color.

42➔ Click on the Placement Properties button and click the Placement tab. Examine the placement properties in this window.

42➔ Choose Curved for the Orientation.

42➔ Check both Above and Below for the Position.

42➔ Fill the button to Remove duplicate labels so that multiple sections of a river won't be labeled multiple times.

42➔ Click OK, OK, and OK.

Point labels are usually placed on top of the point or at the most advantageous position around it. A point label graphic shows the first, second, and third priority locations and can be changed, if desired (Fig. 1.28).

43➔ Open the properties for the Cities layer in the USAGroup layer. Click on the Labels tab.

43➔ Check the box to label the features, and make sure that the Label Field is set to the CITY_NAME field.

43➔ Set the Text Symbol to 8-point Arial.

43➔ Click the Placement Properties tab and examine the options. Leave the default values and click OK to close the Placement Properties window.

Prefer Top Center, all allowed

Fig. 1.28.

44➔ Click the Scale Range button. We already have a scale range set for the cities, so fill the button to use the same scale range as the feature layer.

44➔ Click OK, OK, and OK.

44➔ Zoom in to Maine once more to see the labels. Return to the view of the conterminous states when done. Your group layer should look similar to Figure 1.29.

Now you have learned to use layer files to set and store properties of feature classes. Layers can save time by establishing default symbols for layers and by grouping layers that are commonly used together. In the next chapter, we will learn how to use layers and layer files in ArcMap.

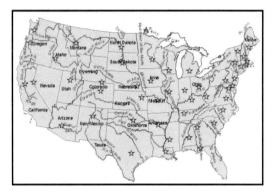

Fig. 1.29. The group layer will look like this after step 44.

Viewing metadata with the Metadata tab

We rely on metadata to provide information when data sets are shared and to help evaluate if the data are suitable for a particular purpose. Learning to read the metadata is the first step.

45➔ In the ArcCatalog folder tree, navigate to the mgisdata\Usa folder, expand the usdata geodatabase, and click on the states feature class to select it.

45➔ Click the Metadata tab.

45➔ Make sure that the Stylesheet drop-down box on the Metadata toolbar says FGDC ESRI. Examine the information.

45➔ Click on the green text **Abstract.** The abstract information folds up and is no longer visible. Click **Abstract** again to display the text once more.

45➔ Scroll down and click **Publication Information.** It tells who created the data and who published it.

6. Who created this data set and when was it published? _____

46➔ Click the Spatial tab at the top of the metadata. It switches to another "page" that contains information about the coordinate system, extent, and accuracy of the data.

46➔ Click the Attributes tab to see a list of the fields in the attribute table, their definitions, and their descriptions.

46➔ Click the AREA attribute and read the information about it. Notice that the description indicates the units are square miles. Had the creator of this data set not entered this information into the metadata, the user might have no idea whether the areas represented square miles or square kilometers.

Metadata is so important that a federal organization, called the Federal Geographic Data Committee (FGDC), has compiled standard rules about what kind of information goes into metadata and how it is organized and stored. These rules make up the FGDC Metadata Standard. The FGDC standard format is a text file with a very specific layout so that different programs can find the information they need.

47➔ Click the Stylesheet drop-down tab in the metadata toolbar and change it from FGDC ESRI to FGDC Classic.

47➔ Scroll down to examine the information, and then return to the top of the document.

Metadata contains seven main sections shown as hyperlinks to the information in the document. The metadata standard has hundreds of "fields" of information, organized into groups and subgroups for easy access. The field names are shown in italics, and the information is shown in plain text.

48➔ Change the Stylesheet to FGDC. This format is similar to FGDC Classic but is a little easier to read.

48➔ Click on the blue hyperlink at the top titled Data Quality Information.

48➔ Read the data quality information for this data set, including the logical consistency, completeness, and positional accuracy.

All of the stylesheets use exactly the same information but present it in a different way. The information is stored in a formatting language called XML, or Extensible Markup Language. It is similar to the HTML used to create Web page documents.

48➔ Change the stylesheet to XML to see what this file looks like. Then set it back to the FGDC ESRI stylesheet.

Working with Internet map services

Let's explore an Internet service site created by ESRI, Inc., the Geography Network.

49➔ Click the Contents tab in the ArcCatalog window.

49➔ Scroll down to the bottom of the folder tree and find the entry labeled **GIS Servers**.

49➔ Double-click the icon labeled Add ArcIMS Server.

49➔ For the URL type **http://www.geographynetwork.com**. Click OK.

49➔ When the Geography Network service appears, expand its plus box in the Table of Contents to view what it contains.

A list of services appears as icons. These entries are similar to other ArcCatalog data sets, and they can be viewed with the Contents and Preview buttons. The only difference is that the information is transmitted via the Internet. Preview works slowly with Internet data, so it is best to keep ArcCatalog in Contents mode until you want to see a specific layer.

50➔ Locate the FEMA_Flood service in the list and click it to highlight it.

50➔ Click the Preview tab. Zoom in to the United States for a closer look.

50➔ Zoom in to the state of New Jersey. Several cities should appear now.

50➔ Zoom in to the Philadelphia area very closely (draw a small box around just the yellow city symbol).

50➔ The FEMA flood zones of the Delaware River should now be visible, color-coded by type in shades of green. Notice that the service is providing data very similar to the layer files you've worked with previously, with different preset symbols, labels, and scale ranges.

The FEMA example demonstrates an image service. Now copy some data from a feature service.

51➔ Locate the EPA Hazards FS feature service and click the plus sign to expand it.

51➔ Preview the Superfund Sites layer. Then preview its table.

51➔ Right-click the Superfund Sites layer and choose Export.

51➔ Navigate to the Usa folder. No geodatabase appears yet because the window is set to store a shapefile.

51➔ Change the Save As Type to File and Personal Geodatabase feature classes.

51➔ Double-click the usdata geodatabase to open it.

51➔ Name the feature class **supersites** and save it. Now it can be accessed any time.

52➔ Navigate to the usdata geodatabase, click on the supersites feature class, and click the Preview tab.

52➔ Change the Preview method to Geography instead of Table.

52➔ Zoom to the full extent of the data set to see all of the sites (about 1400 of them).

➔ Explore more of the Geography Network services, if you have time.

52➔ When finished, right-click the Geography Network service and choose Disconnect.

Some services don't require a GIS program. They are accessible from a Web browser. If you have a high-speed Internet connection on your computer, try exploring the USGS National Map.

53➔ Open your Web browser and type in http://nationalmap.gov and click on the red Go to Viewer button.

53➔ Click the Zoom In tool and zoom in to Colorado.

53➔ Expand the Geology theme and check the box for Active Mines and Mineral Plants. Click Refresh Required in the lower right of the screen.

53➔ Notice the Download icon on the left. You can actually retrieve certain layers using this browser. See the Help if you want to learn more.

➔ Explore the National Map more, if you wish.

This is the end of the tutorial.

➔ Close the Web browser and ArcCatalog.

More skills

Consult the Skills Reference section of this chapter to learn to do the following:

➢ Managing GIS files (creating, deleting, copying, renaming)

➢ Creating new symbols using the Symbol Properties Editor

➢ Searching for data sets in a folder, hard drive, or Internet service

Exercises

> **TIP:** To submit certain answers to your instructor, you must capture an image from the screen and place it in an answer document. Whenever a question has the statement **Capture** in it, capture the answer by making sure the ArcCatalog window is active and simultaneously pressing the Alt and PrintScrn keys. This action places the active window on the Clipboard, and it can then be pasted into a Word or PowerPoint document. For help, ask your instructor.

1. How many feature datasets are there in the oregon geodatabase in the mgisdata\Oregon folder? List their names. How many total feature classes does the geodatabase have? How many each of point, line, and polygon feature classes does it have?

2. What is the coordinate system of the country shapefile in the mgisdata\World folder? Of the parks feature class in the oregon geodatabase?

3. What type of information does the feature class cd106 in the usdata geodatabase have? For what year(s) is the information valid? What was the original source of the spatial data? What are the source scale and the horizontal positional accuracy of this feature class? Could you use it to determine which parcels in your town belong to which district?

4. What is the largest lake in the United States? What is its area?

5. Which state has a county named Itawamba?

6. What is the minimum, maximum, and average number of nozzles in Rapid City gas stations?

7. How many rasters does the mgisdata\BlackHills\rasters folder contain? List them.

8. How many rows and columns does the Landsat image TM_24Sep98_utm have? What is the cell size (including the distance units)? How many bands does it have? What is its coordinate system?

9. Create a layer file showing only the counties in Nevada. Label as many counties as possible as long as the name fits inside it (using 8-point Arial font). **Capture** the ArcCatalog window showing the layer.

10. Consider the watershed feature class in the southdakota geodatabase and the states feature class in the usdata geodatabase. Answer the following questions for each: Who originally created this feature class? Can I publish a map containing this feature class? Under what circumstances, if any, could I give this data to someone else?

Challenge Problem

Search the Geography Network for three different feature services that cover the same geographic area of your choice. Download and save the data and use it to create a group layer file. Place the map in a Word document and include references for the data sets (providers, dates, etc.).

Skills Reference

Starting ArcMap or ArcCatalog 52

Starting ArcToolbox..52

Connecting and disconnecting from folders 53

Connecting to an Internet service 53

Setting options 53

Viewing the contents of a folder 54

Viewing and setting layer properties 54

Examining the coordinate system 54

Previewing layer geography 55

Creating and viewing thumbnails 55

Identifying features 55

Previewing a table 56

Viewing metadata 57

Setting symbols for a layer 57

Creating new symbols 58

Labeling features using dynamic labels 59

Managing ArcCatalog files 60

Searching for data..60

Starting ArcMap or ArcCatalog

1. Look on the computer desktop for an icon named ArcMap or ArcCatalog (Fig. 1.30). Double-click it to start the program.

2. If no icon is present on the desktop, click the Start button on the computer's menu bar. Navigate to Programs > ArcGIS and choose the name of the program desired.

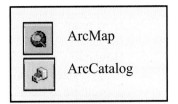

Fig. 1.30. Program icons

3. From ArcCatalog, launch ArcMap by clicking the ArcMap button in the menu bar.

4. From ArcMap, launch ArcCatalog by clicking the appropriate icon in the menu bar.

Starting ArcToolbox

ArcToolbox is a dockable window that sits inside ArcCatalog or ArcMap.

1. Click on the ArcToolbox icon in either ArcCatalog or ArcMap to open the window.

Connecting and disconnecting from folders

In order to access data files from ArcCatalog, you must set up a connection to the appropriate disk or folder. This saves time when frequently accessing data in a subfolder deep below the top. You can set up connections either to a drive letter, such as D:\, or to a subfolder in the drive. Connections can be deleted when they are no longer in use.

 1. To connect to a folder, click on the Folder Connect button in ArcCatalog.

2. Navigate down the directory tree and select a folder to which to connect.

3. Highlight the folder (or drive letter) and click OK (Fig. 1.31).

4. To disconnect from a folder, click the connection in ArcCatalog to highlight it and then click the Disconnect Folder button.

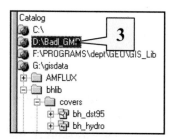

Fig. 1.31. Folder connection

Connecting to an Internet service

Using ArcCatalog, you can connect to and access data from Internet Map Servers. You must know the URL of the service, such as www.geographynetwork.com. Secure services also require a login and password. ArcGIS Servers are similar to ArcIMS servers but may serve data either over a local network or over the Internet.

1. In the left side of ArcCatalog, scroll to the bottom and expand the GIS Servers entry. Double-click the Add ArcIMS Server entry.

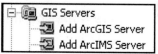

2. Type in the URL of the service in the top box.

3. If the server requires a user login and password, enter these at the bottom of the window and click OK.

Setting options

You can control the way the ArcCatalog performs various actions, displays information, and so on using this dialog box. You can also set defaults for the way tables, images, and other features are displayed.

1. Click Tools in the ArcCatalog menu bar and choose Options.

2. Click the appropriate tab to set the options.

3. When done setting options, click OK.

Viewing the contents of a folder

1. Click on the folder in the tree window.

2. Click on the Contents tab in the content window.

3. Choose one of the display options from the toolbar: Large icons, List, Details, or Thumbnails (Fig. 1.32).

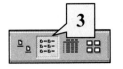

Fig. 1.32. View icons

Viewing and setting layer properties

Data sets, including layers and tables, have properties that can be modified in ArcCatalog. The types of properties will vary with the type of data set.

1. To access the layer/table properties, right-click on the layer/table name in the tree window and choose Properties. Double-clicking the layer/table also opens its Properties.

2. Click the tab containing the properties to change.

Shapefile tabs include: General, XY Coordinate System, Fields, and Indexes.

Geodatabase tabs for feature classes include: General, Fields, XY Coordinate System, Resolution, Tolerance, Domain, Indexes, Subtypes, Relationships, and Representations.

Coverage tabs include: General, Projection, Tics and Extent, and Tolerances.

Subsequent chapters will describe some of these properties and how to set them.

Examining the coordinate system

Every data layer has x-y coordinates that must be in some type of coordinate system. The coordinate system is stored as part of the information about a data set.

1. Right-click the feature class and choose Properties.

2. Click the XY Coordinate System tab (for feature classes) or the Projection tab (for coverages) (Fig. 1.33).

Using the Select, Import, or New buttons in this window will change the coordinate system label attached to this data set and should only be used on feature classes with an unlabeled or incorrectly labeled coordinate system.

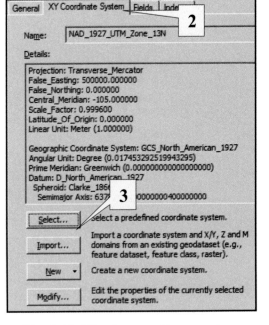

Fig. 1.33. Viewing coordinate systems of feature classes in ArcCatalog

Previewing layer geography

1. Click on the data layer to highlight it in the tree window.

2. Click the Preview tab (Fig. 1.34).

3. Choose Geography from the drop-down menu at the bottom of the Preview window to preview the spatial data, or choose Table to preview the table.

Use the Zoom, Pan, Full Extent, or Identify buttons to explore the geography preview.

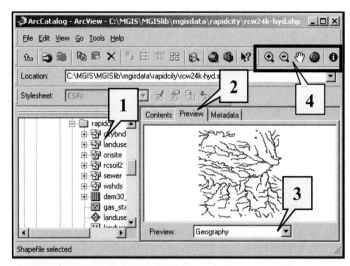

Fig. 1.34. Previewing a layer

Creating and viewing thumbnails

1. Make sure that the layer is highlighted in the tree window and the Preview tab is clicked.

2. If desired, use the Zoom and Pan buttons to modify the appearance of the layer.

3. Click the Thumbnail button in the Zoom/Pan menu.

4. To view all the thumbnails in a folder, click on the folder, make sure the Contents tab is clicked, and choose the Thumbnail display option.

5. To view the thumbnail for a single layer, click on the layer in the tree window to highlight it and make sure the Contents tab is clicked.

Identifying features

1. Click on the Identify tool.

2. Click on the feature to identify.

3. Clicking the identified feature in the dialog box will cause it to flash briefly (Fig. 1.35).

4. In the Identify Results box, click the drop-down list at the top and select the layer(s) to identify.

5. Click on a layer to view the attributes for that feature. Click on a different layer to see its attributes.

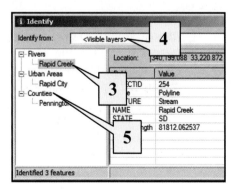

Fig. 1.35. Using Identify

Previewing a table

1. Click on a data layer or table in the tree window, and make sure that the Preview tab is clicked.

2. Choose Table from the drop-down box at the bottom of the Preview window.

Sorting the table

3. Right-click the field name and choose Sort Ascending or Sort Descending from the menu.

Getting statistics on a field

4. To get statistics for a numeric field, right-click on the field name and choose Statistics.

Freezing/unfreezing columns

5. To hold a field at the left side of the table while you scroll to the right, right-click the field and choose Freeze/Unfreeze from the context menu.

6. The field will move to the left edge of the table and remain there as you scroll to the right. More than one field can be frozen at a time.

Finding text in a field

7. To find text or values in a particular field, click on the field name to highlight the field (Fig. 1.36).

8. Shift-click to add additional highlighted fields to search, if desired.

9. Click the Options button in the lower right of the table and choose Find from the menu. (If necessary, enlarge the ArcCatalog window to the right to find the Options button.)

10. Type in the text to find. Modify the search settings, if desired, and click Find Next.

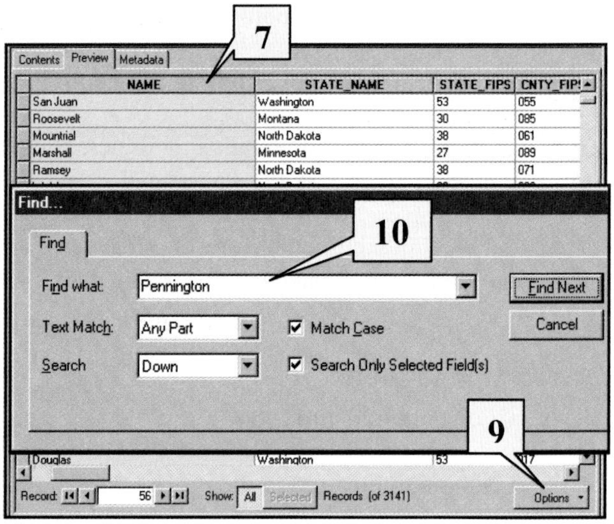

Fig. 1.36. Finding text

Adding/deleting fields from Preview mode

11. To add a field, click on the Options button and choose Add Field. Enter the table name and field type. For more information on adding fields to tables, see Chapter 5.

12. To delete a field, right-click on the field name and choose Delete Field. This action cannot be undone.

Viewing metadata

1. Click on the file you wish to view and click the Metadata tab.

2. Optionally, you can change the way the metadata look by choosing a different stylesheet from the drop-down menu. They are all based on the same metadata file, written in XML; they just present the information differently.

Setting symbols for a layer

You can set symbols a number of ways in ArcMap. Here are a few of the most common ways—all of these assume that all features in the layer are being drawn with the same symbol.

Changing the color of the current symbol

1. Right-click on the layer symbol in the Table of Contents and choose a color.

Changing properties of the current symbol

2. Click on the layer symbol in the Table of Contents to open the Symbol Selector window (Fig. 1.37).

3. Choose a symbol from the scroll box.

4. Modify the symbol's color, size, thickness, outline, or other attributes by setting the options provided. Make additional changes using the Properties button.

5. To load additional symbols in the scroll box, click the More Symbols button and choose from the list of categories.

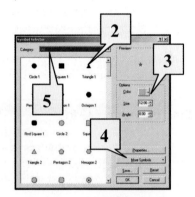

Fig. 1.37. Symbol Selector

6. Use the drop-down Category box to change which categories are currently visible in the box.

7. Click OK when finished modifying the symbol.

Creating new symbols

Symbols are created from one or more layers of symbol objects. For example, the symbol in Figure 1.38 is composed of a blue cross layer and a light-blue circle layer. Users can create new layers, put predefined symbols in them, and modify their colors and other properties to create new symbols. Symbols can also be created from imported bitmap images.

1. Click the Properties button in the Symbol Selector to access the Symbol Property Editor (Fig. 1.38).

2. Add new layers using the + button in the Layer area of the window.

3. Select the type of symbol character to put in the layer.

4. Select the desired character.

5. Modify the size, color, thickness, and other properties of the character.

6. Remove layers or change the order, if necessary.

7. Click OK when finished creating the symbol.

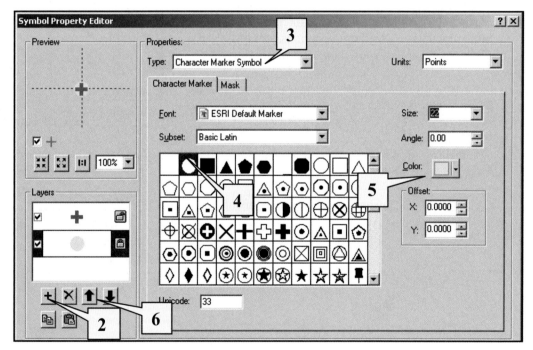

Fig. 1.38. Changing symbol properties

Labeling features using dynamic labels

1. Right-click the layer name in the Table of Contents and choose Properties from the menu.

2. Click the Labels tab.

3. Check the Label Features box (Fig. 1.39).

4. Make sure the method is set to label all the features the same way.

5. Choose the Label Field. Click the Expression button to enter a VBA (Visual Basic for Applications) script.

6. Edit the font settings, or select a predefined text symbol by clicking the
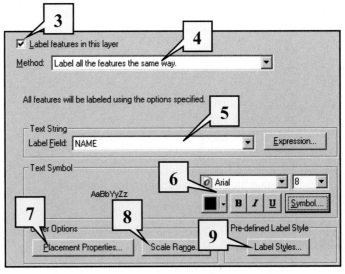
Fig. 1.39. The Label Properties window settings

Symbol button and choosing a predefined symbol style.

7. For detailed control of label placement, click the Placement Properties button.

8. Set the scale range, if desired, by using the label's scale range or by typing in new values. If the map scale is outside the specified range, the labels will not be drawn.

9. Select a label style, if desired. A label style includes BOTH a text symbol and predefined label placement options.

10. Click OK to place the labels.

TIP: Turn labels on and off for a layer by right-clicking the layer name in the Table of Contents and choosing the Label Features option. If the menu choice is checked, the labels are on, and choosing it will turn them off. If it is unchecked, choosing it will turn them on.

Managing ArcCatalog files

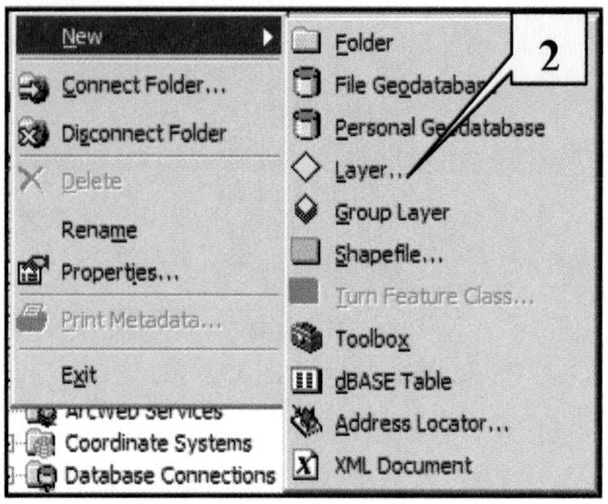

Creating new shapefiles and geodatabases

1. Click on the folder to contain the new file.

2. Choose File > New > and the type of file to create (Fig. 1.40).

3. Follow the instructions in the menu for creating the file. Different file types require different parameters.

4. To delete a file, right-click it and choose Delete. Click Yes to confirm that it should be deleted.

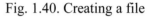

Fig. 1.40. Creating a file

Copy and paste files

5. Right-click the object to be copied and choose Copy.

6. Right-click the folder or geodatabase to place the object in and choose Paste.

Renaming files

7. Click twice slowly on the file name to highlight it. Type in the new name.

Searching for data

The search engine in ArcCatalog locates data sets on a particular disk or server that cover the geographic area specified. The results are written to a catalog entry and can be easily dragged into ArcMap or copied to a different directory.

> **TIP:** The Search option uses the file metadata and may fail to find files if the metadata are missing or incomplete.

 1. Click the Search button in ArcCatalog.

2. Click the **Name & location** tab (Fig. 1.41).

3. To search for particular types of files, choose them from the Type list. Use Ctrl-click to select more than one. To search for all types, don't select any or click Clear.

4. Choose Catalog to search only connected folders, or choose Disk to search the entire disk.

5. Set the disk or other location to search.

6. Click the **Geography** tab. Check to use geographic location in search.

7. If you know the map coordinates to search, type these directly into the boxes.

8. OR, to locate an area to search on a map, use the Map drop-down box at the bottom of the Geography tab to choose a display map to search on. Use one of the default maps or select a different one from the disk by entering <Other>.

9. OR use the buttons to zoom in to the appropriate region and draw a box around the target area.

10. OR select a place name from the list at the top of the tab. The list will depend on the map chosen in step 8. Thus, to search for a county, choose US Counties as the map layer.

11. Choose to search for data entirely within your location or overlapping your location.

12. To search for data for specific dates, click the **Date** tab. Fill out the information.

13. You can also search for explicit fields and values in the metadata by clicking the **Advanced** tab. This option is only recommended for experienced users. However, you can use it to find data produced by certain agencies or having certain keywords. There are many ways to refine a search.

14. In the Geography tab, type in a name under which to save the search or use the default, My Search.

15. When ready, click Find Now. Wait, as the search may take some time to complete. The status bar at the bottom of the window shows the folders being searched. Click the Stop button at any time to quit the search.

16. When the search is complete, click on My Search (or whatever you named it) in the ArcCatalog tree window to expand it and to view the results.

17. The Search routine places links to the original data inside the My Search entry. You can preview the geography and tables of these links just as you would other files and drag and drop them into ArcMap.

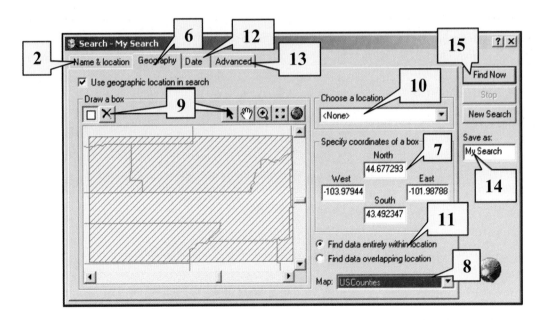

Fig. 1.41. Searching for geographic data

Chapter 2. Mapping GIS Data

Objectives

➢ Understanding map scales and GIS data scaling issues

➢ Knowing types of attribute data and the maps created from each

➢ Understanding how to classify numeric data

➢ Displaying raster images

➢ Understanding how map documents save and represent data

Mastering the Concepts

GIS Concepts

Simply storing GIS data in a computer is no fun. The power of GIS comes from the ability to explore map data, to analyze spatial patterns, and to communicate those findings to others. This chapter will focus on creating maps from the attribute data of a GIS feature class.

Map scale concepts

The act of taking a set of GIS features with *x-y* coordinate values and drawing them on a screen or printing them on a piece of paper establishes a map scale. On a paper map, the scale is fixed at the time of printing. Within a computer system that allows interactive display, the scale changes every time the user zooms in or out of the map.

What is map scale?

Map scale is a measure of the size at which features in a map are represented. The scale is expressed as a fraction, or ratio, of the size of objects on the page to the size of the objects on the ground. Because it is expressed as a ratio, it is valid for any unit of measure. So for a common U.S. Geological Survey topographic map, which has a scale of 1:24,000, one inch on the map represents 24,000 inches

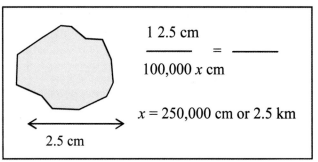

Fig. 2.1. Solving for the size of a lake

on the ground (or one meter represents 24,000 meters on the ground). Imagine a map made to the scale of 1:100,000. You can use the map scale and a ruler to determine the true distance of any feature on the map, such as the width of a lake (Fig. 2.1). Measure the lake with a ruler and then set up a proportion such that the map scale equals the measured width over the actual width (*x*). Then solve for *x*. Keep in mind that the actual width and the measured width will have the same units. You can convert these units, if necessary.

Often people or publications refer to large-scale maps and small-scale maps. A large-scale map is one in which the *ratio* is large (i.e., the denominator is small). Thus, a 1:24,000 scale map is larger scale than a 1:100,000 scale map. Large-scale maps show a relatively small area, such as a quadrangle, whereas small-scale maps show bigger areas such as states or countries.

Scales for GIS data

When data are stored in a GIS, they do not have a scale, technically speaking, because only the coordinates are stored. They acquire a scale once they are drawn on the screen or on a piece of paper. However, the concept of scale does apply to GIS data in the sense that most data layers have an intrinsic scale at which they were created. A 1:1 million scale paper map that is scanned or digitized cannot effectively be used at larger scales. The map in Figure 2.2 shows congressional districts in pink and the state outlines in thick black lines. Notice how the state boundaries are more angular and less detailed than the districts because they were digitized at a smaller scale. Thus, although it is possible to take small-scale data and zoom in to large scales, the accuracy and detail of the data will suffer. The original scale of a map or data set is an important attribute, and it should be included when creating metadata.

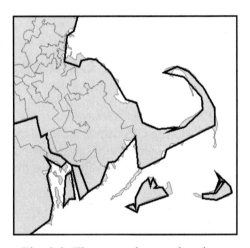

One should exercise caution in using data at scales very different from the original map. Zooming in to a data set may give a false impression that the data are more precise than they actually are. A pipeline digitized from a 1:100,000 scale map has an uncertainty of about 170 feet in its location due to the thickness of the line on the paper. Displaying the pipeline on a city map at 1:60,000 might look fine, but zooming to 1:2000 would not help should you desire to locate the pipeline by digging.

Fig. 2.2. These two layers showing Massachusetts originated from maps with different scales.

From looking at Figure 2.2, one might conclude that it is desirable to always obtain and use data at the largest possible scale. However, large-scale data require more data points per unit area, increasing data storage space and slowing the drawing of layers. Every application has an optimal scale, and little is gained by using information at a higher scale than needed.

Types of maps

Cartographers may choose a variety of strategies for symbolizing features in a map. A data layer may simply be portrayed using one symbol (Fig. 2.3), or different features can be assigned different symbols depending on the value of an attribute field, such as zoning or population. Point data are shown using marker symbols, line features with linear symbols, and polygon features with shaded area symbols (Fig. 2.4). When creating maps, it is important to distinguish between **nominal, categorical**, and **numeric data** because the type of data will affect the kind of map that can be made.

Fig. 2.3. Single-symbol map

Maps for nominal and categorical data

Nominal data names or identifies objects, such as the name of a state or a street. Nearly every feature will have a different name. Nominal data aren't always text. A parcel number or a tax identification number also serves the purpose of uniquely identifying an object. The U.S. Census uses such data in its Federal Information Processing Standards Codes, or FIPS codes. Each state, county, city, block, and so on is assigned a unique identifying number that is consistent from product to product and can be used to link data from different documents together. Nominal data are usually portrayed in a map by labels.

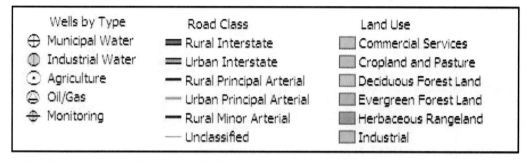

Fig. 2.4. Examples of categorical data representation for points, lines, and polygons

Categorical data separates features into distinct groups or classes. Figure 2.4 shows examples of categorical data for point features (well type), linear features (road class), and polygon features (land use). Other examples might include soil types, ethnic groups, or geological rock formations. Categorical data are often stored as text, but it is also possible to represent the categories with numerical codes, such as Commercial Services = 20 and Industrial = 40. Even though the values are numeric, they still represent categories. Categorical data are usually represented by **unique values maps**, which give each category a different symbol (Fig. 2.5).

Ordinal data is grouped by rank based on some quantitative measure, although the measure may not be linear in scale. For example, cities might be classified as small, medium, or large. Students are assigned grades of A, B, C, D, or F on tests. Soils are sometimes designated as A, B, C, or D depending on their infiltration capacity. Ordinal data must be represented by unique values maps if the values are text, but the colors should be the same shade and give a sense of increase. Numeric ordinal data may be represented either with unique values maps or graduated color maps.

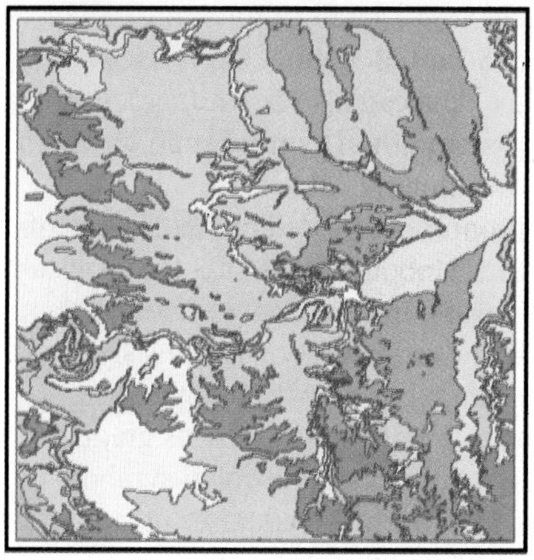

Fig. 2.5. A unique values map based on categorical data representing different geological units

Maps for numeric data

Numeric data store numbers that represent phenomena that are continuous and fall along a regularly spaced measurement scale, such as population or rainfall. Equal changes in interval involve equal changes in the object being measured. For example, the amount of energy to heat a thimble of water from 6 to 7 degrees is the same as heating it from 96 to 97 degrees. **Ratio** data have the property that the measurements are related to a meaningful zero point such that zero indicates the absence of the thing being measured. Precipitation is an example of ratio data; zero precipitation corresponds to a total lack of rain, and two inches of rain is twice as much as one inch. Ratio data support all four arithmetic operations of addition, subtraction, multiplication, and division. **Interval** data have a regular scale but are not related to a meaningful zero point. Temperature data measured in the familiar Celsius or Fahrenheit scale are interval data because a temperature of zero does not correspond to a complete lack of temperature (heat energy). Any measuring scale that can have negative values, such as elevation and pH, must be an interval scale. Interval scales only support addition and subtraction.

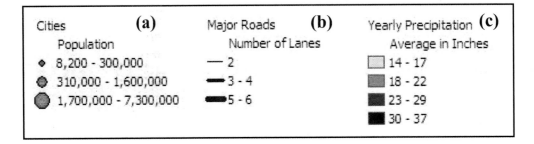

Fig. 2.6. Numerical data representations for (a) points, (b) lines, and (c) polygons

Numerical attribute data for points or lines are typically mapped using variations in symbol size or thickness (Fig. 2.6a and 2.6b). The numerical values are partitioned into distinct ranges of values called classes. These examples are called **graduated symbol** maps. However, the data can also be portrayed using a **proportional symbol map** in which the numeric value is used to proportionally determine the size of the symbol. Instead of a few size classes, the map has a continuous range of symbol sizes. This style is often referred to as an unclassed map. Figure 2.7 compares graduated symbol and proportional symbol maps for the populations of state capitals.

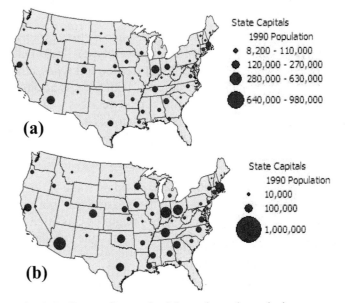

Fig. 2.7. Comparison of a (a) graduated symbol map and a (b) proportional symbol map showing the populations of state capitals

Numerical data for polygons are typically represented by changes in hue, saturation, or intensity of color using shaded symbols (Fig. 2.6c and Fig. 2.8). Such maps are called **graduated color maps**, or **choropleth maps**. Changes are usually represented using a single color that changes in intensity (monochromatic color ramp) for each class. Color ramps with two or more colors may become difficult to interpret unless the symbols are chosen carefully (such as the precipitation map in Figure 2.8, which implies a dry to wet regime).

For some attributes, the size of the feature influences the value of the attribute reported for that feature. For example, counties with larger areas will tend to have more acres of farmland or larger populations. Such data should be **normalized** to the quantity per unit area when they are mapped, in order to prevent the size of the measurement unit from influencing the map appearance. Thus, population data are usually normalized to show the number of people per unit area (population density). Some attributes don't need normalization. In a precipitation map, each value already represents inches of rain falling at any location. The median rent for a county does not depend directly on the area of the county. These attributes would not be normalized.

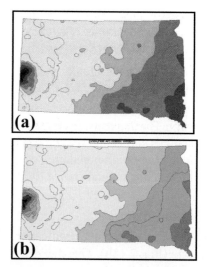

Fig. 2.8. A graduated color map showing annual precipitation with a (a) monochrome color ramp and a (b) two-toned color ramp

Another way to normalize values to the area of a polygon is to create a **dot density map**, which uses randomly placed dots within a polygon to show the magnitude of a value in the attribute table. Each dot represents a certain number, for example, 1 million people. In that case, California would have about 33 dots in it, and South Dakota would barely have 1. In Figure 2.9, one dot equals 100,000 people. Note that the locations of the dots, which are randomly placed, do not necessarily represent the actual distribution of people in the state.

Chart maps

A chart map expands the number of attributes that can be displayed on a map by replacing a single symbol with a chart representing several attributes. The chart could be a pie chart, bar chart, or stacked bar chart. Figure 2.10 shows a pie chart map with the proportions of Caucasians, African Americans, and Hispanics in each state. The pie sizes can be all the same (Fig. 2.10a) or proportional to the sum of the three categories, thus showing the relative number of people in the state as well (Fig. 2.10b). Notice how the pie colors are chosen to facilitate recognition of the classes: cream for Caucasians, dark brown for African Americans, and reddish for Hispanics. Such details help make maps easier for the reader to interpret, and they lessen the need for the reader to look back and forth between the legend and the map.

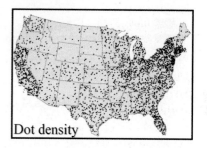

Fig. 2.9. A dot density map showing state populations

Classifying numeric data

Creating graduated color or graduated symbol maps requires dividing a continuous range of data values into groups, each of which receives its own symbol. This process is called **classification**. Different methods of classification are suited to different types of data, and the choice of method affects the appearance of the map and the

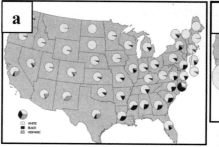
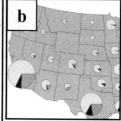

Fig. 2.10. (a) A chart map can represent several attributes, such as the proportion of ethnic groups. (b) The pie size can represent the total population.

message that it portrays. Some common methods are discussed in the following paragraphs, and an example data set of average farm size by state is used to compare the methods (Fig. 2.11).

Natural breaks (Jenks method)

The Jenks method sets the class breaks at naturally occurring gaps between groups of data (Fig. 2.11a). Each class interval can have its own width, and the number of features in each class will vary. The Jenks method works well on unevenly distributed data, such as populations of the capitals shown in Figure 2.7. There are many low-population and medium-population capitals and a few very high-population capitals. The Jenks method works well with almost any data set, making it a natural choice for the default classification scheme in ArcMap.

Equal interval

The equal interval classification divides the values into a specified number of classes of equal size (Fig. 2.11b). This method is useful for ratio data, such as income or precipitation because it gives a sense of regularity to the observed increases. However, it is hard to predict how many features will end up in each class. Notice in the farm size example that nearly all of the states fall into the first class. Compare this map to Figure 2.11a.

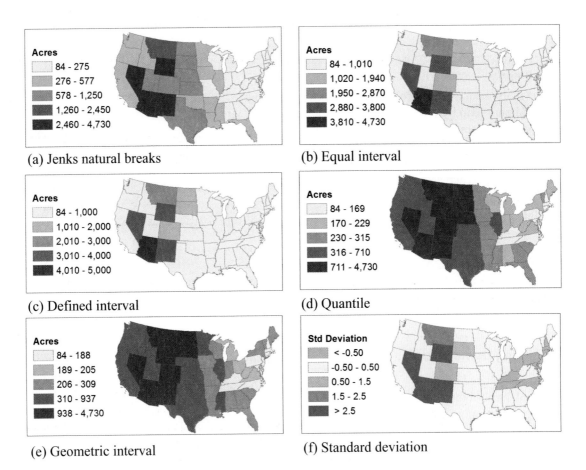

Fig. 2.11. Comparison of different classification methods using the same data set: average farm size in acres by state

Defined interval

A defined interval classification is similar to an equal interval one, except that the user specifies the size of the class interval, and the number of classes then depends on the range of values (Fig. 2.11c). This method will create rounded values in the classes that are easy to interpret. Defined interval maps are ideal when comparing classes composed of percentages, dollars, temperatures, and other values when specific break values are desired (100, 200, etc.). It does, however, suffer the same disadvantages as the equal interval classification.

Quantile

A quantile classification puts about the same number of features in each class (Fig. 2.11d). This method will create a very balanced map with all classes equally well represented, but some of the features in the same class could have very different values, and features in different classes could have similar values. Quantile classifications are best applied to linearly distributed data. Notice that a quantile classification highlights differences between the eastern states that are hidden in the Jenks, equal interval, and defined interval classifications.

Geometric interval

A geometric interval classification bases the class intervals on a geometric series in which each class is multiplied by a constant coefficient to produce the next higher class. A geometric interval classification is designed to work well with continuous data like precipitation and to provide about the same number of values in each class range. In the farm size example, it produces a map similar to the quantile classification (Fig. 2.11e).

Standard deviation

A standard deviation classification apportions the values based on the statistics of the field. The user selects the class breaks as the number of standard deviations, and the data range determines the number of classes needed. This method excels at highlighting which values are typical and which are outliers. In Figure 2.11f, the yellow states are close to the mean farm size, several eastern states appear to have smaller than average farms, and some western states have much larger farms than average. This map is best applied to normally distributed data.

Manual class breaks

Finally, if none of the above options gives the desired map, the user can set class break points manually to any chosen values.

Choosing the classification method

The choice of classification method depends on the mapmaker's purposes and on the type of data. Jenks shows the "nearest neighbors" in a distribution, whereas a defined interval or equal interval map does a better job of portraying relative magnitude. Notice the difference between the population maps in Figures 2.11a and 2.11b. The Jenks map suggests that states with large populations are common; only a careful examination of the legend leads the reader to understand that most states have relatively small populations in comparison to a few with very large populations. Figure 2.11b makes this observation clear at first glance.

Some data possess the quality that the magnitude of the values has an intrinsic meaning. Percentage data, for example, have a physical and a psychological meaning—people have an intuitive understanding of the difference between 50% and 100%. Differences in median rent occur in dollar values to which people can attach meaning. When dealing with such data, it is often wise to use a defined interval map in order to choose classes with logical breakpoints, such as 20%, or 5 inches of rain, or $200, rather than 12.6%, or 1.47 inches of rain, or $187. The reader can more effectively assign meaning to the classes.

The distribution of values has impact also. Jenks and geometric interval classifications are designed to work well with unevenly distributed data. Equal interval, defined interval, and quantile maps can be used with any data but show better results with evenly distributed data. The statistics behind standard deviation maps assume that the data are normally distributed.

Displaying rasters

Raster data occurs as thematic rasters or image rasters. **Thematic rasters** represent map quantities, such as land use or rainfall. Thematic rasters also fall into two categories we have already discussed, discrete or continuous. A discrete thematic raster has coded values that define regions, such as geology or land use. A continuous thematic raster has values that change continuously from one location to another, such as elevation or precipitation. **Image rasters** mainly come from aerial photography and satellites, and the pixels represent different degrees of brightness, typically on a scale of 0–255 DN (digital numbers). A dark shadow would have a low DN, and a bright white cement road would have a high DN. Displaying rasters requires an understanding of these basic data categories and the display techniques applicable to them.

Thematic rasters

Thematic rasters use display options similar to vector features. A discrete raster is best displayed using a **unique values** classification, where each value of the raster receives its own color, just like a unique values map for vector data. The geology map shown in Figure 2.12a provides a good example. The **discrete color** option is similar to unique values but is faster when many unique values are present because it assigns colors randomly.

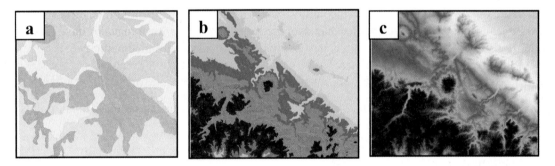

Fig. 2.12. Display methods for thematic rasters: (a) unique values geology; (b) classified elevation; (c) stretched elevation

Continuous thematic rasters can be displayed with two options. The **classified** method divides the values into classes, just like graduated color maps. The same classification methods (Jenks, equal interval, etc.) are available. The elevation map in Figure 2.12b has 12 classes. The **stretched** method scales the image values to a color ramp with 256 shades (Fig. 2.12c). The raster is first subjected to a **slice**, which rescales the image values from 0-255, and then matches them to the available 256 shades in a color ramp.

Image rasters

Image rasters may be displayed by two different methods. The **stretched** method is commonly applied to single-band images. Many images have only 0 to 255 values to begin with, so slicing is often not necessary. However, many images require additional stretch options.

Image data are often normally distributed, with many values in the middle range and fewer at the tails. Figure 2.13a shows a **histogram** of cell values, with the horizontal axis showing the range of values and the vertical axis showing the number of cells for each value. Each cell is assigned a color along a ramp of 256 color values (Fig 2.13a). A stretch ignores the tails of the distribution when assigning the symbols to display the image with improved brightness and contrast (Fig. 2.13b). The two-standard-deviation stretch is most commonly applied, which uses only the values within two standard deviations of the mean. However, other stretches such as a minimum-maximum stretch are also employed. Figure 2.13c shows the image without the stretch applied. Stretches can be applied to thematic continuous rasters as well as images.

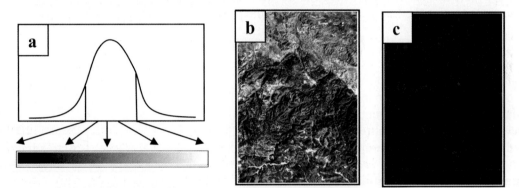

Fig. 2.13. The effects of stretching: (a) stretching an image; (b) standard deviation stretch; (c) no stretch

The **RGB composite** method applies to rasters containing more than one band of data. One band is assigned to each of the three color guns on the computer monitor (red, green, and blue) to produce a color image. If the value in Band 1 is high (close to 255) and the other two bands have low values near zero, a bright red color will result. If red and green are both high and blue is low, a mixture of red and green will create yellow. All the colors on the screen are composed of different proportions of red, green, and blue light, as dictated by the values in the bands. Each band of the composite may be individually stretched. Figure 2.14 shows the same image displayed using the stretched method for a single band (a) versus a true-color composite using three bands (b).

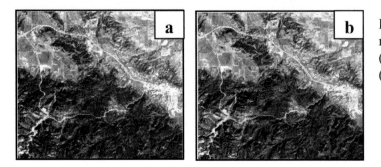

Fig. 2.14. Raster display methods for images: (a) stretched image; (b) RGB composite image

About ArcGIS

Map documents

ArcMap creates map documents, just as Microsoft Word creates Word documents. A map document is a collection of spatial data layers and tables and their properties. When a feature class is added to ArcMap, it becomes a **layer**. Just like the layer files in Chapter 1, each layer has properties that control how it is displayed and used. The only difference is that the layers are stored in memory rather than on the hard drive.

Map documents organize the data using **data frames**, windows that contain groups of layers that are drawn together. Maps can contain multiple data frames (Fig. 2.15). One frame at a time is designated the **active frame**. All commands and actions, such as adding layers or creating a scale bar, occur in the active frame. The blue handles indicate that the frame can be resized on the page.

A map document does not store the GIS data files within it. Instead, the map stores the name and disk location of the spatial data—its **source**. Most changes to the map document do not affect the source data, only its properties within the document. Some functions do change the source data, such as adding attributes or editing boundaries. Users need to learn which are which.

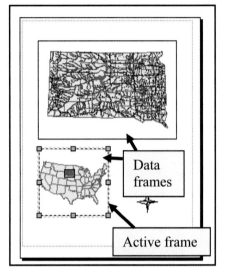

Fig. 2.15. Data frames

Because the source data are stored outside the map document, *changes to source data affect all map documents using the same data set.* Because data are shared in a GIS system, extra precautions are needed to ensure that an editing mistake does not affect dozens of people. Often the source files are write-protected so that only one or two users are authorized to make changes. Systems in which multiple users must edit data typically use an ArcSDE geodatabase that supports versioned editing.

Storing the data separately has several benefits. First, when a shared database is updated, the changes automatically appear in every map document based on that database. Second, it saves space. GIS data sets can be very large, but the map documents that reference them don't have to be. However, the practice has implications for sharing map documents. Giving a colleague a copy of a map document by e-mail only works if the recipient has access to exactly the same data in the same disk location as the original user.

Map documents and pathnames

Map documents keep track of the source files by storing the location of each file as a **pathname**, which lists the successive folders traversed to a data set, each separated by backslashes. The pathname c:\mgisdata\usa\states.shp refers to the shapefile states.shp, which resides in the Usa folder, which in turn resides in the mgisdata folder, which is found on drive C:\ (Fig. 2.16).

The pathname c:\mgisdata\Usa\States.shp is termed an **absolute pathname** because it starts at the drive letter and proceeds downward to the file. When searching for a file using an absolute pathname, the search begins at the drive level. A **relative pathname** is used to indicate that a search should be made starting at the current folder location. A double dot indicates movement up one folder. The pathname ..\Usa\States.shp indicates a search up one folder and then down into the Usa folder and then to the States shapefile (Fig. 2.16).

Map documents can store either absolute or relative pathnames as chosen by the user, with absolute pathnames being the default. Imagine that the map document ex_1.mxd in Figure 2.16 refers to the States file in the Usa folder. When the map document is opened, it searches for States starting at the C:\ drive down through C:\mgisdata\Usa\States.shp in order to locate and draw the states in the map.

However, what if the user decides to move the mgisdata folder to another network drive designated with the letter F:\? When the ex_1.mxd map is opened, it searches down the C:\ path but can no longer locate the mgisdata folder. The links to the data are broken, and in ArcMap, the States layer will appear with a red exclamation point (Fig. 2.17). The user can manually locate the data to fix the links, but this can be time consuming. (ArcCatalog can fix multiple links more efficiently.)

A better solution is to direct the map document to store relative pathnames. In this case, the States pathname is stored as ..\Usa\States.shp, and it no longer matters whether the mgisdata folder exists on the C:\ drive or the F:\ drive. As long as the relative positions of the MapDocuments and Usa folders remain the same, the map document can always locate the data.

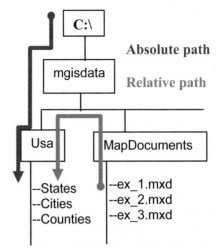

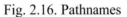

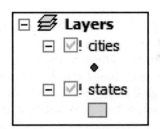

Fig. 2.16. Pathnames

Fig. 2.17. The red exclamation point indicates that the pathname link is broken.

The choice of absolute or relative pathname depends on the situation. If the map documents will always refer to a central data server that all users on a network have access to, then absolute pathnames will more reliably locate the data layers, even when e-mailing documents. On the other hand, if a user plans to move the map documents and data together as a unit, then relative pathnames work better. The map documents on the CD that comes with this book are stored with relative pathnames so that the links to the data layers continue to work no matter where you install the mgisdata folder.

VERY IMPORTANT TIP: Pathnames, including the names of map documents and spatial data sets, should not contain spaces or special characters such as #, @, &,*, and so on. Use the underscore character _ to create spaces in names, if necessary. Although ArcCatalog and ArcMap allow spaces in names, spaces create problems for certain functions. It is wise to avoid them.

Data frame coordinate systems

In Chapter 1, we learned that every data set has a coordinate system in which it stores its x-y coordinates. In older GIS systems, the coordinate systems had to match in order for them to be drawn together in the same map. UTM data could only be shown with other UTM data, State

Plane data with other State Plane data, and so on. If data were in different coordinate systems, they would need to be converted (projected) to the same coordinate system prior to display.

Although it is still true that coordinate systems must match for data to be displayed together, ArcMap can perform the conversion on the fly. This feature allows data to be stored in different coordinate systems yet to be drawn together. The user defines a coordinate system for the data frame in the map document, and all of the data are reprojected to match the data frame (Fig. 2.18). The units defined for the data frame coordinate system, whether they are meters, feet, or degrees, become the **map units** of the data frame and may differ from the stored units in the files. In Figure 2.18, the UTM data use meters to store the *x*-*y* coordinates, the GCS data use degrees, and the State Plate data use feet. The Oregon Statewide Lambert coordinate system is defined using meters, so meters become the map units of the data frame.

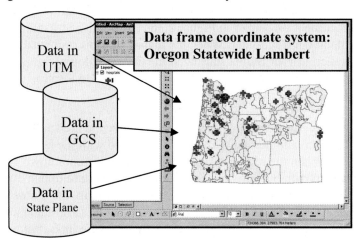

Fig. 2.18. Data in any coordinate system can be displayed together by setting the data frame coordinate system.

Summary

➢ Map scale is the ratio of size in the map to size on the ground. The original scale of a geographic data set determines the scale of its usefulness.

➢ Nominal data name things. Categorical data divide them into a discrete number of groups. Numbers may be categorical when they are used as codes.

➢ Ordinal data rank objects without a regular scale. Interval data are measured on a scale, and ratio data are measured on a scale with a meaningful zero point.

➢ Single symbol maps and labels are used to map nominal data. Unique values maps are used for categorical or ordinal data. Continuous numeric data are displayed using graduated color, graduated symbol, proportional symbol, dot density, or chart maps.

➢ Continuous numeric data are classified before being mapped. Classification methods include Jenks natural breaks, equal interval, defined interval, quantile, geometric interval, standard deviation, and manual. The choice of classification method depends on the type of data and the data distribution.

➢ Rasters may contain thematic or image data. Discrete thematic rasters can be displayed using a unique values method. Continuous thematic rasters may be classified or stretched. Image rasters use a stretched or RGB composite display.

➢ Map documents store references to layers and information about them, such as how to draw them. The actual data exist outside the document and may be used by many people.

➢ Map documents may store either absolute or relative pathnames for data files. Absolute pathnames work better when the data will always be in the same place for every user. Relative pathnames are best if the data will be moved about.

➢ Data frames contain layers to be drawn together, and they also have properties including the coordinate system and the reference scale. Data frames can project data on the fly to any chosen coordinate system.

➢ Map layers store a reference to a data set along with information on how to display the data. ArcMap layers are held in memory, although they may be saved to layer files.

VERY IMPORTANT TIP: Pathnames, including the names of map documents and spatial data sets, should not contain spaces or special characters such as #, @, &, *, and so on. Use the underscore character _ to create spaces in names, if necessary. Although ArcCatalog and ArcMap allow spaces in names, the spaces create problems for certain functions. It is wise to avoid them.

Chapter Review Questions

1. You measure a football field (100 yards long) on a detailed map and find that it is 0.5 inches long. What is the scale of the map?

2. A 1:20,000,000 scale map of the United States displays the interstates with a line symbol that is 3.4 points wide. There are 72 points to an inch. What is the uncertainty in the location of the road due to the width of the line used to represent it? Give the answer in feet and miles.

3. For each of the following types of data, state whether it is nominal, categorical, ordinal, interval, or ratio. Explain your reasoning.

bushels of wheat per county	pH measurement of a stream
vegetation type	state rank for average wage
average maximum daily temperature	number of voters in a district
parcel street address (e.g., 51 Main St)	student grade in a class (e.g., A, B+, etc.)
parcel ID number (e.g., 1005690)	soil type

4. For each of the following attributes, state whether a single symbol, graduated color, or unique values map would be most appropriate. Explain your reasoning.

precipitation	rivers
geologic unit	land use
acres of corn planted per county	household income

5. If mapping the following attributes for counties, indicate which ones would generally be normalized and which would not?

Average daily temperature   Acres of farm land

Number of Hispanics   Median rent

Square miles of park land   Sales tax rate

6. State whether you would use a unique values, classified, stretched, or RGB composite display method for each of the following rasters.

Geologic units   Black and white aerial photo

4-band satellite image   Precipitation

Landslide hazard (high, medium, low)   Slope in degrees

7. Which would take more storage space, a layer file showing all the U.S. counties or a layer file showing all the U.S. states?

8. (a) Jill creates a map document and e-mails it to a colleague in another state. He calls and tells her the document opens, but no map appears. Explain what is wrong. (b) Jill creates a map document using data from a central fileserver and e-mails it to a colleague in her office. Again, no map appears. What is wrong this time?

9. What characters should be avoided when naming GIS files, folders, and map documents?

10. What happens when the data frame is set to a different coordinate system than the data it contains?

Mastering the Skills

Teaching Tutorial

The following examples provide step-by-step instructions for doing basic tasks and solving basic problems in ArcGIS. The steps you need to do are highlighted with an arrow ➜; follow them carefully. Click on the video number in the VideoIndex to view a demonstration of the steps.

➜ Start ArcMap.

1➜ Click the button next to A New Empty Map. Click OK.

> **TIP:** If using ArcGIS on a network, you may see a different configuration of toolbars. Right-click the gray area at the top of the window below the blue window title bar, and you will see a list of toolbars. The following toolbars should be checked: Main, Draw, Standard, and Tools. If you see others checked, click the names to turn them off.

To ensure that moving the mgisdata folder to a different place will not break any data links, we will set the map document to store relative pathnames. We will also make relative pathnames the default for any new map documents that we create.

2➜ Choose File > Document Properties from the main menu bar.

2➜ Click the Data Source Options button.

2➜ Fill the button to Store relative pathnames to data sources.

2➜ Also check the box to make relative pathnames the default for new map documents.

2➜ Click OK and OK. Now as long as the map document stays in the same folder relative to its source data, the links to the data will stay valid.

3➜ Choose Save As from the File menu. Navigate to your mgisdata\MapDocuments folder and save a copy of the map document using the name **ex2mapdoc.mxd**. Then save periodically as you work.

> **TIP:** The names of map documents, spatial data sets, and folders should not contain spaces or special characters such as #, @, &, *, and so on. Use the underscore character _ to create spaces in names, if necessary. Although ArcCatalog and ArcMap allow spaces in names, spaces can create problems for certain advanced functions. It is wise to avoid them entirely.

Working with the map display

In Chapter 1 you created a group layer file of United States cities and counties. Our first act will be to load the layer file into ArcMap.

4➜ Click the Add Data button.

4➜ Navigate to the mgisdata\Usa folder and locate your USAGroup.lyr file. Select it and choose Add. (If you don't have the layer file, open the one in the same folder named USAGroup_sample.lyr.)

The four layers in the group file appear in the **Table of Contents,** which lists the data frames, layers, and tables in the map. It has three tabs that can be selected from the lower-left corner of the Table of Contents. The **Display** tab lists the layers in the order in which they are drawn, from bottom up. The **Source** tab organizes the layers by their source directories and shows tables. The **Selection** tab is discussed in Chapter 4.

5➔ Click each mode in turn and then return to the Display tab.

All of the symbols, labels, and scale ranges are just as you set them when you created the group layer file. Something else will look familiar also. Find the toolbar containing the Zoom In, Zoom Out, Pan, and Full Extent buttons, as in ArcCatalog. This is the Tools toolbar (Fig. 2.19). There are new buttons also.

Fig. 2.19. The Tools toolbar

TIP: Hover the cursor over a tool for a moment to find out the tool name.

6➔ Use the Zoom In tool to draw a box around the lower 48 conterminous states.

6➔ Use the Pan tool to drag the capital of New Mexico to the center of the map window.

6➔ Use the Fixed Zoom In tool to zoom in. Keep clicking the tool until the cities appear on the map. (Recall that you set a scale range for the Cities layer.)

1. Find the scale display. At approximately what scale do the cities appear? _____

6➔ Click the Fixed Zoom Out tool ONCE to zoom back out.

7➔ Right-click the Cities layer in the Table of Contents and choose Properties.

7➔ Click the General tab and read the scale range. It should be the same value it was set to when you created the layer file.

7➔ Close the Layer Properties window.

7➔ Use the Full Extent button to zoom out to all of the United States again.

TIP: You can also set and clear the range scale by right-clicking the layer name in the Table of Contents and choosing Visible Scale Range.

8➔ Use the Go Back to Previous Extent button to return to the close-up of New Mexico.

8➔ Click on the Identify button and click the state capital of New Mexico. Notice that in ArcMap, where many layers can be present, the Identify tool allows you to choose the layer(s) it uses.

2. What is the state capital of New Mexico? _____ What is its population? _____

8➔ Close the Identify window.

8➔ Use the Go to Next Extent button to return to the display of the United States.

Map tips are another nice feature, allowing the user to quickly see the name of a feature.

9➜ Right-click the StateCapitals layer and choose Properties. Click the Fields tab. The Primary Display Field is used for Map Tips; make sure it is set to CITY_NAME.

9➜ Click the Display tab and check the box to Show Map Tips. Click OK.

9➜ Hold the cursor over one of the capitals until its name appears. Try a couple more.

The Find tool in ArcMap has more options than the one in ArcCatalog. It will look for features having a specified string across all the layers in the map or an individual layer. It can also search all fields or confine the search to a single field.

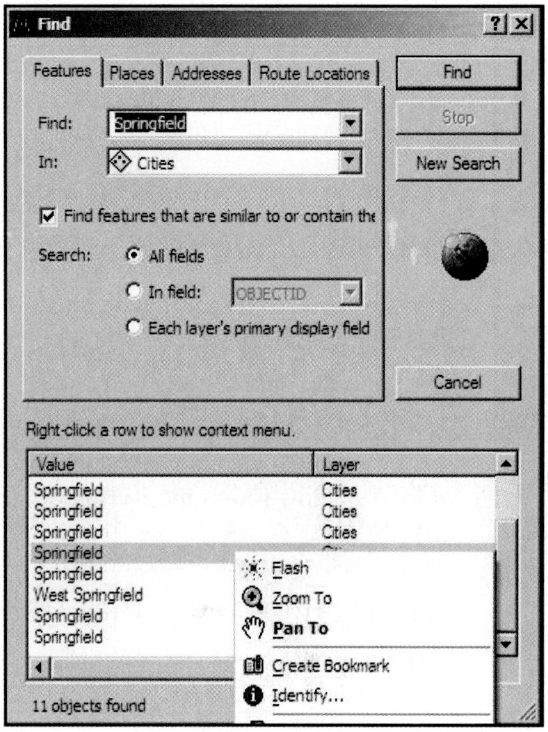

10➜ Click the Find button. Fill out the dialog box as shown in Figure 2.20 to search for **Springfield** in the Cities layer. Click Find.

10➜ It finds many instances of Springfield, a popular city name. Move the Find window off the map so that you can see it, and click on one of the names in the list. The feature flashes so you can see where it is.

10➜ Keep clicking on the cities until you think you have found Springfield, Illinois.

10➜ Right-click that Springfield in the list and examine the context menu. Then choose Identify to check that you have found the correct one.

10➜ Close the Identify window, right-click that Springfield again, and choose Zoom To.

Fig. 2.20. Using the Find tool

10➜ Close the Find window and use the Zoom tools to return to the view of the conterminous states.

Now let's introduce the measuring tool. It is useful for finding areas, lengths, and perimeters in an assortment of different units.

11➜ Locate and click on the Measure tool.

11➜ Click on the lower-left corner of Oregon to start a line. Add a couple of vertices to capture the curve and then double-click on the lower-right corner to finish the line.

11➜ Read the distance in the Measure tool window. Notice that the Measure tool reports the total length as well as the last segment.

12➜ Click the Choose Units button and set the Distance units to miles. Enter the line again.

3. What is the straight line distance between the capitals of California and Arizona? _____

 13➜ Click on the Measure an Area tool. Set the Area units to miles.

13➜ Click around the border of the state of Utah to create a polygon. Double-click to end the polygon and read the measurement.

 13➜Change to the Measure Feature tool and click on Utah to get an exact measurement.

4.   What is the length in kilometers of the Brazos River in Texas? _____ (Keep clicking until you get a line instead of a point.)

13➜ Click the Clear and Reset Results button and close the Measure window.

Data frame coordinate systems

This map document has one data frame named Layers in the Table of Contents. The data frame coordinate system is set to match the coordinate system of the first layer loaded into it. When we loaded our group layer, we also set the data frame coordinate system. What is it?

14➜ Right-click the Layers name in the Table of Contents and choose Properties.

14➜ Click on the Coordinate System tab and examine the information for the data frame.

5.   What is the name of the coordinate system? _____
What is the central meridian? _____ What are the linear units? _____

The **central meridian** of a map is the longitude on which the coordinate system is centered. This central meridian lies in the center of North America. The linear units are the **map units** of the data frame, for example, the units in which the map is drawn. The location display at the bottom of the ArcMap window shows the *x-y* values of the current cursor position.

14➜ Close the Data Frame Properties window.

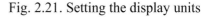

14➜ Move the cursor around the map while examining the *x-y* coordinates in the display.

The values update as the cursor moves. By default, the location **display units** are the same as the map units, in this case, meters. However, we can change the display units, which are a property of the data frame.

15➜ Right-click the Layers name in the Table of Contents and choose Properties.

15➜ Click on the General tab.

15➜ Notice that the Map Units box is dimmed (Fig. 2.21). The map units are determined by the data frame coordinate system and cannot be changed.

Fig. 2.21. Setting the display units

15➜ Set the Display Units to Decimal Degrees and click OK.

15➜ Look at the location display and verify that the units are now shown in degrees of longitude and latitude.

Using data frames

Now we will learn some things about data frames. Starting with the current map, our eventual goal will be to create a printed map showing volcanic hazards in Washington and Oregon. The map will contain two data frames, one major one showing the hazards and one small one showing the location of Oregon and Washington in the United States (a location map).

The map window has two view modes, **Data view** mode and **Layout view** mode. In Data view, only a single data frame is displayed. Layout view shows how the entire map document will look when printed, including titles, scale bars, legends, and so on.

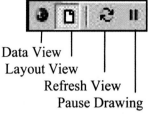

16➡ Click the Layout view icon in the lower left of the display area to switch from Data view to Layout view (Fig. 2.22).

16➡ Click on the Layers data frame name, then click again, and type in a new name for it, Hazards. Click Enter.

16➡ Choose Insert > Data Frame from the main menu bar.

16➡ Click twice slowly on the new frame and name it USA.

Data View
Layout View
Refresh View
Pause Drawing

Fig. 2.22. View buttons

The new data frame appears in the middle of the page, highlighted with blue handles to indicate it is the active frame. The active frame is also shown in boldface type in the Table of Contents. Next, resize both frames so they do not overlap each other on the page.

17➡ Place the cursor inside the USA frame and then click and drag it to the lower right, slightly off the page.

17➡ Click once inside the Hazards frame to make it the active frame. Place the cursor over the lower-left or lower-right corner and click and drag to resize the frame to occupy the top two-thirds of the page.

17➡ Move and resize the USA frame until the page appears similar to Figure 2.23.

18➡ Make sure that the USA frame is the active frame.

18➡ Click the Add Data button. Navigate to the mgisdata\Usa\usdata geodatabase and add the states feature class. It appears in the data frame.

Fig. 2.23. Layout view

18➡ Use the Zoom In tool to zoom to the conterminous 48 states in the USA data frame. Note that the Zoom In tool works just as well in Layout view as in Data view.

19➡ Right-click on the Hazards data frame name in the Table of Contents and choose the Activate command from the context menu. (The blue handles don't move from the USA data frame, but the Hazards frame is active as shown by the bold text.)

19➡ Use the Zoom In tool to zoom in to the extent of Washington and Oregon.

Notice that north is angled about 10 degrees. This effect is a distortion present in the coordinate system for this data set, which is optimized for displaying all 50 states and has a central meridian of −96 degrees. We will modify the central meridian of the data frame to the approximate center of our desired map to remove the distortion.

20➡ Place the cursor near the center of Oregon and read the longitude from the location display. This is the value we will use (about −120.5 degrees).

20➡ Right-click the Hazards data frame name and choose Properties.

20➡ Click the Coordinate System tab and review the information.

20➜ Click the Modify button.

20➜ Find the Central_Meridian box and carefully change the value to −120.5. Don't change anything else. Click OK and OK.

20➜ Adjust the zoom in the Hazards frame, if necessary, to fill the frame with Washington and Oregon.

Next, you will set up an **extent rectangle** to show the location of the Hazards map in the USA data frame. This trick is another data frame property.

21➜ Right-click the USA data frame name and choose Properties.

21➜ Click on the Extent Rectangles tab.

21➜ Click the Hazards frame in the list to the left, and click the arrow to place it in the list to the right. If it is not highlighted on the right, click it once to select it.

21➜ Click the Frame button and change the border to a 2 point black line. Click OK and OK.

Creating a map

Now we will begin working on the map of the volcanic hazards in earnest. Since we will focus on the Hazards data frame for now, we will return to Data view.

 22➜ Click on the Data view button. (You should get the Washington-Oregon map, but if you don't, right-click the Hazards data frame and choose Activate.)

Since we will be changing the original group layer substantially, there is no need to keep it together or to have the scale range or labels on the Cities layer.

22➜ Right-click the USAGroup layer and choose Ungroup. Click once outside all the layers so that none are selected anymore.

22➜ Right-click the Cities layer and choose Visible Scale Range > Clear Scale Range.

22➜ Right-click the Cities layer and choose Label Features. Since they are already checked, this action will turn the labels off.

23➜ Click the Add Data button and navigate to the mgisdata\Usa\usdata geodatabase. Add the volcano feature class and the counties feature class.

23➜ Click twice slowly on the volcano layer and rename it Volcanoes. Click Enter when done typing. Then rename the counties layer Counties.

24➜ Make sure the Display tab at the bottom of the Table of Contents is clicked so that you can rearrange the draw order of the layers.

24➜ Click on the States layer and drag it above the Counties layer.

24➜ Click on the States symbol to open the Symbol Selector. Choose the Hollow symbol and set the line thickness to 2 points and the outline color to black.

25➜ Click on the Rivers symbol and change its thickness to 2 points.

25➜ Click on the Volcanoes symbol and change it to a large dark red triangle.

Customizing dynamic labels

In Chapter 1, you created dynamic labels for the states and counties by setting the layer properties. You can also do it in ArcMap. We will now learn about customizing dynamic labels.

26➔ Right-click the Volcanoes layer, choose Properties, and then click the Labels tab.

26➔ Check the Label features box. Verify that NAME is the Label Field.

26➔ Increase the size of the text to 10-point and make the style Bold. Click OK.

Notice that some of the volcanoes may not have labels because they overlap with names of other volcanoes. The Labeling toolbar offers several convenient functions for working with labels, including viewing unplaced labels.

27➔ Right-click the gray menu area and choose the Labeling toolbar from the list.

27➔ Click the View Unplaced Labels button to show the missing labels in red.

Dynamic labels can be created for different classes of features within a layer. We will label the more active volcanoes with large text and the less active volcanoes with small text. This reduces map clutter and highlights the active volcanoes. The Volcanoes table has a field indicating the number of eruptions.

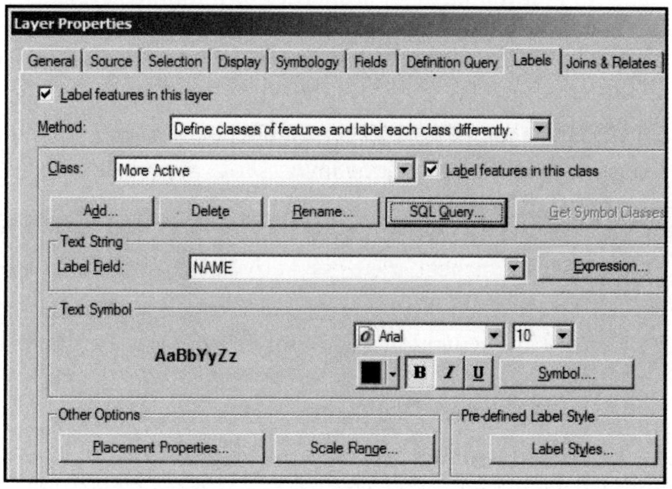

28➔ Open the Labels tab of the layer properties for Volcanoes (Fig. 2.24).

28➔ Change the Method to Define classes of features and label each class differently.

Fig. 2.24. Creating a label class

28➔ The current class is Default. Uncheck the Label Features in this class box. Since we will be creating new classes, we don't want the default class displayed.

28➔ Click Add to add a new class. Name the new class **More Active** and click OK.

Now all of the buttons and properties in this menu refer to the More Active class. Now we need to define the members of the class using a query and set the class symbol properties (Fig. 2.24).

29➔ Click the SQL Query button and enter the expression [KNOWN_ERUP] = > 10.

29➔ Click Verify to check the expression for errors. (If an expression is incorrect, then no labels may appear.) Click OK and OK.

29➔ Check the box to Label Features in the Class.

29➔ Make sure NAME is the label field.

29➔ Set the text symbol to Arial 10-point Bold.

Now the More Active label class is defined. (You could edit other properties for label placement and scale range, but we won't do so at this time.) We now create the other class, Less Active.

30➔ Click the Add button and name the new class **Less Active**. Click OK.

30➔ Enter the SQL expression [KNOWN_ERUP] < 10 for this class, following the steps you just did for the previous class. Make the labels regular 8-point font.

30➔ When finished setting the properties for both classes, click OK to close the property sheet and enact the changes.

TIP: To go back and edit properties for a label class later, access the class by clicking the Class drop-down button in the property sheet and choose the class to edit.

A map with many labeled layers can be cumbersome to adjust if one must keep opening label properties for each layer. The Label Manager facilitates the control of labeled layers and makes it easier to modify the label properties of multiple layers.

 31➔ Click the Label Manager button on the Labeling toolbar.

The Label Manager lists all the label classes from all the layers. Notice the two label classes you just added for the Volcanoes layer, More Active and Less Active. The map is still too cluttered, so turn off the Less Active labels.

31➔ Uncheck the Less Active class and click Apply.

31➔ Click the More Active label class. The settings in the window now apply to this class. Change the label color to dark red to match the volcano symbols. Click Apply.

32➔ Turn off the default labels for the States layer. Click OK.

32➔ When you are finished, your map should look similar to Figure 2.25.

32➔ This is a good place to save your work so far. Choose File > Save from the main menu. Continue to save periodically as you work.

Fig. 2.25. The current map

Creating maps based on attributes

Displaying a data layer with a single symbol is only the beginning. Let's start using attribute fields to display features. Viewing the attribute table is often the best way to begin.

33➔ Add the majroads feature class from the Transportation feature dataset in the usdata geodatabase to the map. Rename it **Highways**.

33➔ Right-click the Highways layer and choose Open Attribute Table. Examine the available fields.

33➔ The HWY_TYPE field appears to be the best choice. Close the table.

6. Is the HWY_TYPE field nominal, categorical, ordinal, ratio, or interval data? _____
What map type should be used to display it? _____

> 34➔ Double-click the Highways layer name in the Table of Contents to open its properties (a nice shortcut). Click the Symbology tab.
>
> 34➔ Choose Categories: Unique Values as the type of map.
>
> 34➔ Choose HWY_TYPE as the Value field.
>
> 34➔ Click Add All Values to display the categories in the legend box.
>
> 34➔ Choose a different Color Ramp—one with dark, strong colors instead of pastels.
>
> 34➔ Click Apply. The markers on the map change, but the Properties dialog box stays open. Move the box aside, if necessary, to view the map.

TIP: If a class has only a few features, it may be grouped with the other small classes into a class called All Other Values. Displaying this class is optional.

Because the lines are thin, it is hard to distinguish the different road classes. Let's change them.

> 35➔ In the Symbology tab, double-click the symbol for Interstate to open the Symbol Selector. Choose the Expressway symbol. Click OK.
>
> 35➔ Double-click the Freeway symbol. Choose the Expressway symbol again and change its color to green. Click OK.
>
> 35➔ Double-click the Arterial symbol and give it the Major Road symbol. Make it green.
>
> 35➔ Change the Unclassified symbol so that it is the same as the Arterial symbol except that the color is gray. Click Apply to see all the changes so far.

Finally, the order of the classes in the legend could be better.

> 36➔ Click on the Interstate class to select it, and click the up arrow to make it first.
>
> 36➔ Move the Freeway class up underneath Interstates. Arterial should be next, and then Unclassified will be last. Use the up and down arrows until the order is right.
>
> 36➔ Uncheck the box to show the <all other values> class, as it is empty.
>
> 36➔ Your final legend should look like Figure 2.26. Click OK.

Symbol	Value	Label
☐ ————	<all other values>	<all other values>
	<Heading>	**HYW_TYPE**
═══════	Interstate	Interstate
═══════	Freeway	Freeway
————	Arterial	Arterial
————	Unclassified	Unclassified

Fig. 2.26. The final legend for the highways should look like this.

7. Is city population data nominal, categorical, ordinal, ratio, or interval? _____
What kind of map should be used to display it? _____

> 37➔ Open the Symbology tab for the Cities layer and choose Quantities: Graduated Symbols as the map type.
>
> 37➔ Choose POP1990 as the Value field. Click Apply.

Notice how the population of the towns is immediately apparent now, with a few larger cities and many smaller ones. We can make some adjustments to this default set of symbols.

38➔ Click on the Template symbol button and choose a light pink color from the Symbol Selector. Click OK and Apply.

38➔ Change the maximum symbol size from 18 points to **10** points. Click the Tab key to enter the **10** value—the Enter key causes the dialog box to close.

38➔ Change the number of classes to four.

38➔ Right-click the legend area and choose Flip Symbols from the context menu. Repeat and choose Reverse Sorting. Now the classes are in reverse order.

38➔ Click the Label column heading and choose Format Labels (Fig. 2.27).

38➔ Click to ensure that the Category is Numeric, choose Number of Significant Digits and set it to 2, and check the Show Thousands Separators box.

38➔ Click OK to close the window, and OK again to apply the changes.

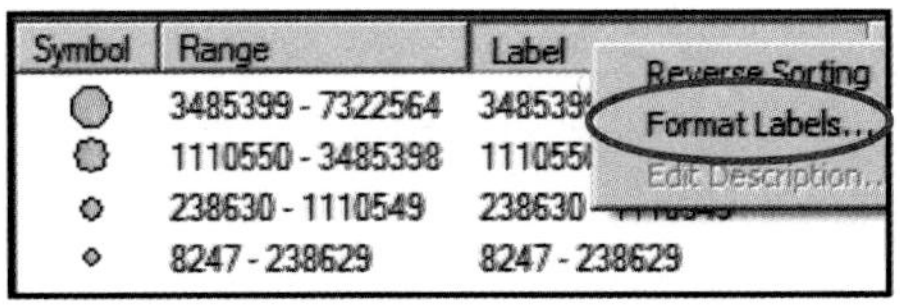

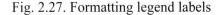

Fig. 2.27. Formatting legend labels

TIP: Using the Format Labels option to make sure the class labels have evenly rounded values is good cartographic practice. Neglecting this detail presents an unprofessional picture.

Next, let's create some maps based on the attributes of a polygon data layer—the counties.

39➔ Open the Symbology Properties for the Counties layer.

39➔ Choose Quantities: Graduated Colors and set POP2000 as the Value field.

39➔ Pick the orange color ramp and click OK. A single color ramp is easier to interpret than a multicolor ramp.

Remember, population data is usually normalized to the area of the measurement unit (in this case, counties). The Normalization option can be used to do this.

40➔ Open the Symbology properties for Counties again.

40➔ Choose Area as the Normalization field.

40➔ Click the Label column heading and choose Format Labels. Choose Numeric and give the values two significant digits and thousands separators, as before. Click OK and OK to implement the symbol changes.

8. Why are population data normalized when mapping counties but not normalized when mapping cities?

Nearly all of the counties fall into the first class. This fact highlights an issue with our symbols. The statistics are based on the entire United States data set, but we are only looking at two states. Therefore the symbols do not emphasize the regional differences. We can fix this by restricting the layer to just Washington and Oregon. We use the Definition Query as we did in Chapter 1.

41➔ Open the layer properties for Counties and click the Definition Query tab.

41➔ Click the Query Builder button and use the buttons to enter the expression:

[STATE_NAME] = 'Oregon' OR [STATE_NAME] = 'Washington'

Remember you can click Get Unique Values to choose values from a list instead of typing them in.

41➔ Remember to Verify the expression before clicking OK and OK.

The map colors have not changed yet because the symbols are based on the old statistics.

42➔ Open the Symbology tab and set the value field to POP2000 again. Also format the labels again to have two significant digits and thousands separators.

43➔ Use the Definition Query again on the Cities layer.

44➔ Reset the symbols with the new statistics for the Cities layer also.

You might think that we could repeat the same steps to eliminate the roads outside the two states, but the road features do not stop at the state boundaries and so cannot be queried that way. However, the data frame can restrict, or **clip**, its layers to the extent specified by another layer.

45➔ Open the data frame properties for the Hazards layer and click the Data Frame tab.

45➔ In the Clip to Shape section, check the Enable box.

45➔ Click on the Specify Shape button and choose Outline of Features. Set the Layer to Counties and click OK.

45➔ Set the Border to a 2-point black line. Click OK. The map should appear like Figure 2.28.

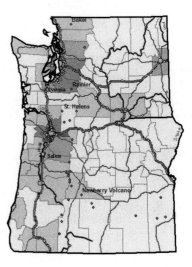

Fig. 2.28. After step 45, the map should look like this.

Classifying data

Up to now, we have been using the default Natural Breaks classification. Let's try some others. However, let's not lose the work done so far. We will create another layer based on the counties and modify its classification scheme.

46➔ Right-click the Counties layer and choose Copy.

46➔ Right-click the Hazards data frame and choose Paste Layer(s). A second copy of the layer appears.

46➔ Rename the copy **Median Rent**.

9. Is median rent data nominal, categorical, ordinal, ratio, or interval?_____ What kind of map should be used to display it? _____ Should it be normalized? _____

47➔ Open the Properties for the Median Rent layer and choose the Symbology tab.

47➔ Keep the Graduated colors map type but change the Value field to MEDIANRENT and set the Normalization back to None.

47➔ Click the Classify button.

Examine the histogram. These data are evenly distributed and in dollars, which gives the values intrinsic meaning. The Defined Interval classification is the best method for this field.

47➔ Set the Classification Method to Defined Interval, and set the Interval Size to 100. Click OK and OK.

48➔ Create another copy of the Counties layer, paste it into the Hazards data frame, and name it **Mobile Homes**.

49➔ Open the new layer's properties and click on the Symbology tab.

49➔ Change the Value Field to MOBILEHOME. The number of mobile homes may be influenced by the area of the county, so we should normalize the data.

49➔ Set the Normalization to the HSE_UNITS field so that we are looking at the fraction of housing units that are mobile homes.

49➔ Click the Classify Button and examine the distribution. These values approach a normal distribution, so it might be helpful to view them by standard deviation.

49➔ Choose the Standard Deviation classification method. Set the Interval Size to 1 Std Dev. Click OK and OK.

Now symbolize the volcanoes. Let us distinguish between those with known historical eruptions and those without. We will use a Manual classification with two classes.

50➔ First, apply the same definition query to the Volcanoes layer so that only volcanoes in Washington and Oregon are included. (Even though they are not displayed, the national statistics are still being used.)

51➔ Open the Symbology properties for the Volcanoes layer. Choose a graduated symbols map with KNOWN_ERUP as the Value field.

51➔ Click the Classify button. Set the number of classes to 2.

51➔ In the Break Values box, enter **0** for the first break value. Set the second break value to the maximum of the data set, **36**. Click OK.

52➔ You can change the labels applied to the data. Click under the Label column on the 0 value. Enter **Inactive**. Click on the 1-36 class beneath it and type **Active**.

52➔ Change the symbol for the Inactive volcanoes to a 12-point green triangle. Change the symbol for the Active volcanoes to an 18-point dark red triangle. Click OK.

Finally, we will use a classification to compare the number of eruptions for each volcano. Since we not interested in the volcanoes with no eruptions, we will use the Exclusion function to eliminate them from the map. Since these data points are unevenly distributed and the hazard is related to the number of eruptions, a proportional symbol map is a good choice.

53➔ Copy the Volcanoes layer and paste the copy into the Hazards data frame. Rename the copy **Known Eruptions**.

54➔ Open the Symbology properties for the Known Eruptions layer.

54➔ Set the map type to Proportional Symbol. Set the Value field to KNOWN_ERUP.

54➔ Click the Exclude button. Enter an expression indicating the volcanoes to be excluded: **[KNOWN_ERUP] = 0**. As always, remember to Verify the expression.

55➔ Leave the units set to Unknown Units. Click on the Min Value button. Choose the Triangle 2 symbol and set the color to pale green. Click OK and OK.

55➔ Turn off the Volcanoes layer to see the new symbols better.

TIP: When a map has overlapping symbols, using a symbol with an outline will help distinguish them more clearly. ArcMap will try to display the largest symbols first.

55➔ The symbols are a little too large for clarity. Open the Symbology properties again.

55➔ Click the Min Value button and change the symbol size to 10 points. The other symbols will be resized proportionally. Click OK and OK.

For one last touch, we will label the largest volcanoes with the number of eruptions since it is not easy to figure that out from the legend.

56➔ Open the Known Eruptions layer properties and click the Labels tab.

56➔ Check the Label features box and choose to label all features the same way.

56➔ Set the Label field to KNOWN_ERUP.

56➔ Set the text symbol to Arial 10-point normal.

56➔ Click the Placement Properties button. Click the Placement tab, if necessary, and fill the button to Place label on top of the point. Click OK and OK.

57➔ Turn off the red unplaced labels by clicking the Unplaced Labels button on the Labeling toolbar.

57➔ Turn off the Mobile Homes and Median Rent layers so that you can see the Counties layer with the population map.

57➔ Your final map should appear similar to Figure 2.29.

10. Examine the map and decide which three volcanoes are the most hazardous, considering the number of eruptions and proximity to high-population areas.

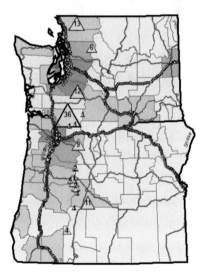

Fig. 2.29. The final volcano hazards map after step 57

Displaying rasters

Two main types of rasters are available for use in ArcGIS, image rasters and thematic rasters. Image rasters store brightness or color values and come in a variety of formats. Photos and satellite images are common examples of images, and scanned maps are another. Thematic rasters portray map data such as soil types, roads, elevation, or precipitation.

TIP: The first time that a raster is displayed, the user is asked if she wishes to build **pyramids**, which help speed the display of the raster on subsequent occasions.

58➔ Choose to start a new map by clicking the New Map File button on the main toolbar. **Save the changes to your previous map**. You will use it again.

58➔ Click the Add Data button and navigate to the mgisdata\BlackHills\rasters directory.

58➔ Add the rasters geolgrid, dem2, hillshd2, and TM_24Sep98_utm.

58➔ Turn off all of the rasters for now.

Displaying thematic rasters

Displaying thematic grid data is not very different from displaying attribute data. You may use a classified, stretched, or unique values display type. Unique values maps are used for grids containing categorical data. Classified and stretched maps are used for continuous grids.

59➔ Turn on the geolgrid raster.

11. Is the geology data nominal, categorical, ordinal, ratio, or interval?_____ What kind of map should be used to display it? _____

Notice that the grid's default display uses unique values—ArcMap recognized that only a few discrete values were present and automatically applied the correct type of display. The user can modify the individual colors or the color scheme.

59➔ Open the geolgrid Properties and click the Symbology tab.

59➔ Set the color scheme to pale pastels.

59➔ Click OK to apply the changes and close the Properties.

59➔ Turn off geolgrid and turn on dem2.

The file is a digital elevation model, or DEM. Each cell represents an elevation value in meters above sea level and represents continuous data.

12. Is elevation data nominal, categorical, ordinal, ratio, or interval?_____ What kind of map type(s) could be used to display it?

60➔ Open the symbology properties for the dem2 raster. ArcMap assigned a Stretched map type to start.

60➔ Use the drop-down box to change the color ramp used to display the image and click Apply. Find one that looks good and keep it. Click OK.

> **TIP:** Rasters may contain areas where the values are unknown and are stored as a zero or as a special data value called NoData. Use the Background Value and NoData settings to give these cells a NoColor shading so that they can't be seen.

Instead of stretching the values in a raster, they can be classified like attribute data for vectors.

61➔ Open the symbology properties for dem2 again.

61➔ Change the display type from Stretched to Classified.

61➔ Choose a rainbow color ramp.

61➔ Click the Classify button and set the classification method to defined interval. Set the interval to 100 meters. Click OK and OK.

61➔ Examine the classification. The colors change every 100 meters, rather like having a color map with a 100-meter contour interval.

Displaying image rasters

Images usually represent brightness of land features as measured by a camera or spectrometer. They can have one band or multiple bands. A single-band image is usually displayed using a grayscale color ramp so that it looks like a black-and-white photo. A single band of a multiband image can also be displayed that way.

62➔ Turn off dem2 and turn on TM_24Sep98_utm.

62➔ Double-click TM_24Sep98_utm to open its Properties sheet.

62➔ Click the Symbology tab and examine its contents.

This image is a Landsat satellite image with seven different bands corresponding to different wavelengths of light. Notice the two ways to display the image. Stretched is appropriate for single-band images or to display one band of a multiband image at a time.

62➔ Click Stretched and select Band 4 as the band to display.

62➔ For the Stretch Type, choose None and click OK.

62➔ Open the Properties and change the Stretch Type to Standard Deviations. Click OK.

RGB composites display three bands at a time. The computer screen has a display gun with three colors: red, green, and blue. In a color composite, one band of the image is assigned to each gun, and the combination creates a full-color picture.

63➔ Open the Symbology properties, if necessary, and choose RGB Composite.

63➔ Click Apply and examine the image.

Currently in this image, the first, second, and third bands are assigned to red, green, and blue, respectively. In a Landsat image, Band 1 records the blue light from the earth's surface as measured by the satellite sensor, Band 2 records green, and Band 3 records red.

63➔ To see the image in true color, use the drop-down boxes to assign Band 3 to the red channel, Band 2 to green, and Band 1 to blue. Click Apply.

Band 4 of Landsat measures near-infrared reflectance. Vegetation reflects this wavelength of light strongly. Foresters and range managers often use a 4–3–1 band combination to assess the presence and health of vegetation.

> 63➜ Assign Band 4 to red, Band 3 to green, and Band 1 to blue. Click OK.

Different band combinations highlight different features in Landsat images. The combination of 7–4–1 is often used to contrast vegetation with urban areas or dry plains. Try it and experiment with some other band combinations.

> 64➜ Turn off the TM_24Sep98_utm raster and turn on the hillshade raster.

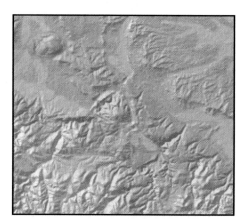

A hillshade map is created from an elevation map by calculating the brightness and shadows induced on the surface by an illuminating source. A display feature called Transparency can be combined with a hillshade to produce informative backgrounds and can be applied to raster or vector layers.

> 64➜ If necessary, click and drag the geolgrid layer until it is above the hillshd2 layer in the Table of Contents. Turn it on.
> 64➜ Open the geolgrid Properties and click the Display tab.
> 64➜ Set the Transparency value to 40 percent and click OK.
> 64➜ Examine the map (Fig. 2.30). Zoom in and examine an area in more detail.

Fig. 2.30. Transparent geology displayed over a hillshade raster

This is the end of the tutorial.

> ➜ Close ArcMap. You do not need to save your changes.

More skills

Consult the Skills Reference section of this chapter to learn to do the following:

> ➢ Group layers together for display.
>
> ➢ Create and use bookmarks.
>
> ➢ Manage toolbars in ArcMap and ArcCatalog.
>
> ➢ Drag and drop files from ArcCatalog to ArcMap.

Exercises

Use ArcMap and the data in mgisdata\Sdakota\southdakota.mdb to answer the questions.

1. Find the Mount Rushmore Memorial in the interest feature class and zoom in to it. Then answer the following questions.

    In which county is it?

    What is the name of the river running north of it?

    What is the name of the closest town?

    How far from the town is the memorial in miles?

    What are the names of the two tourist spots to the immediate northwest of the memorial?

2. Create a map of South Dakota majroads and counties. Make the outline of the counties medium gray so that they are less prominent. Change the symbols of the roads to look like this legend. **Capture** the map.

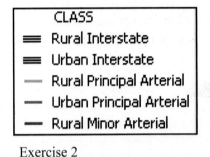

Exercise 2

3. What is the driving distance across South Dakota along Interstate 90 in miles? In kilometers?

4. Create a map of South Dakota counties, rivers, and lakes (hydroply). Label the rivers with blue italic text that curves along them. **Capture** the map.

5. Create a map of South Dakota cities and counties. Label the large cities with more than 25,000 people with 12-point bold labels and the other towns with 8-point labels. Add the roads and zoom in to the Sioux Falls area. **Capture** the map.

6. Use the fedlandp feature class to create a map showing the federal ownership of the land using the colors and legend labels shown here (i.e., the legend should look like this one). Make BIA orange, BLM lavender, BRec blue, DoD dark purple, FS light green, FWS red, and NPS dark green. Make the counties white and turn off all the other layers. **Capture** the map and the legend from the Table of Contents.

7. Create a map showing each watershed subregion in a different color. Add a dot density map showing the density of farms in each county, using black 2-point dots, with the ratio 1 dot = 30 farms. Which watershed subregion appears to have the highest density of farms? **Capture** the map.

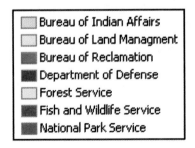

Exercise 7

8. Create a map of South Dakota counties showing the median home value in each county. What is the best classification method for this map? **Capture** the map and the legend.

9. Create a graduated color map of the *density* of farms in each county. What is the best classification method for this map? Make sure the legend values have only two decimal places. **Capture** the map and the legend.

10. Use the hillshade and dem2 grids in the BlackHills\rasters folder to create a color-shaded relief map of the central area of the dem2 grid so that the color represents elevation. **Capture** your map.

Challenge Problem

Begin with a hypothesis that counties with a high proportion of mobile homes also have lower median rents. Create a map layout to test this hypothesis for a single state. Use the data in mgisdata\Usa\usdata geodatabase.

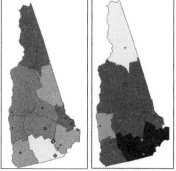

> ➢ Choose one state and make sure your map shows only the counties and cities for that state. Display the cities by population.

> ➢ Create a map layout with two data frames, each occupying half of the page. Arrange them to best show the state maps side by side. Put your state in each one and make sure they are at roughly equal scales.

> ➢ In one data frame, symbolize the number of mobile homes in each county. (Do you need to normalize? By what?) In the other frame, show the median rent for each county. Use the best classification method.

> ➢ **Capture** the page layout and the Table of Contents for your instructor. Discuss the results and confirm or reject the hypothesis for that state.

Skills Reference

Starting ArcMap 95

Managing toolbars 96

Adding data 96

Dragging and dropping files to ArcMap 96

Saving an ArcMap document 97

Opening an ArcMap document 97

Controlling display of layers 97

Setting Map Tips 98

Grouping layers for display 98

Using the Zoom/Pan tools and bookmarks 99

Measuring features 100

Finding features by searching attributes 100

Setting data frame and layer properties 101

Clipping to a layer 101

Examining the coordinate system...101

Setting the data frame coordinate system..102

Creating a map based on an attribute 103

Classifying attributes for a map 104

Modifying the appearance of the legend 105

Creating a map layer or map group layer 106

Starting ArcMap

ArcMap can be started several ways.

1. Click the Start button on the Start Menu bar of the computer, navigate to Programs > ArcGIS > ArcMap, and click it.

2. Find the ArcMap icon on the desktop and double-click it (Fig. 2.31).

3. Click the appropriate button in ArcCatalog (Fig. 2.31).

4. As ArcMap starts, it displays a dialog screen (Fig. 2.32). Choose to start with a new empty map, open a map template, or open an existing map.

5. To open existing maps, click Browse for maps to search for a map on the disk. Alternatively, click one of the recently used maps shown in the list.

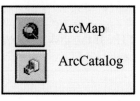

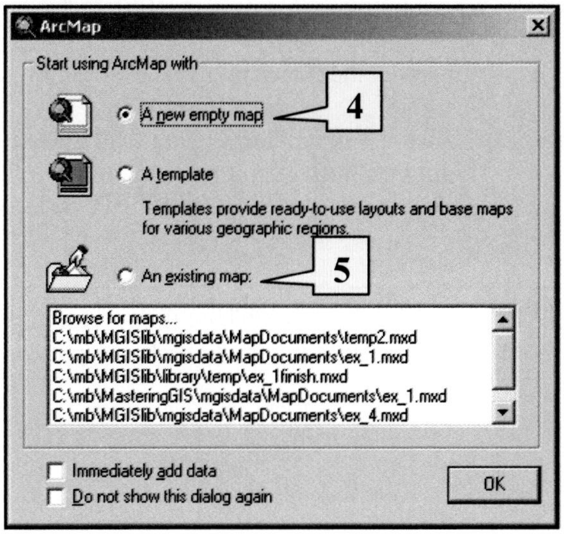

Fig. 2.31. Program icons

Fig. 2.32. ArcMap start-up dialog

Managing toolbars

1. To move a toolbar to another location on the GUI, click and hold on its handle (a shadowed line on its left edge) and drag it to the desired position (Fig. 2.33). This is a docked toolbar.

2. To move a toolbar off the GUI entirely, use the handle to drag it off the GUI. It then becomes a floating toolbar. It can be docked again by dragging it to a location on the GUI.

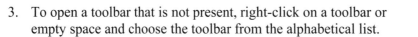

Fig. 2.33. Handle

3. To open a toolbar that is not present, right-click on a toolbar or empty space and choose the toolbar from the alphabetical list.

4. To close a toolbar, click on the X box in its upper-right corner. Alternatively, right-click on a toolbar or empty space and choose the toolbar to close from the list.

Adding data

1. Click the Add Data button on the main toolbar, or choose File > Add Data from the main menu bar.

2. Use the Add Data dialog box to navigate to the folder containing the data to open.

3. Click on the data layer to select it. Select multiple layers by holding down the Control or Shift key while clicking additional data sets.

4. Click Add. The data will be added to the active data frame.

TIP: If the disk or directory containing the data does not appear in the list, you probably need to add a connection to it using the Add Connection button, as described in Chapter 1.

Dragging and dropping files to ArcMap

After searching through data in ArcCatalog, you can easily drag them into ArcMap for use.

1. Position the ArcCatalog and ArcMap windows so that both are visible.

2. Click the file in ArcCatalog that you want to open in ArcMap. Hold down the mouse button and drag it into the ArcMap window (you must be in the map window or the Table of Contents—not on a menu bar). Release the mouse click to drop the file into ArcMap.

TIP: If ArcMap is not visible on the screen, drag the file onto the ArcMap icon on the computer's Start Menu bar. Hold it there until ArcMap opens and then drag it into ArcMap.

Saving an ArcMap document

1. Click on the Save button, or choose File > Save from the menu bar.

2. If the document has not previously been saved, a dialog box appears for the user to select a location and a name for the map document. Navigate to a directory in which to save it and then enter a name for the document.

3. Choose File > Save As from the menu bar to save the map document under a new name, leaving the original version unchanged. Specify the location and name as for step 2.

> **TIP:** The names of map documents, spatial data sets, and folders should not contain spaces or special characters such as #, @, &, *, etc. Use the underscore character _ to create spaces in names, if necessary. Although ArcCatalog and ArcMap allow spaces in names, spaces can create problems for certain advanced functions. It is wise to avoid them entirely.

Opening an ArcMap document

1. Click the Open button, or Choose File > Open from the main menu bar.

2. Use the dialog box to navigate to the directory containing the file.

3. Click on the file to select it and choose Open.

4. If the currently opened map has unsaved changes, you will be prompted to save them before opening the new document. ArcMap can only open one map document at a time.

5. To create a new empty map document, click the New Map File button, or choose File > New from the main menu bar.

Controlling display of layers

1. To turn a layer on or off, click inside the check box.

> **TIP:** If the check box is dimmed, then the current map scale is outside the display range set for the layer. Try zooming to another scale or reset the scale range in the layer properties.

2. To change the draw order, click and hold on the layer name and drag it higher or lower in the list. Release the mouse button to finish.

> **TIP:** If the layers don't seem to go anywhere when you drag them, check to make sure that the Display tab in the lower-right corner of the Contents box is clicked. You cannot move layers when the Source tab is clicked.

3. Hide the legend for a data layer by clicking the minus sign in the box next to the layer. Click the plus sign to expand it again.

Setting Map Tips

1. Right-click the name of the layer in the Table of Contents and choose Properties from the context menu.

2. Click the Fields tab and set the Primary display field for the text to appear in the tip.

3. Click the Display tab and check the box to Show Map Tips.

TIP: If the Show Map Tips box is dimmed, then you must create a spatial index for the layer using ArcCatalog.

Grouping layers for display

Grouped layers can be turned on and off with one mouse click.

1. Click the first layer in the group to highlight it, and then click the other members of the group while holding down the Ctrl-key. When you are done, all the group layers should be highlighted in blue.

2. Right-click a member of the group and choose Group (Fig. 2.34).

3. Click on the New Group Layer name once to highlight it and then click on it again to type in a name for it. Press Enter to finish.

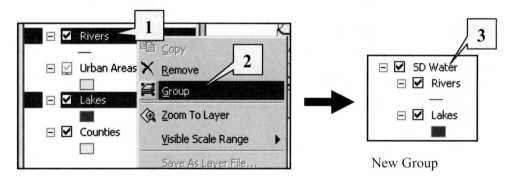

Fig. 2.34. Creating a group layer

Groups have joint properties, such as setting scale ranges for the entire group at once. You can also add group members directly from the disk without needing to add them to the map document first. All layers in a group must be in the same data frame.

4. Right-click on the Group name and choose Properties from the context menu.

5. Click the General tab to set scale ranges for the group.

6. Click the Group tab and click Add to add more members to the group from the disk. Use Remove to remove members from the group. The Properties button is used to specify different properties for individual group members.

7. Click the Display tab to set transparency, brightness, or contrast for the group.

Using the Zoom/Pan tools and bookmarks

1. To zoom in to an area, click the Zoom In tool. Position the cursor to the upper left of the target area and then click and drag a box to encompass the area desired. Release the mouse button to finish.

2. To zoom out, click the Zoom Out tool and click once at the center zoom point. Draw a large box to zoom out slightly or a small box to zoom out a great deal.

3. To zoom in by a fixed amount while remaining centered within the current extent, click the Fixed Zoom In tool.

4. To zoom out by a fixed amount while remaining centered within the current extent, click the Fixed Zoom Out tool.

5. To move the map, click the Pan tool, click on the map, and drag it to the desired area.

6. Click the Previous Extent button to return to the extent just before the current one.

7. Click the Next Extent button to return to the extent after the current one (this button is dimmed unless the Previous Extent button has been clicked).

8. Click the Full Extent button to view the full area of all the layers in the data frame.

9. To set a **bookmark**, zoom to the desired area and select Bookmarks > Create from the main menu. Type in a name for the bookmark. (In version 9.2 choose View > Bookmarks > Create.)

10. To zoom to an existing bookmark, click Bookmarks and choose the desired bookmark (Fig. 2.35).

11. To delete an existing bookmark, choose Bookmarks > Manage. Select the bookmark name by clicking on it, and then click the Delete button.

Fig. 2.35. Zooming to a bookmark

12. To zoom to the reference scale, right-click the frame name and choose Zoom to Reference Scale from the menu.

Measuring features

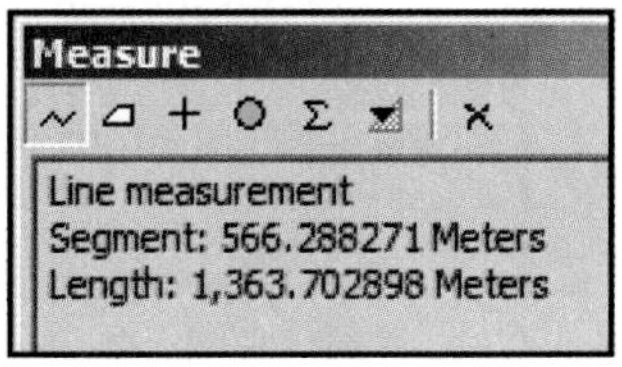

1. Click the Measure tool. By default it measures distance.

2. Click on the map to start the line. Click along the path to be measured. Double-click to end the line. Both total length and the length of the last segment are shown (Fig. 2.36).

Measure

Line measurement
Segment: 566.288271 Meters
Length: 1,363.702898 Meters

Fig. 2.36. The Measure tool

3. Click the Measure Area button to measure a polygon area. Click to create vertices for the polygon and double-click when finished.

4. Click the Measure Feature button, and click on a feature to get its measurement.

5. Use the Choose Units button to set the units for length and area.

6. Use the Snap to Features button to make the cursor coincide with a feature when it gets close to it. This option is useful for precise measurements between or along features.

7. Use the Show Total button to keep a running count of all features measured.

8. Use the Clear/Reset Results button to clear running totals and reset the measurements.

Finding features by searching attributes

1. Click the Find tool and type the search text in the first box (Fig. 2.37).

2. Choose the layer or set of layers to be searched.

3. Check the box for an exact match or for features that are similar to or contain the string.

4. Choose to search all fields, a specific field, or the primary display field.

5. Click Find to begin the search. Records that match will be listed below.

6. Right-click one of the found records to display a context menu with options for flashing, zooming to, identifying, bookmarking, or selecting the found feature (Fig. 2.38).

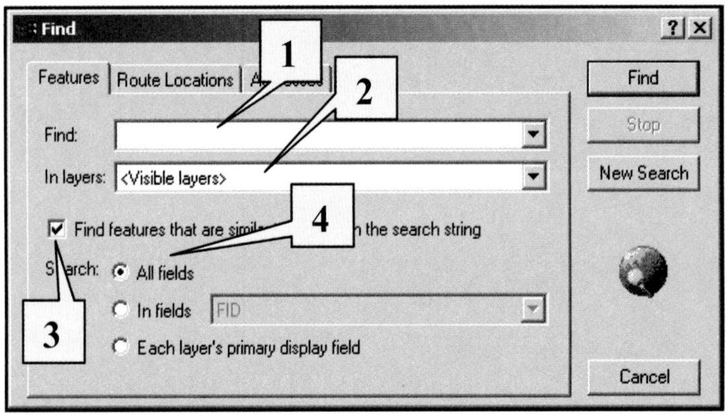

Fig. 2.37. Finding features

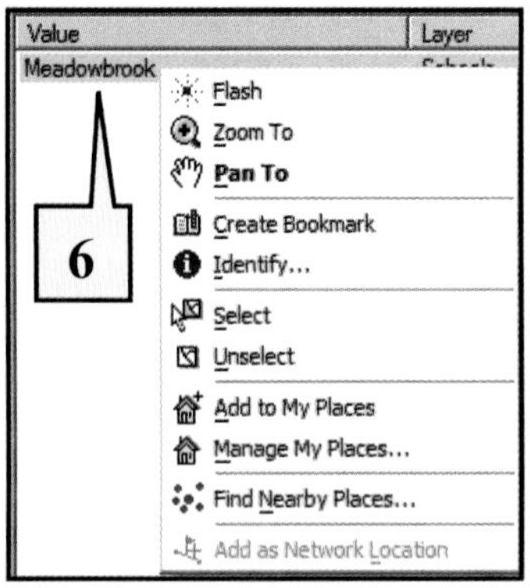

Fig. 2.38. The Find list

Setting data frame and layer properties

1. Right-click the frame or layer name and choose Properties from the context menu.

2. Click the tab containing the properties to set.

3. Enter the appropriate information and settings.

Clipping to a layer

1. Open the Data Frame Properties and click the Data Frame tab.

2. Check the Enable box under Clip to Shape (Fig. 2.39).

3. Click the Specify Shape button.

4. Choose to Outline Features and select the layer defining the outer boundary.

5. Alternatively, enter coordinates for a custom rectangle.

6. Select a border to outline the clip feature, if desired.

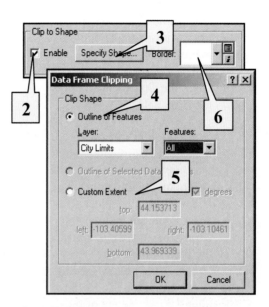

Fig. 2.39. Clipping layers in a data frame

Examining the coordinate system

Every data layer has *x-y* coordinates, which must be in some type of coordinate system. The coordinate system is stored as part of the information about a data set. You can access this information in ArcMap. The information about the coordinate system and the units reflects what is stored in the file, not the coordinate system of the data frame.

1. Right-click the layer name and choose Properties.

2. Click the Source tab.

3. Read the coordinate system information (Fig. 2.40).

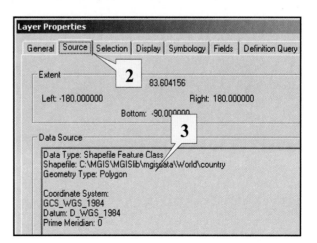

Fig. 2.40. Viewing a coordinate system in ArcMap

Setting the coordinate system of a data frame

Use this feature to display data sets with different coordinate systems in the coordinate system of your choosing. The data sets are projected on the fly to whatever coordinate system you specify. The original data are not affected.

1. In ArcMap, right-click the data frame name and choose Properties from the context menu.

2. Click the Coordinate System tab (Fig. 2.41).

3. Use the folder tree provided to select a coordinate system.

4. Alternatively, click the Import button to select a spatial data set that has the coordinate system to apply to the data frame.

5. If desired, click Modify to alter the parameters of the existing coordinate system.

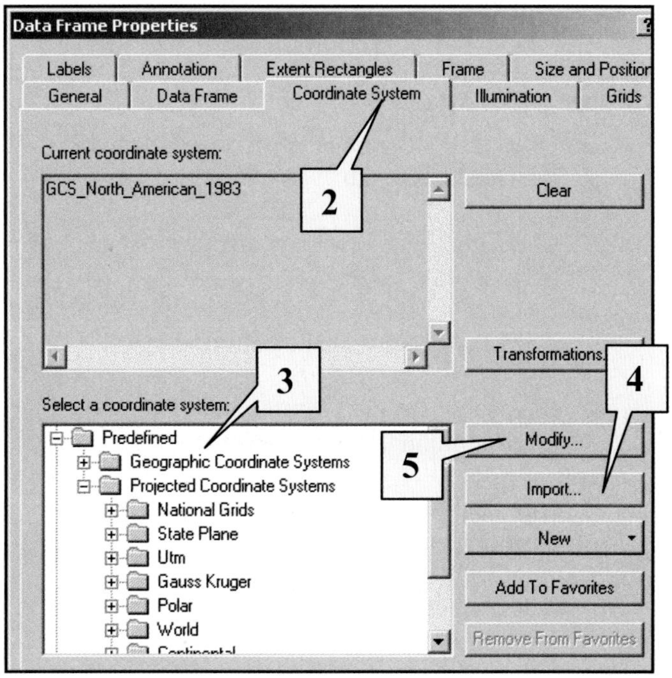

Fig. 2.41. Setting the coordinate system of a data frame

Creating a map based on an attribute

1. Right-click the layer name and choose Properties from the context menu.

2. Click the Symbology tab (Fig. 2.42).

3. Choose the type of map desired. OR, use the Import button to bring in the symbology of an existing layer file and apply it to this data set.

4. Select the field to base the map on.

5. Choose a Color Scheme.

6. If no values are displayed in the window yet, click Add All Values to view the categories.

7. For Graduate Color or Symbol maps, choose a normalization method and field if desired.

8. Double-click a symbol in the list to change its properties using the Symbol Selector.

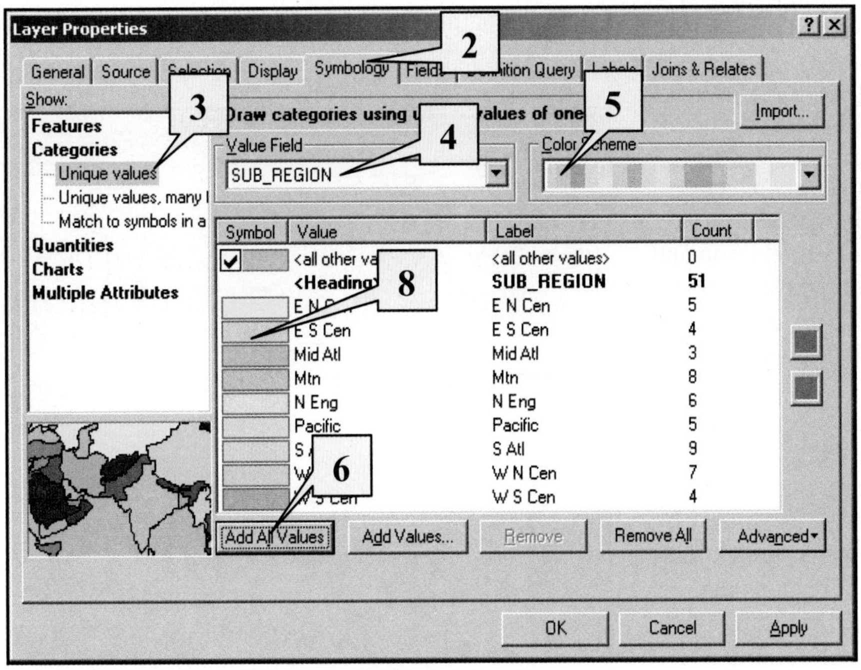

Fig. 2.42. Making a unique values map

TIP: If a class has only a few features, it may be grouped with the other small classes into a class called All Other Values. For example, when mapping land use there might be many residential and industrial parcels but only a few zoned for parks. To display the parks in the regular legend, click the Add Values button and choose it to be shown in the regular legend. If the All Other Values entry displayed in the legend is not desired, uncheck the box next to it.

Other map types, such as dot density, have different settings in the Symbology tab. Experiment with setting these properties or look up the recommended settings in the Help files.

Classifying attributes for a map

For Quantities map types, ArcMap applies a default classification strategy of Jenks Natural Breaks with five classes. Use the following procedure to modify the classification.

1. Open the Symbology tab in the Layer properties. Make sure one of the quantities maps is selected for the map type and that a field has been chosen for the map.

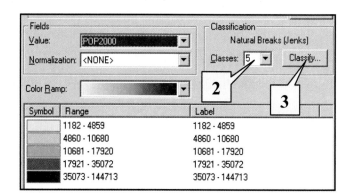

2. Use the drop-down box to change the number of classes, if desired (Fig. 2.43).

3. For other changes, click the Classify button to open the Classification dialog box.

Fig. 2.43. Classification box in the Symbology tab

4. Change the classification method and number of classes, if desired (Fig. 2.44).

5. To exclude certain records, such as those containing zeros or NoData values, click the Exclusion button and enter an expression defining the records to be excluded.

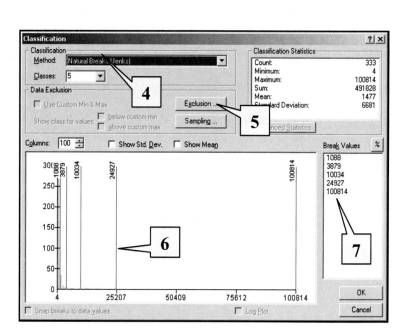

6. The Sampling button is used to reduce the number of features used to calculate statistics, which is useful when data sets are very large. The default sample is 10,000 records.

Fig. 2.44. Setting class breaks

7. To manually set class breaks, either click and drag on the blue lines in the graph, or type the break values directly into the box on the right. When either of these actions is performed, the classification method automatically changes to Manual. The Statistics shown in the upper-left corner can be helpful when assigning classes manually.

8. Click OK when done setting the classes.

Modifying the appearance of the legend

Right-clicking the legend area provides a context menu with several options for modifying the legend (Fig. 2.45).

1. Use Flip Symbols to reverse the order of the symbols only.

2. Select a symbol by clicking on it (or multiple symbols using Ctrl-click) and choose Properties for Selected Symbols to change their appearance with the Symbol Selector. Use Properties for All Symbols to launch the Symbol Selector and modify the appearance of all of the symbols at once, such as changing them all from red to green.

3. Use Reverse Sorting to sort the classes and symbols in the reverse direction.

4. Select one or more classes and delete them with Remove Class(es).

5. Select two or more classes and merge them with Combine Classes.

6. Click Format Labels to apply numeric formatting options to the labels, such as changing the number of decimal places or including thousands separators.

7. Click Edit Description to enter or change a descriptive headline that appears at the top of the legend.

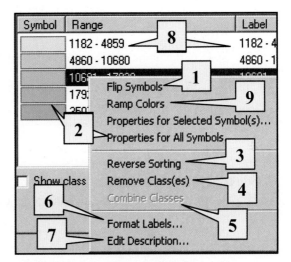

Fig. 2.45. Modifying the legend

8. Click on any Range (except the lowest value of the first range) or Label in the legend to type in new values.

9. To create a new color ramp, set the first and last symbols to the desired end colors, and then choose the Ramp Colors option.

10. When making graduated symbol maps, change the template symbol (the same symbol that increases in size with each class) by clicking the button marked Template.

Creating a map layer or map group layer

A map layer is a set of stored characteristics for displaying a spatial data set.

Creating a layer file

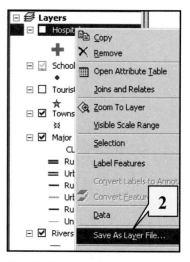

1. Set the symbology, labeling, and other display properties of the data layer using the techniques learned in this chapter.

2. Right-click the data layer name in the Table of Contents and choose Save As Layer File from the context menu (Fig. 2.46).

3. Specify the name and location of the file to be stored.

4. To use the layer file, open it using the Add Data button, just as for any other layer.

Fig. 2.46. Creating a layer

Creating a map group layer

1. Set the symbology, labeling, and other display characteristics of all the data layers to be included in the group.

2. Click the first group member in the Table of Contents to highlight it. Use Ctrl-click to also highlight the other layers in the group (Fig. 2.47).

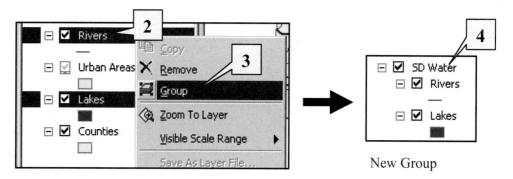

New Group

Fig. 2.47. Creating a group layer

3. Right-click one of the group members and choose Group from the context menu.

4. Click on the group name to highlight it and then click it again to type in a name for it.

5. Right-click the group name and choose Save As Layer File from the context menu.

6. Enter the location and file name of the group layer and click OK.

Chapter 3. Presenting GIS Data

Objectives

➢ Learning basic elements of creating good maps

➢ Choosing appropriate coordinate systems for mapping

➢ Learning to label features with annotation

➢ Creating map layouts and printed maps

Mastering the Concepts

GIS Concepts

GIS analysis often results in sharing information with others in the form of maps or reports. Whether you are creating a large poster-style map, a page-sized map, or a report, a few guidelines help in devising a map design that expresses the essence of the data and gets the message across. This chapter introduces some basic ways to communicate ideas to others, but it cannot cover the full range of concepts included in the art and science of cartography. Taking a cartography class or reading some map design texts are good ways to improve your skills.

Basic elements of map design

A map can be a work of art as well as a vehicle of information, and these two aspects support each other. By designing the map carefully and using some basic guides to style, you can maximize the transfer of information and also present yourself as a competent professional. A poorly designed map gives the opposite impression.

When pen and ink were involved, people spent more time on design to make a good map with minimum effort. Now that computers can quickly create and modify maps, the design step is often ignored. People may not even take advantage of the computer's wonderful ability to try out different designs, and many bad maps result. High-quality maps require planning, attention to detail, and aesthetic sense. The basic process follows these steps:

➢ **Determine the objectives of the map.** What is being communicated? Who is likely to use the map and under what circumstances? The use of the map will guide some important design decisions, such as how big the text must be. A map being projected on a screen to a large audience needs simplicity, large text, and bold colors. A highway travel map needs more details.

➢ **Decide on the data layers to be included.** Computers provide a big advantage here because one can easily add or remove layers as the map takes shape and can change a design decision if it does not appear to work in practice.

➢ **Plan a layout** that includes all the frames and other **map elements** (legends etc.) needed. Following some simple guidelines about placing the elements yields more elegant, pleasing maps.

> **Choose colors and symbols** that create the right effect and maximize readability. Symbols with psychological relevance can be used to make maps easier to interpret, such as making water features blue. The choice of symbols can enhance a message or hide it.

> **Create the map,** paying attention to the details, colors, layout, and aesthetic qualities.

Determine the map objectives

Maps serve a variety of purposes. A one-page map can present information in a report or a paper. A poster-sized map can display important information at a public meeting or an information booth. Some maps are presented to the public and must be meticulously planned and executed. Others are rough copies to take into the field for collecting more data. Some questions to help guide in map planning are suggested in the following paragraphs.

Who will be using the map? Maps for the general public should be simple and carefully explained using legends and supporting text. These maps can benefit from orthophotos or shaded relief backgrounds that show land surfaces and help the reader easily identify landmarks and other features of interest (Fig. 3.1). Consideration should be given to possible human disabilities in mapmaking, such as poor eyesight or color blindness. Maps for professional interest may be more complex and make more assumptions about the readers' level of knowledge.

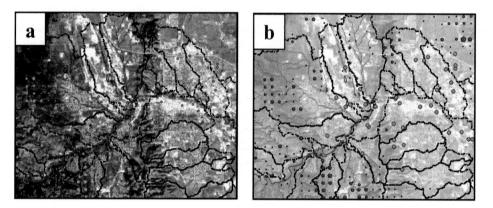

Fig. 3.1. Watersheds, septic systems (orange), and gas stations (green). (a) The satellite image dominates. (b) A transparent image lets the other symbols show up better.

Under what circumstances will the map be used? A page-sized map in a report should be legible to most readers with normal eyesight in good light. A travel map might be used at night or while driving a car, so enhanced readability helps its purpose. A poster map should be tested to ensure that all its text is legible to a person standing several feet away—an 18-point font is considered a minimum size for this type of map. A map used for field work might have many details to help the user find the right locations easily. One issue that is seldom considered is the distribution of the map. Maps in reports or articles are often read as a facsimile—consider designing the map so that its main features are apparent even as a black-and-white copy. Designing a good black-and-white map is even more challenging than using all of those beautiful colors.

What objectives should the map achieve? Maps do not always simply convey information. The design can contribute to the messages intended by the map creator or can detract from it. For example, a city map showing point-source pollution sites in big red symbols suggests danger and may unnecessarily alarm nearby residents. Such a map might cause unwanted publicity and action by interest groups out of proportion to the actual risk. On the other hand, the map could alert the city council to a real danger in order to encourage appropriate action.

How sensitive is the map information? Are privacy issues involved? Many GIS mapmakers, carried away with the ability to create dozens of maps, do not stop to consider the possible effect of certain types of information, especially when layers are put together. For example, laying a fault map on top of the city parcels layer might have serious consequences for the property values of parcels near faults. The public may not be aware that most faults are inactive and do not pose a threat. Maps of fossil sites or archeological digs must ensure that the exact locations of sensitive material remain out of the public domain. One must also be cautious in putting private information on public maps, such as the names of parcel owners or property values.

Choosing the map layers

The objectives of the map usually dictate the choice of the main layers, but the addition of other layers can often improve a map (or diminish it). Generally, the more ancillary information on a map, such as roads, towns, landmarks, topography, and vegetation features, the easier it is to find locations on the map and relate it to what we know. A geologic map without topography or roads does little to help a geologist evaluate hazards in a development area. However, too much information can drown the message or make the map difficult to interpret.

A good design rule of thumb is to present the most important layers with the clearest and largest symbols so that the critical aspects of the layer are emphasized. Minor layers can be presented with less obtrusive symbols and smaller text. One should not have to hunt to be able to see the major patterns of import in the map. Figure 3.1a attempts to show the distribution of septic systems and gas stations with respect to watersheds in Rapid City. The satellite image in the background helps people relate the watersheds to known landmarks and neighborhoods, but it interferes with the interpretation of the map. A redesigned map in Figure 3.1b uses the transparency function to tone down the satellite image and changes the size and color of the marker symbols for enhanced visibility. Now one can easily identify which watersheds are most at risk of contamination.

Planning the layout

The basic additions to a map include a title, legend, indication of north, and scale. Other elements include graphics, pictures, neatlines, and graticules (latitude-longitude markers).

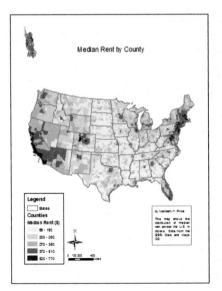

Balance and alignment are very important in designing the map. Balance means that the elements are evenly arranged on the page and are a good size relative to each other. Filled space should be evenly distributed with empty space. Take a look at the map shown in Figure 3.2. All of the basic elements are there, but notice the poor layout. The elements are crowded into the lower-left corner. Their edges are uneven. Alaska is cut off in the corner. The legend is large compared to the map, and the legend boxes are large compared to the legend text. The scale is almost as long as the map. The elements are not aligned but form arbitrary stair-step lines along the map edges. This is a poorly designed map. It looks like

Fig. 3.2. A poorly designed layout

someone made it at 2:00 a.m. the morning before the report was due. However, a few simple changes can dramatically improve its appearance (Fig. 3.3).

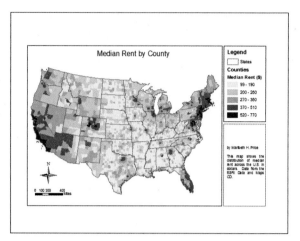

➢ Choose a landscape layout to complement the east-west length of the United States and reduce blank space. This choice also increases the size of the map for better visibility.

➢ Distribute the elements evenly on the page.

➢ Reduce the size of the legend and scale to emphasize the importance of the map.

Fig. 3.3. A revision of the map in Figure 3.2 enhances its professionalism.

➢ Use **neatlines** (boxes around map elements) to enclose and balance the space. Make sure that the neatline edges align with each other.

Choosing symbols

Symbols can highlight or obscure the meaning of a map. Here are some important guidelines when choosing symbols:

➢ Natural earth tones, such as browns, greens, blues, and tans, are more pleasing and easy on the eye than strident reds, hot pinks, and wild purples. Such colors should be saved for special emphasis or contrast and should occupy a small area of the map.

➢ Light pastels are best for large areas and should take up most of the map. Dark or bold colors should be used sparingly to highlight important features.

➢ Take advantage of the psychological aspects of symbols. Blues and greens are cool and calming and are easily understood when applied to water and vegetation, respectively. Reds and oranges tend to denote danger, heat, and agitation. Browns are steady and solid or can suggest dryness when combined with greens for vegetation.

➢ Use symbols that mimic the phenomenon being mapped. Water should usually be blue. A reader viewing the map in Figure 3.4 would probably at first glance interpret the Gulf of Mexico as land. NDVI maps showing the amount of vegetation are often displayed with browns and tans in low-vegetation areas, light greens in moderate areas, and dark greens in heavy areas, mimicking the actual appearance of the ground conditions.

Fig. 3.4. Blue makes features look like water.

➢ Maps showing numeric values, such as county population, city size, or road traffic loads, should have symbols that follow an easily recognized pattern. Increases for polygons are best shown using a single or dichromatic color ramp. For lines, increasing thickness can illustrate larger values, and increasing marker size is typically used for points.

> Avoid "vibrating" color patterns, such as thick alternating white and black lines (Fig. 3.5), because they are tiring and unpleasant to look at. They are called moiré patterns. However, thinning the lines, increasing the separation, and making the white background transparent creates an effective symbol to show the extent of urban area (Fig. 3.6).

> Emphasize important information by using dark and bold colors, large sizes, or high contrast with surrounding map elements. Deemphasize less important information using pale colors, turning black to gray, or using transparency.

Fig. 3.5. A moiré pattern

Fig. 3.6. Eliminating the white background provides a useful see-through pattern.

Examples

In Figure 3.7, we see the population density and highways of the San Francisco area. In Figure 3.7a, a rainbow color ramp was used, but the sense of increase is not obvious. The lower-density counties are dramatized in dark red, drawing inappropriate attention to them. The preponderance of mixed bold colors is also tiring to look at. This map is aesthetically poor and difficult to interpret.

In Figure 3.7b, the rainbow color ramp is replaced with a graduated set of browns, making it obvious which counties have high populations. The color balance seems far more restful and pleasant. The highways are shown in a contrasting green for good legibility, and the urban highways, which are shorter and more difficult to spot, have been put in bright green to enhance their visibility. The major interstates are shown as thick lines and the highways as thin lines, delineating the difference at a glance. In Figure 3.7c, the county outlines are light gray to deemphasize them relative to the roads. The final map is clearer, more attractive, and less cluttered than the original.

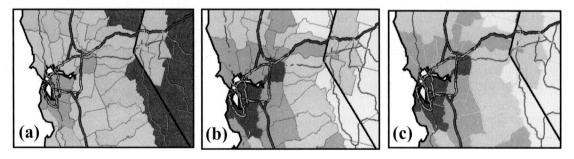

Fig. 3.7. Map of county populations near San Francisco. (a) A map with poorly chosen symbols. (b) A monochromatic ramp highlights the patterns instead of obscuring them. (c) Gray county boundaries emphasize the more important roads.

This world population map illustrates other guidelines (Fig. 3.8a). The ocean is blue, but the mapmaker has disregarded the guideline to reserve dark, bold colors for highlighting important areas. The red-orange graduated color serves to highlight the population but contrasts noisily with the dark blue. The latitude-longitude lines, which are not the focus of the map, crisscross the population data and create a busy feeling. Let's see how a redesign can improve this map.

In Figure 3.8b, a light blue color for the oceans instantly creates a more relaxed, agreeable atmosphere. The latitude-longitude lines have been deemphasized by making them gray and drawing them behind the continents instead of on top of them. The graduated color shades contrast nicely with the ocean, are brown/orange to suggest earth, and are predominantly light. The darker populous countries are naturally emphasized.

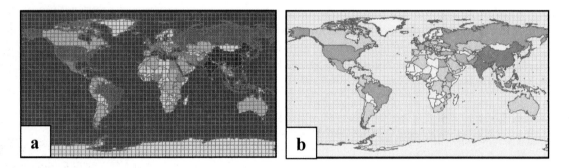

Fig. 3.8. A map of the world. (a) Dark colors and busy lines give this map a heavy and chaotic feel. (b) The pastel colors and deemphasized graticules are far more pleasant.

The layout of the legend often reveals the difference between an amateur and a professional map. The legend in Figure 3.9a illustrates the kind of effort to avoid. No self-respecting cartographer would use the population classes shown here, a jumble of random numbers. The file name cities.shp makes a poor heading—the reader does not care what the name of the file is. The POP1990 label might not be understood by some people as population data. The outline crowds the text. This legend demonstrates some classic mistakes made by beginners or an amateur in a hurry.

The redesign in Figure 3.9b has replaced the layer and attribute names with more understandable labels. The class numbers have been formatted to three significant digits, and the thousands separator has been included for additional clarity. Finally, the cartographer increased the gap between the legend neatline and its contents. This legend is clear, balanced, and professional.

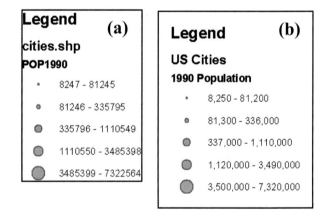

Fig. 3.9. (a) A poor legend; (b) a better legend

Choosing coordinate systems

In Chapter 1, we introduced the idea of a coordinate system and learned that data may be stored in different coordinate systems such as UTM or State Plane. In Chapter 2 we discovered that all layers within a data frame are projected on the fly to match the coordinate system of the data frame set by the user. But what coordinate system should be set? The choice of the coordinate system affects the appearance and use of a map. Figure 3.10 shows the United States using two alternative coordinate systems. Why do they look so different?

The map in Figure 3.10a is based on the familiar latitude-longitude coordinate system, often called a geographic coordinate system, or GCS. This system is a three-dimensional coordinate system based on a sphere, and the latitude-longitude degrees represent angular measures, not distances. Every GCS uses a datum, which adjusts the spherical GCS to fit better on the not-perfectly-spherical earth. Coordinate system labels include the datum being used, with **North American Datum** 1983 (NAD 1983) or World Geodetic System (WGS 1984) being two of the most common ones.

When degrees of latitude and longitude are displayed by a GIS, they are treated as if they were planar coordinates instead of spherical coordinates. This practice introduces a distortion to the features and elongates them in the east-

Fig. 3.10. An (a) unprojected and (b) projected coordinate system

west direction (Fig. 3.10a). Thus, although most people are familiar with the latitude-longitude values and they may be a convenient way to store coordinates, they are a poor choice for maps.

The map in Figure 3.10b utilizes a **projection**, which takes the 3-D angular coordinates of a GCS and uses mathematical equations to displace them to a cylindrical, conic, or planar surface and then unwraps them to a plane (Fig. 3.11). During this process, the angular units in degrees are converted to planar units of feet or meters.

Four properties of map features may be distorted by projections: *area, distance, shape,* and *direction.* Usually map projections reduce or eliminate certain distortions at the expense of the others. Maps based on cylindrical projections (Fig. 3.11a) typically preserve direction and shape at the expense of distance and area. Notice that the longitude lines point north-south, indicating the direction correctly, but Alaska and Greenland are enlarged. Maps based on conic projections typically preserve area and distance at the expense of direction and shape (Fig. 3.11b). The longitude lines no longer point north-south, but the areas are better preserved. The map in Figure 3.10b is in a conic projection. Maps based on azimuthal projections typically preserve areas and distances also (Fig. 3.11c).

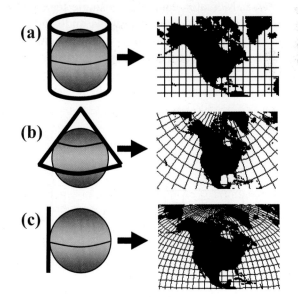

Fig. 3.11. Projections: (a) cylindrical; (b) conic; (c) azimuthal or planar

Choosing a suitable coordinate system is part of developing a good map. National-scale and world-scale maps are impossible to portray without distortion, so the map's purpose drives the choice. If the project includes measuring distances and areas, it is important to choose equidistant or equal-area map projections. If the maps are used for navigating by compass, maps that correctly display direction are important. The cylindrical Mercator projection has been popular for world maps (Fig. 3.12a), but recently critics have charged that this projection enlarges the size of the prosperous countries in Eurasia and North America with respect to developing countries nearer the equator. Attempts have been made to popularize equal-area projections that don't exaggerate the importance of the northern hemisphere (Fig 3.12b).

In Chapter 11 you will learn more about projections, but for now a few simple guidelines can help with the choice. Often the name of the coordinate system is a helpful clue, containing "equidistant" or "equal-area," which indicates those properties are preserved. "Conformal" indicates that shape is preserved. Generally, conic maps tend to preserve distance and area. Mercator and other cylindrical projections tend to preserve shape and direction. Regional data at the state, county, or local scale generally use UTM or State Plane coordinate systems that preserve all four properties. The inside front cover of this text summarizes some of the more common map projections along with the properties and uses of each.

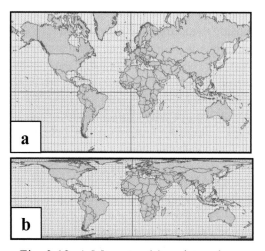

Fig. 3.12. A Mercator (a) and equal-area cylindrical (b) projection

The map projection may influence some of the map elements placed in a layout. A conic projection showing a large area does not have a single north direction, and a single north arrow cannot represent it. A map that distorts area or distance, such as Mercator, cannot use a single scale bar to represent distances. In these maps, a **graticule grid** is employed, which shows latitude and longitude markings (Fig 3.13a). Two other types of grids are often used on maps. A **measurement grid** shows the map units present in the coordinate system (Fig. 3.13b). A **reference grid** shows letters and numbers delineating each grid square (Fig. 3.13c). A reference grid is commonly found in a telephone directory in conjunction with a street index, allowing streets to be located within squares of the map.

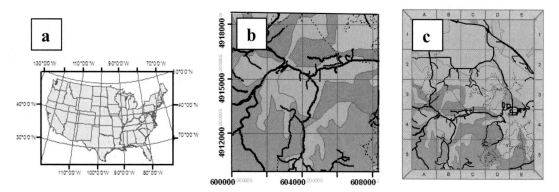

Fig. 3.13. Data frame grids: (a) graticule grid; (b) UTM measurement grid; (c) reference grid

About ArcGIS

Concepts

Maps and reports in ArcGIS

ArcGIS provides two basic tools for creating maps and reports. The layout mode in ArcMap facilitates map design by incorporating data frames and map elements such as legends, north arrows, scale bars, text, and more. Each map document can contain one map layout. Map documents may also be saved as templates, which can be used to re-create a design again and again or to give an organization's maps a common format and look.

Reports may be designed using either the simple report utility packaged within ArcGIS or the more sophisticated Crystal Reports software, which is purchased separately. Creating a report includes choosing fields to present, selecting statistics to report, and setting up the report layout and formatting. Reports can be included in map layouts or printed separately (Fig. 3.14). ArcGIS also supplies a basic **chart** tool to create a variety of different types of graphs, such as bar charts, pie charts, and scatter charts. Like reports, graphs can be created and included in layouts (Fig. 3.14).

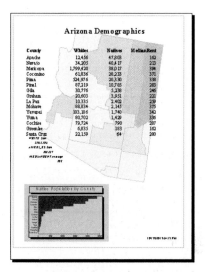

Fig. 3.14. A layout with a report and a chart

Assigning map scales

GIS spatial data acquire a scale as soon as they are drawn on a computer screen or printed on paper. The **map extent** is defined as the range of x-y values currently displayed in the data frame. Zooming in reduces the map extent; zooming out enlarges it. The map scale on a computer screen is determined by the current map extent and the size of the data frame on the screen. The printed scale is determined by the map extent and the size of the data frame on the paper. Assigning the map scale of a layout can follow one of three strategies.

Automatic scaling is used most frequently (Fig. 3.15a). The user sets the size and position of the data frame on the layout page and zooms in or out of the data until the desired map extent is achieved. The user does not try to impose a specific scale but uses his or her eye and aesthetic sense to balance the map frame and other map elements.

Fixed scale is employed when the printed map must have a specific scale, such as 1:24,000. In this mode, the features always appear at the same size, and changing the data frame size changes the map extent and may crop the map if the user is not careful (Fig. 3.15b). Zooming in or out of the map is not permitted, but panning is.

Fixed extent establishes the current map extent and prevents it from changing. The data frame can be resized, which changes the scale of the map but not its extent (Fig. 3.15c). The aspect ratio of the data frame is also fixed. All of the Zoom and Pan tools are unavailable in Fixed Extent mode.

Using Fixed Scale or Fixed Extent establishes the map scale in Layout view and for the printed map. In changing from Layout view to Data view, the scale of the map shown on the screen can change. However, the scale on the page remains as set.

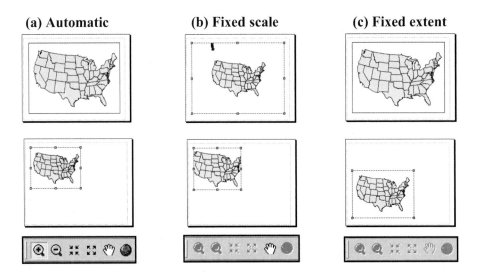

(a) Automatic   **(b) Fixed scale**   **(c) Fixed extent**

Fig. 3.15. Scaling options for data frames, shown in Layout mode. The effects of resizing the data frame differ in each case. The Zoom/Pan tools are only active in Automatic mode.

Setting up scale bars

Most maps should contain either a scale bar or a statement of the scale, such as "This map is shown at a scale of 1:62,500." ArcMap can automatically construct a suitable scale and resize the it when the map scale changes. The user has to set several options, which requires some knowledge of terminology.

The **divisions** are the sections into which the scale is divided. The scale bar in Figure 3.16 has four divisions. The first division may be subdivided, yielding what are called **subdivisions**. This scale bar also has four subdivisions. The **division units** are the units of measurement used to divide the scale bar (miles in this case). The **division value** is the length of each division, expressed in division units. This scale bar has a division value of 0.5 miles.

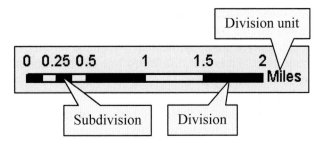

Fig. 3.16. Terminology of a scale bar

The division units, the map scale, the division size, and the length of the scale bar are interrelated. Zooming, for example, changes the map scale and requires the scale bar to change also. Three options control how the scale bar will respond to changes in map scale.

Adjust Width keeps the division size and number of divisions the same. If the map scale changes, the entire scale bar gets larger or smaller. This method keeps nicely rounded numbers for the divisions but may result in an unacceptably small or large scale bar if the scale changes dramatically.

Adjust Division Value keeps the scale bar roughly the same length on the page and with the same number of divisions but changes the size of the divisions. If the map scale changes, the division size of the scale in Figure 3.16 might go from 0.5 miles to 0.6 or 0.4. This option preserves the width of the scale bar but may result in less evenly rounded division sizes and labels.

Adjust Number of Divisions leaves the division value constant and adds or subtracts full divisions if the scale changes. For minor changes in scale, the bar will increase or decrease slightly in size. For dramatic changes, the scale bar will remain about the same size but may have more or fewer divisions. The bar in Figure 3.16, for example, might become 1 mile long or 3 miles long, but the division size would stay 0.5.

Labeling and annotation

Labeling is commonly employed on map layouts. In the previous two chapters, you learned to automatically label many features using **dynamic labeling**. ArcMap provides two additional kinds of labels. **Interactive labels** are used to place a few labels one at a time. **Annotation** is used to gain exact control of labels. We briefly summarize the key features of each label type.

Interactive labels

Interactive labels are constructed by the user, one at a time, to label map features or provide titles and supporting text for a map layout. Several different tools are provided to create standard, splined, callout, or multiline labels or to label a feature using the value from its attribute table (Fig. 3.17). All interactive labels exist on the map as simple graphic elements. They remain where they are placed but are inefficient when many labels must be made.

Fig. 3.17. Labeling tools

Dynamic labels

Dynamic labels are created from an attribute and are redrawn every time the user changes the map view. They can be turned on or off as needed. Dynamic labels utilize a label-placing algorithm that omits overlapping labels. The number of labels displayed can vary, depending on the zoom scale and label size.

Although largely automatic, dynamic labels provide some control options. The *Placement Properties* help to manage the overlap and optimization of the labels to give the best possible result. The *Conflict Detection* options can handle priorities between different layers. *Label weights* control which layers take priority; for example, states might have a higher priority than cities and be preferentially labeled in case of a conflict between the two. *Feature weights* specify the layers whose features cannot be overlapped by labels. Finally, label *buffers* prevent new labels from encroaching on the edges of existing labels.

Because the labels are redrawn each time, printing maps with dynamic labels can bring surprises. What shows on the screen does not necessarily appear the same way on the map. For many purposes a few extra or missing labels do not matter. If they do, however, converting labels to annotation provides greater control over label placement.

Annotation

Annotation provides precise control of each label. Ordinary dynamic labels can be converted to annotation. The labels that fit on the map are automatically placed, and the overlapping labels are directed to an overflow window for interactive placement. Once placed, it always appears in the same place. Annotation may be stored in three ways.

➢ You can store annotation in the map document. They essentially become interactive labels except for the fact that they were created from dynamic labels. This annotation can only be used in a single map document.

➢ You can store annotation as a feature class in a geodatabase and use it in other map documents. Editing must be used to modify it, so we leave this method to Chapter 13.

➢ You can store annotation as a feature class in a geodatabase and link it to the point, line, or polygon features being annotated. Feature-linked annotation is automatically created, moved, and deleted as the features are edited, but it is only available to ArcEditor or ArcInfo users.

Reference scale

When labels or annotation are created, they are assigned a size in points. For a given map, the size of the text relative to the features, such as the counties shown in Figure 3.18a, is determined by the text point size and the current map scale. But what happens if you zoom in? By default, the labels (and symbols) stay the same size when the user zooms in or out (Fig. 3.18b). Twelve-point text remains 12-point text regardless of the scale of the map.

This behavior changes if the user sets the **reference scale**, which is a property of the data frame. The reference scale is the scale at which the labels appear at their assigned size. If a user places 12-point labels on a map with a scale of 1:100,000, then the labels will appear 12 point in size only as long as the map scale remains 1:100,000. If the user zooms in, the labels and symbols increase in size (Fig. 3.18c). If the user zooms out, they decrease in size.

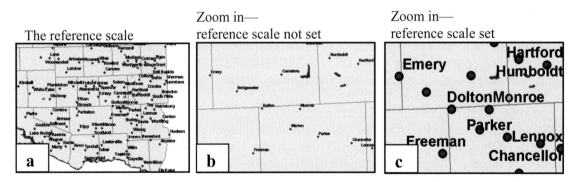

Fig. 3.18. (a) Text and symbols appear at their assigned size at the reference scale. (b) If the reference scale is not set, the symbols will always appear the same size. (c) If the reference scale is set, then zooming in or out changes the size of the text and symbols.

TIP: Be sure to understand the difference between the terms *map scale*, *reference scale*, and *visible scale range*. Map scale determines the ratio of features on the map to features on the ground. The reference scale is the scale at which symbols and text appear at their assigned size. The visible scale range determines the range of scales at which a layer will appear.

Summary

➤ For best results, making maps requires planning, knowledge of some basic guidelines, and attention to detail.

➤ The purpose and audience of a map influence the design process. However, most maps should contain at least a title and contact information, scale indication, north arrow, and an appropriate legend.

➤ Maps should contain sufficient detail for interpretation, without excessive detail that detracts from the purpose.

➤ Layouts should be designed with attention to symmetry, size, and balance of the elements. Neatlines and shades can be used to help distribute space well. Alignment of objects along page margins is an important detail.

➤ Careful choice of symbols enhances the map. Guidelines for symbols include using primarily pastels and earth tones except when emphasis is desired, using larger and bolder symbols for the more important features of a map, using colors and symbols that naturally reflect the nature of the features, and avoiding clutter and clash.

➤ Maps should use projected coordinate systems. All projections introduce distortions in area, distance, direction, or shape. Different projections preserve some properties at the expense of the others.

➤ The choice of projection depends on the application and should be chosen to minimize distortion in the properties important to the map function, such as using an equidistant map to measure distances.

➤ Maps are created in Layout view where all data frames are visible. The Layout tool bar has its own zooming tools that act on the map page instead of within the data frame.

➤ Scaling of the map is controlled by the data frame properties. Three scaling methods may be used: automatic scaling, fixed scale, or fixed extent.

➤ Three types of labeling are available. Interactive labels are used to place any kind of text. Dynamic labels are automatically generated for entire layers at a time. Annotation is created from dynamic labels but can be precisely controlled.

Additional reading on cartography

MacEachren, Alan M. 1995. *How Maps Work: Representation, Visualization, and Design.* New York: The Guilford Press.

Monmonier, Mark. 1996. *How to Lie with Maps.* Chicago: The University of Chicago Press. Second edition. First published 1991.

Robinson, Arthur H., Joel L. Morrison, Phillip C. Muehrcke, A. Jon Kimerling, and Stephen C. Guptill. 1995. *Elements of Cartography.* New York: John Wiley & Sons. Sixth edition. First published in 1953.

Chapter Review Questions

1. List four questions about map objectives that would influence the design of a map.

2. What factors should be considered in evaluating the balance of a map?

3. What types of colors generally work best for maps? How can the psychology of colors be used to enhance a map's meaning?

4. List three common pitfalls that amateurs make when creating legends.

5. What is a geographic coordinate system, and why is it a poor choice for creating maps?

6. What four properties are distorted by map projections? Which tend to be preserved by conic projections? What distortions are present in UTM and State Plane projections?

7. Examine the map projections on the inside front cover. List which projection(s) might be suitable for a (a) map of a county, (b) map of the United States, (c) United States map used to calculate travel distances, and (d) United States map used to calculate areas.

8. When does a north arrow not point up? When should a north arrow not be used?

9. If you have an ArcView license and wish to create and use annotation in different map documents, how would you need to store it?

10. What is the difference between the map scale, the scale range, and the reference scale?

Mastering the Skills

Teaching Tutorial

The following examples provide step-by-step instructions for doing basic tasks and solving basic problems in ArcGIS. The steps you need to do are highlighted with an arrow ➜; follow them carefully. Click on the video number in the VideoIndex to view a demonstration of the steps.

➜ Start ArcMap. Choose to Browse for maps on the splash screen and open the ex2mapdoc_sample.mxd from the mgisdata\MapDocuments folder.

➜ Use Save As to rename the document and remember to save frequently as you work.

1➜ Switch to Layout view by clicking the Layout button.

1➜ Click the Full Page button in the Layout menu, if necessary, to see the entire page.

Setting up the map page

This map document comes from the tutorial in Chapter 2, but we will use this specific one to make sure that each tutorial step works properly. First, we are going to change the page layout and turn on the grid so we can align the objects more easily.

1➜ Choose File > Page and Print Setup from the main menu bar.

1➜ Examine the options to see what these are.

Look at the two sections labeled Paper and Map Page Size. The Map Page Size section controls the size and layout of the map itself. Always set this section first. If you are planning to create an 11-inch by 17-inch map, then set the Map Page Size settings to reflect that. *Always design a map at the size at which you intend to print it.*

The Paper section is used to modify the printer settings at the time of printing. It should not be set until you are ready to print. In many cases, the settings will be the same as the Map Page Size. However, after you created an 11-inch by 17-inch map, you might decide to print it on a letter-size page for a report. Setting the Paper section makes it possible.

1➜ Uncheck the box to Use the Printer Paper settings if it is checked.

1➜ Check the button to scale the map elements proportionally to changes in page size to help in resizing the frames later.

1➜ Fill the buttons to change the Map Page Size to Landscape orientation. Click OK.

> **TIP:** Checking the box to use the Printer Paper settings is not recommended by this author. It can cause undesired changes in the layout if you switch printers.

2➜ To set the grid, choose Tools > Options from the main menu and click the Layout View tab.

2➜ Notice the options for setting rulers: the page units (currently inches), the grid, and snapping.

A grid is a set of points a specified distance apart (now set to 0.25 inches). When snapping is on, objects within the snap tolerance (currently 0.2 inches) will be automatically snapped to the closest grid location. This feature simplifies aligning map elements.

2➔ Check the box to Show the grid.

2➔ Check the Snap Elements to Grid box and make sure the tolerance is 0.2 inch.

2➔ Click OK to close the Options box.

2➔ Open the Layout toolbar, if necessary, and position it in a convenient place.

The Layout toolbar (Fig. 3.19) provides basic zoom tools, which look similar to those on the regular Tools menu, but they affect the view of the page rather than a data frame. These Zoom/Pan tools have no effect on the printed map scale or on the view in the data frame.

Fig. 3.19. The Layout toolbar zoom tools

Now let's see how snapping works.

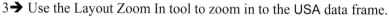

3➔ Use the Layout Zoom In tool to zoom in to the USA data frame.

3➔ Click the Select Elements tool on the Drawing or Tools toolbar. Click the USA frame to select it, and drag it to a new location. Notice that the frame corner snaps to a grid point when it gets close.

3➔ Click the Zoom Whole Page button on the Layout toolbar to view the entire page.

Next we will set up the basic layout design. Be sure to provide ample margins to avoid cropping when the map is printed.

4➔ Use the Select Elements tool to resize and move the two data frames until they look approximately as shown in Figure 3.20.

Although it is simple to arrange the data frames by clicking, dragging, and resizing, sometimes precise sizes and locations are needed. Set these options using the data frame properties Size and Position tab.

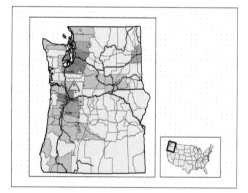

Fig. 3.20. Arrange the frames like this.

5➔ In the Table of Contents, right-click the Hazards data frame and choose Properties.

5➔ Click the Size and Position tab.

5➔ Set the X and Y positions to 0.75 (inch). The page has coordinates of inches, with 0.0 at the lower left. This step will give the lower-left corner of the data frame exactly 0.75-inch margins.

5➔ In the Size window, uncheck the As Percentage box. Set the frame size to 6 inches wide and 7 inches high. Click OK.

TIP: If the map does not redraw completely, use the Refresh button, located next to the Layout and Data view buttons, to redraw it.

6➜ Use the Select Elements tool and the snapping grid to place the USA frame to fit 0.2 inches to the right of the Hazards frame and with a 0.75 margin on the bottom.

6➜ Open its properties and make it exactly 2 inches high.

7➜ Click on the Hazards frame to make sure it is the active frame.

7➜ Use the Zoom/Pan tools on the standard Tools toolbar (NOT the Layout toolbar) to center the Hazards map and make it fill the frame.

The tools on the standard Tools toolbar operate inside the data frame, whereas the Layout tools operate on the layout. It takes a little practice to always use the right one the first time. To see the difference, do the following exercise:

7➜ Still using the Pan tool, move the map inside the Hazards data frame around a little, as you did before. (Put it back in the starting spot before you stop.)

7➜ Now switch to the Layout Pan tool, put the cursor inside the Hazards frame, and move the map again. Do you see the difference?

1. Which scaling method is being used for this data frame? _____ How do you know?

Setting the map projection

We have not yet considered the issue of the map projection for this layout. Recall that both data frames were originally in the same coordinate system, but we modified the central meridian of the Hazards data frame to ensure that north is up. The USA frame shows the direction distortion. It would look better if these two matched.

8➜ Right-click the Hazards frame and open its properties. Click on the coordinate system tab and read the information. Recall that we changed the central meridian from −96 to −120.5 to remove the direction distortion for Washington and Oregon. Write the present coordinate system and central meridian here:

_____

8➜ Open the properties for the USA frame and examine the coordinate system. Write the coordinate system and central meridian here:

_____

It might seem logical to modify the USA frame coordinate system to match the Hazards coordinate system. Try it.

8➜ Click the Modify button in the USA frame coordinate system properties and carefully change the central meridian to −120.5. Click OK and OK.

We have improved the direction distortion in the western United States, but now the east is skewed worse. The Equidistant Conic coordinate system tends to preserve distance and area but distorts shape and direction. We need a projection that preserves direction. Let's look.

2. Which of the projections from the front of the book preserves direction? _____

9➜ Open the USA data frame coordinate system properties again.

9➔ Under Select a coordinate system, click the plus sign to expand the Predefined folder. Navigate to the Predefined > Projected Coordinate Systems > Continental > North America.

9➔ Choose one of the USA Contiguous projections and click Apply. Then try another.

Notice that the direction distortion does not change much, regardless of which projection is chosen. These projections are all conic. We only want a location map and won't be measuring distances, so we might consider a Mercator projection that preserves direction. This list has none, but we can find one in the World folder.

10➔ Click the minus signs to collapse the North America and Continental folders, and click the plus sign to expand the World folder.

10➔ Find the Mercator projection and select it. Click the Modify button.

10➔ Change the central meridian from 0 to –96 to center the map on the United States. Click OK and OK.

10➔ Click Yes to use the coordinate system. You'll learn more about this warning in Chapter 11.

10➔ Use the Zoom and Pan tools to adjust the map inside the data frame.

Labeling features interactively

Dynamic labels are useful for labeling entire layers. Sometimes it is useful to create **interactive labels**, which are simple graphics placed on the map. They are created and controlled using the Draw toolbar, usually located at the bottom of the ArcMap window. If placed in Layout view, the labels are placed on the layout page and appear only in Layout view. If placed in Data view, the labels are part of the map and always appear.

Several styles of interactive labels are available. They are accessed through a drop-down button. When the arrow on the button is clicked, the full set of buttons is shown, and the user can pick one. The selected button is then shown on the toolbar until a different button is selected.

11➔ Click the New Text tool on the Drawing toolbar. The black triangle indicates that more versions of this tool are accessible by clicking the drop-down arrow.

11➔ Click offshore of the map and type Pacific Ocean. Press Enter.

11➔ Click once on the text to select it. Notice the blue dashed line that appears around the text, indicating that it is currently selected and will be affected by changes in its font or other properties.

11➔ Use the Draw toolbar buttons to set the font size to 20 and make the text Italic and blue. Click and drag the text to place it better in the ocean. It's OK if it does not fit right now.

TIP: To work with graphics, use the Select Elements tool from the Draw toolbar. To delete text, click the text to select it and press the Delete key. Use the Select Elements tool to draw a box around many text items, or hold down the Ctrl-key to select multiple text boxes.

12➜ Let's try a different text tool. Click on the black arrow next to the New Text tool and select the New Splined Text tool.

12➜ Click to enter vertices of a line that curves along the coast of Oregon and Washington from south to north. Double-click to end the line.

12➜ A small text box appears. Type in **Pacific Ocean** and press Enter. The text appears splined along the line you entered.

12➜ Change the text size to 24, the style to Italic, and the color to blue.

12➜ Select and delete the earlier horizontal label. Select and move the curved label if its position needs adjusting.

Most of the interactive labels require you to type the text to be displayed. The Label tool, however, can read the label from an attribute field in the feature class table. It uses the primary display field, which is one of the layer properties.

13➜ Right-click the StateCapitals layer and choose Properties. Click the Fields tab and verify that the Primary Display Field is set to CITY_NAME. Click OK.

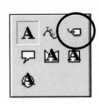

13➜ Try to choose the Label tool from the interactive labels drop-down button. It is dimmed. The Label tool must be used in Data view.

13➜ Click the Data view button. Activate the Hazards frame, if necessary.

14➜ Choose the Label tool. The Label Tool Options window appears.

14➜ Keep the default options to automatically find the best label placement.

14➜ Fill the button to Choose a style for the text label. Scroll down and choose the Capital style from the list.

15➜ Click on the capital of Oregon and watch the label appear as "Salem". If you accidently clicked the wrong feature, delete the label and try again.

15➜ Click on the capital of Washington (Olympia).

15➜ Use the Select Elements tool to position the labels to be clearly seen.

15➜ Close the Label Tool window and return to Layout view.

TIP: Interactive labels and graphics are collectively termed Elements. To Select All Elements or Unselect All Elements, use the functions in the Edit pull-down menu on the main menu bar.

Working with annotation

Although dynamic labels work fine for map layouts, we can convert them to annotation to gain precise control. Examine the labels showing the number of eruptions for each volcano. They would look better in the triangle centers. We can achieve that effect with annotation.

> **TIP:** Annotation is created for the data frame and includes all the labels currently set for the data frame. You must turn off any layer labels that you don't want converted.

We want to create annotation for Known Eruptions, and we would also like to label the active volcanoes. We already set up labels for the Volcanoes layer, which are not currently being displayed but which we can use.

16➜ Add the Labeling toolbar, if necessary, and click on the Label Manager button.

16➜ Click Clear All to turn off all the labels.

16➜ Turn the Known Eruptions labels back on.

16➜ Turn on the Volcanoes labels and make sure More Active is the only class checked.

16➜ Click on the More Active label class. Click the SQL Query button and change the expression to volcanoes with >= 5 eruptions. Click OK.

16➜ Click OK to make the changes and close the Label Manager.

17➜ Turn on the Volcanoes layer.

17➜ Examine the map to make sure you set everything correctly. You should have labels in the triangles, large brown labels for the volcanoes, and no river labels.

18➜ Right-click the Hazards data frame and choose Convert Labels to Annotation.

18➜ Fill the button to create annotation in the map.

18➜ Notice the reference scale value listed in the upper left corner. This is the scale at which the labels will appear at their assigned point size.

18➜ Choose to create annotation only for the features in the current extent.

18➜ Keep the box checked to convert unplaced labels to unplaced annotation. These labels will be put in an overflow window for individual placement later.

18➜ Click Convert to create the annotation.

View the new annotation. It looks similar to the dynamic labels, but the labels are now individually adjustable. If you had unplaced labels, they were put in an overflow window (Fig. 3.21). Your list might be slightly different than the one in the figure. We can place these labels manually.

19➜ In the overflow window, right-click the label for Baker and choose Flash Feature. Repeat, if necessary, until you see it flash. (You may need to move the overflow window to see it.)

19➜ Right-click the Baker label again and choose Add Annotation. The label is added to the map. The placement is not perfect, but we will adjust it in a moment.

20➜ Add the other volcano names to the map.

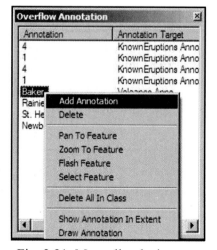

Fig. 3.21. Manually placing overflow annotation

20➔ We don't want the eruptions labels for the smaller triangles. For each piece of annotation that is a number, right-click it and choose Delete. Then close the Overflow Annotation window.

To adjust the annotation, we need to be in Data View.

21➔ Click the Data View button. If you get the USA frame, right-click the Hazards frame name and choose Activate.

21➔ Click the Select Elements tool, if necessary. Click on the label for Baker to select it. Adjust the position of the label so that it is not cropped at the edge of the map.

21➔ Click on the 13 inside the Baker symbol to select it. Use the mouse or the arrow keys to move it to the center of the triangle.

21➔ Carefully select and delete any annotation in the smaller triangles with less than 5 eruptions.

21➔ Adjust the remaining labels for best effect.

3.   When you returned to Data view, the labels got larger. What does this tell you about annotation? _____

Now is a good time to adjust the layers and colors to reduce clutter and improve legibility. Notice how the interstate symbols dominate the map. Although it is helpful to have the roads to see which travel routes are at risk of blockage during an eruption, they can be toned down.

22➔ Turn off the Cities layer.

22➔ Click on the Interstate symbol of the Highways layer in the Table of Contents to open the Symbol Selector.

22➔ Change the symbol to Double, Plain (near the bottom). Set the color to Spruce Green (hold the cursor over a color to see its name).

23➔ Change the symbol for the Freeways to Double, Plain with Dark Olivinite color.

23➔ Right-click the Arterial symbol and change the color to Spruce Green.

23➔ Right-click the Unclassified symbol and change the color to Burnt Umber.

24➔ Open the Symbology properties for the Counties layer.

24➔ Click on the Symbol heading above the shades and choose Properties for All Symbols. Change the Outline color to Gray 20%. Click OK and OK.

25➔ Turn off the Volcanoes layer. The labels disappear. Turn them back on. We want the labels, but not the symbols. This is not hard to do.

25➔ Open the Symbology Properties for Volcanoes. Right-click the Active class and select Remove Class(es). Repeat for the Inactive class. Click OK.

Finally, let's investigate the reference scale.

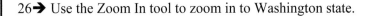

26➔ Use the Zoom In tool to zoom in to Washington state.

Notice that the annotation appears larger because it had a reference scale assigned to it when it was created. However, the symbols do not scale, and the eruption numbers don't fit in the triangles any longer. We need to set the reference scale for the symbols separately.

26➜ Return to the previous extent showing both states.

26➜ Right-click the Hazards data frame name and choose Reference Scale > Set Reference Scale.

26➜ Zoom in again and watch the symbols scale this time.

26➜ Return to the previous extent.

TIP: To set an exact reference scale such as 1:24,000, right-click the data frame name, choose Properties, click the General tab, and type a specific reference scale in the appropriate box.

Some annotation properties can be modified, such as placing a scale range to show it only at certain scales. It can also be removed from the map if no longer needed.

27➜ Open the Hazards data frame properties and choose the Annotation Groups tab.

27➜ Examine the options.

27➜ Select one of the annotation groups and notice that you could now click Remove to get rid of it. Don't do it though; just choose Properties.

27➜ Examine the settings and then close both windows without making changes.

Adding a legend to the map

Now that the map is in good shape, it is time to continue with creating the layout.

28➜ Click the button to return to Layout view. Choose the Select Elements tool.

28➜ Right-click the Hazards frame name and choose Activate. The legend is always added to the active frame.

28➜ Choose Insert > Legend from the main menu bar. A Legend Wizard will appear.

29➜ Examine the list of layers and make sure you want them all. Volcanoes has only labels, so select it on the right and click the < button to remove it from the list.

29➜ Now establish the order of layers. To move a layer, click on it to select it and click the Up or Down arrow until it is in the right location. Put Known Eruptions first, Counties next, then Highways, then StateCapitals, then Rivers, then States. Click Next.

30➜ Leave the Legend Title as is, except click the button to Center it. Click Next.

30➜ Choose the 1.0 Point border for the Legend. Make sure that the Gap is set to 10 points. Click Next.

30➜ To choose a different style patch, click the Rivers layer and choose the Flowing Water line. Click Next.

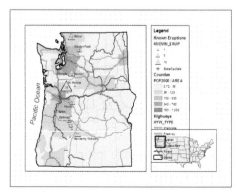

Fig. 3.22. Adding the legend

30➔ This section gives very detailed control of the spacing between different elements of the legend. It is fine to leave the defaults on this step. Click Finish.

30➔ The legend appears in the middle of the map. Click and drag it next to the map frame and even with its top (Fig. 3.22).

If you made a mistake creating the legend, don't worry. You can go back and change its properties. Note that the legend is longer than the space available. We can make changes to help it fit better. The Known Eruptions layer has two headings. We only need one.

31➔ Right-click the Legend and choose Properties. Click the Items tab.

31➔ Click on Known Eruptions and click the Style button.

31➔ Choose the item style shown in Figure 3.23, called "Horizontal Single Symbol Layer Name and Label".

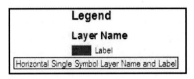
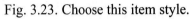

Fig. 3.23. Choose this item style.

31➔ Click OK and Apply and examine the change in the legend.

31➔Click on Highways and change its item style also. Click OK and OK.

Legend entries for layers can be given different styles with different elements present. Figure 3.24a shows these elements and their terms. Layer names and headings are edited in the Table of Contents (Fig. 3.24b). Labels can be edited on the layer's Symbology tab, and descriptions are created by right-clicking the class in the Symbology tab (Fig. 3.24c).

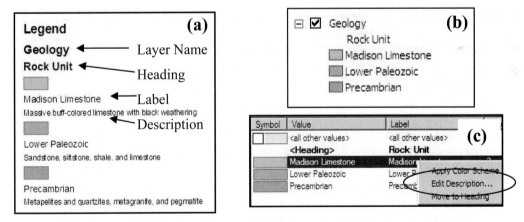

Fig. 3.24. Editing legends. (a) Legend items and their terms. (b) Edit layer names and headings in the Table of Contents. (c) Edit labels and descriptions on the Symbology tab.

32➔ Click twice on the POP2000/AREA heading under the Counties layer in the Table of Contents and change it to **People per sq. mile**.

32➔ Click twice on the StateCapitals layer name in the Table of Contents and change the name to **State Capitals** so it looks better in the legend.

TIP: If you did not know the units of the AREA field, you could look it up in the metadata.

We have shortened the legend, but it still does not fit. Let's use two columns instead of one.

33➔ Double-click on the Legend to open its Properties.

33➔ Click the Items tab, if necessary. Move the Properties box to see the Legend.

33➔ Select the Highways and check the box to Place it in a new column. Click Apply.

That looks like it will work, but we want to align the edges of the legend with the location map below it. The location map width is 3.25 inches.

34➔ Click the Size and Position tab in the Legend Properties window.

34➔ Leave the Height alone but set the width to 3.25 inches. Click OK. Make sure the vertical edges of the legend align with the location map edges.

34➔ Move the legend into place, using the snap grid so that it is 0.2 inches from the location map and the Hazards frame.

Placing a scale bar on the map

The scale bar is placed in the active frame, so make sure that the correct frame is active before creating the scale bar. Our scale bar will refer to the Hazards data frame.

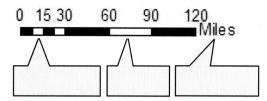

Fig. 3.25. Terminology of a scale bar

4. To review the scale bar terms, write the correct terms in the boxes of Figure 3.25

35➔ Make sure the Hazards frame is active.

35➔ Choose Insert > Scale Bar from the main menu.

35➔ Choose the Alternating Scale Bar 1 style bar.

35➔ Click Properties, click the Scale and Units tab, and make sure that the Division Units are set to miles.

35➔ Click OK and OK. Move the scale bar down to the lower-right corner of the Hazards frame, on top of Oregon.

35➔ Use the Layout Zoom In tool to zoom in to the scale bar to examine it more closely.

The scale bar might show up rather short with overlapping labels, or it might be long with uneven units such as 2.6 miles. If it appears unsuitable, try resizing the scale bar to improve it.

36➔ Click the Select Elements tool.

36➔ Click the right boundary of the scale bar (a double horizontal arrow will appear) and drag it to the left to decrease the length of the scale bar.

36➔ Repeat until the scale bar is about 120 miles long.

Unless you were lucky, you probably ended up with some fairly uneven labels instead of nice, round numbers. We can modify the scale bar to get exactly the properties we want.

37➔ Double-click the scale bar to open its properties, and click the Scale and Units tab.

Notice that the division value is dimmed out. In order to set it explicitly, we must change the way ArcMap adjusts the bar when resizing.

37➔ Under When Resizing, set the drop-down box to Adjust Number of Divisions.

37➔ Set the Division value to **50** miles.

37➔ Set the number of subdivisions to **2**. Click OK and OK.

37➔ Examine the new scale bar. Resize it several times. Make it 150 miles long.

The size changes in even increments of division units, 50 miles. This method is a better way to handle your scale bars to prevent uneven division units.

Adding other map elements

Many other elements can be added to the map using the Insert menu. Keep in mind that the elements will be inserted into whichever data frame is currently active. If you make a mistake and insert an element into the wrong data frame, simply click the Delete button to get rid of it.

Adding a north arrow

38➔ Click the Zoom Whole Page button to see the entire page again.

38➔ Choose Insert > North Arrow from the main menu bar.

38➔ Click the symbol you like best.

38➔ Click the Properties button and set the color to a dark blue. Click OK and OK.

38➔ Click the Select Elements tool and move the north arrow to the lower-right corner of the Hazards frame.

TIP: Some map projections do not preserve direction, and the north arrow should point at the appropriate angle instead of vertically. ArcMap does not tilt the arrow automatically. Use the Properties tab to set the north arrow to the correct angle.

The United States location map might look good with a graticules grid on it.

39➔ Right-click the USA data frame name in the Table of Contents and choose Properties. Click on the Grids tab.

39➔ Click New Grid. Choose the Graticule grid. Click Next.

39➔ Choose Graticule and labels. Accept the default intervals. Click Next.

39➔ Accept the defaults on the axes and labels. Click Next.

39➔ Accept the defaults on the graticules. Click Finish and OK.

The grid appears, but notice that the labels extend into the other map frame. At this point a design decision is in order. You can resize the USA data frame so that the labels fit into the space available, or you can remove the grid. Since this frame merely contains a location map, having a scale or north direction or grid is not important. Make your own decision to keep the grid or not.

40➔ If keeping the grid, resize the USA data frame so that the grid labels fit comfortably in the space provided.

40➜ If removing the grid, open the Properties for the USA data frame, select the Graticule grid, and click Remove. Click OK.

Adding titles and text

41➜ Choose Insert > Title from the main menu bar. A centered text field appears.

41➜ Type in **Pacific Northwest Volcanic Hazards** and press Enter.

41➜ Double-click the title to open its properties. Edit the text, placing a line return after Northwest. Press OK when finished.

41➜ Using the Drawing toolbar, set the font size to 18 and make it Bold. Center it above the legend, aligning it with the top of the Hazards data frame.

42➜ Choose the Insert > Text choice in the main menu bar.

42➜ Type **by Your Name** and press Enter.

42➜ Move the text to a centered position below the Title.

43➜ To add wrapped text, click the New Rectangle Text tool (one of the label tools) and click and drag it to create a box inside the legend, aligning the edges with the data frame and the legend.

43➜ Double-click inside the new text box to open its properties and type the following text into the box without using any Enter keys: **ESRI Data and Maps [DVD]. (2006) Redlands, CA: ESRI, Inc. [October, 2008]**.

43➜ Make the text left-justified.

43➜ Click the Frame tab and set the border to <None>. Click OK.

43➜ Set the font size to 8 points. Adjust the location to fit neatly in the legend space. The legend should look like Figure 3.26.

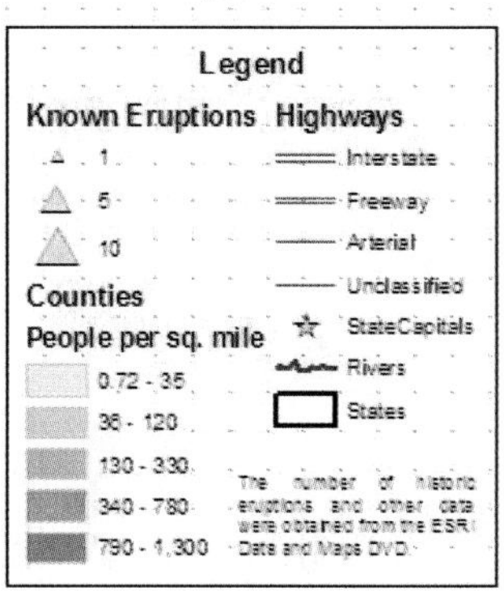

Fig. 3.26. The final legend

Adding neatlines and shades

44➜ Choose Insert > Neatline from the main menu bar.

44➜ Fill the button to place it around all elements.

44➜ Set the gap to 10 points (the gap controls the distance between the elements and the neatline).

44➜ Choose the Triple Graded border from the drop-down box.

44➜ Choose the Linear Gradient fill from the Background drop-down box (Fig. 3.27). Click OK in all boxes.

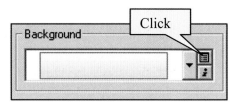

Fig. 3.27. Changing the color of the gradient

TIP: To change the gradient color, click the small icon to the upper right of the background symbol box. Click Properties and then click Change Symbol. Click Properties again and choose the color ramp. Click OK in all boxes.

Reviewing and printing the map

Now for the finishing touches.

45➔ Use Tools > Options > Layout View, if necessary, to turn the grid snapping off.

45➔ Examine the map one last time for balance and aesthetics and fix any problems with alignment of boxes, text, or other issues.

45➔ Change the color of the United States map to harmonize it with the layout.

45➔ The final map should look similar to Figure 3.28.

46➔ It is always wise to SAVE the map document before you print.

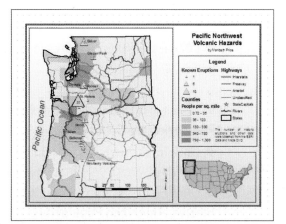

Fig. 3.28. The final layout

46➔ To print, choose File > Print from the main menu bar.

46➔ Click Setup to change the printer, if necessary. Adjust the other settings, if necessary. When ready, click OK.

TIP: Colors shown on the computer screen often look somewhat different than colors printed on paper. Plan to print at least one draft of the map to make sure that the colors look acceptable. Colors can also vary from printer to printer.

Maps can also be saved as image files. The PDF format is a popular way to distribute maps electronically to others. A JPEG or GIF can be easily imported into word processing documents or slide shows.

47➔ Choose File > Export Map.

47➔ Choose the location and name for the image file.

47➔ Set the Save As type to JPEG.

47➔ Click on the Options drop-down flag, if necessary, to show the Options settings.

47➔ Check the resolution. The minimum resolution for printing should be 200 dpi. It could be smaller for a Web or screen image.

47➔ If planning to place the map in another document such as PowerPoint, check the box to Clip Output to Graphics Extent.

47➔Click Save.

This is the end of the tutorial. You may continue on to do the optional tutorial on creating graphs.

➜ If stopping, exit ArcMap. You do not need to save your changes.

Making a simple graph (optional)

Imagine that you have been commissioned to write a report on the Native American population in Arizona. You need to create a graph showing the Native American population of each county.

➜ Open the map document ex_3b.mxd in the MapDocuments folder.

➜ Use Save As to rename the document and remember to save frequently as you work.

48➜ Choose Tools > Graphs > Create from the main menu. The Create Graph Wizard appears.

48➜ Fill out the boxes as shown in Figure 3.29. Click Next.

49➜ Enter the title of the graph as **Native Population by County**.

49➜ Click on the Left axis tab and title it **County**.

49➜ Click on the Bottom axis tab and title it **Native American Population**.

49➜ Uncheck the Legend box so the final graph is not cluttered with one.

49➜ Click Finish. The bar graph appears in a separate window. Adjust its size, if necessary, so that all the counties are labeled.

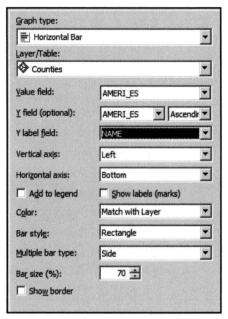

Fig. 3.29. Creating a bar graph

After the graph is made, you can still modify its properties.

50➜ Place the cursor on the window bar of the graph. Right-click it and choose Advanced Properties.

50➜ Click the Walls entry and select the Back tab. Set the color to white. Click OK and close. The final graph should appear similar to Figure 3.30.

50➜ Take a moment to see what other properties can be set for the graph.

51➜ Right-click the top window bar of the graph again and choose Add to Layout.

51➜ Close the graph window. Adjust the size and position of the graph on the layout, if necessary. (Open the properties Size and Position tab and uncheck Preserve Aspect Ratio to be able to resize it fully.)

Fig. 3.30. A native population graph

134

51➜To open the graph window again, choose Tools > Graphs > Native Population by County from the main menu bar.

➜ Use File > Save As to save the map document in the MapDocuments folder under a new name, such as Arizona.mxd.

This is the end of the optional tutorial.

➜ Exit ArcMap.

More skills

Consult the Skills Reference section of this chapter to learn to do the following:

➢ Add pictures or graphics to a layout.

➢ Set the data frame to use the Fixed Scale or Fixed Extent scaling method.

➢ Use a map template to produce a quick layout of a map.

Exercises

Create an attractive map layout showing median rent in the counties of all 50 states of the United States. Use additional data frames to show Alaska and Hawaii individually at larger scales, each with its own appropriate coordinate system and scale. Label the coordinate systems used in each data frame on the map.

Consider: Ideally, each data frame should use the same classification breaks in the legend so that you need only put the legend on the map once. How do you make sure this happens?

In your work, pay close attention to details of layout, colors, formatting, balance, and all the other aspects we have learned about. Make the map as aesthetically pleasing and professional as you can. Be sure to include all the required map elements, as well as your name. Be as creative as you want.

Export the map as a pdf file and print it for your instructor.

Challenge Problem

Create a scatter plot graph showing the relationship between median home value and median rent. Print it separately from the map.

Skills Reference

Composing the data frames 136

Setting up the map page 137

Using the Layout toolbar..138

Setting the scale or extent..139

Labeling features interactively 139

Setting the reference scale 141

Converting dynamic labels to annotation...142

Adding a north arrow 143

Adding titles and text 143

Adding graphics to layouts..143

Adding a legend...144

Changing the appearance of a legend...144

Adding a scale bar 147

Adding neatlines, backgrounds, and shadows.......................................148

Adding pictures 148

Creating a map from a template 149

Printing a map 150

Exporting a map as a picture file..150

Creating a simple graph...151

Composing the data frames

The map layout has at least one data frame that can be dragged and resized on the page.

1. Make sure that the Layout view button at the bottom left of the display window is clicked (Fig. 3.31).

2. Choose the Select Elements tool from the Drawing toolbar or the Zoom/Pan toolbar.

3. Click on a data frame to activate it. Blue handles and dashed lines will appear to indicate that it is active.

4. Click and drag on the data frame to move it to a new location (Fig. 3.32).

5. Click and drag a side handle to increase or decrease the size in one direction.

6. Click and drag on a corner handle to increase or decrease the size in two directions.

Layout view

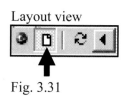

Fig. 3.31

Fig. 3.32. Click and drag the active frame to move it or change its size.

136

Setting up the map page

1. Choose File > Page and Print Setup from the main menu bar.

2. Check the printer and use Properties to change it, if necessary (Fig. 3.33).

3. Set the printer paper size, source, or orientation options, if necessary.

4. Set the map page width, height, and orientation.

5. To use the settings from the current printer for the map page size, check the box. Otherwise, set the page size. See the tip below.

6. If you plan to change the map size and want the map elements to be resized along with the page, check the box.

7. If desired, check the box to show the printer margins in the layout.

8. Examine the page preview and adjust settings, if necessary.

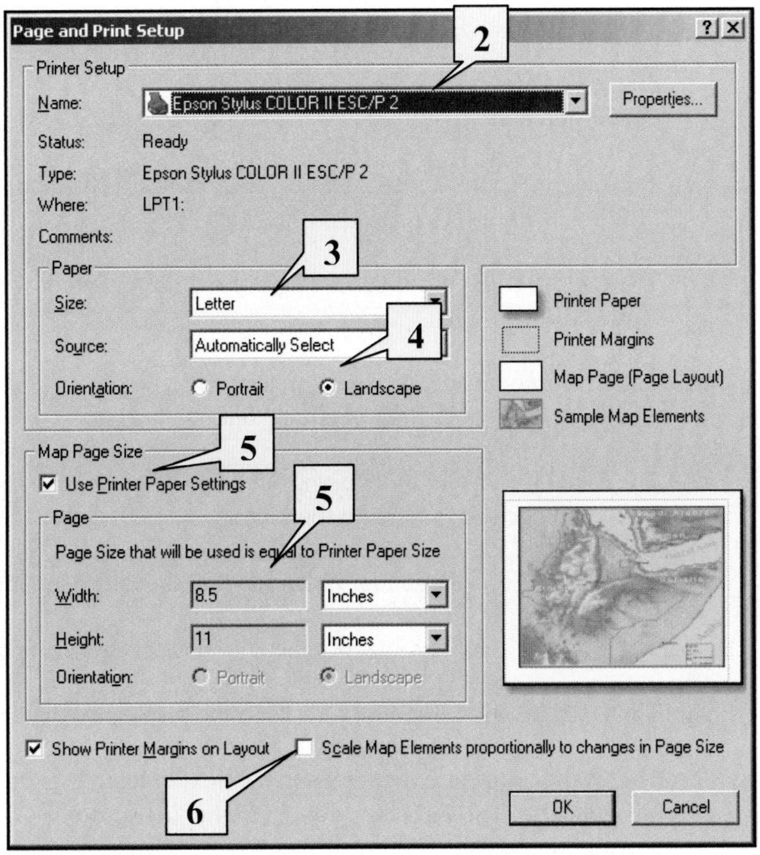

Fig. 3.33. The Page and Print Setup window

TIP: Using the Printer Paper Settings may cause a map document to issue a warning when it is opened on a system where the original printer is not available or to make unwanted changes to the layout when the printer is changed. The author does not recommend using this check box.

Using the Layout toolbar

The Layout toolbar provides tools for zooming around the layout page (Fig. 3.34). It has no effect on the zoom in the data frame. The function of the buttons, from left to right, is as follows:

1. The Layout Zoom In tool is used to enlarge a portion of the layout. Click the tool and then click and drag a box around the desired area.

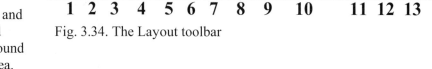

Fig. 3.34. The Layout toolbar

2. The Layout Zoom Out tool zooms out a specified distance from the layout, centered on the clicked point.

3. The Layout Pan tool moves the layout within the window. Click and drag on the layout to move it.

4. The Fixed Zoom In tool zooms in a specified amount, centered on the current display, when you click anywhere on the layout.

5. The Fixed Zoom Out tool zooms out a specified amount, centered on the current display, when you click anywhere on the layout.

6. The Zoom Whole Page tool zooms so that the entire layout page can be seen.

7. The Zoom to 100% tool shows the layout at the same scale as it will be printed.

8. The Go Back to Extent tool returns to the previous extent.

9. The Go Forward to Extent goes to the next extent. This button is only available if you have clicked the Go Back to Extent tool at least once.

10. The Zoom Control box sets a particular percent enlargement for the layout.

11. The Toggle Draft Mode allows you to display each element as a simple labeled box. This feature can make setting up the layout easier without waiting for each element or data frame to redraw each time a change is made.

12. The Focus Frame mode switches to Data View but lets you keep editing any text placed on the layout. It's sort of a hybrid between Data View and Layout View.

13. The Change Layout button launches the Template window so you can add or change the layout of the map using a predefined template.

Setting the scale or extent

The Data Frame properties tab provides three options for setting the map scale of the frame: automatic, fixed scale, or fixed extent. Automatic scaling is the default. Fixed scale or fixed extent will deactivate the Zoom/Pan tools for the frame.

1. Right-click the data frame name to open its properties. Click the Data Frame tab. Choose the desired scaling method (Fig. 3.35).

Setting a fixed scale

2. Fill the second button and type the desired scale in the box. Click OK.

Setting a fixed extent

3. First, use the Zoom/Pan tools to set the Data view to the desired extent . Alternatively, enter the *x-y* coordinates of the desired extent.

4. In the Data Frame tab, fill the Fixed Extent button. The boxes will be filled with the current *x-y* extent of the data frame. Change the *x-y* coordinates, if desired. Click OK.

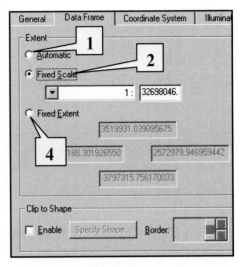

Fig. 3.35. Scaling options in the data frame properties window

Labeling features interactively

1. Look for one of the labeling tools on the drawing toolbar. Click on the black arrow for a drop-down menu to choose one of the tools.

Adding text to the map

1. Click the Add Text tool.

2. Click the desired location on the map and enter the text. Press Enter when finished.

3. Newly entered text is always selected, as shown by the dashed blue box around it. At this point you can change its font, size, or color using the menus on the Drawing toolbar, or you can use the cursor to click and drag it to a new location.

TIP: To delete text, click the black arrow on the Drawing toolbar, click the text to select it, and press the Delete key. Use the Select tool to draw a box around many text boxes, or hold down the Ctrl-key to select multiple text boxes for deletion.

TIP: To delete all graphic elements including text, choose Edit > Select All Elements from the main menu bar and press the Delete key. Be careful! This option also deletes any annotation saved in the map document.

Labeling a feature with its attribute

1. If necessary, right-click the layer name, choose Properties, and click the Fields tab. Set the Primary Display Field to the desired attribute to appear in the label. Click OK.

 2. Choose the Label tool from the labeling drop-down button. The Label Tool Options window will appear.

3. Choose whether to find the best placement (e.g., the center of a feature) or to place the label where you click. Also choose whether to use the symbol properties already set for the layer or to choose a new symbol style now. Click the X box to close the window.

4. Click on the feature to be labeled. ArcMap will label the topmost layer clicked if more than one layer is present.

> **TIP:** The text symbol is set for the layer using the Layer Properties. Modify the default symbols by right-clicking the layer name, choosing Properties, clicking the Labels tab, and setting the text symbol.

Splining text from an attribute along a line

Splining text makes it follow along a linear feature such as a road or stream.

1. Set the font size and style options desired using the Drawing toolbar.

 2. Choose the Spline Text tool from the labeling drop-down button.

3. Click vertices to define the line along which the text will appear. Double-click to end the line.

4. Type the text in the box and press Enter.

5. Use the Drawing toolbar to modify the position or font characteristics as needed.

Adding a callout label

A callout places text in a box with a pointer to the feature of interest.

1. Choose the Callout Text box tool from the labeling drop-down button.

2. Click on the feature to be labeled and drag the cursor to define the direction and length of the callout pointer.

3. Enter the text into the box and press Enter.

4. Click and drag on the text box to change its location, if desired. Click and drag on the blue dot to change the location of the pointer. Use the tools on the Drawing toolbar to modify the callout's font, style, size, and so on.

Creating wrapped text boxes

1. Choose one of the wrapped text tools: the New Polygon text, the New Rectangle text, or the New Circle text.

2. For the circle or rectangle text tool, click and drag to draw the circle/box, releasing the mouse when it reaches the desired size and shape. For the polygon tool, click on each vertex to define the desired shape. Double-click when finished.

3. Click in the empty shape to open the text Properties box (Fig. 3.36). Type in the text to be displayed. Do not use the Enter key unless you wish to enforce a new line within the text.

4. Set the symbol, spacing, or other options, if necessary.

5. Use the other tabs to change the margins, columns, frame border, size, position, or area background of the text box as needed.

6. Click OK to place the text.

7. To modify the text later, double-click to open its properties box.

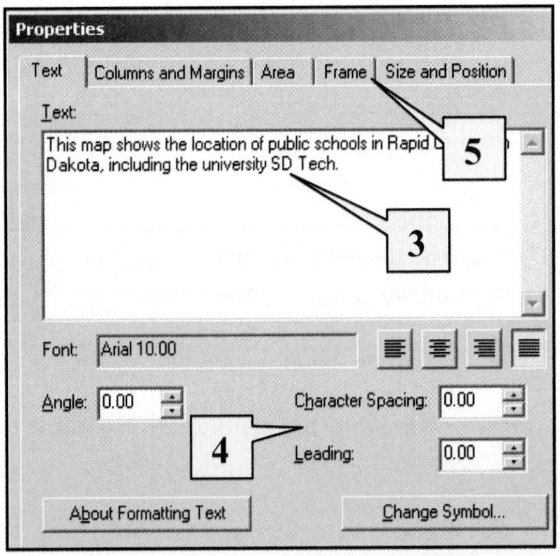

Fig. 3.36. Wrapped text properties window

Setting the reference scale

1. Use the Pan/Zoom or Bookmark tools to zoom to the desired scale.

2. Right-click the data frame name in the Table of Contents and choose Reference Scale > Set Reference Scale.

TIP: To set an exact scale value such as 1:24,000, right-click the data frame name, open the data frame Properties, click the General tab, and type a specific reference scale in the appropriate box.

3. To remove the reference scale, right-click the data frame name in the Table of Contents and choose Reference Scale > Clear Reference Scale.

4. To zoom to the reference scale, right-click the data frame name and choose Reference Scale > Zoom to Reference Scale.

TIP: Annotation, once created, retains its original reference scale set when it was created, even if the reference scale of the data frame is changed later.

Converting dynamic labels to map annotation

These directions show how to create annotation stored as text graphics in a map. Consult the Help files for information on creating annotation as a geodatabase feature class.

1. Use the Layer Properties to create dynamic labels for all the desired layers. Take care in setting the properties, weights, and so on because these will control the labels that appear.

2. Annotation will be created for all layers in the data frame with dynamic labels turned on. Turn off any dynamic labels for layers not to be converted.

3. Right-click the data frame name and choose Convert Labels to Annotation. The dialog box will appear (Fig. 3.37a).

4. Examine which layers will be converted to ensure they are the desired ones.

5. Choose to create annotation in the map.

6. Check the box to Convert unplaced labels if you want to place overlapping labels interactively.

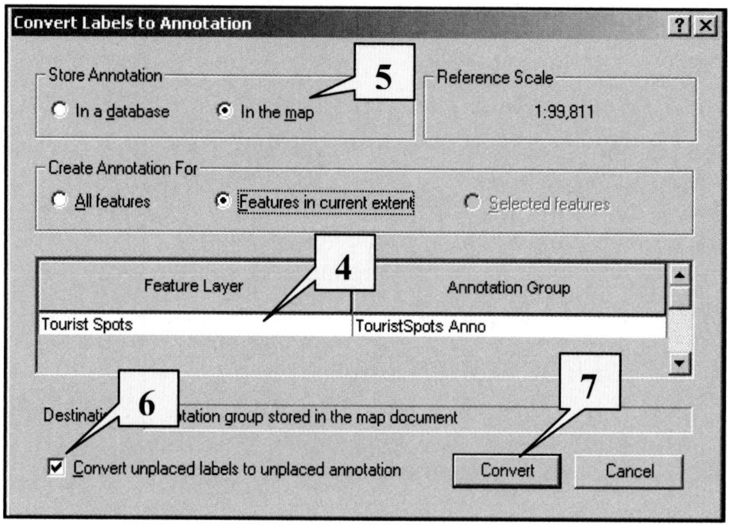

Fig. 3.37a. Creating and placing annotation

7. Click Convert to create the annotation.

Unplaced labels will be placed in an overflow window if you checked that option. The next step is to place these labels interactively (Fig. 3.37b).

8. Right-click a label in the overflow list and choose a method to locate it (flash, pan, zoom, etc.).

9. Choose Add Annotation to add the label to the map. Click and drag it to adjust its location, if necessary.

10. If you decide not to place the label, delete it from the list using Delete.

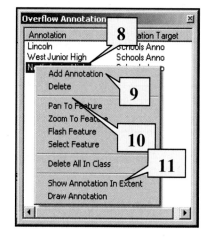

Fig. 3.37b. Placing overflow annotation

11. Use Show Annotation In Extent to list only the labels in the current view. Place all of these before zooming to another location—it saves time.

Adding a north arrow

1. Choose Insert > North Arrow from the main menu and choose the desired symbol.

2. Click the Properties button to modify the symbol. Click OK to place it.

3. If necessary, click and drag it to the desired location, or resize it.

TIP: In some map projections, north is not straight up. Set the north arrow marker angle manually to point true north, if necessary.

Adding titles and text

1. Choose Insert > Title or Insert > Text from the main menu bar. These commands are similar except that the default font size is larger for the Title.

2. Enter the text in the box on the map and press Enter.

3. The text remains selected. If necessary, change its size or font using the Drawing toolbar or click and drag it to a new location.

4. To modify the text after it has been created, double-click it with the Select Elements tool, or right-click it and choose Properties. You can change the text, set the font and size, or specify a particular position for the text.

Adding graphics to layouts

The Drawing toolbar (Fig. 3.38) provides functions for creating and modifying objects on a layout. These objects may also be created within a data frame itself, in which case they will be scaled if the map changes size. If they are in the layout, they will be unaffected by scale changes.

 The Drawing toolbar contains common functions found in other programs. A small black triangle on a button indicates that the button contains a menu.

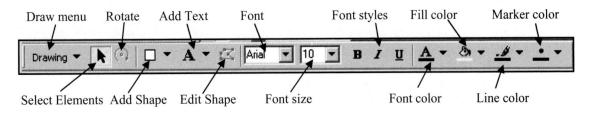

Fig. 3.38. The Drawing toolbar

Adding a legend

1. Use the Select Elements tool to click on the data frame containing the layers to appear in the legend. The legend is always created from, and placed in, the active frame.

2. Choose Insert > Legend from the main menu bar.

3. Choose which layers will be included in the legend. To add a layer, click it in the box on the left and click the > button. To remove a layer from the legend, click it in the box on the right and click the < button. Choose the number of columns in the legend. Click the Preview button to examine a preview of the legend. Click Next.

4. Modify the legend title text and formatting to desired settings. Click Next.

5. Specify a legend border, background, and drop shadow, if necessary. Click Next.

6. Click a layer to modify its symbol size and patch style. Click Next.

7. Modify the spacing, if necessary (usually not necessary). Click Finish.

8. The legend appears in the map, selected. Click and drag the legend to the desired location and resize it, if desired. Resizing will change the size of the text and boxes.

TIP: To edit the individual elements of a legend, right-click the legend and choose Convert to Graphics. To work with each element separately, choose Draw > Ungroup from the Drawing toolbar.

Changing the appearance of a legend

Right-click the legend and choose Properties. The window has four tabs: Legend, Items, Frame, and Size/Position.

The Legend tab

1. Change the title of the legend (Fig. 3.39), its position, or its symbol.

2. The Patch box specifies the default size and symbol styles for the entire legend and can be changed here. (The patch style for individual layers can be changed in the Items tab.)

3. Adjust the default spacing between the elements in the legend by typing in the desired values, in points.

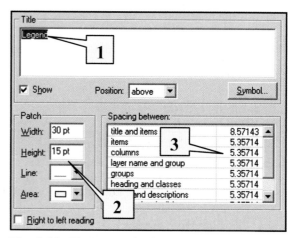

Fig. 3.39. The Legend tab

The Items tab

This tab gives you individual control of every layer in the legend.

1. Change which layers are displayed by adding or removing them from the Legend Items box (Fig. 3.40).

2. To modify other properties of a single layer, click on it to select it.

3. Change the position of the layer in the list by clicking on the Up or Down arrows.

4. To place the selected layer in a new column, check the box.

5. Set the number of columns for this layer to span in the legend.

6. To change the style of the legend layer, click the Style button. The Legend Item Selector window appears.

7. Choose one of the predefined styles. To further modify the chosen style, click the Properties button. The Legend Item window appears.

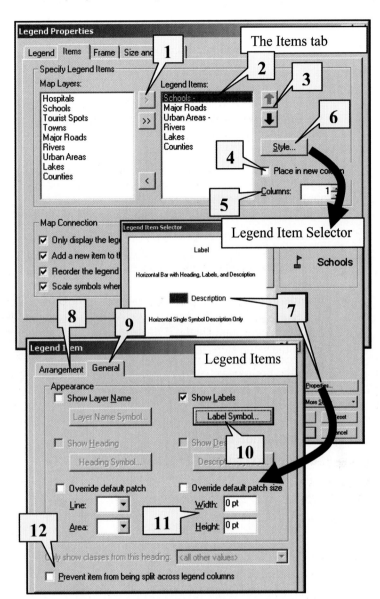

Fig. 3.40. The Items tab and its submenus

8. Click the Legend Item window Arrangement tab to change the relative position of the patch, the layer name, and the description.

9. Click the Legend Item window General tab to set specific properties of the item style.

10. Change the font style of the text by clicking the Label Symbol button.

11. Change the patch shape or the patch width and height, if desired.

12. You can also prevent the legend item from being split across a column break.

The Frame tab

The Frame tab can set a border around the legend and/or give it background shading or a drop shadow (Fig. 3.41).

1. Choose a border style using the drop-down box and change its color, if desired.

2. To modify the existing border, such as changing its pattern or color or thickness, click the Border Selector button or the Edit Border button.

3. Set the gap distance between the map and border to the desired X and Y values.

4. Choose to round the corners by entering a percentage greater than zero; the higher the percentage, the more rounding.

5. Set the background shade, if desired, using the same steps as for the border.

6. Set the Drop Shadow desired, using the same steps as for the border.

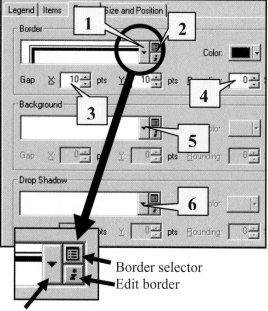

Fig. 3.41. The Frame tab

The Size and Position tab

The Size and Position tab lets you specify an exact position and size for the legend in page units, which are inches by default (Fig. 3.42).

1. Set the position in page units, relative to the lower-right corner of the page.

2. The anchor point indicates which part of the legend sits the XY distance from the corner. To place the lower-right corner of the legend at 3 inches from the left and 3 inches from the bottom, enter 3,3 and click the lower-left anchor point. To center the legend at 3,3, enter 3,3 and click the center anchor point.

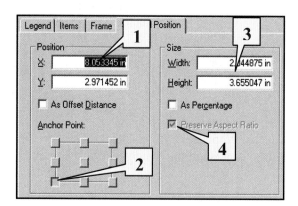

Fig. 3.42. The Size and Position tab

3. Set a specific width and height of the legend, if desired, either in page units or as a percentage of the page size.

4. If Preserve Aspect Ratio is checked, the shape of the legend will remain constant if it is resized.

Adding a scale bar

1. Using the Select Elements tool, click to activate the desired frame. The scale bar will be placed in the active frame and will be sized according to the scale of the active frame.

2. Choose Insert > Scale bar from the main menu. The Scale Bar Selector window will appear (Fig. 3.43).

3. Choose the desired scale bar.

4. To modify the scale bar, click the Properties button to open the Scale Bar window (Fig. 3.43).

5. The Numbers and Marks tab controls the spacing of numbers and marks on the bar. These values can usually be left as defaults.

6. The Format tab controls the font of the scale text and the style of the scale bar. The defaults are usually fine.

7. The Scale and Units tab controls the length and divisions of the scale bar (Fig. 3.44).

8. Set the units for the scale bar to miles, kilometers, or some other unit.

9. Choose the When resizing… option. One or more of the input boxes here may be dimmed, depending on which option is chosen.

10. Set the division value, the number of divisions, and the number of subdivisions, as applicable.

11. Check the box to place the subdivisions before the zero point rather than in the first division.

12. The chosen units will be labeled on the scale bar (e.g., Miles). Choose the label position, change the label text and font symbol, and set the gap between the scale bar and the label, if desired.

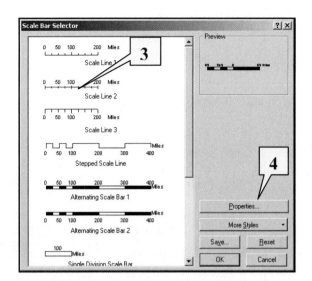

Fig. 3.43. The Scale Bar Selector

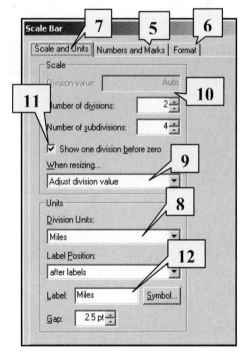

Fig. 3.44. Scale bar properties

TIP: To change the properties of a scale bar after it has been created, double-click the scale bar using the Select Elements tool, or right-click it and choose Properties.

Adding neatlines, backgrounds, and shadows

Neatlines are lines that enclose one or more map elements. They are used to balance the space and provide alignment between irregular map elements for a more aesthetically pleasing map. Most objects in layouts, including data frames, legends, and scale bars, have a tab in their properties to set up borders, backgrounds, and drop shadows. To access these tabs, open the element's properties, by either double-clicking it or right-clicking it and choosing Properties.

To create a neatline or shaded box as a separate object on its own accord, do the following:

1. Choose Insert > Neatline from the main menu bar. The Neatline window appears.

2. Choose the desired Placement option (Fig. 3.45).

3. Set the border, background, and drop shadow styles.

4. To further modify the available styles, click the Border Selector or Edit Border buttons as shown.

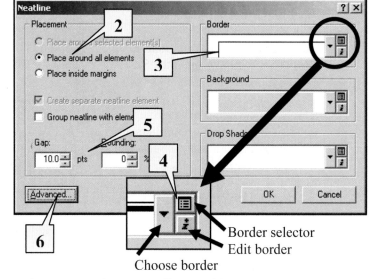

Fig. 3.45. Setting up neatlines and boxes

5. Set the gap between the neatline and the elements and enter rounding, if desired. The higher the percentage, the more rounding occurs.

6. The Advanced button allows customized editing of the symbols, gaps, and rounding for each border, background, and shadow.

Adding pictures

1. Choose Insert > Picture from the main menu bar.

2. Navigate to the folder containing the picture and click on it to select it.

3. Click Open to add the picture.

4. Resize and/or move the picture to the desired location using the Select Elements tool.

TIP: Information on allowed image formats can be found in the online help index by typing the entry "rasters" and choosing the subheading "formats, supported."

Creating a map from a template

A map template is a set of data frames, titles, styles, and other map elements that are already formatted and ready to receive the data in the data frame(s). Use a map template to quickly create a map in a standard format. You can save any map as a template to create a similar map again.

1. Click the Change Layout button in the Layout toolbar.

2. Click one of the tabs to see a choice of templates (Fig. 3.46). Templates you have created will be stored in the My Templates tab.

3. Click on a template to see a preview of it.

4. Use the Browse button to navigate to another directory containing more templates saved elsewhere.

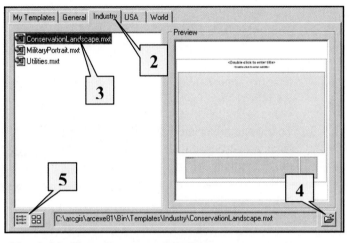

Fig. 3.46. Choosing a map template

5. Click the Thumbnail button to see icons of all the templates; click the List button to go back to the original list view.

6. Click the desired template and choose Next or Finish.

TIP: If the template has more than one data frame, it will prompt the user to assign the data frames in the map document to the data frames in the template.

---

7. If the data frames need assigning, click each of the frames in the list on the left (Fig. 3.47).

8. Use the Move Up and Move Down buttons to put them in the same order as the numbered frames in the new layout.

9. Click Finish.

10. Finally, change any titles or other map elements in the template that need to be customized.

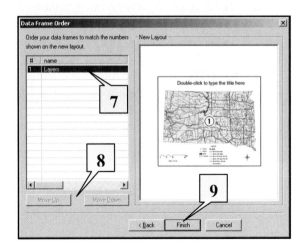

Fig. 3.47. Assigning data frames to the frames in the template

Printing a map

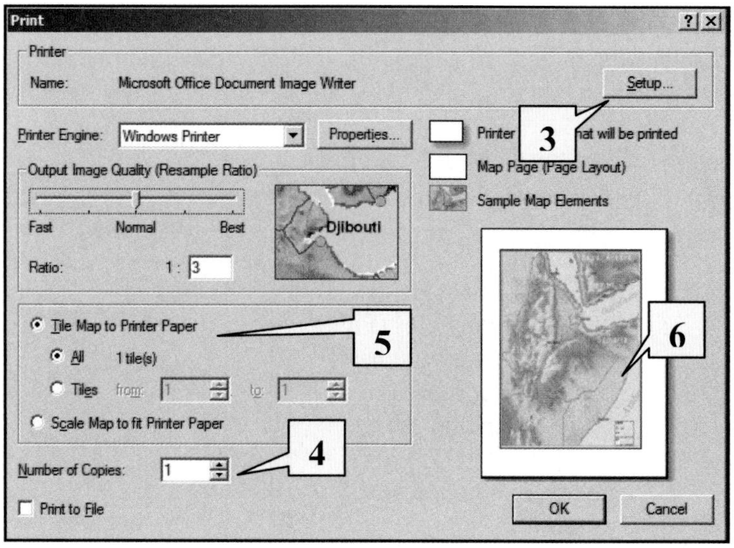

1. To preview a map to see how it will look on paper, choose File > Print Preview from the main menu bar.

2. To print, Choose File > Print from the main menu. The Print window appears (Fig. 3.48).

3. To change the printer or its properties, click the Setup button.

4. Set the number of copies to print.

Fig. 3.48. The Print window

5. Choose the desired tiling options if the map is larger than the printer paper.

6. Preview the layout placement on the page. Click OK.

Exporting a map as a picture file

You can export a map as an image file to put it on a Web page or inside another document or as a PDF to share it with others.

1. Choose File > Export Map from the main menu.

2. Navigate to a folder for the saved picture file (Fig. 3.49). Choose the type of file and enter a name for it.

3. Click the gray arrow for more export options, such as the resolution. If planning to enlarge the map before printing, increasing the default resolution may be necessary.

4. To avoid having a white border around the picture, check the box to Clip Output to Graphics Extent.

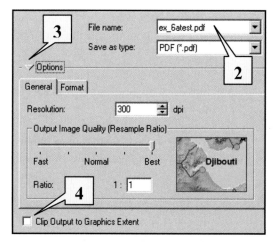

Fig. 3.49. Exporting a map as a picture

5. Click Save.

150

Creating a simple graph

This example shows the steps to create a horizontal bar graph. Other graph types will have slightly different options. Have fun experimenting.

1. Choose Tools > Graphs > Create from the main menu bar. The Create Graph Wizard will appear (Fig. 3.50).

2. Choose the graph type.

3. Choose the layer/table and set the Value field to be graphed.

4. To sort the bars, make the Y field the same as the Value field and choose Ascending or Descending.

5. Choose the X or Y label field.

6. Uncheck the Add to Legend box if you don't need a legend.

7. Set the bar color. Use Match to Layer to make it the same as the map. Choose Custom to make all the bars one color of your choosing. Choose Palette to make every bar a different color.

8. Click Next.

9. Give the graph a title and footer.

10. Select a title and position for the legend if you have one.

11. Click the Left tab and provide a title for that axis.

12. Click the Bottom tab and enter a title.

13. Click Finish.

14. To place a graph on the layout, open the graph. Right-click the blue bar at the top of the graph and choose Show on Layout.

15. To edit the graph properties, right-click on the top blue bar of the graph window and choose Properties or Advanced Properties.

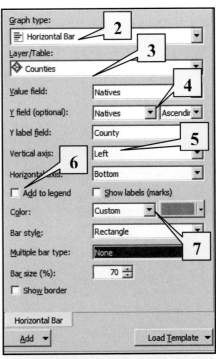

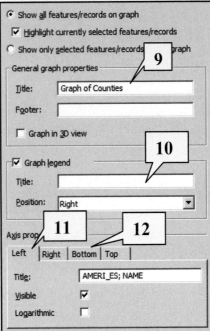

Fig. 3.50. The Graph Wizard

Chapter 4. Attribute Data

Objectives

- Understanding how tabular data are stored and used
- Understanding the links between database management systems and tables
- Using queries to select records of interest
- Understanding joins and cardinality concepts
- Summarizing tables to get statistics on groups
- Learning how to define fields
- Editing and calculating fields in tables

Mastering the Concepts

GIS Concepts

Overview of tables

What is a table?

A **table** is a data structure for storing multiple attributes about a location or an object. It is composed of rows, called **records**, and columns, called **fields** or **attribute fields**. Figure 4.1 shows an example of a table containing attributes of counties.

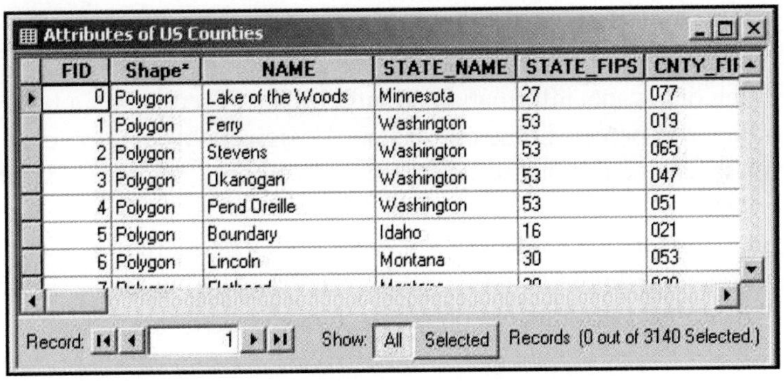

Attributes of US Counties

FID	Shape*	NAME	STATE_NAME	STATE_FIPS	CNTY_FI
0	Polygon	Lake of the Woods	Minnesota	27	077
1	Polygon	Ferry	Washington	53	019
2	Polygon	Stevens	Washington	53	065
3	Polygon	Okanogan	Washington	53	047
4	Polygon	Pend Oreille	Washington	53	051
5	Polygon	Boundary	Idaho	16	021
6	Polygon	Lincoln	Montana	30	053

Record: 1 Show: All Selected Records (0 out of 3140 Selected.)

Fig. 4.1. A table with information about U.S. counties

Tabular data files fall into two main categories, **attribute tables** and **standalone tables**. An attribute table, such as the one shown in Figure 4.1, contains information about features in a geographic data set. In an attribute table there is always one and only one row of information for each feature. In a georelational data model the row is linked to the spatial feature in a separate file using a unique ID number called a Feature ID, or FID (Fig. 4.1). In an object-oriented data model the file stores both the attributes and the *x-y* coordinates in the same data file, although the coordinates are not visible in the table. Attribute tables have an Object ID, or OID instead of a FID. In contrast, a standalone table simply contains information about one or more objects in tabular format instead of having information about map features. A standalone table might come from a text file, an Excel® spreadsheet, a global positioning system data file, or a database. Standalone tables exist independently of a geographic data set and may be only incidentally related to map features. They also have an OID rather than an FID.

Database Management Systems

GIS tables share a past history and many current properties of the large database programs that are routinely used in many commercial, governmental, and academic settings. Nearly every GIS system uses an underlying database management system (DBMS) to store its data. These programs are designed to store, manipulate, analyze, and protect tabular data of all kinds. Governments use them to store information about citizens, parcels, taxes and more. Companies use them to store information about customers. Universities use them to manage information about students, classes, and faculty. Three types of databases have traditionally been used.

A **flat file database** simply stores rows of information in a text or binary file. Finding information requires parsing the table and selecting the records of interest. They are simple but not very efficient.

A **hierarchical database** has multiple files, each of which contains different records and fields. Parent tables can be linked to child tables through a specified field called a key. For example, a table of college classes might be linked to a table of students in each of the classes through a course ID number. Relationships between tables are fixed, which makes looking up information quick. However, the relationships are inflexible, designed to permit only a small set of operations.

A **relational database** also has multiple tables stored as files. However, the relationships are not defined ahead of time. Instead, the user can temporarily associate two tables if they share a common field. This process is called **joining** the tables, and the common field becomes the key. This database model is extremely flexible and is overwhelmingly the preferred choice for GIS systems. It is called a relational DBMS, or RDBMS.

Queries on tables

Often one wants information about a subset of records in a table, for example, knowing the number of parcels in the city that are designated as commercial. To determine this information a **query** can be performed on the table. In a query, a **logical expression** is used to specify certain criteria (e.g., zoning = commercial), and then the software searches the table and finds the records (parcels) that match the criteria. Those records are returned as a **selected set**. Selected records can become the input to another action, such as printing them, exporting them to a new file, or executing a GIS function on them.

Most databases use a special language called Standard Query Language (SQL) to write and execute queries. A land-use table might contain a field with the zoning code. If the field containing the zoning was named ZONE, and the code for commercial property was 492, the logical expression in SQL might look like this:

> SELECT * FROM landuse WHERE [ZONE] = 492

In this example landuse is the name of the table, [ZONE] is the field name, and [ZONE] = 492 is the logical expression stating the criteria to be met. SQL queries can have multiple lines and include many criteria.

> SELECT * FROM landuse WHERE
> [ZONE] = 492 AND [VALUE] > 300000

This query would find all commercial parcels with a value greater than $300,000. This chapter presents methods for performing queries on tables. Chapter 5 examines queries in more detail.

Joining and relating tables

In a RDBMS and in a GIS, tables are commonly combined, or **joined**, in order to bring different types of information together. The tables are combined using a common field called a **key**. When a join is performed, the two separate tables become one and contain the information from both tables (Fig. 4.2). The join is a temporary relationship and may be removed when it is no longer needed. Joins are especially useful when one needs to link a standalone table, such as a report of agricultural yields by county obtained from the Internet, to a spatial data file so that the information in the table can be mapped.

Joins have a direction. The information from one table is appended to the information from the other table. The table containing the information to be appended is called the **source table**. The table that receives the appended information is called the **destination table**. In Figure 4.2, the destination table, Attributes of US States, is a feature attribute table. The source table containing demographic data is a standalone table. When the two tables are joined, the demographic data is appended to the US States attribute table. Often joins are used to bring data from standalone tables into an attribute table for mapping or analysis.

Join direction matters. Once the demographics are joined to the states shapefile table, then the information could be used to make a map of the population data. If, however, the join had been performed in the opposite direction, with the demographics as the destination and the states as the source, then the resulting table would be a standalone table. In this case, a map showing the demographics could not be made because the demographics data is not part of the feature class.

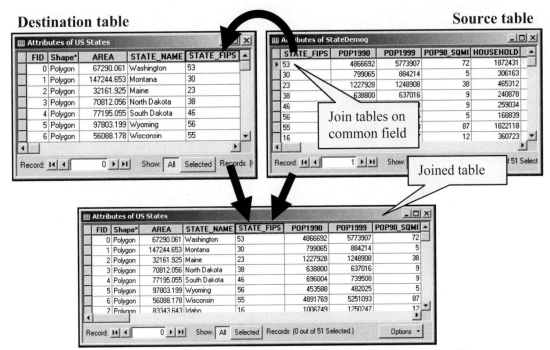

Fig. 4.2. Joining on a common field. These two tables are joined into one using the common field STATE_FIPS. The join field appears only once in the joined table. The data from the source table are placed into the destination table.

Cardinality

When joining, the **cardinality** of the relationship between the tables must be considered. The simplest kind of relationship is *one-to-one*, in which each record in the destination table matches exactly one record in the source table. In Figure 4.2, each state has one corresponding record of demographic data. In a *one-to-many* relationship, each record in the destination table could match more than one record in the source table. For example, a store location could have many employees. In a *many-to-one* relationship, many records in the destination table would match a single record in the source table, such as many cities falling within one state. Finally, a *many-to-many* relationship indicates that multiple records can appear in both tables. For example, a student may take more than one class, and most classes have more than one student.

The direction of the join must be taken into account when ascertaining the cardinality of a relationship. The destination table is the point of reference and comes first; that is, the relationship cardinality is reported as {destination} to {source}. Imagine two tables containing *states* and *counties*. If one performs a join with *states* as the destination table and *counties* as the source, the cardinality is one-to-many because each state contains many counties. If the join is reversed and *counties* is the destination table, then the cardinality becomes many-to-one because there are many counties in one state.

> **TIP:** Putting the destination table first when stating cardinality is only a convention but one adopted by ESRI in its publications and Help documents. Although to some readers it may initially seem backwards, sticking to this convention will cause the least confusion in the long run. It may help to always imagine the destination table on the left, as in Figure 4.3.

The cardinality of a relationship dictates whether the tables can be joined. The Rule of Joining stipulates that there must be one and only one record in the output table for each record in the input table. Consider the four tables shown in Figure 4.3.

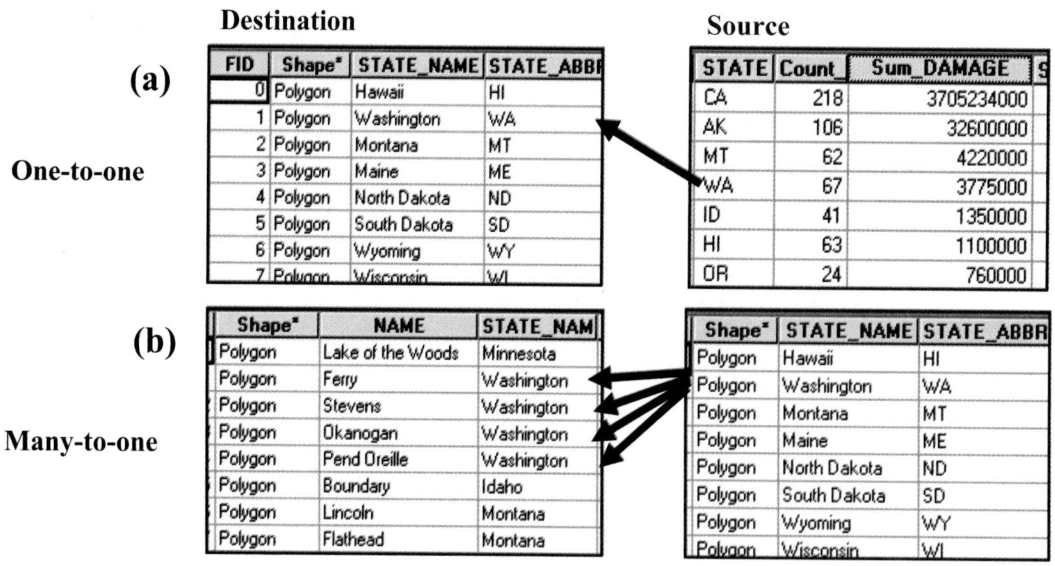

Fig. 4.3. A cardinality of one-to-one or many-to-one permits tables to be combined without violating the Rule of Joining. (a) One-to-one cardinality. (b) Many-to-one cardinality.

One-to-one cardinality. In Figure 4.3a, a table containing the number of earthquakes and total damage in each state (source) is being joined to a states attribute table (destination), using the common field provided by the state abbreviation. Each state occurs once in each table so that there is no ambiguity about matching the source records to the destination records.

Many-to-one cardinality. In Figure 4.3b, a table containing state information (source) is being joined to a table of counties (destination) using the common field provided by the state name. Although there are many counties in each state, there is only one record for each state in the source table and no ambiguity in linking the source records to the destination records. The state record does get used more than once, but it still does not violate the Rule of Joining.

One-to-many cardinality. Figure 4.4 shows the reverse of the join in Figure 4.3b. Now states are the destination table and counties are the source table. Many county records in the source table match each state record in the destination table. The Rule of Joining is violated, and it becomes ambiguous which county record should be matched to the state. For this reason, one cannot perform a join if a one-to-many relationship is present. Instead, we perform a different operation called a **relate**.

Destination

Shape*	STATE_NAME	STATE_ABBR
Polygon	Hawaii	HI
Polygon	Washington	WA
Polygon	Montana	MT
Polygon	Maine	ME
Polygon	North Dakota	ND
Polygon	South Dakota	SD
Polygon	Wyoming	WY
Polygon	Wisconsin	WI

?

Source

Shape*	NAME	STATE_NAM
Polygon	Lake of the Woods	Minnesota
Polygon	Ferry	Washington
Polygon	Stevens	Washington
Polygon	Okanogan	Washington
Polygon	Pend Oreille	Washington
Polygon	Boundary	Idaho
Polygon	Lincoln	Montana
Polygon	Flathead	Montana

Fig. 4.4. A one-to-many relationship violates the Rule of Joining because more than one record in the source table matches a record in the destination table.

In a relate, the two tables are still associated by a common field, but the records are not joined together. The two tables remain separate. However, if one or more records are selected in one table, then the associated records can be selected in the other table. For example, selecting the state of Washington in the states table allows the selection of all the counties in Washington in the related table, as shown by the red boxes in Figure 4.4. Many relationships in the world have cardinalities of one-to-many, such as a school to its classes or a well to its yearly water quality tests, and relates provide valuable support in dealing with those features and their attributes.

To summarize, a join may be used to combine tables whenever a one-to-one or many-to-one relationship exists between the destination table and the source table. If the relationship is one-to-many or many-to-many, then a relate must be used.

Summarizing tables

The Summarize function is a powerful form of statistics applied to databases. The Statistics function calculates statistical values for all of the selected values in a field. Summarize first combines the records into groups based on a specified attribute field and then calculates statistics for each group. Because many statistics for many fields can be returned from a Summarize, this command produces a new table. This table can subsequently be joined to another table. The

Chapter 4

combination of the Summarize function followed by a join works well for taming unwieldy one-to-many relationships.

We might use Summarize to gather information about the damage and deaths caused by earthquakes. Figure 4.5 contains a table showing major historical earthquakes in or near the United States. Each earthquake is listed with information including the state in which it occurred, the number of

STATE	DEPTH	DEATHS	DAMAGE	MAG	MMI	LOCATION
MO	0	7	0	7.88	12	New Madrid, Missouri
SN	0	51	0	7.36	12	Northern Sonora, Mexico
AK	0	0	0	8.15	11	Yakutat Bay, Alaska
AK	0	0	0	8.26	11	Southeast Alaska
AR	0	7	0	7.68	11	Northeast Arkansas
CA	20	3000	52400000	7.80	11	Near San Francisco, California
CA	16	12	6000000	7.48	11	South of Bakersfield, California
CA	8	65	50500000	6.62	11	North of San Fernando, California

Fig. 4.5. Earthquake table

deaths it caused, the total damage caused, and the Richter scale (MAG) and Modified Mercalli Intensity (MMI) measuring the energy of the quake. Note that this relationship is one-to-many, with each state containing many earthquakes. We would like to know which states have suffered the most from historical earthquakes. In particular, we would like to know the total deaths, total damage, and average Richter and MMI scales for each state. We summarize on the state field, dividing the earthquakes into groups by state, and select the statistics we want to calculate. Separate statistics are then calculated for each group (state).

Attributes of quakesum

OID	STATE	Count	Sum_DAMAGE	Sum_DEATHS	Average_MAG	Average_MMI
6	CA	218	3705234000	3777	5.2575	7.422
0	AK	106	32600000	125	6.5042	3.7358
26	MT	62	4220000	32	2.9737	5.9677
54	WA	67	3775000	15	3.5894	5.8955

Fig. 4.6. Summarize produces a file which contains the summary field (STATE), the number of earthquakes in each state, and the requested statistics.

The quakesum.dbf output table shown in Figure 4.6 has one record for each state. It also has a Count field indicating the number of earthquakes in each state. The output table also includes one field for each of the statistics we requested. To better interpret the data, we sorted the table by the damage field in descending order to find the most grievously affected states. Our next step might be to join the quakesum.dbf table to the US States layer and create a map showing the total damage or total deaths for each state (Fig. 4.7). Using Summarize, we collapsed a one-to-many relationship between states and earthquakes into a one-to-one relationship between states and earthquake statistics, allowing the data to be mapped.

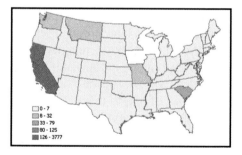

Fig. 4.7. Earthquake deaths by state

Field types

Creating database tables always begins with an analysis of the fields that the table will contain. Unlike a spreadsheet, in which any cell can contain any type of data, a database table must always contain the same type of information in each field. Each field must be defined, and the type of its

158

contents must be established, before any data are entered. Furthermore, once a **field definition** is set, it cannot be changed. The most common field types are numbers, strings, dates, and Boolean fields. Fields have defined parameters as well, including length, precision, and scale.

For example, the field *Street* would be a string field because street names typically contain letters. The **field length** defines how many letters can be stored in the name. If the *Street* field is given a length of 10, then any name longer than 10 letters will be truncated to the first 10 letters. Thus, "Elm Street" fits perfectly (including the space), but "Maple Street" would be truncated to "Maple Stre" to fit in the field. It is very important to ensure enough space in fields to hold the longest possible case.

When defining a number field, the user must designate a storage width (**precision**) and the number of decimal places (**scale**). For example, a field with a precision of 5 and a scale of 0 could store a number between –9999 and 99999 (the minus sign takes one space). A field with a precision of 5 and a scale of 2 could store a number between –9.99 and 99.99 (the decimal place also takes one space). When designing a database, try to use the smallest field widths that will store every possible value because extra spaces increase the size of the database.

Numbers can be stored in a variety of ways. Most databases offer several formats for storing numeric values, and these formats can differ from database to database. The next section describes the common options in general terms.

ASCII versus binary

A **byte** is the basic unit of storage space for a computer—it is composed of a string of eight digits (bits) which may be zeros or ones. These zeros and ones represent a number in base 2; such numbers are called binary numbers. A single byte can store a binary value from 0 (00000000) to 255 (11111111). Two bytes can store values up to $2^{16} - 1$, or 65,535. The more bytes allotted to a value, the higher the number that can be stored. Computers can store numbers and letters several different ways.

All text is stored as sequences of characters using a special code called **ASCII** in which every number, every letter, and every symbol (such as $) is assigned a single-byte code between 0 and 255. Then to store the word "cat", the computer stores the code for "c", the code for "a", and the code for "t". Thus, it takes three bytes to store "cat". The word "horse" requires five bytes. Numbers can also be stored using ASCII by storing the ASCII one-byte code for each numeral. Thus, it requires three bytes of data to store "147" and five bytes to store "147.6 ". This scheme is a simple and standard way to store information. Text files and HTML files, among others, are stored in ASCII.

Another scheme involves storing **binary** data. In this case a number is stored in base 2 directly rather than being assigned one byte per character. The number "16" would be stored in binary as "00001000" in a single byte of information. Binary is more efficient than ASCII. The number 33456 would require five bytes in ASCII but only two bytes in binary. It is also faster to compute with binary values because the computer by design does all its calculations in base 2. If the number is already stored in base 2, the computer does not need to convert it before calculating. Thus, it is often advantageous to use binary storage when it is appropriate. Many types of files use a binary encoding scheme, including spreadsheets, word processing documents, and shapefiles. Raster and image data are also stored as binary data.

Precision

Very large numbers, such as 1,000,000,000,000, require many bytes to store, as do very small numbers such as 0.0000000000001. People often use scientific notation when dealing with very large or very small numbers; these two would be written as 1.0×10^{12} and 1.0×10^{-13}. Computers can also use scientific notation, usually called exponential or floating-point data. This notation stores values composed of a mantissa (the decimal part of the number) and an exponent. For example, the number 123456789 stored as an exponent would consist of a mantissa of 1.23456789 and an exponent of 8, yielding 1.23456789×10^{8} (often written as 1.23456780e08). The computer usually truncates the mantissa at a certain level of precision—when storing values in the trillions, differences of tenths or hundredths are often of little interest. The number might become 1.2345e08. A **single-precision** floating-point field stores up to eight significant digits of information in the mantissa. A **double-precision** field stores up to 16 significant digits. Floating-point data types are much more flexible than numeric or binary types because they can store either very large or very small values using the same field.

Different types for different data formats

Every attribute must be defined before use; that is, the field type must be specified (text or numeric) and the field properties set. Once a field is defined, the definition cannot be changed. If a mistake is made defining the field, one must usually delete the field and redefine it. Different GIS systems may have different allowable data types, although most use the basic categories of text, integers, and floating-point values. Table 4.1 shows the specific data types used in ArcGIS shapefiles and geodatabases (coverages have different data types; see the ArcGIS Help for more information). In Version 9.3, geodatabases also have a Raster and GUID type for advanced users.

Table 4.1. Field data types available for feature classes (*geodatabases only)

Field Type	Explanation	Examples
Short	Integers stored as 2-byte binary numbers *Range of values −32,000 to +32,000*	255 12001
Long	Integers stored as 10-byte binary numbers *Range of values −2.14 billion to +2.14 billion*	156000 457890
Float	Floating-point values with eight significant digits in the mantissa	1.289385e12 1.5647894e–02
Double	Double-precision floating-point values with 16 significant digits in the mantissa	1.12114118119141e13
Text	Alphanumeric strings	'Maple St' 'John H. Smith'
Date	Date format	07/12/92 10/17/63
BLOB*	Binary large object; any complex binary data including images, documents, etc.	

About ArcGIS

Tables in ArcGIS

Tables in ArcGIS may come from any of the underlying RDBMS programs supported by ArcGIS. The tables may be in different formats for storage, but the table itself always looks the same and has the same functions so that users don't need to learn different commands for working with different file types.

In the georelational shapefile model, the table is stored in a dBase format file and includes a unique feature identification number (FID) that links the spatial and the attribute data. In the object-oriented geodatabase model and in any standalone table, each record contains an Object ID (OID) analogous to the row number in a spreadsheet. Both shapefiles and geodatabase feature classes contain a Shape field that represents the *x-y* coordinate values of the feature. Geodatabase tables for line and polygon feature classes also have Shape_Area and/or Shape_Length fields, which keep track of the areas and lengths of features.

Figure 4.8 shows some terminology associated with tables. A table consists of rows and columns. A row is called a record, and it contains information about a single object or feature. A column is called a field, and it stores one type of information. Each field has a name shown in the top row. Field names must contain 13 or fewer characters and should contain only letters, numbers, and underscores. An alternative name, called an **alias**, can be used to temporarily give a field a more descriptive name that does not have to follow the naming rules. For example, the somewhat perplexing field MEDREN could be given a more understandable alias such as MedianRent. Aliases form part of a layer definition. If defined in ArcMap, they persist only inside that map document. If defined as part of a layer file, they can be used in different map documents by loading the layer file instead of the feature class.

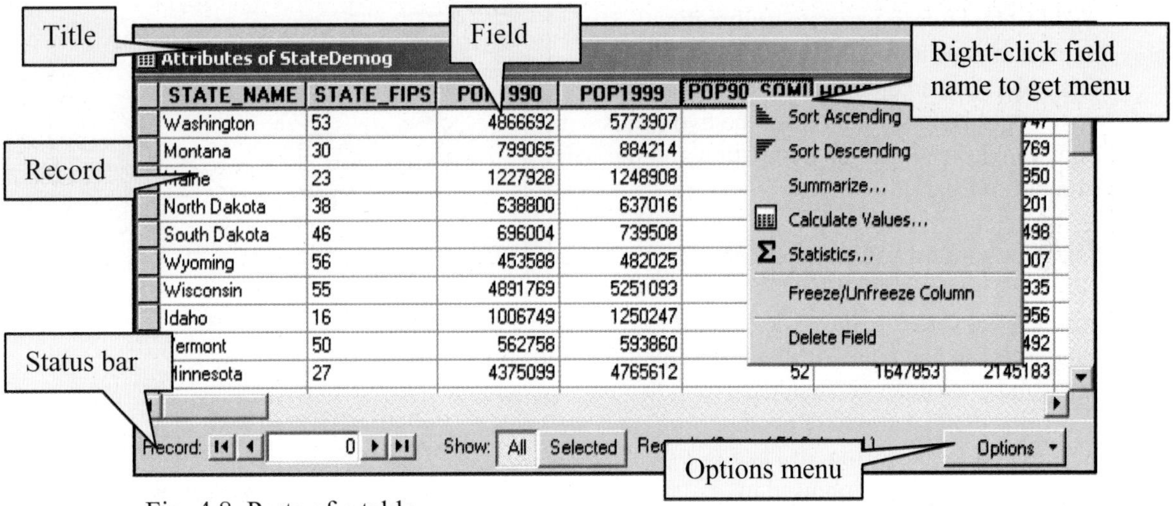

Fig. 4.8. Parts of a table

In an attribute table, some fields are required and are created and updated by the program that creates the geographic data set. For example, a shapefile has a feature ID (FID) field and a Shape field. An ArcInfo coverage will have the cover# and cover-id fields, plus additional fields depending on the types of features in the attribute files. **These fields should never be altered by the user.** Likewise, users of shapefiles must make sure that they never delete records in an attribute table unless the accompanying features in the spatial data set are also deleted. Editing features only within an editing session in ArcMap will meet this criterion, as does deleting rows in a geodatabase feature class.

TIP: Tables with fields containing *x-y* coordinates can be used to create feature classes.

Table formats

As we learned in Chapter 1, the different GIS data formats use different underlying databases. The coverage model uses a RDBMS called INFO. Shapefiles use a dBase format file. Personal geodatabases also use the database underlying Microsoft Access, called Jet. SDE geodatabases use a large-scale commercial RDBMS such as Oracle or SQLServer. ArcGIS can read comma-delimited text files (fields separated by commas) and tab-delimited text files (fields separated by tabs). Text formats cannot be edited, but they can be converted to dBase format.

Microsoft Excel spreadsheets are a common source of data for GIS. ArcGIS can open spreadsheets as read-only tables. However, since Excel does not enforce field names or field content the same way that databases do, not all Excel spreadsheets are suitable. The Skills Reference section includes a summary of how to prepare spreadsheets to work properly.

Editing and calculating fields

ArcGIS offers two ways to change the values in a table, by typing the information directly into the fields or by calculating the value of a field. Typing information into fields must be done during an edit session in ArcMap. Calculating fields can be done inside or outside an edit session and uses the Field Calculator to generate arithmetic expressions using fields as variables. For example, the percentage of Hispanics in each state could be calculated from the two fields containing the total state population and the number of Hispanics.

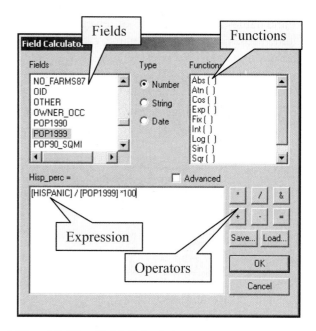

Fig. 4.9. The Field Calculator

The first window in the Field Calculator (Fig. 4.9) contains the fields in the table that can be used to create the expression. The functions box contains different functions that can be used in expressions. The functions displayed will depend on whether Number, String, or Date is selected. The expression used to calculate is entered in the large box at the bottom. The operators (+, *, −, /, &, =) appear to the lower right. The Advanced button allows more complex Visual Basic expressions to be entered. (Visual Basic is a programming language that comes with ArcGIS.) A complicated expression can be saved and loaded for use another time.

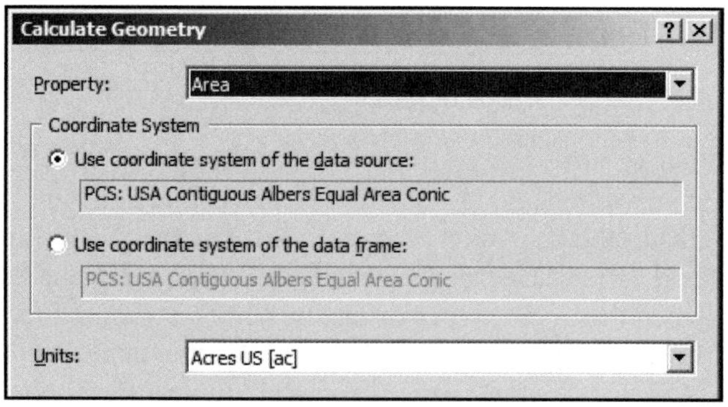

Fig. 4.10. The Calculate Geometry function

The Calculate Geometry function (Fig. 4.10) can add information on feature areas, perimeters, lengths, or *x-y* centroids to a table. The field to contain the information must already exist. The user may choose which coordinate system and output units to use in the calculation.

Summary

➤ Tables consist of rows and columns of information. A row is associated with one feature in the real world and includes columns of information called fields.

➤ Tables associated with spatial data sets are called attribute tables and contain records, one for each feature in the data set. Standalone tables are not associated with map features.

➤ Relational database management systems construct temporary links between data tables and are the preferred model for GIS software.

➤ Table queries allow the user to select certain records based on one or more criteria. Once selected, these records may be viewed, exported, or analyzed.

➤ Tables may be joined or related on a common field in order to access information in one table from another. Joins may be performed on tables with one-to-one or many-to-one cardinality. Relates must be used on tables with one-to-many or many-to-many cardinality.

➤ The Statistics function calculates basic statistical values for all selected records in a table. The Summarize command generates statistics about groups of features defined by a field.

➤ Fields must be defined before they can receive data. Different data formats have different ways of defining fields, and fields are converted when the data formats are changed.

➤ ArcGIS tables are interfaces designed to display and manipulate tabular data from a variety of sources, including dBase files, INFO files, geodatabases, SQL queries, or comma-delimited text files.

➤ New fields may be added to tables by defining names and field types. New tables may have data entered by typing in values or by using the Calculate function. The Calculate Geometry function can put areas and lengths in attribute tables for the corresponding features.

TIP: Field names must have 13 or fewer characters; may include letters, numbers, and the underscore character (_); and should not contain spaces or special characters such as @, #, !, $, or %. Field names must also start with a letter, not a number.

VERY IMPORTANT TIP: Do not use programs other than ArcGIS to delete or edit attribute tables that are associated with shapefiles. You may corrupt a shapefile and make it unusable.

Chapter Review Questions

1. Describe the difference between an attribute table and a standalone table.

2. Which type of database management system are GIS systems based on? How does this type of system differ from other DBMS types?

3. List the types of data sources from which tables may display data.

4. Describe how storing the number 255 in ASCII would differ from storing it as a binary representation.

5. Choose the best field type for each of the following types of data in a geodatabase:

> population of countries in the world
> precipitation in inches
> number of counties in a state
> highway name
> distances between U.S. cities, in meters
> birthdays

6. What is the cardinality of each of the following relationships?

> students to college classes
> states to governors
> students to grades
> counties to states

7. Describe the differences between a join and a relate.

8. You have a table of states and a table of airports, both with a state abbreviation field. Can you join them if states is the destination table? If airports is the destination table? Explain your answer.

9. Describe the difference between using Statistics and Summarize functions on a field.

10. For each of the following problems, using data sets for the United States, state whether using a query, the Statistics function, or the Summarize function would be the best approach to solving it.

> _____ Find all towns with more than 20,000 people.
> _____ Find the total number of volcanoes in each state.
> _____ Determine the total amount of damage caused by earthquakes in the U.S.
> _____ Find the states in which Hispanics exceed the number of blacks.
> _____ Find out which subregion of the country has the most Hispanics.

Mastering the Skills

Teaching Tutorial

The following examples provide step-by-step instructions for doing basic tasks and solving basic problems in ArcGIS. The steps you need to do are highlighted with an arrow ➔; follow them carefully. Click on the video number in the VideoIndex to view a demonstration of the steps.

➔ Start ArcMap, if necessary. Navigate to the MapDocuments folder in the mgisdata directory and open the map document ex_4.mxd.

➔ Use Save As to give the document a new name and save frequently as you work.

1➔ Right-click on the US States layer and choose Open Attribute Table from the context menu.

Changing the appearance of a table

First, let's experiment with changing the appearance of this table. Notice how the STATE_NAME field is much wider than its data.

2➔ Narrow the STATE_NAME field to a more suitable width by clicking and dragging the right border of the field.

The ObjectID field information contains no interesting information, so hide it from sight.

3➔ Right-click the US States layer and choose Properties from the context menu.

3➔ Click the Fields tab.

3➔ Select the ObjectID field and uncheck the box next to its name.

Notice that the STATE_ABBR field is also too wide for its data but that narrowing it will cut off the field name.

3➔ Click on STATE_ABBREV in the Alias column, type in the alias **ABBREV**, and also give the STATE_NAME field the alias **NAME**.

3➔ Click OK to close the Layer Properties window and see the changes.

Scroll over to the right of the table and look at all the different fields of population data from the U.S. Census. It is difficult to match the values to the right state, however, because the state names quickly scroll out of sight.

4➔ Freeze the NAME column by right-clicking on the NAME field and choosing Freeze/Unfreeze Column.

4➔ Now scroll again and see how much easier it is to interpret the data.

TIP: An unfrozen column will not go back to its original place in the table. To move it back, remove the table from the map document and add it again.

Now let's use the Sort function to obtain information about the largest and smallest states, using the POP2000 field.

> 5➜ Right-click on the POP2000 field name and choose Sort Descending from the context menu.

1. Which state has the lowest population? _____

TIP: Use Advanced Sorting to sort on more than one field at a time.

Notice that the AREA field has decimal values in it, which makes it hard to compare the sizes of different states. We can format the field so that no decimals are displayed.

> 6➜ Right-click on the US States layer and choose Properties from the context menu. Click the Fields tab.
>
> 6➜ Click the ellipses in the Number Format column in the row for AREA.
>
> 6➜ Fill the button that says Number of Decimal Places and choose 0. Click OK and OK.

You can also change a field's appearance or turn it off or on directly from the table instead of opening the table properties.

> 7➜ Right-click the HOUSEHOLD field and choose Turn Field Off. (You must go into the table properties to turn it back on.)
>
> 7➜ Right-click the POP2000 field and choose Properties.
>
> 7➜ Note that you can give the field an alias, turn it off, set the primary display field, and format its numeric values from this window.
>
> 7➜ Click on the ellipses next to the Numeric format entry.
>
> 7➜ Set the number of significant digits to 3 and check the box to show thousands separators. Click OK and OK.

Using queries on tables

Now let's use the Select By Attributes function to select the states that had more than 5 million people in the year 2000.

> 8➜ Click the Options button in the table and choose Select By Attributes.
>
> 8➜ Enter the query [POP2000] > 5000000 and click Apply.
>
> 8➜ Click on the Show Selected button to view only the selected records.

2. How many states were selected? _____

Notice that when you select the states in the table, the associated features on the map are also selected and highlighted.

> 8➜ Click the Show All button. (You cannot clear a selection when only the selected records are being displayed.)

8➜ Clear the selected set by clicking the Options button on the table and choosing Clear Selection from the context menu.

> **TIP:** Queries on geodatabases use brackets around the field names as in [ZONE]. Queries on shapefiles use double quotes around the field names, as in "ZONE".

Calculating statistics for fields

Now it is time to calculate and view some population statistics for the states.

9➜ Right-click on the POP2000 field name and choose Statistics from the context menu.

9➜ Notice the statistics that are calculated and spend some time examining the frequency distribution.

3.   What is the population of the largest state? _____

4.   Why are there 51 "states" listed in the statistics? _____

9➜ Use the drop-down list in the Statistics box to select the POP1990 field.

9➜ Close the Statistics window.

> **TIP:** The frequency diagram cannot be placed into a map layout as a graph. However, you *can* capture the window on the screen by holding down the Alt key and pressing the PrintScrn key on the keyboard. This will place the graph on the clipboard so that it can be pasted into the Windows Paint program and saved as a JPEG file. The JPEG can then be placed in a layout.

The Statistics command calculates using all records, unless a subset of records has been selected, in which case it uses only the selected subset to calculate the statistics.

10➜ Use the Select By Attributes option to select all states with a year 2000 population greater than 5 million. Close the Select By Attributes window.

10➜ View the Statistics for the POP2000 field again. Look at the Count statistic and observe that fewer states are included this time—statistics are calculated only for the selected records. Close the Statistics window.

10➜ Clear the selected records by clicking on the Options button in the table and choosing Clear Selection.

10➜ Close the US States table.

11➜ Open the table for the 108th Congress layer.

11➜ Examine the fields, noting the field PARTY.

11➜ Close the table. Turn off US States and turn on 108th Congress.

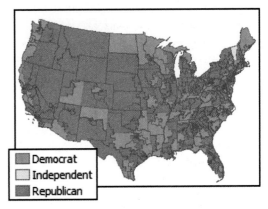

Fig. 4.11. Congressional districts

11➔ Open the Symbology properties for 108th Congress and create a unique values map of the PARTY field showing Democrats as blue, the Republicans as red, and the Independent district (Vermont) as light gray (Fig. 4.11).

Now let's find out which districts may have changed the party affiliation of the representative since the 106th Congress.

12➔ Add the cd106 feature class from the mgisdata\Usa\usdata geodatabase and move it underneath the 108th Congress layer. Rename it **106th Congress**.

12➔ Open the table for 106th Congress by right-clicking on the layer name and choosing Open Attribute Table.

12➔ Examine the fields.

Notice that in this table the PARTY field is absent and that the congressperson's party affiliation is listed after his or her name in the NAME field. The two tables can only be compared if both contain a PARTY field, so you'll need to create one for the 106th District table.

13➔ With the table open, click on the Options menu and choose Add Field.

13➔ Type in the field name **PARTY**, choose Text as the field type, and enter a Length of **11**. Click OK.

TIP: Field names must have 13 or fewer characters; may include letters, numbers, and the underscore character (_); and should not contain spaces or special characters such as @, #, !, $, or %. Field names must also start with a letter, not a number.

Next we will select the records with Democratic representatives and use Calculate to enter **Democrat** in the PARTY field.

14➔ Click on the Options button and choose Select By Attributes from the menu.

14➔ Enter the expression [NAME] LIKE '*(D)*'. Click Apply.

14➔ Scroll through the table and notice that all the Democrats are now selected.

LIKE is an operator that searches for matching subsets of characters in a field. The * character is a wild card meaning any one or more characters. (Shapefiles use % as the wild card.) This expression finds all records which contain (D). Since the parentheses are part of the expression, it will ignore other occurrences of D, such as in Danforth. Also notice the brackets around the field name and single quotes around text strings—these are required.

15➔ Right-click on the PARTY field and choose Field Calculator from the context menu. (Click Yes if you receive a warning message.)

15➔ In the Field Calculator, enter "**Democrat**" in the expression box. Make sure you enclose it in *double* quotes. Click OK.

TIP: Yes, it's confusing. You must put single quotes around strings when *selecting* them but double quotes around strings when *calculating* them.

16➜ The Select By Attributes window is still open. Change the D in the expression to an R and click Apply again.

16➜ Use Field Calculator again to place "Republican" in the Republican records.

17➜ Repeat the query and calculation one more time using I for Independents.

17➜ Click Options and choose Clear Selection. Close the tables and windows.

The fields are calculated, and now we can compare the changes. To do so, we need to join the two tables together and execute a query on both PARTY fields to see if any have changed. We'll use 108th Congress as the destination table.

18➜ Right-click the 108th Congress layer and choose Joins and Relates > Join.

18➜ Choose to Join attributes from a table (Fig. 4.12).

18➜ Choose DISTRICTID as the name of the field to base the join on.

18➜ Choose 106th Congress as the table to join.

18➜ Choose DISTRICTID as the second field to base the join on.

18➜ Click OK to join and say Yes when it asks whether to index the join field (indexing reduces search time on large tables).

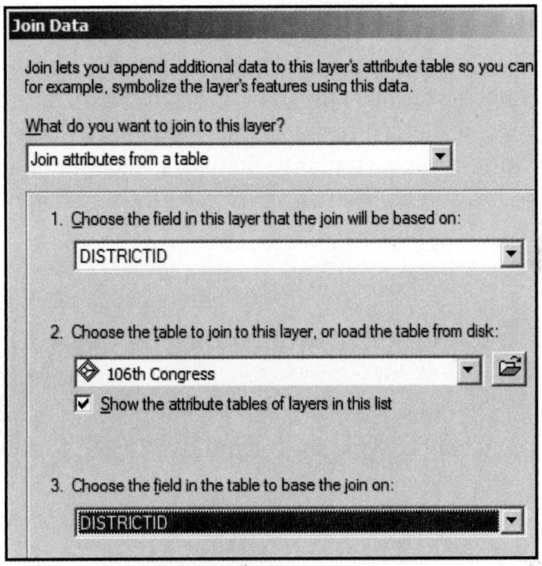

Fig. 4.12. Joining tables

19➜ Open the 108th Congress table.

The joined fields appear to the right of the table. (In version 9.2, the table name prefixes the field names. In version 9.3, the prefix is there, but each field is automatically given an alias without it.) Some of the records contain <Null> in their cd106 fields because a matching district-id could not be found in the source table.

19➜ Choose Options > Select By Attributes.

19➜ Enter the expression cd108.PARTY <> cd106.PARTY to find out which records changed. (Notice the query does not use brackets for joined field names.)

19➜ Close the query window and the table.

19➜ Right-click the 108th Congress layer and choose Selection > Create Layer from Selected Features.

20➜ The new layer appears near the top of the Table of Contents as 108th Congress selection. Rename it Changed Party. Turn off the 106th Congress layer.

20➔ Click on the Changed Party layer symbol to open the Symbol Selector and change it to 10% Crosshatch (near the bottom of the scroll window). This pattern shows the change but allows the original color to show through (Fig. 4.13).

21➔ Right-click the 108th Congress layer and choose Joins and Relates > Remove Join(s) > Remove All Joins.

21➔ Right-click the 108th Congress layer and choose Selection > Clear Selected Features.

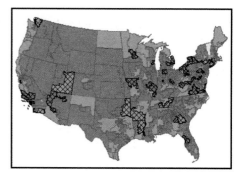

Fig. 4.13. Map of changed districts

Summarizing tables

Recall that the districts table has a field called STATE_ABBR, indicating the district's home state. Some states have only one district; others have more. How many districts does each state have? The Summarize function can yield this information. Summarize is similar to Statistics, except that it groups records based on one field (in this case the state abbreviation) and then calculates statistics for each group, including the number of districts in each group (state).

22➔ Open the attribute table of 108th Congress.

22➔ Right-click on the STATE_ABBR field and choose Summarize.

The Summarize command always counts the number of features in each group and will report it in the output table so that you don't need to request any other statistics.

22➔ Type the name of the new table to be created, **dists_per_state**, making sure it will be saved in the Usa folder. Click OK.

22➔ Click Yes when it asks to add the table to the map document.

TIP: Since this table was created to answer a transitory question, it does not seem useful to make it a permanent addition to the usdata geodatabase. Thus, we simply save it in the folder. Also, although it seems easy to take the default table names, you may have trouble later remembering what the table contains. Take a few moments to give the table a descriptive name.

---

5. Is this new table an attribute table or a standalone table? _____

23➔ Click on the Source tab at the bottom of the Table of Contents, if necessary, to make sure that the new table is visible.

23➔ Close the 108th Congress table.

23➔ Open and sort the new dists_per_state.dbf table by number of districts.

23➔ Close the dist_per_state.dbf table and turn off the 108th Congress, 106th Congress, and the Changed Party layers.

In other cases, the goal is to generate some statistics about each group in addition to simply counting the records. We'll perform the example given in the Concepts section—determining the total deaths, damage, and average magnitude of earthquakes in each state.

24➔ Add the quakehis feature class to the map document from the usdata geodatabase.

24➔ Open the quakehis table and examine the fields, noting the state abbreviation field and the fields for deaths, damage, and magnitude (MAG).

Since we want to summarize the earthquakes according to the states in which they occur, the state abbreviation field is chosen as the field to summarize by.

25➔ Right-click the STATE field and choose Summarize from the context menu (Fig. 4.14).

25➔ Verify that the field to summarize is set to STATE.

25➔ Locate the DAMAGE field in the list, expand it, if necessary, and choose Sum as the statistic to calculate.

25➔ Locate the DEATHS field and choose Sum as the statistic to calculate.

25➔ Locate the MAG field and choose Average as the statistic to calculate.

25➔ Enter a location and name for the output table, calling it **quakesum** and placing it in the Usa folder.

25➔ Click OK and say Yes when asked whether to add the table to the map.

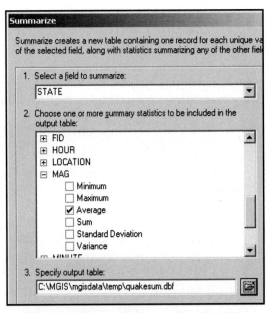

Fig. 4.14. Summarizing on the STATE field in the earthquake table

Now the new standalone table has been created and added to the map document.

26➔ Close the quakehis table.

26➔ Right-click the new quakesum table and open it.

26➔ Examine the table.

26➔ Sort the Sum_DEATHS field in descending order.

6.   Which state has the most deaths caused by earthquakes? _____

26➔ Close the quakesum table and turn off the quakehis layer before going on.

Joining tables

A map of the states with graduated color symbols showing the number of districts in each state would be a helpful addition to this map document. However, notice that the new dist_per_state table is a standalone table and its information can't be portrayed on a map. Joining it to the US States attribute table using the common field STATE_ABBREV solves the problem. In this join, the dist_per_state table is the source, and the US States table is the destination.

27➔ Right-click on the US States layer and choose Joins and Relates > Join.

27➔ In the Join dialog box, enter Join attributes from a table in the drop-down box.

27➜ Choose ABBR as the field in US States to base the join on.

27➜ Choose dists_per_state as the table to join.

27➜ Choose STATE_ABBR as the field in the source table to join on.

27➜ Click OK to finish the join.

7. What is the cardinality of this join? _____

27➜Open the US States table and scroll to the right to find the new fields added as a result of the join.

Now let's make this table easier to read. Notice that the OBJECTID and STATE_ABBR fields were added during the join, but we don't need them. Also, the field containing the number of districts is called Count_STATE_ABBR—hardly a descriptive name.

28➜ Open the Properties for the US States layer and click the Fields tab. Scroll down to the bottom of the field list to find the fields from the joined dists_per_state table.

28➜ Hide the ObjectID and STATE_ABBR fields.

28➜ Give the Count field the alias NumDistricts. Click OK.

29➜ Close the US States table.

29➜ Create a graduated color map showing the number of representatives for each state (Fig. 4.15).

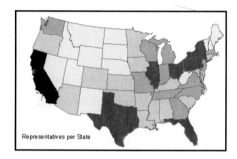

Representatives per State

Fig. 4.15. Your map after step 29

Relating tables

Next, it would be convenient to obtain information on which representatives come from which subregions of the country, as presumably they would have similar issues to confront with their constituencies. However, the SUB_REGION field is in the US States table, not the districts table. Setting up a relationship between the US States table and the 108th Congress table will provide the combination of information that is needed. The goal is to select a subregion of the country and obtain a list of representatives from that region.

8. What is the cardinality of this relationship?

_____

9. Which table is the source table, and which is the destination table?

_____

30➜ First remove the previous join by right-clicking on the US States layer and choosing Joins and Relates > Remove Join(s) > Remove All Joins.

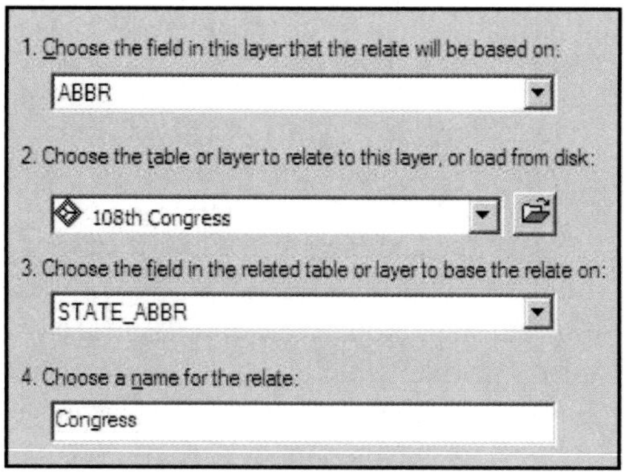

1. Choose the field in this layer that the relate will be based on:

ABBR

2. Choose the table or layer to relate to this layer, or load from disk:

108th Congress

3. Choose the field in the related table or layer to base the relate on:

STATE_ABBR

4. Choose a name for the relate:

Congress

Fig. 4.16. Setting up a relate

30➜ Right-click on the US States layer and choose Joins and Relates > Relate.

30➜ Fill out the fields as shown in Figure 4.16. Click OK.

Now imagine that a group called Daughters of New England is trying to contact all of the representatives in New England. With the related tables, it is easy to fulfill their request for a list of all the New England reps.

31➜ Open the US States table and the 108th Congress table and position them so that both can be seen.

31➜ Click the Options button on the US States table and use Select By Attributes to select the states in the New England (N Eng) subregion. (Click Get Unique Values to select the N Eng value from the list.)

31➜ Close the Select By Attributes window when done.

Now that the New England states are selected, the relate gives easy access to the representatives from these states.

31➜ Click on the Options button in the US States table and choose Related Tables > Congress: cd108.

31➜ In the Districts table, click the Show Selected button to view the matching representatives from New England.

10. How many representatives come from New England states? _____

Exporting the selected records from the representatives table will provide a new file that can be given to the Daughters of New England group.

32➜ In the Congress table, click the Options button and choose Export.

32➜ Make sure it says to export the selected records, and type in NE_reps.dbf as the name of the table. Save it in the Usa folder.

32➜ Say Yes to add the table to the map, and close the open tables.

32➜ Open the new NE_reps table and verify that it contains the correct representatives. (You may need to click the Source tab in the Table of Contents.)

> **TIP:** ArcMap has no direct way to export only certain fields from a table. However, one may export an entire table and then delete the unwanted fields in ArcCatalog.

33➜ Close NE_reps table.

33➜ Right-click the NE_reps table in the Table of Contents and choose Remove.

33➜ Remove the relate by right-clicking the US States layer and choosing Joins and Relates > Remove Relate(s) > Remove All Relates.

33➜ Choose Selection > Clear Selected Features from the main menu.

> **TIP:** Exporting a joined table will put both the destination and the source table information in the new table as a permanent copy of the joined tables.

Analyzing tabular data

Let's explore the distribution of Hispanics in the United States. Although the HISPANIC field in the US States table could be used to create a graduated color map showing the number of Hispanics in each state (Fig. 4.17), the map is difficult to interpret because the population of Hispanics is typically larger in states with larger populations. The percentage of Hispanics in each state provides a more useful statistic. Creating a new field and calculating the percentage of Hispanics solves the problem.

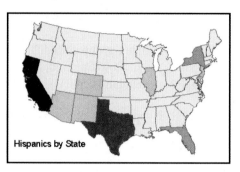

Fig. 4.17. Hispanics by state

34➔ Open the US States attribute table.

34➔ Click on the Options button and choose Add Field.

34➔ Name the field **HISP_PERC** and select Float as its type. Click OK.

TIP: When setting numeric fields in shapefiles, you may take the precision and scale defaults, or you may specify values. The precision refers to the number of digits a field can hold, including negative signs and decimal points. The scale indicates the number of decimal places. Precision and scale cannot be set for geodatabases; they have default values.

Now calculate the percentage of Hispanics.

35➔ Right-click on the new HISP_PERC field (located to the far right of the table) and choose Field Calculator from the context menu.

35➔ Enter the expression [HISPANIC] / [POP2000] * 100 and Click OK.

TIP: Do all of the records equal zero? You might have accidentally selected a single record by clicking on the table. Check the bottom status bar to see if any records are selected. If so, choose Options > Clear Selection from the table and try the calculation again.

36➔ Now create a map showing the percentage of Hispanics in each state (Fig. 4.18). Looks different, doesn't it?

Finally, we will gather some information about Hispanics in the different subregions of the United States, in particular the total number in each subregion and the average percent in each subregion. This is another job for Summarize.

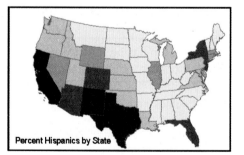

Fig. 4.18. Percent Hispanics by state

37➔ In the US States table, right-click on the SUB_REGION field and choose Summarize.

37➔ For statistics, click on the plus sign next to the HISPANIC field and choose Sum for the statistic.

37➔ Click on the plus sign next to the HISP_PERC field and choose Average for the statistic.

37➔ Enter the name **SubRegHisp** for the table to be produced, and make sure it is being saved in the Usa folder. Click OK.

37➔ Click Yes to add the table to the map document.

37➔ Close the US States table.

37➔ Click the Source tab, if necessary, locate the new table, and open it.

11. Which subregion has the greatest number of Hispanics? _____ ·•

Finally, for a report you need to determine the density of Hispanics in people per square kilometer in the counties. The county table has an area field and population density field, but these are based on the full population and square miles. You'll need to create and calculate two new fields, the area in kilometers and the density of Hispanics. Because you're calculating areas, make sure the data frame coordinate system is set to an equal-area projection.

38➔ Close the SubRegHisp table.

38➔ Open the data frame properties, click the Coordinate System tab, and set the coordinate system to USA Contiguous Albers Equal Area Conic (Continental folder).

39➔ Open the Counties attribute table.

39➔ Click the Options button and choose Add Field.

39➔ Name the field **Area_km** and set its Type to Float.

40➔ Add another new Float field named **Hisp_Dens**.

Now we will calculate the new values in the fields.

41➔ Right-click the Area_km field and choose Calculate Geometry. Click Yes to calculate outside an edit session.

41➔ Set the Property to Area, use the coordinate system of the data frame, and set the units to square kilometers. Click OK.

42➔ Right-click the Hisp_Dens field and choose Field Calculator. Enter the expression [HISPANIC] / [Area_km] and click OK.

Working with Excel data

We will do a brief exercise on using Excel data in ArcGIS and displaying locations in a table as points on a map.

43➔ Click on the New Map File button to open a new map document. Save the changes to the previous one.

43➔ Click the Add Data button to add the counties feature class from the mgisdata\Oregon\oregon geodatabase.

44➔ Click the Add Data button again, navigate to the mgisdata\Oregon folder, and double-click the ORstations.xls workbook. Select the ORstations$ worksheet and click Add.

44➜ Repeat the previous step to add the ORprecipnormals$ worksheet from the ORprecipnormals.xls file.

45➜ Open the ORstations$ table and examine the fields. Note the LAT and LON fields containing the station location in decimal degrees.

45➜ Open the ORprecipnormals$ table and examine the monthly and annual precipitation normals for each station.

12. What is the common field in these two tables? _____

Any table with *x-y* coordinate locations can be displayed as a point layer called an event theme. You need to know the coordinate system of the *x-y* locations in the table. In this case the units are degrees of latitude and longitude, so we know that it is a geographic coordinate system. A look at the Web site where these data were downloaded from the National Climatic Data Center (NCDC) tells us that the GCS uses the NAD 1983 datum.

45➜ Close both tables.

46➜ Right-click the ORstations$ table name and choose Display XY Data.

46➜ Set the X Field to the LON field.

46➜ Set the Y Field to the LAT field.

46➜ Click the Edit button to set the coordinate system.

46➜ Choose Select. Choose Geographic Coordinate Systems > North America > North American Datum 1983.prj and click Add, OK, and OK.

Now we will add the precipitation values to the newly created event theme and export the result to the geodatabase to save it permanently as a feature class.

47➜ Right-click the new ORstations$ Events layer and choose Joins and Relates > Join.

47➜ Enter STATION NAME as the field to base the join on, the ORprecipnormals$ table as the table to be joined, and STATION as the second field.

47➜ Open and examine the joined table. Close it when finished.

48➜ Right-click the event theme and choose Data > Export Data.

48➜ Choose to use the same coordinate system as the data frame so that the stations will be stored in a projection (Oregon Statewide Lambert).

48➜ Click the Browse button and navigate to the mgisdata\Oregon folder. Change the Save As Type to File and Personal Geodatabase feature classes so that the oregon geodatabase appears. Open it.

48➜ Name the output feature class **precip**. Click Save and OK.

48➜ Choose Yes when asked to add the feature class to the document.

49➜ Open and examine the table of the precip layer. Close it when done.

49➜ Open the Properties of the precip layer.

49➜ Click the Symbology tab and create a graduated symbol map of the annual precipitation in the ANN field.

50➜ Right-click the event theme layer, which you no longer need, and choose Remove.

50➜ Click Add Data and add the gtoposhd raster from the oregon geodatabase.

50➜ Turn off the counties layer. Admire the map (Fig. 4.19).

50➜ Close ArcMap. Save the map document if you wish.

This is the end of the tutorial.

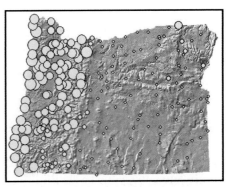

Fig. 4.19. Oregon precipitation

More skills

Consult the Skills Reference section of this chapter to learn to do the following:

➢ Preparing Excel workbooks for use in ArcMap

➢ Creating new tables

➢ Creating and using reports based on a table

Exercises

Open the ex_4.mxd map document in the MapDocuments folder of the mgisdata directory and answer the following questions.

1. How many counties in the United States have the name Washington? What is their total 2000 population? Which one has the largest area?

2. Calculate the percentage of the population in each state that is African-American. (Use the BLACK field for the African-Americans and the POP2000 field for the total population in each state.) Which state has the largest *number* of African-Americans? Which state has the largest *percentage* of African-Americans?

3. Which subregion of the country has the greatest number of African-Americans? Which subregion has the highest average percentage of African-Americans in a state?

4. Add the table popestmt from the mgisdata\Usa\usdata geodatabase to the map document. This table contains population estimates by county between 1990 and 1998. Examine the table. Which fields could you use to join this table to the US Counties table?

   In the US Counties table, examine the three fields STATE_FIPS, CNTY_FIPS, and FIPS. How are these fields related? Join the tables.

5. Use the POP1990 and POP1995 fields from the popestmt table (now joined to the US Counties table) to select the counties that *lost* population between 1990 and 1995. How many are there?

    In what area of the country are these counties predominantly located? Make a map showing the "losing" counties. **Capture** the map. (**Hint:** Right-click the US Counties layer and choose Selection > Create Layer from Selected Features. Name it **Losing Counties**.)

6. Use Summarize to determine the number of counties in each state which lost population. Which three states had the most losing counties, and how many losing counties did each have? (**Hint:** If Summarize doesn't work, try removing the join first.)

7. Imagine that you are the PR director of a group seeking aid for Native American artists, and you wish to contact the *Democratic* representatives in the 108th Congress who are from states in which the number of Native Americans and Eskimos (AMERI_ES field in US States) exceeds 100,000. Which states are included in the list? How many representatives will you need to contact?

8. Using the Major Cities layer, determine how many people in the United States live in state capitals. (**Hint:** Look for the designation "State Capital" within the FEATURE field.) What is the smallest, largest, and average population of a state capital?

9. Which state has the highest percentage of people who are married? Which one has the highest percentage of people who are divorced?

10. Which subregion of the United States has the largest number of counties? How many total people did that subregion have in the year 2000? Create a map of the counties showing to which subregion each belongs. **Capture** the map.

Challenge Problem

Imagine that you are planning to open a dating service in Texas. Your preliminary research indicates that such a service would expect to gross $3 for every divorced person, $2 for every single (never married) person, and $1 for every widowed person in a typical county. Calculate the projected gross income for each county in Texas and create a map layout showing the projected income for each county. (Make sure your map shows only Texas counties.) What is the total projected gross income for the entire state? Is there a large cluster of high-income counties that might be a good place to put your office?

Skills Reference

Adding or removing a table 179

Opening a table 179

Modifying a table's appearance..180

Selecting records using attributes 182

Getting statistics for a field 183

Summarizing on a field 184

Exporting a table 184

Adding or deleting fields...185

Calculating fields 187

Editing fields in a table...187

Creating joins and relates 188

Creating a new table..189

Deleting a table..189

Opening Excel Data...190

Adding or removing a table

Adding a table means putting it into a map document in ArcMap so that it may be used. Attribute tables are added by adding the spatial data set with which it is associated. Adding a standalone table is done using the Add Data button, just as for spatial data sets.

1. In ArcMap, click the Add Data button or choose File > Add Data from the menu bar.

2. If necessary, use the Add Connection button in the Add Data dialog box to add a folder connection to the disk area containing your data.

3. Browse to find the tabular data file. Click to select it and click the Add button. (Shortcut: Double-click the desired file.)

4. To remove a table, right-click the table name in the Table of Contents and choose Remove.

Opening a table

1. To open an attribute table, right-click on the spatial data set *name* (not the icon) and choose Open Attribute Table from the context menu (Fig. 4.20).

2. To open a standalone table, first make sure that the table is visible by clicking on the Source tab at the bottom of the Table of Contents.

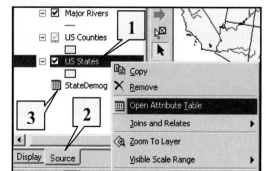

Fig. 4.20. Opening a table

3. Right-click on the table name and choose Open from the context menu.

Modifying a table's appearance

You can change certain ways in which the table displays data. Note that none of these operations modifies the actual data on the disk, only the way in which the data are presented.

Adjusting field width

1. Use the cursor to hover over the right edge of the field to resize until it changes to a bar with a double arrow.

2. Click and drag the edge to the desired width and then release the mouse button.

Sorting on a field

Right-click on the field name to sort by and choose either Sort Ascending or Sort Descending. Use Advanced Sorting to sort on more than one field at a time.

Displaying Selected or All Records

A table can show all records with the selected ones highlighted, or it can show only the selected records.

Click on the Selected button at the bottom of the table (Fig. 4.21). Notice that it also reports how many records are currently selected.

Fig. 4.21. Showing selected records only

Hiding fields or creating aliases

Fields can be made invisible to hide them from display. Aliases can be used to give fields more descriptive names that need not follow the restricted naming conventions for fields.

1. If the table is an attribute table, right-click on the data set name and choose Properties from the context menu. If it is a standalone table, right-click on the table name and choose Properties.

2. Click the Fields tab (Fig. 4.22).

3. To create an alias, type a new name in the Alias column.

4. To hide a field, uncheck the box.

5. To format a numeric field, click the ellipses (see the next section).

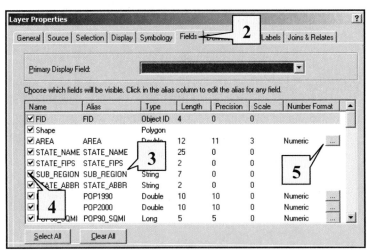

Fig. 4.22. The Fields tab of the Layer Properties window

Formatting field values

You can also change the alignment, decimal places, and other formatting characteristics which determine how a numeric field is displayed.

1. Open the Field Properties tab as described in the previous section.

2. Click on the ellipses in the Number Format column.

3. Choose the numeric category, such as currency or percentage (Fig. 4.23).

4. Specify the number of decimals or significant digits, the alignment, and other options.

5. Click OK.

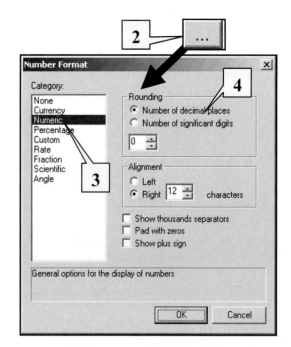

Fig. 4.23. Formatting field values

Freezing or unfreezing columns

Freezing places a column at the left edge of the table and keeps it always in sight as the user scrolls through the table. More than one column may be frozen.

1. With the table open, right-click on the field name of the field to freeze and choose Freeze/Unfreeze Column from the context menu.

2. To unfreeze it, right-click on a frozen field heading again and choose Freeze/Unfreeze Column again.

TIP: An unfrozen column will not go back to its original place in the table. To move it back, remove the table from the map document and add it again.

Changing table colors and fonts

The user can change how a table looks by changing its font and the colors that are used to highlight selected records.

1. With the table open, click the Options button in the lower area of the table and choose Appearance.

2. Click the color squares to choose a new color from the drop-down palette for the selection and highlight colors.

3. Choose a font, font size, and color.

Selecting records using attributes

A query selects records that meet certain criteria, such as cities having a population greater than 5 million.

1. With the table open, click the Options button and choose Select By Attributes. A dialog box will appear (Fig. 4.24).

2. Choose the selection method. For a discussion of the different methods, see Chapter 6.

3. Enter a field name from the list into the large query box by double-clicking on the field name.

4. Enter an operator by clicking on it once.

5. To choose a value from the table, click Get Unique Values and then choose the desired value from the list.

6. Alternatively, enter a value by typing it into the query box. If it is a text string, put single quotes around it.

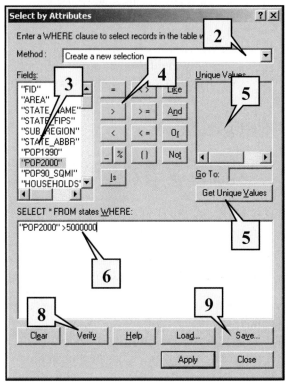

Fig. 4.24. Selecting by attributes

7. Multiple queries use criteria from more than one field, such as finding all cities that are capitals AND have more than 5 million people. The field name must be repeated for each query (see the examples that follow).

8. Choose to Verify your query if you are uncertain whether it is correctly formulated.

9. Use the Save button to save a query that is long and complex and frequently needed. It will be saved with an .exp extension. Load it again later using the Load button.

10. Click Apply to execute the query. The selected records will be highlighted in the table.

11. To clear the selected records, click on the Options button in the table and choose Clear Selection from the context menu.

Examples of valid queries:
        "POP1990" > 1000000
        "STATE_NAME" = 'Alabama'
        "STATE_NAME" = 'Alabama' OR "STATE_NAME" = 'Texas'
        "POP2000" >= "POP1990"

Getting statistics for a field

Statistics are useful for learning more about the data in fields. Statistics may be calculated for a selected set of records or for the entire table.

1. If desired, use one of the selection methods described in Chapter 6 to select a subset of records requiring statistical analysis.

2. With the table open, right-click on the field on which to get statistics and choose Statistics from the context menu.

3. Examine the statistics and frequency diagram in the Statistics dialog box (Fig. 4.25). If the table has records selected, then statistics will be calculated only for the selected records.

4. Look at statistics for a different field by choosing it from the Field drop-down box.

TIP: The frequency diagram cannot be placed directly into a map layout as a graph. However, you *can* capture the window on the screen by holding down the Alt key and pressing the PrintScrn key on the keyboard. This will place the graph on the clipboard so that it can be pasted into the Windows Paint program and saved as a JPEG file. The JPEG can then be placed in a layout.

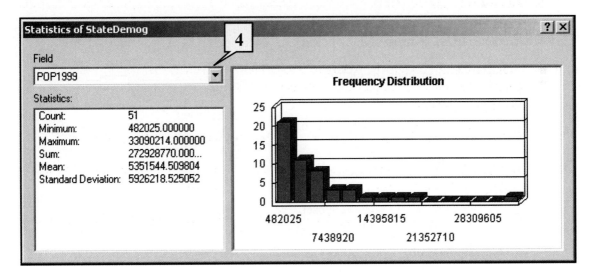

Fig. 4.25. The Statistics window

Summarizing on a field

Summarize generates statistics for groups of features in a given field, such as finding the average magnitude of earthquakes occurring in each state.

1. With the table open, right-click on the field on which to generate statistics and select Summarize from the context menu.

2. In the Summarize dialog box, verify that the chosen field appears in the first box (Fig. 4.26).

3. In the second box, choose one or more statistics to calculate. Click on the plus sign to expand the list of statistics for a field, and then check the box next to the desired statistic(s).

4. Specify the name of the output table. Click on the Browse button to change the directory where the file will be saved. Click OK.

5. Click Yes to add the table to the map.

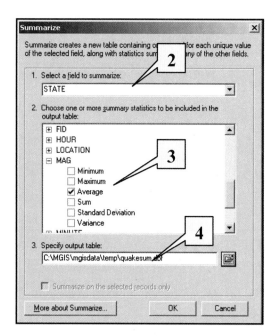

Fig. 4.26. The Summarize window

Exporting a table

Exporting a table means saving it under a new name. There are two ways to export a table.

1. If the table is open, click the Options button and choose Export from the menu. If it is a standalone table, you can also right-click the table name in the Table of Contents and choose Data > Export.

2. To export the entire table, choose All records from the first drop-down box. To export only a previously selected subset, change the drop-down box to say Selected records (Fig. 4.27).

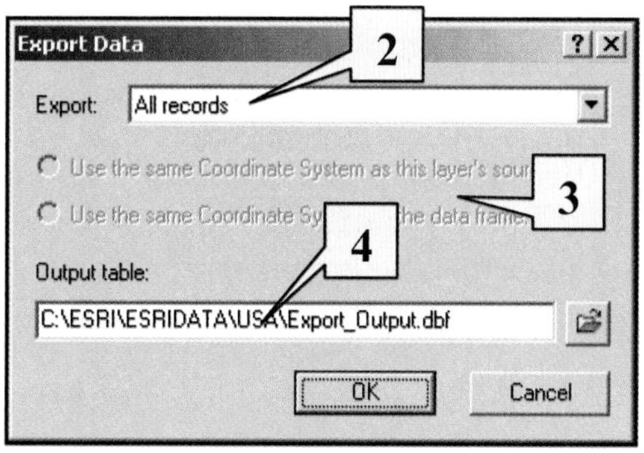

Fig. 4.27. The Export Data window

3. The two choices regarding the coordinate system are dimmed because you are exporting a table instead of a spatial data set.

4. Choose the directory and name of the output file by typing it in or clicking the Browse button. Click OK.

TIP: Exporting a joined table will put both the destination and the source table information in the new table.

Adding or deleting fields

Fields may be added to tables in two ways. The first way uses ArcCatalog to enter and define new fields. This way is most efficient when you are entering many fields at once. You can also add a field during an ArcMap session.

Managing fields of shapefiles or DBF files in ArcCatalog

1. If the table is currently open in ArcMap, save the map document and exit ArcMap. Tables cannot be modified when they are part of an ArcMap session.

2. Navigate to the table in the left side of the ArcCatalog window.

3. Right-click the table and choose Properties from the context menu.

4. Click the Fields tab (Fig. 4.28).

5. Scroll down the list of fields, if necessary, to the first empty space at the bottom.

6. Click on the empty space and type in the name of the new field.

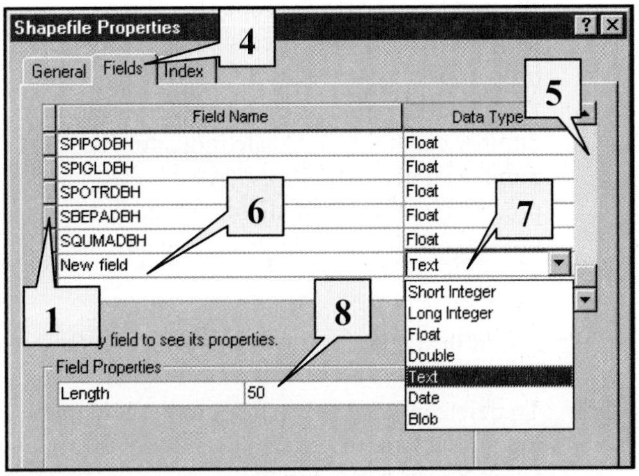

Fig. 4.28. Adding shapefile fields in ArcCatalog

7. Click on the Data Type box to the right of the new field, and a drop-down list of field types will appear. Choose the appropriate type.

8. View the field property information in the lower part of the dialog box. Beginners are advised to leave these properties alone for the most part. However, the length, precision, and scale properties may be set, if necessary. Precision determines the number of digits a number can hold, and scale indicates the number of decimal places. (A value with a precision of 6 and a scale of 2 could hold numbers between –99.99 and 999.99.) Only shapefile fields can have the precision and scale set; these values have defaults in geodatabase tables.

9. Repeat steps 5 through 8 until all the fields have been added.

10. Click OK or Apply to enter the changes in the Properties dialog box.

IMPORTANT TIP: Note that field names must have 13 or fewer characters; may include letters, numbers and the underscore character (_); and should not contain spaces or special characters such as @, #, !, $, or %. Field names must also start with a letter, not a number.

---

11. To delete a field, follow steps 1–4 to bring up the table's Fields tab in the properties menu. Select the field by clicking on the gray tab next to its name, and press the Delete key. Warning: You cannot undo this step.

Managing fields in ArcMap

1. With the table open, click on the Options button and choose Add Field from the menu.

2. Type in the name of the field (Fig. 4.29).

3. Choose the type of field from the drop-down box.

4. View the field property information in the lower part of the dialog box. Beginners are advised to leave these properties alone. However, the length, precision, and scale properties may be set, if necessary. Precision determines the number of digits a number can hold, and scale indicates the number of decimal places. (A value with a precision of 6 and a scale of 2 could hold numbers between –99.99 and 999.99.)

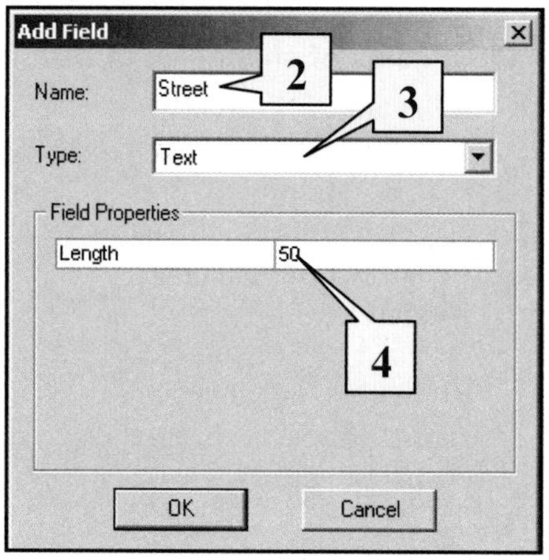

5. Click OK. The field will be added to the end of the table.

Fig. 4.29. Adding a field in ArcMap

6. To delete a field, right-click on the field name and choose Delete Field from the context menu. Because this action cannot be undone, you will be prompted to confirm the action. Click Yes to delete the field.

IMPORTANT TIP: Note that field names must have 13 or fewer characters; may include letters, numbers, and the underscore character (_); and should not contain spaces or special characters such as @, #, !, $, or %. Field names must also start with a letter, not a number.

TIP: You cannot add or delete a field while you are in an editing session; you must stop editing first. If the Add/Delete options are dimmed in the menu, check to see if you are editing.

Calculating fields

Use the Field Calculator to quickly enter information into many records at once, either the entire table or a selected set. The information entered can be as simple as a single number, or it could be a complicated mathematical expression. Calculations may be done outside of an editing session, but then they cannot be undone.

1. With the table open, right-click on the field to be calculated and choose Field Calculator. The Field Calculator will appear (Fig. 4.30).

2. Begin entering the expression in the box to the lower right. Double-click a field name to enter it in the box. Click on operators to include those. Double-click functions to enter them. Numbers may be entered from the keyboard.

3. For example, the expression in Figure 4.30 was created by double-clicking [Hispanic] in the Fields box, clicking the / operator, double-clicking [Pop1990], clicking the * operator, and typing 100.

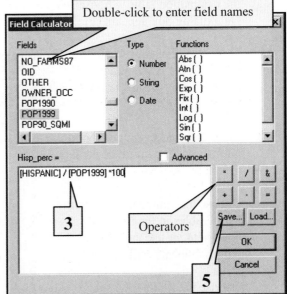

4. Click OK.

Fig. 4.30. Calculating a field

5. Complex expressions that are used frequently may be saved as *.cal files and loaded later. Press Save, navigate to the desired directory, and enter a file name. To reload the expression, click Load, navigate to the file location, and select the desired expression file.

Editing fields in a table

Fields can only be edited during an editing session. For more information, consult Chapter 11.

1. If not already in an editing session, click the Editor Toolbar button in the main ArcMap button bar to open the Editing Toolbar.

2. Choose Editor > Start Editing from the Editor Toolbar.

3. Open the table, if it is not already open, by right-clicking it and choosing Open.

4. Click in the field space of the record to edit and start typing in the changes.

5. Choose Editor > Save Edits from the Editor Toolbar to save during an edit session. When finished typing the changes, choose Editor > Stop Editing and answer Yes when asked whether to save your changes.

Creating joins and relates

Joins and relates are used to link information from two different tables, based on a field that is common to both. The fields may be numeric or string fields, and they may have different names.

In a join, information from the source table is appended to the destination table. The destination table will then have its original fields plus the fields from the source table. The source table field names will be prefixed with the source table name.

1. In the Table of Contents, right-click on the spatial data set or the table that is to become the destination table. Choose Joins and Relates > Joins from the context menu. The Join Data dialog box appears (Fig. 4.31).

2. Make sure that Join Attributes from a Table is the selection action in the first drop-down box.

3. Enter the field in the destination table that will be used as the key to join the records.

4. Choose the source table from the list of tables in the map document, or use the Browse button to locate a file on disk.

5. Choose the field in the source table that will be used as the key.

6. Choose to keep unmatched records (in which case the appended fields will contain <Null> values) or drop the unmatched records from the table.

7. Click OK.

8. To remove a join, right-click the destination table or layer in the Table of Contents and choose Joins and Relates > Remove Join(s). Then select a specific join to remove, or remove all joins on that table.

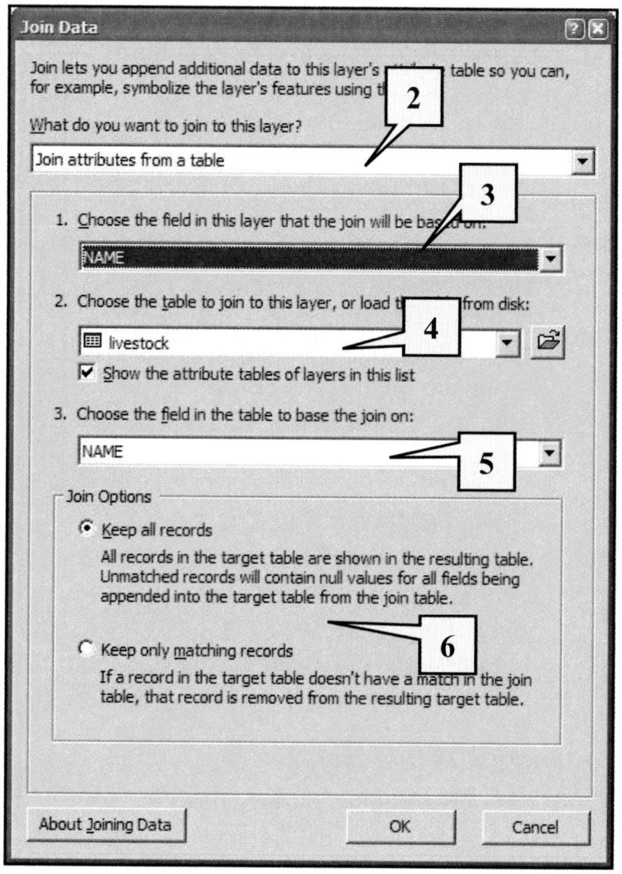

Fig. 4.31. The Join Data window

Creating a new table

New standalone dBase tables must be created in ArcCatalog. Attribute tables are generated automatically when the program creates the associated spatial data set.

1. In ArcCatalog, click on the directory or workspace that is to contain the new file. Make sure you are not inside a coverage directory.

2. Choose File > New > dBase Table from the ArcCatalog menu bar. The new table will appear in the directory with the name New_dBase_Table.

3. Type in the name of the new table. Make sure the table name does not contain spaces or special characters such as @, #, %, ^, or &. Underscore characters are permitted.

4. To add fields to the table, see the instructions under "Adding or deleting fields" in this chapter.

Deleting a table

Standalone tables should be deleted using ArcCatalog.

1. Make sure that the table is not open in a current ArcMap session; otherwise, you will not be able to delete it.

2. In ArcCatalog, navigate to the directory where the table is located.

3. Right-click the table name and choose Delete from the context menu. This action cannot be undone.

VERY IMPORTANT TIP: Do not use Windows™ to delete an attribute table that is associated with a shapefile. You will corrupt the shapefile and make it unusable.

Opening Excel Data

Preparing the data

Using Excel data is not difficult, but it requires careful preparation. First, make sure that the spreadsheet meets the following requirements.

- The first row of the worksheet contains the field headings.
- Field headings must start with a letter, must have no more than 13 characters, and must not contain spaces or shift characters such as %, $, #, and so on.
- No blank lines occur between the headings and the data or within the data rows.
- The spreadsheet contains no formulas.
- The bottom of the spreadsheet has no extraneous data, such as column totals.
- There are no merged or split cells, and every column contains consistent data types.
- Numeric columns have only numeric values in them, without text characters like "x" or "n/a" to indicate missing values. If present, missing values should be replaced by a numeric NoData marker value such as –99.
- Text fields contain no commas, unless all the text is enclosed in quotes.
- It is helpful if each column has been specifically formatted as text data or as numeric data with a specified number of decimal places.

An Excel workbook may have multiple worksheets inside. You must open each one separately. If they have not been named they will appear as Sheet1$, Sheet2$, and Sheet3$, with Sheet1$ containing the data. If they have been named, the names will appear.

1. Click the Add Data button and navigate to the folder containing the Excel document.

2. Double-click the desired spreadsheet document to look inside it. Select the named sheet you want, or select Sheet1$ if the worksheets have not been named (Fig. 4.32). Click Add.

3. Use the spreadsheet as if it were any other table (except it is read-only).

4. To convert it to dBase format or save it in a geodatabase, right-click the spreadsheet table and choose Export. See the Skill Reference entry in this chapter on Exporting.

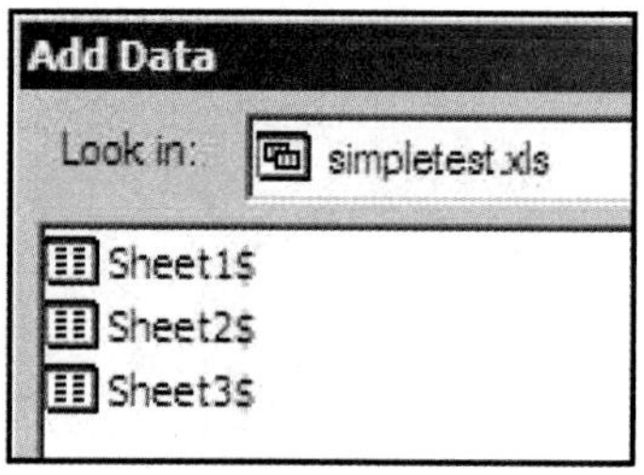

Fig. 4.32. An Excel workbook with multiple worksheets

TIP: You must CLOSE the spreadsheet in Excel before you can open it in ArcMap.

You can also save a spreadsheet in .csv or .dbf format and open it from ArcMap that way.

Chapter 5. Queries

Objectives

➢ Understanding queries and how they are used

➢ Selecting features based on attributes using SQL and Boolean operators

➢ Selecting features based on their spatial location with respect to other features

➢ Applying selection options, including the selectable layers and the selection method

Mastering the Concepts

GIS Concepts

About queries

Queries are a fundamental tool in GIS, used in nearly every analysis. They serve a critical function in data exploration or in picking one's way through a data set looking for patterns. They can also serve as the preparatory step of extracting certain features prior to performing another function. Some common applications involving queries include the following:

- **Selecting features of interest.** Queries can be used to search a table and to find which features meet certain criteria. How many houses currently for sale fall into my price range? Combined with the statistics function, this tool becomes powerful. What is the average cost of three-bedroom homes in this town?

- **Exploring patterns.** Creating a map from selected features and examining their spatial distribution can be illuminating. Where do the wells with high values of contaminants occur? Do they show a pattern relative to known point source pollution emissions? Are they widely spaced or clustered together?

- **Isolating features for more analysis.** Select out the Democrats and calculate an entry in the Party field. Extract the primary roads into a separate layer and create polygons around them at a specified distance (buffering).

- **Exploring spatial relationships.** Which parcels lie in the floodplain? Which cities are close to a volcano? Where do roads cross unstable shale units?

- **Creating raster queries.** Although we have only discussed vector queries, cells of a raster can be queried also. Which cells have a slope greater than 10 degrees? Which cells show an increase in vegetative greenness as measured by satellite images taken in 1998 and 1999? Given two rasters showing land use in 1970 and 2000, where has land use changed?

Formally speaking, a query extracts features or records from a feature class or from a table and isolates them for further use, such as printing them, calculating statistics about them, editing them, graphing them, creating new files from them, or doing more queries on them. In the simplest kind of query, you look at a map or a table and use the mouse to select the desired record(s). After selecting features with a query, the selected features are highlighted on the screen, and the corresponding records in the table are highlighted as well.

Queries fall into two main categories, **attribute queries** and **spatial queries**. An attribute query uses records in the attribute table to test a condition, such as finding all cities with population greater than 1 million people. Attribute queries are sometimes called aspatial queries because they do not require any information about location. A spatial query uses information about how features from two different layers are located with respect to one another. A spatial query might choose cities within a certain county, rivers that are completely within a state, or hospitals within 20 miles of an airport. Attribute queries can be performed on either attribute tables or standalone tables. A spatial query requires a spatial data layer. Spatial and attribute queries may be combined, such as selecting cities with more than 50,000 people who are within 50 miles of an airport. The spatial and attribute queries must be performed separately but can be done in any order.

Attribute queries

In this method, the user specifies a certain condition based on fields in the attribute table and selects the records that meet that criteria, such as choosing all states with a population greater than 5 million. We already introduced selecting by attributes from tables in Chapter 4. In this section we extend selection to features and discuss the more advanced selection options.

An attribute query must be based on an **expression**. Expressions are written in a format called **Structured Query Language** (SQL), which is used by most major databases to perform selections. SQL expressions can be complex and sophisticated, or fairly simple.

SELECT *FROM cities WHERE "POP1990" >= 500000

SELECT *FROM counties WHERE [BEEFCOW_92] < [BEEFCOW_87]

The SELECT *FROM and WHERE keywords must be present as dictated by the syntax rules of SQL. The lowercase words (e.g., cities) give the name of the table from which to select, and the terms in double quotes or brackets are the field names. SQL recognizes a variety of conditional operators, including <, >, and =. The operators AND, OR, and NOT can also be used in building expressions. This same syntax would be used in any SQL-compliant database, so SQL is a transportable language. However, minor differences between database SQL varieties can be found, such as whether field names are enclosed in brackets or quotes.

Boolean operators

When a query includes more than one condition, set theory and comes into play. Consider the following queries with two conditions:

SELECT *FROM accounts WHERE "Cust" = 'COM' AND "Balance" > 500

SELECT *FROM accounts WHERE "Cust" = 'COM' OR "Cust" = 'GOV'

SELECT *FROM accounts WHERE [Balance] > 500 AND [Balance] < 1500

The first expression selects records from a customer accounts table that are held by commercial customers and have a balance over $500. The second expression finds the accounts held by commercial and governmental customers. The third expression finds accounts with balances between $500 and $1500.

TIP: SQL requires that a field name must appear in every condition of the expression, even if it is the same field. SQL will not correctly evaluate the expression "Cust" = 'COM' AND 'GOV'.

AND, OR, XOR, and NOT are called **Boolean** operators, and they evaluate two input conditions, which may be either true or false, and return a set of records meeting those conditions. Figure 5.1 represents these operators graphically using **Venn diagrams**. The largest circle labeled T represents all the records in the table. Circle A represents the subset of records meeting condition A, circle B represents the subset meeting condition B, and the blue region indicates the records selected by the different Boolean operators.

Examine the first expression in the examples above. The first condition, "Cust" = 'COM', constitutes condition A; condition B would be "Balance" > 500. A AND B finds the accounts for which both input conditions are true (commercial customers and balance over $500). If either condition is false, the record is not selected.

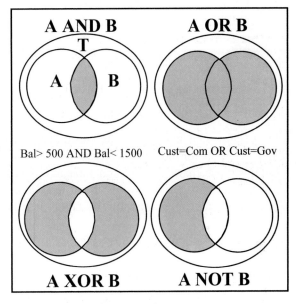

Let us examine the other operators using the same conditions. The expression A OR B selects the records for which either condition is true, finding all the commercial customers and all the accounts with balances over $500 (commercial or not). A NOT B finds all the commercial accounts but excludes the ones with balances over $500, instead finding the commercial customers with balances under $500. The "exclusive or" operator XOR is not included in the SQL calculator in ArcGIS but

Fig. 5.1. Boolean operators

is another common Boolean operator used to find records for which one of the conditions holds but not both at the same time. XOR would find all the commercial customers except those with high balances, plus all the high balances except those held by commercial customers.

Novices often confuse the AND and OR operators when writing expressions, for example, in selecting both the commercial and the governmental accounts. In English we say, "select the commercial and the governmental accounts," and it seems natural to use the expression "Cust" = 'COM' AND "Cust" = 'GOV'. However, this expression yields no records because a customer account cannot be both commercial and governmental at the same time. The OR operator should be used.

The AND operator is used to find records within a numeric range. Consider the third expression, "Balance" > 500 AND "Balance" < 1500. If a record with "Balance" = 1000 is being considered, it would test true for both conditions and be selected. Using OR is incorrect; it would return ALL of the values instead of those inside the range. A record with "Balance" = 300 would test true for the second condition and so be included. A record with "Balance" = 3000 would test true for the first condition and also be included.

SQL does accept three or more Boolean conditions. Typically, parentheses must be used to enforce the correct order of evaluation. Compare these two expressions for selecting parcels:

("LUCODE" = 42 AND "VALUE" > 50000) OR "SIZE" > 50

"LUCODE" = 42 AND ("VALUE" > 50000 OR "SIZE" > 50)

In the first expression, all parcels > 50 hectares are chosen regardless of their zoning or value, and additional parcels are added if they are code = 42 and high in value. In the second expression all of the parcels must have a zoning of 42 and can be either high in value or large in size.

Searching for partial matches in text

When selecting strings, one must allow for extraneous spaces and case mismatches. For example, the string "Maple St " with a space following St would not match the string "Maple St" or "Maple St." or "Maple st". Often databases have little inconsistencies like this one because it is very difficult to standardize the formats perfectly, and, of course, people make typing mistakes. Most GIS systems use an operator such as LIKE or CONTAINS that allows searches for a particular substring within another string.

An operator such as LIKE is often used instead of EQUALS (=) when selecting on string fields. LIKE searches for the specified set of characters within the field and returns any record that contains those characters. Most databases have case-sensitive searches. When using LIKE, a wild card character (% for shapefiles; * for geodatabases) is used to stand for other letters that may appear. Consider selecting cities based on their NAME field. The two queries

"NAME" LIKE 'New %' *(shapefile)*

"NAME" LIKE 'New *' *(geodatabase)*

would select all cities beginning with the word New, such as New London and New Haven, but would not include Newcastle because there is no space in the latter. The expression

"NAME" LIKE '%new%'

would return only lowercase 'new' in the middle of a word, such as Kennewick and Pinewood.

Spatial queries

Spatial queries are a powerful tool unique to GIS because they select based on spatial relationships, such as finding wells within five miles of a river or finding parcels inside a floodplain. A spatial selection uses two layers and one spatial condition. The features of the layer being selected are compared spatially with the features of the second layer to see which ones meet the criteria, and features meeting the criteria are selected.

Because feature classes vary in precision and geometric accuracy, it can often happen that two objects, which coincide in the real world (e.g., a stream gage and a stream), will not match exactly when their *x-y* coordinates in the GIS are compared. It is helpful to be able to specify a **buffer** when evaluating a spatial condition so that the features need to be close but not exactly matching. If the user specified a buffer of 10 meters, then the stream gage would be considered to intersect the stream if they were located within 10 meters of each other. Specifying a suitable buffer requires some knowledge of the geometric accuracy of the two feature classes being compared.

Spatial queries test for basic spatial relationships including **containment**, **intersection**, and **proximity**. Consider two feature classes, A and B. The spatial operators shown in Figure 5.2 test for the relationship of each feature in A to each feature in B and return the features from A that meet the criteria with respect to B.

Containment tests for whether one feature includes another wholly or partly. Six of the operators in Figure 5.2 test containment. The most stringent is that the feature from A equals the feature from B exactly (i.e., they have precisely the same geometry). This corresponds to the operator *are identical to*. The test A *completely contains* B returns all features in A that fully surround the features in B, such as a county that completely contains a lake. A *contains* B if only part of the lake lies inside the county. The *are completely within* operator is the inverse of *completely contain*, returning the features of A that lie completely inside B, such as the parcels that lie completely inside the flood plain. The *are contained by* operator is less stringent and would include the parcels that lie wholly or partly inside the floodplain. The *have their center in* operator tests whether the center of a feature in A lies inside a feature in B.

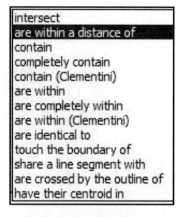

Fig. 5.2. Spatial criteria operators

The Clementini operators apply to the *contain* and *are within* conditions. These operators treat the boundary of a polygon as distinct from its inside and outside, with the result that it would not select features if only the boundary matches between the two layers. Otherwise, they return the same result as the *are within* and *contain* operators.

Intersection is the most generic operator and returns any feature in A that touches, crosses, or overlaps any part of feature in B. The *are crossed by the boundary of* operator is a special case of intersection that returns features in A that are crossed only by the boundary of features in B, such as the parcels that are crossed by the city limits.

Proximity tests how close features in A are to features in B. The most generic test, *are within a distance of*, selects features in A that are within a certain distance of B, such as returning all parcels within two miles of a school. Adjacency is a special case of proximity where the distance goes to zero and the boundaries of the features actually touch each other. These conditions are covered by the *share a line segment with* and *touch the boundary of* operators.

TIP: A complete description of the different spatial operators with graphic examples can be found in the ArcGIS Help by searching in the Index for *Select By Location, described.*

Figure 5.3 shows the results of applying some of the operators to two different feature classes. Sometimes a problem calls for testing spatial relationships using a *subset* of features rather than the entire layer. For example, a user might want to select cities within 50 miles of Interstate 80, rather than cities within 50 miles of any interstate. This procedure begins with an attribute query to isolate the desired interstate, I-80 and then a spatial query to select the cities. In Figure 5.3, the user first selected the state of Texas and then selected the rivers that intersected Texas.

A spatial query can also test spatial relationships within a single layer, such as finding all the parcels that are adjacent to city park parcels or finding all the restaurants that are within two miles of a favorite restaurant.

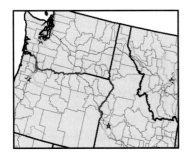

Select counties that *contain* state capitals.

Select counties *within 200 miles of* Denver.

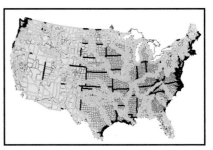

Select counties that *intersect* rivers.

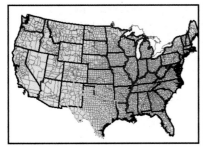

Select rivers that *intersect* Texas.

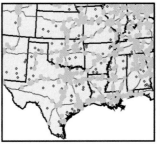

Select cities that *are within 20 miles* of an interstate.

Select rivers that *touch the boundary of* states.

Fig. 5.3. Examples of selecting a layer based on its spatial relationship to another layer

About ArcGIS

ArcMap offers three ways to select features. **Interactive Selection** uses a pointer to select features on the screen. Attribute queries are performed using **Select By Attributes**. Spatial queries are executed using **Select By Location**. After a query, the selected features are highlighted in the map and in the table.

Selection states

ArcGIS follows an important rule when processing features or records in any of its functions or tools: **After a query is performed and a subset of features or records is selected, any subsequent operation on that layer or table honors the selected set.** You have already seen this rule in action. When you selected the Democratic representatives in Chapter 4 and calculated a "D" in the PARTY field, the calculation was performed only on the selected records, the Democrats. If you were to select the New England states and calculate statistics, only the New England states are included in the statistics. If you selected the parcels in a floodplain and then executed the Buffer tool on the parcels layer, only the parcels in the floodplain would be buffered. Once a query has been executed on a layer, the layer acts as if the unselected features do not exist.

A layer or a table can exist in one of three selection states determined by two factors, whether a query has been applied and how many features are selected as a result of that query. The selection state impacts the behavior of geoprocessing and other commands applied to the layer.

By default, a layer has no query applied; that is, it has "no selection." In this state, all of the features are available for processing by a command. If a Statistics command is requested in this state, all of the features would be included in the calculations. In Figure 5.4, the layers shown in normal-face type are in the default state with no selection applied, as they would be shown in the Selection tab in the Table of Contents. Observe that none of these layers is highlighted in the map and that in tables these layers would appear as having zero records selected.

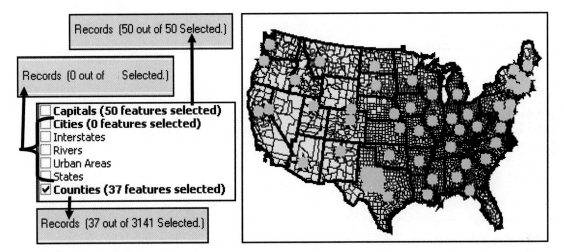

Fig. 5.4. Selection states, as displayed in the Selection tab, the map, and the layer tables

Executing a query on a layer pushes it into a new state. The layer has a selection applied, and the number of features or records selected may vary from zero to the full set of features. The layers in boldface in Figure 5.4 have selections. An interactive query on Counties yielded a block of counties selected in Texas, with a total of 37 features selected. A Select All command applied to the Capitals caused all 50 of its features to be selected. A Select By Attribute query for Cities with elevations greater than 10,000 feet resulted in zero features selected, since none of the cities met that criteria. These states are reflected in the map by the selection highlights and in the table by the number of records selected.

For the three layers with selections, the result of a command, such as Statistics, reflects the number of features selected. The statistics for Counties would be based on 37 records, the statistics for Capitals would reflect all 50 records, and the statistics for Cities would be based on no records. Any other function applied in ArcMap or in ArcToolbox on these layers would also restrict the operation of the function to the selected features or records.

It is important to be aware that the terms "no selection" and "no records selected" have different meanings. In the first case, no query has been applied, and all records would be used. In the second case, a query has been applied but no records have been selected, so no records would be used. Also note that the state of no selection and the state of all records selected would have the same result in a processing step—all of the records would be used, even though the number of records shown as selected in the table is zero in one case and all in another.

Interactive selection

In an interactive selection, you use the mouse to click on one or more objects in the map or table until you have all the desired records. For example, to get all the states beginning with A, you could sort the table by state name and then find and click on each record containing states starting with A. Or, if you wanted the states that have a Pacific Ocean coastline, you could look on the map, hold down the Shift key, and choose California, Oregon, Washington, and Alaska. You can also create a graphic, such as a box, a shape, or a line, on the screen and use the graphic to select the objects it touches (Fig. 5.5a). The Selectable Layers option can turn off selection for one or more layers. In Figure 5.5b, the same rectangle has been used as before, but the user has set the selectable layers to states only.

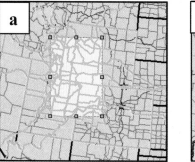 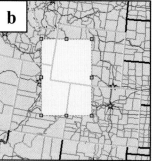

Fig. 5.5. (a) Selecting states and counties using a rectangle. (b) Selecting with only states selectable.

TIP: The Selectable Layers option only applies to interactive selection with the mouse. It does not affect selecting by attributes or selecting by location as described in later sections because those options always occur on the specified layer(s).

Selecting by attributes

SQL expressions in ArcMap are constructed using the fields and the buttons accessible in the Select By Attributes window (Fig. 5.6). The window menu automatically enters the first part of the expression (SELECT *FROM *layer* WHERE) for you, so in subsequent examples we will drop this part and stick to the part needing to be entered. Notice a few additional options that can be used in this window.

The *Query Wizard* button provides a series of step-by-step menus and explanations to help beginners set up a query. Experienced users will usually find it faster to use the Query window directly.

The *Get Unique Values* button adds the values from the table to the window where they may be selected by the user instead of being typed in.

Fig. 5.6. The Select By Attributes window for creating queries

The *Verify* button checks an SQL expression before executing it.

The *Save* and *Load* buttons store queries for use again later or to record what criteria were used, especially for long complex queries with many conditions. Queries are saved in text files with the extension .exp and can be edited with a simple text editor. If saving many queries, it is helpful to place them in a special folder and give them expressive names.

The *Clear* button erases old or invalid queries and starts again fresh.

TIP: Expressions for shapefiles and coverages enclose the field names in quotes. Expressions for feature classes in a geodatabase enclose the field names in brackets.

Selecting by location

The Select By Location window provides the means to set up a spatial query. It reads something like a sentence from top to bottom, such as "I want to select features from the layer Cities that are contained by the features in the layer Rivers," using the Rivers features that are already selected and applying a buffer of 25 miles (Fig. 5.7). You may find it helpful to read through queries to verify that the selection is accurate.

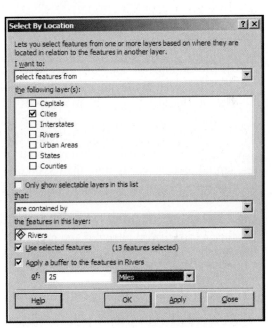

Fig. 5.7. The Select By Location window

Choosing the selection method

The selection method applies to all three types of selections (interactive, selecting by attribute, and selecting by location). These methods give greater flexibility in defining selection sets and enable selection using multiple steps. When performed sequentially, complex expressions involving many criteria are often less confusing than when executed by entering a long expression full of ANDs and ORs and trying to figure out where to put the parentheses.

ArcMap has four methods for selecting features (Fig. 5.8). The methods may be specified in two places, from the main Selection menu when doing interactive selection or within the Select By Attributes or Select By Location windows.

Create New Selection always selects the specified records regardless of what else is already selected. If other records are selected, they will be replaced with the new selected records. For example, if you have selected green jelly beans and then choose to select red jelly beans, then you will only have red jelly beans selected. This method is the default.

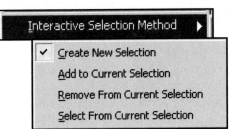

Fig. 5.8. Selection methods

Add to Current Selection yields the records that fit the query, in addition to any that are currently selected. Thus, if you have green jelly beans already selected and then select the red ones, then you will have both red and green jelly beans selected. Using this method is equivalent to doing a single selection step with a double expression using the OR operator, as when selecting beans where "color" = red OR "color" = green.

Remove From Current Selection takes away the records that meet the selection criteria. If you already have red, green, and yellow jelly beans selected and you "choose" green jelly beans, then you will have red and yellow jelly beans left selected. In this option, what you choose gets deleted from your selected set, so you must think slightly backward on this one.

Select From Current Selection chooses the records that fit the criteria, provided that they are already selected. If you have red, green, and yellow jelly beans selected and choose to select the green ones, you will have the green ones selected. If you choose to select orange ones, then you will end up with *nothing* selected because orange jelly beans were not in the original selection. Performing several selections in a row with this method is similar to performing a series of AND expressions.

A real-life example of applying these selection methods might proceed as follows: Imagine selecting all residential parcels with a VALUE greater than $100,000 that intersect the 100-year floodplain. Because this problem involves selecting both by attribute *and* by location, it occurs in two steps. First, use Select By Location to choose all parcels in the floodplain, using Create New Selection as the method. Then use Select By Attributes to choose LU-CODE = Residential AND VALUE > 100000. However, using Create New Selection as the method would yield ALL residential parcels worth more than $100,000, not just those in the floodplain. You need to specify Select From Current Selection in order to obtain the desired records.

One other selection option, *Switch Selection,* replaces the selected set with the unselected records. In other words, if you have red and green jelly beans available and the red ones are selected, and if you do a switch selection, then you will have green ones selected.

Creating layers from selected features

Often after performing a query it is useful to preserve the selected features as a separate layer in the Table of Contents. In Figure 5.9, the user selected the southeastern states and then chose to create a layer from the selected features, resulting in the new **states selection** layer shown above the original **states** layer.

Creating a selection layer has several advantages. The layer can be given its own symbols and can be displayed separately from the original layer. If the selected set is the focus of a complicated analysis, then the layer preserves the selected features for future reference and eliminates the risk of accidentally clearing the selection and having to perform it over again. Selection layers can be used to input the same set of features to several different tools or commands. They are helpful for viewing and recording intermediate results in a long and complex series of queries. The layer can also be saved as a layer file for use in other map documents.

Fig. 5.9. The South Atlantic states were selected and used to create a layer.

Summary

➢ Queries extract a subset of records or features from a data set based on a set of one or more conditions.

➢ Attribute queries extract features based on conditions from fields in the attribute table, such as cities with population greater than 1 million. They are written in SQL and use operators such as =, >, <, AND, and OR. Partial matches to text strings are evaluated with the LIKE operator.

➢ Spatial queries extract features based on criteria of how two layers are spatially related, using tests to evaluate containment, intersection, and proximity/adjacency.

➢ ArcGIS always uses only the selected features in a layer when a function, command, or tool is applied to the layer. If no query has been applied, all features are used. If a query has been applied, only the selected features are used. If a layer has a selection applied that resulted in no (zero) selected features, then no features will be used.

➢ An interactive query uses the Select Features tool and lets the user pick certain features by finding them on the screen and clicking them. Attribute queries are implemented using Select By Attributes, and spatial queries use Select By Location.

➢ The Selectable Layers option restricts the operation of the Select Features tool to the specified layers. By default all layers are selectable.

➢ Four selection methods add greater flexibility to queries. These methods include creating a new selection, adding to the current selection, removing from the current selection, and selecting from the current selection.

➢ Creating a new layer from a set of selected features allows the selection to be stored, displayed, and passed on to tools or commands.

Chapter Review Questions

1. What is a query?

2. Write a valid SQL expression to select cities between 1000 and 10,000 people using a field called POP2000.

3. Write a valid SQL expression to select all counties whose names begin with the letter Q.

4. Let T be a table containing all students attending a community college in New York. Let A be the subset of students living in New Jersey. Let B be the students with a GPA greater than 3.0. The query A AND B yields 200 records. The query A OR B yields 1100 records. The query A NOT B yields 400 records. Construct a Venn diagram for the sets, labeling each section with the number of students. How many students live in New Jersey? How many students have a GPA greater than 3.0?

5. From the information in Question 4, can you determine the number of students attending the community college? If yes, state how many. If not, explain why.

6. What does it mean to set the selectable layers? What is the default setting?

7. Imagine that you have some trail mix composed of peanuts, raisins, almonds, cashews, dried cranberries, and chocolate candies colored red, green, yellow, and orange. Imagine that you apply the following set of "queries" to the trail mix:

    Create new selection all candies
    Add to selection cashews
    Remove from selection red and green candies
    Select from selection all nuts and candies

    What do you have selected now? _____

8. For each of the following queries, state whether it is correct syntax or incorrect. If incorrect, explain why.

    a. [ZONE] = 'COM' AND [ZONE] = 'RES'

    b. [COVTYPE] = 'SPRUCE' AND [CROWNCOV] > 50

    c. [POP2000] > 2000 OR [POP2000] < 9000

    d. [INCOME] < 100000 AND [INCOME] > 50000

9. Explain the difference between having no selection and having no selected features. Which one would result in all statistics being zero if the Statistics command were chosen for the table?

10. List some advantages of creating a new layer from the selected features?

Mastering the Skills

Teaching Tutorial

The following examples provide step-by-step instructions for doing basic tasks and solving basic problems in ArcGIS. The steps you need to do are highlighted with an arrow ➔; follow them carefully. Click on the video number in the VideoIndex to view a demonstration of the steps.

➔ Start ArcMap and open the document ex_5.mxd in the MapDocuments folder.

➔ Use Save As to rename the document and remember to save frequently as you work.

Before we begin, we will add the Clear Selected Features button to the general toolbar. Although you can access this function through the Selection menu, it is faster if you place the button on the toolbar (Fig. 5.10).

Fig. 5.10

1➔ Choose Tools > Customize from the main menu bar.

1➔ Move the window if necessary in order to see the Tools toolbar (the one with the Zoom/Pan tools on it).

1➔ Click the Commands tab in the Customize window.

1➔ Scroll down the list of Categories on the left to find Selection, and click on it to highlight it.

1➔ Locate the Clear Selected Features command on the right, click on it, and drag it to the general toolbar, just below the Select Features tool.

1➔ To have this tool appear in all map documents by default, change the Save In file to Normal.mxt (a template containing your map document preferences).

1➔ Close the Customize box.

Add to toolbar

Using interactive selection

Now, let's experiment with selecting features with the mouse.

2➔ Choose Bookmarks > Texas from the main menu bar. (If using version 9.2, look in View > Bookmarks.)

2➔ Click on the Select Features tool, and then click on the star that represents Austin, the capital of Texas.

Your results may differ slightly, but you should now have several features selected, including Texas, one or more counties near the capital, and an interstate highway (Fig. 5.11). Any layer near the clicked point was added to the selected set.

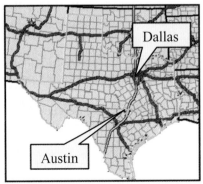

2➔ Try clicking several other spots to see what gets selected.

2➔ Click and drag a box around the Dallas area and notice that anything in the box or anything that goes through the box is selected.

Fig. 5.11. Interactive selection with all layers selectable

Setting the Selectable Layers offers control over what gets selected by a click.

> 3➜ Choose Selection > Set Selectable Layers from the main menu bar.

1. Which of the layers are currently selectable? _____

> 3➜ Click the button to Clear All, and then check the Interstates box. Click Close.
>
> 3➜ Draw a box around Dallas again. This time, only the interstates are selected.

The Table of Contents in ArcMap contains a tab to facilitate the viewing and the management of layer queries. The Selection tab offers a faster way to change the selectable layers.

> 3➜ Click the Selection tab at the bottom of the Table of Contents.
>
> 3➜ Uncheck the Interstates layer, and check the Counties layer to make it selectable.
>
> 3➜ Draw the box again. Now only the counties are selected.

Notice that when a new selection is made, the selection on the other layers disappears. The Selection tab keeps track of how many features are currently selected for each layer. Next we will experiment with other ways to interactively select multiple features from one layer.

> 4➜ Choose Bookmarks > Michigan from the main menu.
>
> 4➜ Click the Display tab and turn off Interstates for now.
>
> 4➜ Click one county in Michigan with the Select Features tool.
>
> 4➜ Hold down the Shift key and click another county. It is added to the selected set.
>
> 4➜ Add three more counties using the Shift-click method.
>
> 4➜ Now hold down Shift and click one of the counties that are *already selected*. It will be *removed* from the selected set.

Imagine that you want to efficiently select all the counties in the northern portion of Michigan that lie between Lake Superior and Lake Michigan.

> 5➜ Click and drag a box that goes through as many counties as possible in northern Michigan (Fig. 5.12). Do it several times, if necessary, to get a good result. You will not be able to get all the counties, however.

The Interactive Selection Method makes it easier to add to an already selected group.

> 6➜ Choose Selection > Interactive Selection Method > Add to Current Selection.

Fig. 5.12. Counties of northern Michigan

> 6➜ Click on each of the remaining northern Michigan counties. Each one clicked becomes selected with the previous ones, without needing to hold down the Shift key.
>
> 6➜ Continue until all of the northern Michigan counties are selected.

7➔ Now add a few southern Michigan counties by drawing a box around them.

7➔ Choose Selection > Interactive Selection Method > Remove from Current Selection.

7➔ Click on or draw a box around the southern Michigan counties to remove them from the selected set.

At this point, you should have all the northern counties selected and none of the southern ones (Fig. 5.12). Recall that the Statistics command shows information about the selected features.

8➔ Right-click the Counties layer and choose Open Attribute Table.

8➔ Right-click the POP2000 field and choose Statistics.

2. How many counties are selected and what is the total number of people in them?

_____

8➔ Close the Statistics box and the attribute table.

Now let's get back to testing the different selection methods.

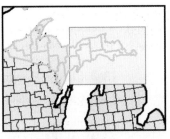

9➔ Choose Selection > Interactive Selection Method > Select from Current Selection.

9➔ Use the Select Features tool to draw a box that encloses the northern peninsula, covering the area shown in Figure 5.13. Be sure to include some of the southern counties in the box.

Fig. 5.13. The box

Notice how the northern counties touching the box were selected, but the southern counties were not because they were not already part of the selected set. Now, let's use a graphic.

10➔ Set the Interactive Selection method to Create New Selection.

10➔ Click the tool to Clear Selected features on the Tools toolbar.

10➔ Click the Down arrow next to the Shape tool menu on the Drawing toolbar and choose the New Circle tool (Fig. 5.14).

10➔ Click on the state capital, Lansing, and drag to make a circle (Fig. 5.14). Watch the radius measurement in the lower left corner of the window and try to get it close to 100,000 meters. The completed circle will be selected, as shown by the blue-handled box around it (Fig. 5.14).

10➔ Choose Selection > Select by Graphics. Any counties touching the circle are selected (Fig. 5.14).

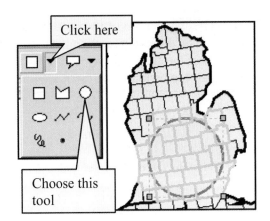

Fig. 5.14. Selection using a graphic

11➔ Try using the rectangle and line tools to make selections. Try to draw a polygon that selects all of the northern counties in one step.

When finished selecting, delete the graphic(s).

12➔ Click the Select Elements tool on the Drawing toolbar.

12➔ Hold down the Shift key and click each graphic to select it.

12➔ Click the Delete key to get rid of them.

The Selection tab has a context menu that performs common query functions. Use this context menu to clear the current Counties selection.

12➔ Click the Selection tab in the Table of Contents.

12➔ Right-click the Counties layer and choose Clear Selected Features.

Selecting by attributes

Chapter 4 demonstrated one way to select by attributes from the Options button in the table. Another method does not require opening the table, and it can be used on any layer.

13➔ Choose Bookmarks > USA from the main menu.

13➔ Choose Selection > Select By Attributes from the main menu.

The Select By Attributes window is identical to the one accessed by the table Options menu, except that you must set the layer to select in the first drop-down box.

14➔ Use the drop-down box to choose Counties as the layer to select.

14➔ Confirm that the selection method is Create New Selection.

14➔ Enter the expression [POP2000] > 1000000 (1 million) by double-clicking POP2000 until it is entered in the expression box, clicking the > button, and typing in **1000000**.

TIP: SQL does not gracefully accept extra spaces, parentheses, and other minor syntax errors, which can be difficult to spot and correct. Most users will find it faster to use the buttons and the windows of the query window rather than trying to type the expressions. If you have trouble making an expression work, then clear the expression box and reenter the expression.

14➔ Click Apply and move the Select By Attributes window to show the map and the window at the same time.

3. How many counties in the Unites States had more than 1 million people in 2000? _____

One advantage of using this window is being able to query another layer from the same window.

15➔ Click the Select By Attributes box again to activate it.

15➔ In the Select By Attributes box, change the layer being selected to States. Keep the Create New Selection method.

15➜ Change the expression to [POP2000] > 5000000 (5 million). Click Apply.

Notice that both counties and states are now selected, even though Create New Selection was the selection method. This happened because States and Counties are separate layers and are treated separately in this query box. Also notice that the states were selected even though they are not currently a selectable layer. The selectable layer applies only to interactive selection, and it is ignored by the Select By Attributes and Select By Location functions.

Now let's try a selection with double criteria, all counties that lost population between 1990 and 2000 and have more males than females.

16➜ Clear all the selected features using the Clear Selected Features button.

16➜ Click the Select By Attributes box to activate it, if necessary.

16➜ Set the layer to Counties.

16➜ Enter the expression [POP2000] < [POP1990] AND [MALES] > [FEMALES] in the query box. Click Apply.

4. How many counties with more males than females lost population between 1990 and 2000? _____ Where are they mainly located? _____

Now from these, select those counties that have fewer than 10,000 people. Be careful to ensure that the new set of counties is pulled only from the currently selected ones.

17➜ In the Select By Attributes box, change the selection method to *Select from current selection*.

17➜ Click the Clear button to erase the old expression.

17➜ Enter the expression [POP2000] < 10000. Click Apply.

Watch as a few of the counties disappear from the selected set.

5. How many counties remain selected? _____

17➜ Close the Select By Attributes window.

17➜ Clear the selected features.

TIP: The Clear Selected Features tool on the toolbar and in the Selection menu clears the selection from ALL layers. To clear the selection for a single layer, right-click the layer in the Selection tab and use the Clear Selected Features command in the context menu.

Selecting by location

Next we will try several examples of selecting by location. First we will select all the counties that are adjacent to rivers.

18➜ Click the Display tab and turn on the Rivers layer.

18➜ Choose Selection > Select By Location, and adjust the window to show both the Selection window and the map.

18➔ In the first drop-down box, choose the selection method *I want to select features from* (Fig. 5.15).

18➔ Choose Counties as the layer to select from.

18➔ Choose the selection criterion *intersect*.

18➔ Choose Rivers as the layer the selection is based upon.

18➔ Click Apply and view the map.

6. What percentage of counties in the United States are intersected by rivers?_____

Read the window just as though it were a sentence in order to check the logic. In this example, the sentence says, "I want to select features from Counties that intersect the features in Rivers." Next, let's select the counties that contain state capitals.

19➔ Change the window to read, "I want to select features from Counties that *completely contain* the features in Capitals."

19➔ Click Apply.

Fig. 5.15. The Select By Location window

7. Which of these counties containing capitals has the smallest population? _____ What is the capital and which state is it in? _____

Now let's do an example with points and lines by selecting the cities close to interstate highways.

20➔ The cities are not visible because the scale range is set. Right-click Cities and choose Visible Scale Range > Clear Scale Range from the context menu.

20➔ Turn on the Interstates layer to see the results more clearly.

20➔ Change the Select By Location window to read, "I want to select Cities that are within a distance of Interstates."

TIP: The layer from the previous query, Counties, has remained checked in this list. Uncheck it before doing the query, or both counties and cities will be selected. The previous layers always remain checked in this window, so get in the habit of checking the list before each query.

20➔ Check the box to apply a buffer to Interstates and set the distance to 20 miles.

20➔ Click Apply and view the results, closing the selection window.

8. What percentage of U.S. cities is within 20 miles of an interstate highway? _____

(To count all records, click the "Go to end record" button on the right.)

In the previous example, we compared *all* the features in one layer to *all* the features in the other. You can also select features based on a subset of the comparison features. For example, let's select all the rivers that pass through Texas. We do this by selecting Texas first, using either interactive selection or selection by attributes, and then performing a Select By Location.

21➜ Clear all selected features.

21➜ Turn off the Cities, Capitals, and Interstates.

21➜ Set the Selectable Layers to States only.

21➜ Click the Select Features tool and select the state of Texas.

22➜ Open the Select By Location window.

22➜ Change the window sentence to read, "I want to select features from Rivers that intersect the features in States."

22➜ Check the box, if necessary, to use the selected features in States.

22➜ Click OK and view the results.

9. Which rivers intersect Texas? _____

Spatial queries may also be performed within a single layer, using it both as the target and as the criterion layer. The features comprising the criteria must be selected first. For example, select all cities within 400 miles of Kansas City, MO.

23➜ Clear all selected features.

23➜ Turn off the Rivers and turn on the Cities.

23➜ Open the Select By Attributes window and make sure the selection method is set to Create a new selection.

23➜ Make sure that the box to show only selectable layers is unchecked.

23➜ Select Cities as the layer to select.

23➜ Enter the expression to select Kansas City. Remember that text queries must have the text enclosed in single quotes.

23➜ Click Apply and close the Select By Attributes window.

24➜ Open the Select By Location window.

24➜ Change the window sentence to read, "I want to select features from Cities that are within a distance of 400 miles from Cities."

Notice that Cities is used twice in the window and that the Use Selected Features box is automatically checked when the selection and criterion layers are the same.

24➜ Click Apply to make the selection, and close the selection window.

24➜ View the map.

Attribute and location queries gain power when combined sequentially. Let's select all the rivers that intersect states in the Mountain subregion. We do this by first selecting the mountain states using an attribute query and then selecting the rivers that intersect these states.

25➔ Clear all selected features. Turn off Cities and turn on Rivers.

25➔ Choose Selection > Select By Attributes.

25➔ Set the layers to States, and enter the expression [SUB_REGION] = 'Mtn'. Use the Get Unique Values button to find the 'Mtn' entry.

25➔ Click Apply and close the Attribute Query window.

25➔ Open the Select By Location window, if necessary.

25➔ Set the parameters to select features from Rivers that intersect the selected features of States.

25➔ Click Apply and view the results (Fig. 5.16).

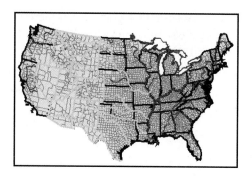

Fig. 5.16. Rivers that cross the Mountain subregion

Notice that many of the rivers extend far to the east to join the Mississippi. Because they do pass through the mountain states, though, they are included. You could try selecting only the rivers that are completely within the mountain states instead.

26➔ Change the selection criterion to *are completely within* and click Apply.

10. Now which rivers are selected? _____

Now hardly any rivers are selected. Ideally, you would like to select the *portions* of rivers that fall inside the mountain states. However, Select By Location can only work on entire features. If a river feature crosses the state boundary without splitting into two features, then the entire river is selected, both inside and outside the state. Unless the data creator specifically forced the features to break at state boundaries, a spatial query will find portions outside the state. This weakness of spatial queries can be troublesome. Chapter 7 presents techniques for splitting features when they cross other features.

Combining attribute and location queries can create sophisticated selections. Imagine that a meat packing company is considering adding the capacity to slaughter bison and wants to survey potential customers between the packing plants in Pierre and Denver to see if the demand for bison meat justifies the added expense of handling these more difficult animals. The company decides to target the survey to counties that are within 300 miles of both Denver and Pierre and have more than 10,000 people.

27➔ Clear all the selected features.

27➔ Turn on the Capitals layer. To make it easier to find the capitals, right-click the Capitals layer and choose Properties. Click the Display tab and turn on Map Tips.

27➔ Use the Selection tab to make Capitals the only selectable layer. Click on the Select Features tool and click on Denver in Colorado to select it.

27➔ Open the Select By Location window, if necessary, and set it to select all Counties within 300 miles of the selected features of Capitals (Denver). Click Apply.

28➔ Right-click the Capitals layer in the Selection tab and choose Clear Selected Features from the context menu. By clearing from the context menu, only the features in Capitals are cleared; the counties stay selected.

28➔ Choose the Select Features tool, hold down the Shift key, and click on Pierre in South Dakota to select it. If you forget to Shift, the county selection will be cleared.

28➔ In the Select By Location window, change the selection method to *Select from the currently selected features*. Make sure the Use Selected Features box is checked and click Apply.

28➔ Close the Select By Location window.

At this point a patch of counties between Denver and Pierre should show as selected (Fig. 5.17). Now we select from this set the counties with more than 10,000 people.

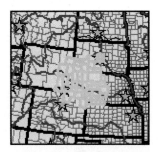

Fig. 5.17. Counties within 300 miles of Pierre and Denver

29➔ Open the Select By Attributes window.

29➔ Set the layer to select to Counties.

29➔ Set the selection method to *Select from current selection*.

29➔ Enter the expression [POP2000] >10000 and click Apply.

29➔ Close the Select By Attributes window.

29➔ Right-click the Capitals layer in the Selection tab and clear the selected features. The map should now look similar to Figure 5.18.

29➔ Right-click the Counties layer in the Selection tab and choose Create Layer from Selected Features.

29➔ Switch to the Display tab and name the new layer **Packing Survey**. Change the symbol, if necessary, to show up well against the background.

29➔ Clear all selected features.

Fig. 5.18. Target counties for the packing survey

11. How many counties will be in the survey? _____ What is the total number of people who live in these counties? _____

Exporting data

One common operation performed after selecting data involves using the **Export** function to create a new feature class from the selected set. Imagine that you want to create series of maps that show only the state of Georgia and want to have the state by itself in a shapefile for repeated use.

30➔ Use Select By Attributes to select the counties in the state of Georgia. (Make sure you set the selection method back to Create a new selection.)

31➔ Right-click the Counties layer and choose Data > Export Data.

31➔ Choose to export the selected features (Fig. 5.19).

31➔ Fill the button to use the same coordinate system as the source data.

31➔ Click the Browse button and save a shapefile named **georgiacnty** in the mgisdata\Usa folder. Click OK.

31➔ Click Yes to add the new shapefile to the map.

31➔ Clear all the selected features.

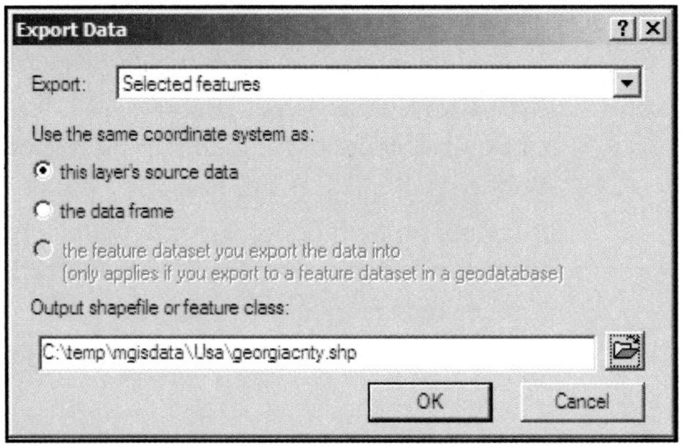

Fig. 5.19. Exporting a map layer

TIP: You can also use Export to project from one coordinate system to another by setting the data frame coordinate system to the desired projection and choosing to use the data frame coordinate system.

This is the end of the tutorial.

➔ Exit ArcMap and save your changes.

More skills

Consult the Skills Reference section of this chapter to learn to do the following:

➢ Changing the selection color and other selection options

Exercises

Use the files in ex_5.mxd to answer the following questions.

1. How many states have counties named for Thomas Jefferson (i.e., how many Jefferson Counties are there)? Which state has the Jefferson County with the most people in the year 2000?

2. How many counties in the United States have more men than women? What percentage of the counties do they represent?

3. How many cities in the United States have more Hispanics than African-Americans, median rents greater than $500, and a population between 250,000 and 500,000? List them (there are fewer than 10). (**Hint:** Do your queries in more than one step.)

4. What percentage of counties in the United States has a river in or next to them? What percentage of the U.S. population lived in these counties in 2000?

5. How many counties have more than 800,000 people and also contain a state capital? List the states these counties are in.

6. How many cities are within 50 miles of a volcano? (Add the volcano feature class from the Usa\usdata geodatabase). What is the total number of people living in those cities?

7. How many other volcanoes are there within 300 miles of Crater Lake, a volcano in Oregon? How many of these volcanoes are also within 50 miles of an interstate?

8. How many cities in the West South Central subregion of the United States are less than 200 miles from Oklahoma City? **Capture** a map showing your selected cities.

9. How many urban areas are *more* than 50 miles from an interstate highway? Which state has the most such areas and how many does it have?

10. Imagine that you are looking for a nice place to live. You like being in a small town in a fairly rural area but within easy driving distance of an urban area for fun. You would prefer a city with between 20,000 and 40,000 people, no more than 50 miles from an urban area, in a county with a population density of fewer than 25 persons per square mile. List the names and states of the cities in the United States that meet all these criteria. (There are fewer than 10.)

Challenge Problem

Congress has awarded FEMA 10 million dollars to help large cities prepare for earthquakes. The cities that qualify for the funding must have more than 500,000 people and be less than 50 miles from one or more earthquakes exceeding 6.0 in magnitude. The bill stipulates that the funding is to be divided among the qualified cities in proportion to their population. Create a feature class that contains only the qualified cities and has a table field listing the amount of funding to be given to each. **Capture** a view of the table showing the city name and the funding amount.

Skills Reference

Setting the selectable layers 214

Selecting features interactively 215

Selecting features using shapes 215

Clearing a selection 216

Using the Selection tab..216

Selecting by attributes 217

Selecting by location 218

Creating a layer from selected features 219

Changing the selection options..219

Setting the selectable layers

1. Choose Selection > Set Selectable Layers from the main menu bar.

2. Add or remove layers from the list of selectable ones by clicking inside the check box to the left of the layer (Fig. 5.20).

3. Click Select All to be able to select all the layers.

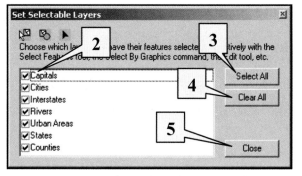

Fig. 5.20. Setting the selectable layers

4. Click Clear All to remove all the checks in order to easily check one or two layers without having to turn all the others off individually.

5. Click Close when finished. The selectable layers remain in effect until changed.

NOTE: You can also set the selectable layers in the Selection tab. See the Using the Selection Tab section.

TIP: The Selectable Layers setting does not apply to selections based on attributes or location.

Selecting features interactively

1. Set the Selectable Layers, if necessary.

2. If necessary, change the selection method by choosing Selection > Interactive Selection Method and picking one of the methods (Fig. 5.21).

3. Click the Select Features tool. Click on a feature in the map to select it. It will be highlighted in the current selection color (blue by default).

✓	Create New Selection ———	**2**
	Add to Current Selection	
	Remove From Current Selection	
	Select From Current Selection	

Fig. 5.21. Choosing a selection method

4. To add a feature to the selection, hold down the Shift key and click the feature to be added. To remove features from the selection, hold down the Shift key and click the feature to remove.

5. To add a group of features using a rectangle, click on one corner of the rectangle, drag the mouse to the desired size, and release the mouse button. All features touching the rectangle will be selected.

TIP: Hold down the Shift key while drawing the rectangle to add a group of features to the current selection.

Selecting features using shapes

1. Change the Selectable Layers and/or the Selection Method, if necessary.

2. Choose one of the Shape tools from the Drawing toolbar by clicking on the Down arrow and selecting one of the tools (Fig. 5.22).

3. Draw the shape on the map using mouse clicks. Double-click to finish the shape.

4. Choose Selection > Select by Graphics to make the selection.

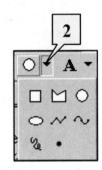

Fig. 5.22. Choosing a shape

Clearing a selection

There are several ways to clear a selection.

To clear features from ALL layers

Choose Selection > Clear Selected Features from the main menu bar.

or

Click the Clear Selected Features tool, if you have added it to a toolbar.

To clear features from a single layer only

Right-click the layer in the Display or Source tab and choose Selection > Clear Selected Features.

or

Right-click the layer in the Selection tab and choose Clear Selected Features.

TIP: Clearing a selection is a common task, and the Clear Selected Features button is a helpful addition to the general toolbar (with the Zoom/Pan tools). To add this button to the toolbar, right-click the gray menu area and choose Customize from the context menu. Click the Commands tab and then click the Selection section. Drag the Clear Selected Features icon to the toolbar (next to the Select Feature tool is a good spot).

Using the Selection tab

The Selection tab in the Table of Contents facilitates viewing and managing selections (Fig. 5.23).

1. Check or uncheck the layer boxes to make layers selectable or unselectable.

2. View the selection state of each layer; normal text indicates the default state of no selection applied and bold text indicates that a query is currently applied and that all features, a subset of features, or zero features are selected.

3. Right-click a layer to open the context menu and choose one of the functions.

See the discussion in the Concepts section regarding the different possible selection states.

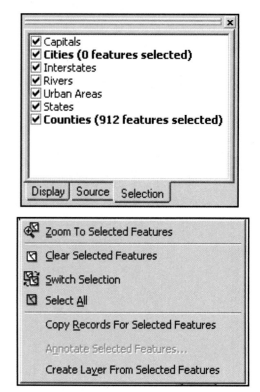

Fig. 5.23. The Selection tab and its context menu

Selecting by attributes

1. Choose Selection > Select By Attributes from the main menu bar.

2. For a detailed walk-through of the process, click the Query Wizard. Otherwise, fill out the boxes in the window (Fig. 5.24).

3. Choose the layer whose features are to be selected. Check the box to show only the selectable layers in the list, if desired.

4. Choose the selection method.

5. Enter the expression. Double-click a field from the list to enter it in the expression box. Click a condition (such as = or >) to enter it in the box. Type the condition value.

6. To enter a condition value from the table instead of typing it, click the Get Unique Values button to show the values, and click on the desired value to enter it in the box.

7. Click Apply to execute the expression and create the selection.

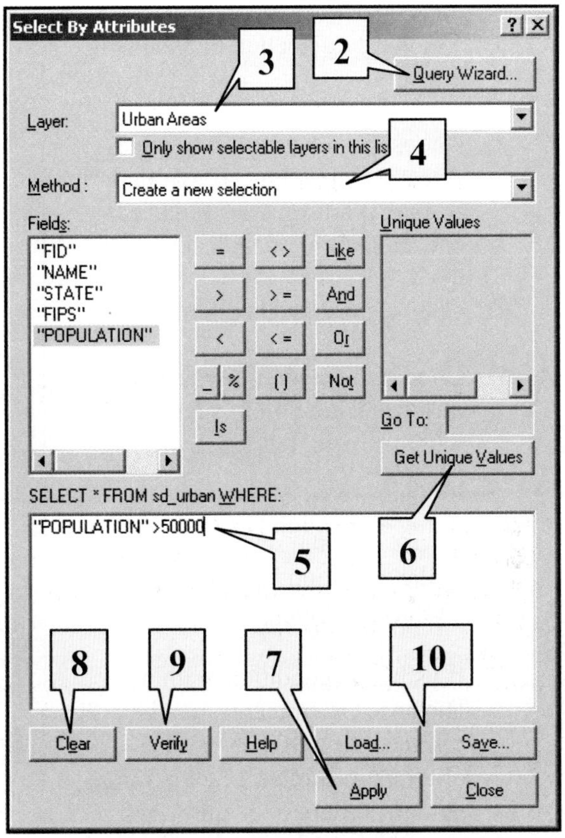

Fig. 5.24. The Select By Attributes window

TIP: Many users find that typing in or editing the expression by hand tends to generate errors. Use the boxes and the buttons provided to enter expressions for quickest results.

TIP: When using multiple conditions, the field name must appear in every condition. To find rents between 500 and 1,000 dollars, for example, enter [RENT] > 500 AND [RENT] < 1000. You cannot enter [RENT] > 500 AND < 1000.

Other options in this window

8. Click Clear to clear the expression and begin again.

9. Click Verify to check the expression syntax without actually executing it.

10. Use the Load and the Save buttons to save expressions to use again later.

11. Click Close to exit the window without executing the expression.

Selecting by location

1. Choose Selection > Select By Location from the main menu bar.

2. Choose the selection method (Fig. 5.25).

3. Choose the layer(s) from which to select features. Be careful—layers from previous queries remain checked until you uncheck them.

4. Choose the spatial condition.

5. Choose the layer to use as part of the spatial condition (such as within a certain distance of Rivers).

6. To use only the selected features of the condition layer, check the Use selected features box.

7. If using the condition Within Distance Of, or increasing the distance over which the other criteria are applied, enter a buffer amount and distance units, and check the Apply a buffer box, if necessary.

8. Click Apply to make the selection.

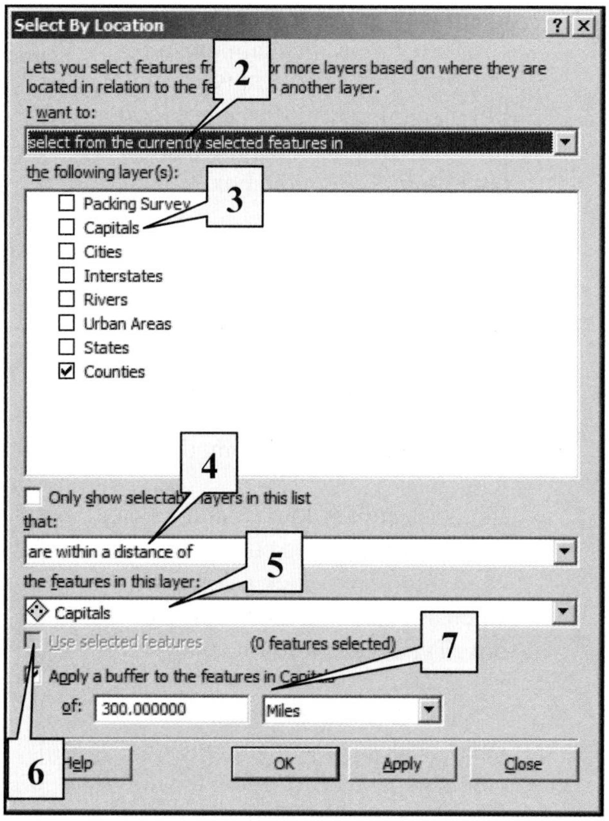

Fig. 5.25. The Select by Location window

TIP: To use the same layer as the selection layer and the condition layer (e.g., to select all restaurants within two miles of the restaurant you own), you must first use an attribute or an interactive query to select the feature of interest—your restaurant. Then do the Select By Location and specify the same layer for step 3 and step 5.

Creating a layer from selected features

1. Perform attribute or location queries to select the desired records.

2. Right-click the layer name and choose Selection > Create Layer from Selected Features from the context menu.

3. The new layer appears at the top of the Table of Contents.

4. Click on the new layer name twice to give it a new name.

Changing the selection options

1. Choose Selection > Options from the main menu bar.

2. The first box controls how shapes select features when using the Select Feature tool or the Select by Shape tool (Fig. 5.26). By default, any feature partially or completely in the shape is selected. You may also choose to select features completely within the shape or to select features within which the shape is completely contained.

3. The second box sets how close you need to click to a feature before it is selected. If you often have trouble selecting points, increase this distance.

4. This box changes the selection highlight color.

5. In the fifth box, you may alter the threshold that determines how many records will be modified by a Switch Selection command. If the number exceeds the threshold, a warning prompt will be issued. This option prevents you from mistakenly doing a switch that will take a very long time to complete.

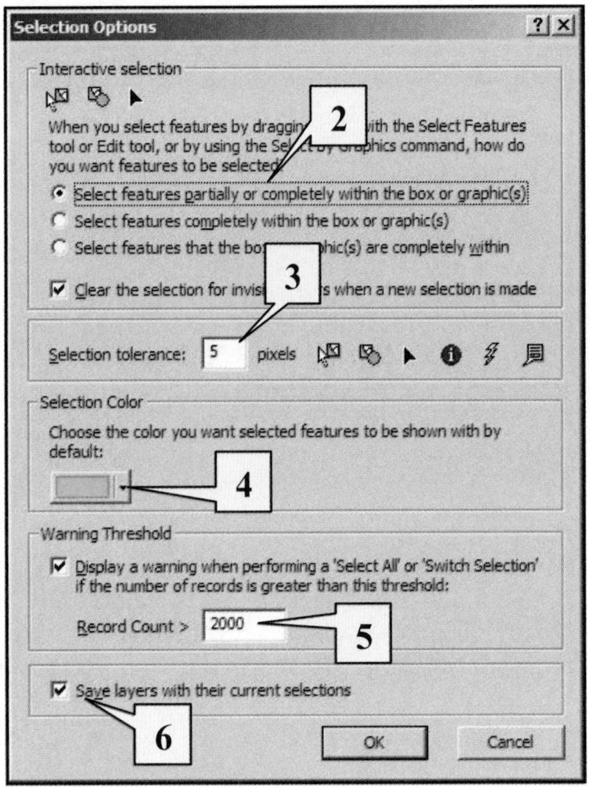

Fig. 5.26. Setting the selection options

6. Finally, you can choose whether to save layers with the current selections (the default) or to clear the selections of all layers when saving the map document.

Chapter 6. Spatial Joins

Objectives

➢ Learning the purpose and capabilities of spatial joins

➢ Correctly setting up spatial joins based on cardinality and feature type

➢ Learning to solve problems with spatial joins

Mastering the Concepts

GIS Concepts

What is a spatial join?

Chapter 4 presented attribute joins performed on tables—for example, joining information in the livestock.dbf table to a map of counties to create a graduated color map of cattle populations. This join was based on a common field, the FIPS code, and had a cardinality of one-to-one. The join resulted in combining the two tables as if they were one table. The destination table received the information from the source table.

A spatial join is similar to an attribute join, except that instead of using a common field to decide which rows in the table match, the *location* of the spatial feature is used. The spatial join uses either a containment criterion (one feature inside the other) or a proximity criterion (one feature close to another).

Like attribute joins, spatial joins designate a source feature class and a destination feature class. Unlike an attribute join, which appends the source attributes to the existing destination table, a spatial join creates a new feature class. It retains the features from the destination layer and appends the attribute information from the source layer (Fig. 6.1). The two original feature classes are unaffected. The destination feature class determines the type of features in the output feature class. If an airports feature class is joined to a cities feature class with cities as the destination, then the output feature class contains cities.

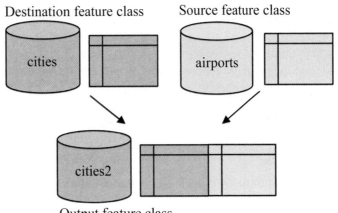

Fig. 6.1. A spatial join keeps the features from the destination layer and appends the attributes from the source layer.

Distance joins

A distance join uses a proximity criterion to link one feature and its attributes to another based on whether one feature is closest to another. Figure 6.2a shows the details of the join between airports and cities diagrammed in Figure 6.1. The source feature class, airports, has been joined to the destination feature class, cities. The output feature class contains cities. Each city has been given the attribute information from the airport that lies closest to it, and a new field has been added to record the distance. The attribute table contains two parts, the original data from cities and the joined data from airports. So Corvallis Municipal is the closest airport to Adair Village, and Eastern Oregon Regional is the closest airport to Adams.

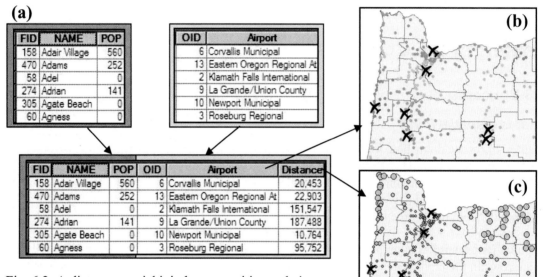

Fig. 6.2. A distance spatial join between cities and airports. (a) Joined tables. (b) Unique values map based on the airport name. (c) Graduated symbol map based on distance.

Two maps have been created from the new joined layer. Figure 6.2b is a unique values map based on the airport name, so each dot gets a color based on the closest airport. The colors indicate which cities are served by each airport, assuming that people will drive to the closest one. The second map is a graduated symbol map based on the distance field, with the larger circles indicating greater distance from the airport (Fig. 6.2c). The units of the distance field are always given in the stored coordinate system units. These data are in Oregon Statewide Lambert coordinate system, and the units are meters.

Inside joins

In an inside join, the records of the feature classes are joined based on whether one feature is inside another (wholly or partly). Figure 6.3 shows a point layer containing septic system locations and a polygon layer showing geological units. Imagine that a community has a porous and fractured geologic unit that serves as a groundwater aquifer, and the city water supply comes from several wells that tap this unit. Outside the city limits, extensive development has occurred in the past 20 years. The city has become concerned that contamination from faulty septic systems may enter city water supplies through this porous unit. Assessing the threat requires identifying the number of septic systems that occur in the outcrop area of the aquifer.

A spatial join solves this problem perfectly. The septics are the destination layer and become the output feature class. Each septic system is evaluated to find the geology polygon within which it falls, and the attributes from that polygon are appended to the septic system's record in the table. As shown in Figure 6.3, septic system 345 falls inside the Minnelusa Formation polygon in the geology layer, so the geology polygon information is appended to its row in the table. Septic system 348 falls inside a gravel deposit unit, and it gets the attributes from the gravel polygon.

Spatial joins may be performed on any two spatial data layers. A user can join points to points, polygons to polygons, lines to points, and nearly any combination of the three types of data. The output layer will always have the same feature type as the destination layer.

Cardinality

Cardinality is an important issue for spatial joins, just as it is for attribute joins. Because records in tables are being matched together, the Rule of Joining must be fulfilled in spatial joins also. Each feature record in the destination table must have one and only one matching record in the source table. This condition is met if

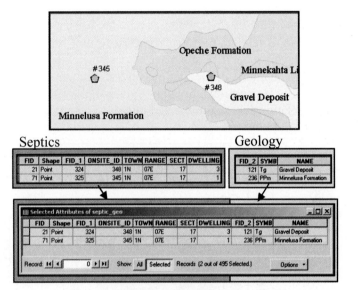

Fig. 6.3. A spatial join gives each septic system the attributes of the geology polygon within which it lies.

the cardinality of destination to source is one-to-one or many-to-one. In attribute joins, we had to use a relate if the Rule of Joining was not fulfilled. With spatial joins, we must use a **summarized join** if we encounter a one-to-many relationship.

Recall that the Summarize function calculates statistics for groups of records in a table. It uses one field to divide the records into groups, and then it calculates statistics for other fields for each group. In a summarized join, each feature in the destination layer is matched to many features in the source. Statistics are calculated for that group of features, and the result is appended to the feature record.

In Figure 6.2, imagine reversing the join so that airports is the destination layer and cities is the source layer. Each airport has many cities that are closer to it than to any other airport. Instead of attaching the single closest city, a summarized join finds all of the cities closer to the airport and calculates one or more statistics, for example, the sum of the city populations. Then for each airport we would know the total number of people being served by that airport

OID	NAME	Count_	Sum_POP_98
5	Mahlon Sweet Field	31	204048
6	Corvallis Municipal	33	148118
7	Bend Municipal	21	38589
8	Roberts Field	32	24893
9	La Grande/Union County	59	54115
10	Newport Municipal	22	24955
11	Aurora State	58	437083
12	Portland International	69	870598

(a)

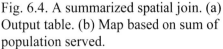

(b)

Fig. 6.4. A summarized spatial join. (a) Output table. (b) Map based on sum of population served.

(i.e., the sum of the populations of the cities that are closer to that airport than to any other). Figure 6.4a shows the output table of the joined layer with a Count field representing the number of cities and the Sum_POP_98 field representing the total people in those cities. So Portland International has 69 cities close to it with a total of 870,598 people. Figure 6.4b shows a proportional symbol map based on the Sum_POP_98 field to represent the total potential population served by the airport.

Types of spatial joins

Spatial joins fall into four main types according to the cardinality of the relationship between the joined layers (simple or summarized) and the choice of spatial criteria (inside or distance). Figure 6.5 shows a matrix of the four possible combinations. A simple join may be used whenever the cardinality is one-to-one or many-to-one so that the Rule of Joining is maintained. In a one-to-many relationship, a summarized join must be employed.

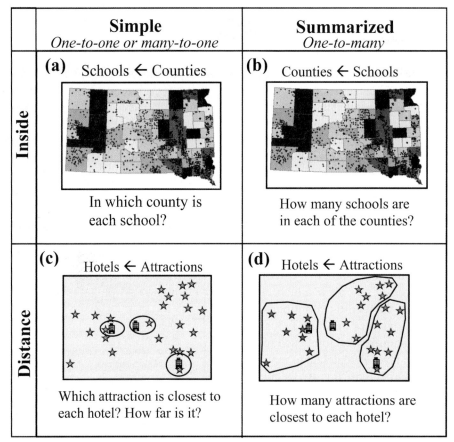

Fig. 6.5. Matrix of spatial joins resulting from different choices of destination table, spatial condition (inside or distance), and cardinality

Consider the simple join types first. In Figure 6.5a, schools and counties are being joined, with schools as the destination layer (some call it the target layer). The output layer contains schools, and each school will have the attributes of the county it falls inside. With the output layer, one could answer the question, "In which county does a particular school lie?" This example is a *simple inside join*. The previous example of the septic systems is also a simple inside join.

If, however, the destination layer is reversed, the cardinality of counties to schools is one-to-many, and a simple join is not possible. A summarized join may be employed in this case. The summarized join groups the schools together based on which county they are in and generates a single record of statistics for the group. This single summarized school record can then be appended to the matching county in the destination table. A Count field is always generated, containing the total number of schools in each county. Figure 6.5b shows this as a *summarized inside join*.

Distance joins also come in simple or summarized varieties. Consider evaluating the desirability of several hotels based on their distance to local tourist attractions, using a layer of hotels and a layer of attractions with the yearly number of visitors in each one. What is the cardinality of this relationship? This question seems confusing because there are several hotel sites and many attractions. In a distance join, it is the question that dictates the cardinality. If we ask, "Which attraction is closest to the hotel?" in order to find out whether that attraction has a large visitor pool, we have specified a one-to-one criterion because only one attraction can be *closest* to each hotel. A *simple distance join* suffices, and each hotel appears in the output table along with the attributes of the attraction closest to it and the distance between them. In Figure 6.5c, the circles connect each hotel site with its closest attraction.

A different question may be asked: "How many tourist attractions are closer to this hotel site than they are to any other?" In this case, we are interested in evaluating the richness of choice of attractions for each hotel or perhaps the total combined visitor pool from all the closest attractions. In this case, each hotel is connected with many attractions, as shown in Figure 6.5d, and summary statistics for the group, such as the number of attractions and the sum of the visitor pools, may be added to the record for each site. This would be a *summarized distance join*.

Feature geometry and spatial joins

The available join types will depend in part on the geometry of the features being joined. When joining points to points, for example, an inside join type does not apply. Thus, only two options are available when joining points to points, simple distance or summarized distance. Table 7.1 lists all of the possible geometry combinations and the join types that can be applied to each. (In this table we break the convention of putting the destination first in order to match the descriptions in the spatial join window in ArcGIS.)

Let us examine some other combinations of geometry types. Consider this problem: Many counties in South Dakota do not have a hospital. The state emergency planning office wishes to know the hospital closest to each county and how far away it is. They require a list of counties, each with the closest hospital attached to it. Thus, counties is the destination layer, and hospitals is the source layer. According to Table 6.1, joining points to polygons requires either a summarized inside or a simple distance join. In this case, the

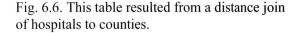

NAME	NAME_1	Distance
Shannon	Battle Mountain National Sanitarium	0.474633
Fall River	Battle Mountain National Sanitarium	0
Pennington	Bennett Clarkson Hospital	0
Lincoln	Canton-Inwood Hospital	0
Union	Canton-Inwood Hospital	0.216911
Clay	Canton-Inwood Hospital	0.336159
Custer	Custer State Hospital	0

Fig. 6.6. This table resulted from a distance join of hospitals to counties.

simple distance join is the correct choice. After the join, each county has a field indicating the name of the closest hospital (Fig. 6.6) as well as the distance from the county to the hospital.

Table 6.1. Join types are available for each combination of feature geometries in a spatial join. The second feature type is the destination layer in each case.

Geometry Type	Join Type	Example
Points to Points	Simple distance	Find the hospital closest to each town.
	Summarized distance	Find all the towns closer to one hospital than to any other hospital.
Lines to Points	Simple distance	Find the water main closest to the proposed building site.
	Summarized inside	Find the total voltage of all electric lines meeting at a substation.
Polygons to Points	Simple inside	Find the soil type that underlies each gas station.
	Simple distance	Find the lake that is closest to each campground.
Points to Lines	Simple distance	Find the elementary school that is closest to each residential street.
	Summarized distance	Find the total number of septic systems closer to a particular stream than to any other stream.
Lines to Lines	Summarized inside	Find the number of roads that cross each river.
	Simple inside	Give a section of hiking trail the attributes of the road it follows for a short distance.
Polygons to Lines	Summarized inside	Give a stream the average erosion index of the soil types it crosses.
	Simple distance	Find the lake closest to a hiking trail or the national park within which a road lies.
Points to Polygons	Summarized inside	Find the total number of schools and students in a county.
	Simple distance	Find the town that is closest to a lake. A point inside a polygon is given a distance of zero.
Lines to Polygons	Summarized inside	Find the total number of rivers crossing a state.
	Simple distance	Find the carrying capacity of the closest power lines to an industrial site.
Polygons to Polygons	Summarized inside	Find the total population of all counties that intersect part of a watershed.
	Simple inside	Find the county within which a lake falls completely.

In Figure 6.7, a unique values map was created based on the hospital name. In the northeast part of the state, it is easy to see that the purple counties are closest to the hospital in the purple area, the green counties to the green hospital, and so on. The long, hatchet-shaped county in western South Dakota (Pennington County) has several hospitals. Since all three lie inside Pennington, they technically have a distance of zero, and Pennington's closest hospital was arbitrarily assigned based on which hospital was found in the table first. The lighter pink counties in the central part of the state were assigned to a different hospital in Pennington County and appear in a different color. The map in Figure 6.8 shows each county displayed according to its distance to the closest hospital.

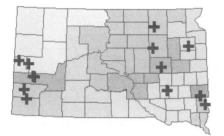

Fig. 6.7. Counties with the same color are closest to the same hospital.

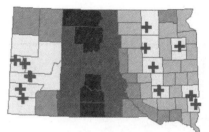

Fig. 6.8. Counties colored according to distance from the closest hospital

Figure 6.9 demonstrates an example of joining points to lines to predict the susceptibility of streams to contamination from septic systems. In this analysis, we make the assumption that the more septic systems that are close to the stream, the greater the susceptibility of the stream to contamination. The point locations represent the centers of one-mile by one-mile sections, and the size of the symbol indicates the number of septic systems in the section. We need to join each stream line to the closest septic systems, so streams is the destination layer. Joining points to lines requires either a simple distance join or a summarized distance join. In this case a summarized distance join, which sums the total septic systems that are closest to each stream, is correct. The map shows the results: the thicker the line symbol of the stream, the more septic systems are closest to that stream and the higher the susceptibility of the stream to contamination.

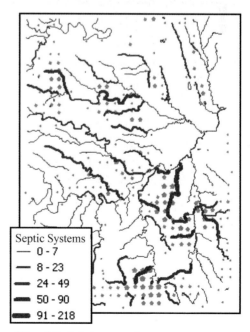

Fig. 6.9. Joining septic systems to streams to evaluate stream susceptibility to contamination

Coordinate systems and distance joins

Distance joins should always be performed on layers with projected coordinate systems. If a join is performed on a layer with a geographic coordinate system (GCS) and units of decimal degrees, two problems arise. First, the distances reported in the table will be in decimal degrees, as shown in the table in Figure 6.6. Degrees cannot be easily converted to miles or kilometers because the conversion factor changes with latitude. Second, the result could be invalid. The distance algorithm relies on the relative *x-y* coordinates of the features to calculate distances and assumes a Cartesian coordinate system to do so. Degrees of latitude and longitude are spherical coordinates, not Cartesian, and the relative distances calculated may be incorrect.

Figure 6.10 shows two cases of a summarized distance join of tourist attractions to hotels. In Figure 6.10a, the join was done with both layers in a GCS. Notice that the three attractions in the northeast corner were assigned to Hotel C based on their distance from it in decimal degrees. In Figure 6.10b, both layers were projected to UTM prior to joining, and the attractions were assigned instead to Hotel B. Since a UTM projection preserves distance within the zone, we know that the second example gives the correct spatial distances between the attractions and the hotels and is the valid result. Notice that the spatial distribution of attractions is elongated in Figure 6.10a, as a result of being displayed in a GCS.

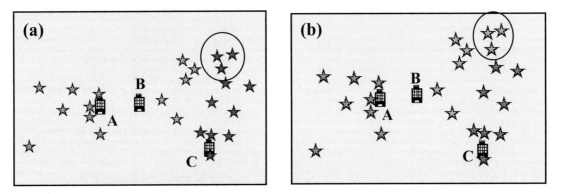

Fig. 6.10. The effect of the coordinate system on a spatial distance join. (a) Three attractions are incorrectly joined to Hotel C (purple stars) when the input feature classes have a GCS. (b) When the feature classes are projected to UTM before joining, the attractions are correctly assigned to Hotel B (green stars).

It is not sufficient to set the data frame to a projected coordinate system; the layers themselves must be projected. Furthermore, the projected coordinate system should be one that preserves distance over the region of analysis, such as equidistant conic. Otherwise, the distance analysis may again be incorrect.

About ArcGIS

Spatial joins are performed on two input feature classes and result in a new feature class that may be stored as a shapefile or geodatabase feature class. The original inputs are unchanged. Spatial joins are initiated using the same method as attribute joins.

Choosing the join type

As in an attribute join, the process begins by deciding which layer is the destination. The user right-clicks the destination layer, chooses to join the data based on spatial location (Fig. 6.11), and enters the desired source layer.

The join menu always offers two ways to decide how to match the fields, taken from the four types shown in Figure 6.5. The Join Data window in Figure 6.11 illustrates the process of joining schools to counties, with counties as the destination layer as described in Figure 6.5a. Each county can have more than one school, so it is a one-to-many join. There are always two options in the window based upon the choices listed in Table 6.1. The user must choose between them based on the desired result.

For this example, Option 1 offers an *inside summarized join,* which gives each county a statistical summary of the attributes of all the schools inside it, such as the total number of schools or the sum of the students. Option 2 specifies that a *simple distance join* should be performed, which gives each county the attributes of the closest school. In this case, the second option is nonsensical. All of the schools are inside the county and would be assigned a distance of zero, so it avails little to find the "closest" one. In this example, the summarized join is the appropriate choice.

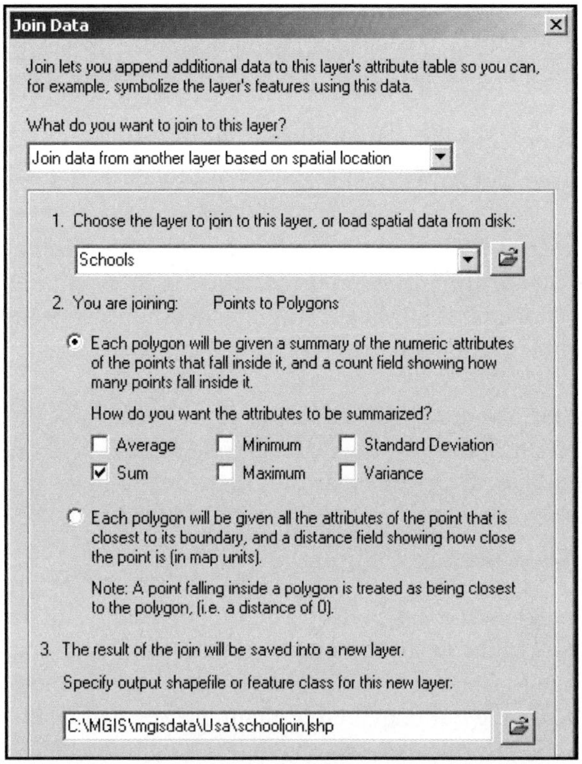

Figure 6.12 shows the result of the join. The resulting layer is a map of counties with the original county attributes plus a field called Count_ that contains the number of schools in the county. From this field, the graduated color map was created, showing the number of schools in each county. When summarizing, the user can choose from several statistics, such as minimum, maximum, average, and so on. All numeric attributes in the source table are summarized using the chosen statistics and are placed in the output table. String fields cannot be summed or averaged, so they are not included in the output table.

Fig. 6.11. A one-to-many cardinality is handled by either a summarized join or a distance join.

Setting up a spatial join

Performing the spatial join itself is a simple process. However, determining that a spatial join is required and identifying the destination table and the type of join can challenge beginners. This section presents a series of questions to be answered when setting up a spatial join to help produce the correct result. Too many students of GIS simply try random combinations until they hit on the right one—this process is designed to give the right answer the first time.

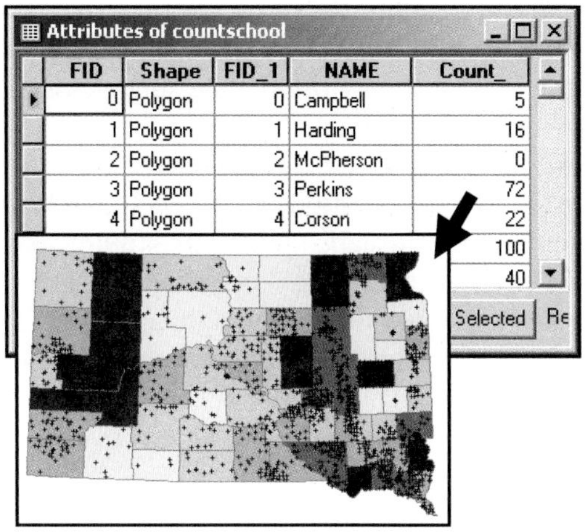

When faced with a suspected spatial join problem, first make a simple sketch of the relationships between the layers to be joined. Then answer this series of questions, designed to lead to the correct solution:

Fig. 6.12. Using a spatial join of schools to counties, one can create a graduated color map showing the number of schools in each county.

229

What should the final output layer or table look like?

Which is the destination layer?

Should a distance join or an inside join be used?

What is the cardinality of the join?

Should a simple join or a summarized join be used?

Let us demonstrate this process by using the problem of estimating stream susceptibility to contamination from septic systems, as shown in Figure 6.9. Recall that each point represents a section and is attributed with the total number of septic systems in the section. First sketch the problem and then go on to the questions.

What should the final output layer/table look like? The desired result is a layer of streams. In the attribute table, each stream must be assigned the total number of septic systems that are closer to it than to any other stream. Imagine the fields in the table, with the stream followed by the sum of the number of septic systems. Once the output is envisioned, it is usually easy to see which feature class is the destination layer.

Which is the destination layer? The features in the output layer are always the same features as those in the destination layer. If streams is chosen as the destination, then the output layer will contain streams. If the septics layer is chosen, then the output layer will contain septic points. Imagining our output table again, we see that what we really want is streams, each of which has septic information assigned to it. Thus streams is the destination layer.

Should a distance join or an inside join be used? Since we're looking for septic systems closest to each stream, this is clearly a distance join.

What is the cardinality of the join? Assignment of cardinality depends on the destination layer, in this case streams, so consider the relationship between streams and the septic systems. Since one stream can have multiple septic points closer to it than to any other stream, this is a one-to-many cardinality.

Should a simple join or a summarized join be used? Since the cardinality is one-to-many, a simple join cannot work. Because the goal is to sum all the septic systems closest to the streams, rather than simply finding the single closest septic to the stream, a summarized join is indicated, with Sum as the statistic.

Now that the questions are answered, the problem setup becomes clear: We need to do a summarized distance join with streams as the destination layer and septics as the source layer. Even experienced users may find that following this suggested procedure when setting up a join makes it easier to find the right approach. In the next few examples, we will apply this process to set up and solve three different spatial join problems.

Problem 1

Number of earthquake deaths in congressional districts

Imagine that a representative from one of the congressional districts in California is sponsoring legislation to provide earthquake emergency planning funds and is looking for support for the bill. She is planning to throw a big party and invite all of the reps from districts with a significant number of earthquake deaths. She asks one of her staffers, who is a GIS specialist, to draft a list of names for the invitation list. The staffer might approach the problem as follows, using the data in the mgisdata directory.

STOP! Write the answers to the questions and then read on to see if you analyzed the problem correctly.

What should the output layer/table look like? _____

Which is the destination layer? _____

Should a distance join or an inside join be used? _____

What is the cardinality of this join? _____

Should a simple join or a summarized join be used? _____

The earthquake table in quakehis.shp has a state attribute but none for congressional district. The only way to associate the number of earthquakes with districts is to perform a spatial join. The goal is a layer of districts with a table containing the number of earthquake deaths for each one. Thus, the districts will be the destination layer. The relationship between districts and earthquakes is potentially one-to-many, so a simple join is out of the question. Since the staffer wants to know the total deaths from quakes in each district, the summarize option will be the best, and the proper statistic is Sum. We will do a summarized inside join, with districts as the destination layer.

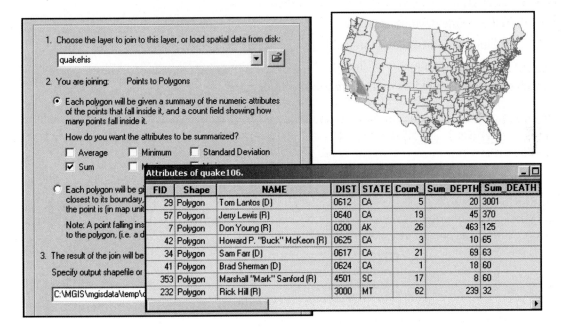

Fig. 6.13. Using a summarized spatial join to find the congressional districts that have had more than 10 earthquake-related deaths

Figure 6.13 shows the spatial join window filled out for this problem, as well as the resulting table. The table and the map show the selected records from the query—many of the districts are in California, as one might expect. After performing the join, the staffer uses an attribute query to select all of the districts that have had 10 or more earthquake-related deaths. Finding only 12 districts on the list and knowing that the boss wants a BIG party, the staffer then uses Select By Attributes to find all the districts that have had ANY earthquake deaths. This brings the total up to 29 districts. The guest list should be even bigger, so the staffer decides to use Select By Location to also select any districts that touch the districts with earthquake deaths (Fig. 6.14). This brings the guest list to 91, and the staffer can make up the invitations.

Problem 2
Pollution risk of rivers based on county populations

Imagine that Lindsey must do an analysis estimating the risk of pollution to rivers in the United States. She has no detailed information of the sources and types of pollutants, but she can make use of the observation that, generally speaking, large numbers

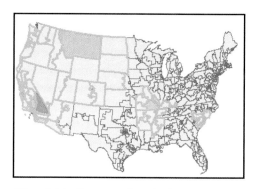

Fig. 6.14. Congressional districts with earthquake deaths or that touch a district with earthquake deaths

of people are generally correlated with high risks of pollution. She decides to use the population of counties as an estimate of pollution risk. For each river, she wants to find the total number of people living in counties that contain the river or lie next to it.

STOP! Write the answers to the questions and then read on to see if you analyzed the problem correctly.

What should the output layer/table look like? _____

Which is the destination layer? _____

Should a distance join or an inside join be used? _____

What is the cardinality of this join? _____

Should a simple join or a summarized join be used? _____

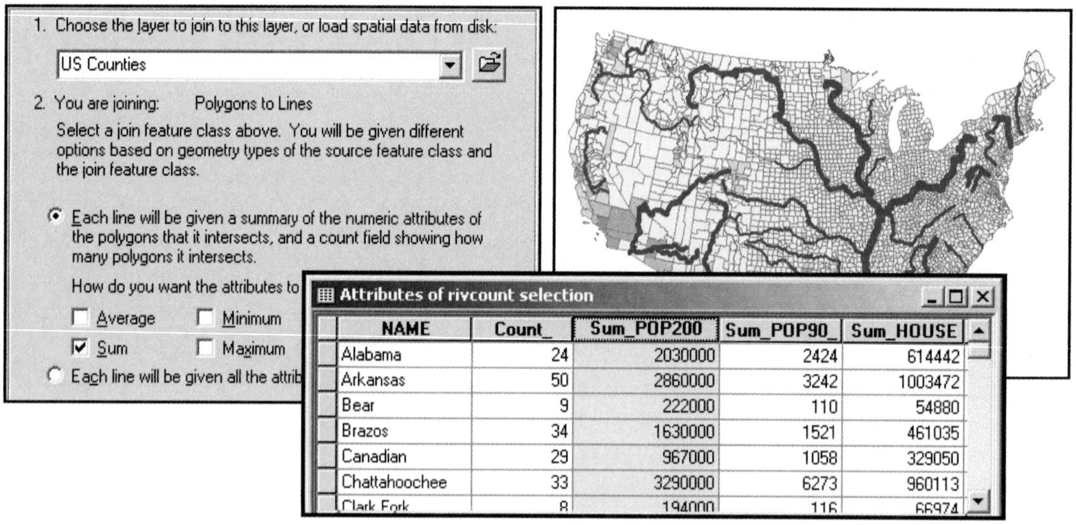

Fig. 6.15. Estimating pollution risk for rivers based on adjacent county populations

The final output layer should contain rivers, with a table listing each river and the total population of counties adjacent to that river. Thus, rivers must be the destination layer. Since the counties must actually touch the river, this is an inside join rather than a distance join. The cardinality is one-to-many since one river can have many counties touching it. An inside summarized join must be used with rivers as the destination layer, and the SUM statistic should be requested.

Figure 6.15 shows the spatial join window filled out for the join. The output table shows the river name, the number of counties adjacent to it, and the sum of each numeric field, including the POP2000 field. The figure also shows a graduated symbol map based on the Sum_POP2000 field, in which the width of the river represents the total county population living adjacent to it. Notice that ALL of the numeric fields are summed, not just the POP2000 field. One drawback to summarized spatial joins is that the chosen statistic is applied to all the fields, yielding potentially large attribute tables.

Problem 3
Closest volcano to a city

In this final example, imagine that a professor who specializes in volcanoes has built a Web site about them. He would like to put a table on his Web site so that schoolchildren all over the United States could enter the name of the city they live in and get an information page about the volcano that is closest to them. Since the professor knows nothing about GIS, his graduate student, Cody, gets to make the table.

STOP! Write the answers to the questions and then read on to see if you analyzed the problem correctly.

What should the output layer/table look like? _____

Which is the destination layer? _____

Should a distance join or an inside join be used? _____

What is the cardinality of this join? _____

Should a simple join or a summarized join be used? _____

The final goal is a cities layer with fields indicating the closest volcano and the distance to it. Thus, Cody realizes, Cities is the destination table, and a distance join to the volcanoes layer will provide the necessary information. Because the closest volcano is the target, this is a simple distance join.

1. Choose the layer to join to this layer, or

volcano

2. You are joining: Points to Points

○ Each point will be given a summary
the points in the layer being joined t
count field showing how many point

How do you want the attributes to b

☐ Average   ☐ Minimum
☐ Sum   ☐ Maximum

◉ Each point will be given all the attributes of the point in the layer
being joined that is closest to it, and a distance field showing
how close that point is (in map units).

Attributes of cityvolc

CITY_NAME	STATE_NAM	STATE	NAME	TYPE_1	Distance
Central Islip	New York	NM	Raton-Clayton	Volcanic field	31.191047
Central Manchest	Connecticut	NM	Raton-Clayton	Volcanic field	32.012169
Centralia	Washington	WA	St. Helens	Stratovolcano	0.939807
Centralia	Illinois	NM	Raton-Clayton	Volcanic field	15.100639
Centreville	Virginia	NM	Raton-Clayton	Volcanic field	26.747356
Ceres	California	CA	Red Cones	Cinder cones	1.906377
Cerritos	California	CA	Lavic Lake	Volcanic field	1.687508
Chalmette	Louisiana	NM	Raton-Clayton	Volcanic field	15.527910

Fig. 6.16. A table showing each city in the United States and its closest volcanoes, including the distance to the volcano (in decimal degrees)

Figure 6.16 shows the spatial join window settings to produce this analysis and the result of the spatial join. The table shows the city name (sorted alphabetically), the state name, the state where the volcano is located, the volcano name and type, and the distance to the volcano.

Notice the field name TYPE_1 in the output table. It happens that the Cities layer has a TYPE field, and so does the Volcano layer. The output table cannot have two fields with the same name, so during the join the repeated names are automatically converted to unique names. The Cities type field remains TYPE, but the Volcanoes type field becomes TYPE_1. (The city TYPE field was hidden from display to make the figure clearer, but it remains in the table.) Usually the source table field is the one that is renamed.

Also notice that the distance from Champaign, Illinois, to the Raton-Clayton volcanic field in New Mexico is approximately 15. These units are clearly not miles or kilometers. In fact, they are degrees, indicating that the coordinate system of the original data is a GCS. Recalling that distance joins should always be done on projected data, Cody realizes that he must repeat this analysis in order to get the correct result. He needs to first project the Cities and Volcanoes layers into a projected coordinate system, such as Equidistant Conic, *before* performing the spatial join. Then the distance units will be given in meters, and the actual miles can be calculated.

> **TIP:** Always check the distance values after performing a distance join to make sure that the values are in projected units.

Summary

➤ A spatial join combines the records of two feature tables based on the location of the features. A new feature class is created by a spatial join.

➤ An inside join uses the criterion that one feature falls inside another or, in the case of points and lines, on top of each other. A distance join matches the destination layer feature to the record of the closest feature in the source layer. A distance field reporting the distance between the joined features is added to the table.

➤ A simple join may be used whenever the cardinality of the layers is one-to-one or many-to-one. A summarized join generates summary statistics for all the source features matching the destination features and is used when the cardinality is one-to-many or many-to-many.

➤ Four types of spatial joins exist based on the combination of the criterion and the cardinality of the relationship: simple inside joins, simple distance joins, summarized inside joins, and summarized distance joins.

➤ Distance measurements are reported in map units of the input layers.

➤ Distance joins should only be performed with layers having projected coordinate systems that do not distort distances. Using layers with a geographic coordinate system may yield incorrect results.

➤ In the joined table, fields with identical names will be renamed in the output file so that all field names are unique, such as NAME to NAME_1. Usually the source field is renamed.

➤ Use the following series of questions to help set up a spatial join properly:

What should the output layer/table look like? _____

Which is the destination layer? _____

Should a distance join or an inside join be used? _____

What is the cardinality of this join? _____

Should a simple join or a summarized join be used? _____

Chapter Review Questions

1. What primary characteristic distinguishes a spatial join from an attribute join?

2. What two options may be used to handle one-to-many relationships in a spatial join?

3. If a polygon feature type is joined to a line layer, with the lines as the destination table, what will the feature type of the output layer be?

4. How many output fields will result if a summarized join is specified and a single statistic (e.g., Sum) is selected?

5. Why should distance joins always be performed on layers with a projected coordinate system? What additional consideration should be added?

6. What happens if the two input layers in a join each have a field with the same name?

For the following spatial join problems, answer the series of questions in the text and then state the type of join that should be used: simple inside, simple distance, summarized inside, or summarized distance.

7. Determine the number of parcels within each of the Tallahassee watersheds.

8. Find the closest school for each house in a realtor's database.

9. Find the land-use zoning type associated with each well in Atlanta.

10. Determine the number of counties and the total number of people served by each airport in the United States.

Mastering the Skills

Teaching Tutorial

The following examples provide step-by-step instructions for doing basic tasks and solving basic problems in ArcGIS. The steps you need to do are highlighted with an arrow ➔; follow them carefully. Click on the video number in the VideoIndex to view a demonstration of the steps.

A simple join

➔ Start ArcMap and open the map document ex_6a.mxd.

➔ Use Save As to rename the document and remember to save frequently as you work.

Spatial joins produce new feature classes. We only need these for practice and don't want them to become part of our permanent feature classes in the source geodatabase. So, for this lesson, we'll create a new personal geodatabase to store the outputs.

1➔ Open ArcToolbox, if necessary, and open the Data Management > Workspace > Create Personal GDB tool.

1➔ Specify the mgisdata\Sdakota folder as the output location. (You'll need to navigate to the mgisdata folder to be able to select Sdakota as the output location.)

1➔ Name the output geodatabase **chap6results**. Click OK.

1➔ Close ArcToolbox.

> **TIP:** When you see the word **STOP**, pause a moment and set up the problem using the questions presented in the Concepts section. Then read on to see if you analyzed the problem correctly.

What should the output layer/table look like?

Which is the destination layer?

Should a distance join or an inside join be used?

What is the cardinality of this join?

Should a simple join or a summarized join be used?

First, we would like to create a list of the attractions, sorted by the counties in which they lie, so that tourists visiting a county know what attractions are available. **STOP** and think it through.

The desired output is a table of attractions with a county field for each one. A simple inside spatial join is indicated, with the tourist attractions as the destination table.

2➔ Right-click the Tourist Spots layer and choose Joins and Relates > Join from the context menu.

2➔ In the top drop-down box, choose to join from another layer based on spatial location (Fig. 6.17).

2➔ Choose Counties as the layer to join.

2➔ Choose to join the point to the polygon that it falls inside.

2➔ Click the Browse button and change the Save As Type to File and Personal Geodatabase feature classes.

2➔ Navigate to the chap6results geodatabase and enter **tour_county** as the name of the feature class to be created. Click Save and Click OK.

Notice that the new feature class appears at the top of the Table of Contents.

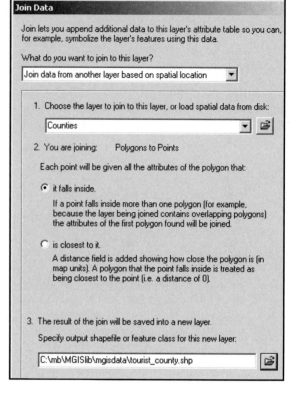

Fig. 6.17. The Join Data window for spatial joins

3➔ Turn off the original Tourist Spots layer.

3➔ Right-click the tour_county layer and choose Open Attribute Table.

3➔ Scroll to the right to find the fields from the Counties table, including the county name and population information.

Both original layers had a NAME field, but they cannot both have the same name in the output file. To distinguish them, the new table was given a NAME_1 field for the county names and this field received the alias counties_NAME. This feature makes it easier to distinguish common field names and which table they came from. (In version 9.2, no alias is created, and the field is called NAME_1.)

1. What other two fields received new names in this file? _____

4➔ Close the attribute table.

4➔ Open the tour_county Symbology properties and create a unique values map based on the counties_NAME field so that the tourist attractions are colored the same when they are in the same county.

4➔ Click the minus sign next to the tour_county layer to collapse the long list of county names in the Table of Contents.

Suppose we now wish to determine the number of tourist attractions in each county. Now that each attraction has a county, a Summarize will perform this task admirably.

5➔ Right-click the tour_county layer and choose Open Attribute Table.

5➔ Right-click the counties_NAME field and choose Summarize. You do not need to choose any statistics (the Count field will be automatically generated).

5➔ Click the Browse button and set the output file type to File and Personal Geodatabase tables. Navigate to the chap6results geodatabase and name the output tour_sum. Click Save.

5➔ Click OK and choose to add the new table to the map.

5➔ Close the tour_county table and open the new tour_sum table.

5➔ Sort the Count_NAME_1 field in descending order. (Notice that it reverted to the actual name of the summarize field, NAME_1, instead of the alias counties_NAME.)

2. Which three counties have the most tourist attractions? How many does each have?

_____

A summarized join

The previous problem used two steps to determine the number of tourist attractions in each county, first a simple join to assign a county to each attraction and then a Summarize to sum the attractions for each county. However, we could have found the answer in one step by setting up a slightly different join. **STOP** and think it through.

In this case, the output table should list the counties, each of which has a field indicating the total number of attractions in it. Thus, an inside summarized join is indicated, with Counties as the destination table. We will try it this way now.

6➔ Close the tour_sum table.

6➔ Right-click the Counties layer and choose Joins and Relates > Join.

6➔ Set Tourist Spots as the layer to join to.

6➔ Choose the option to summarize the points within the polygons.

6➔ There are no numeric fields in the Tourist Spots table and no need to choose a statistic. Leave all the boxes unchecked.

6➔ Name the output feature class tour_county2 and put it in the chap6results geodatabase. Click OK.

7➔ Open the tour_county2 attribute table and scroll right to the last fields.

This time the output layer contains the county polygons instead of tourist spots because Counties was the destination table. One new field was added, Count_, indicating the number of tourist spots found in each county.

7➔ Sort Descending on the Count_ field and verify that the result is the same as before.

7➔ Close the tour_county2 table.

This exercise highlights the function of a summarized spatial join, which takes a one-to-many relationship and uses Summarize to reduce many records into a single record for joining to the destination. The first join had a many-to-one cardinality, and a simple join was used followed by a Summarize. The second join had a one-to-many cardinality, and the summary was built into the analysis.

Distance joins

Point-to-polygon spatial joins are common analysis problems. Let us apply one to the problem of mapping the service areas of hospitals in South Dakota, defined as the counties that are closer to that hospital than to any other hospital. **STOP** and think it through.

To map the service areas, we need a layer of counties with an attribute indicating the closest hospital for each. Then a unique values map can be created from the hospital names. Thus, Counties is the destination layer, and this problem requires a simple distance join to find the single closest hospital.

> 8➔ Click the Display tab. Turn off both tour_county layers and turn on the Hospitals layer.
>
> 8➔ Right-click the Counties layer and choose Joins and Relates > Join.
>
> 8➔ Set Hospitals as the layer to be joined.
>
> 8➔ Choose the option to join the closest hospital to the county boundary.
>
> 8➔ Name the output file **hospcount** and click OK.

Now we can analyze the results.

> 9➔ Open the hospcount attribute table and examine all of the fields. Most of these fields came from the Counties layer. Scroll to the end to find the hospital fields added by the spatial join.
>
> 9➔ Note that the hospital name field is called hospitals_NAME now. Close the table.

> 10➔ Move the Hospitals layer over the hospcount layer in the Table of Contents.
>
> 10➔ Use the Distance field in the hospcount layer to create a graduated color map showing the distance from each county to a hospital, as in Figure 6.8 earlier in the chapter.
>
> 11➔ Right-click the hospcount layer, copy it, and paste it into the South Dakota data frame.
>
> 11➔ Use the hospcount layer copy to create a unique values map based on the hospitals_NAME field, showing the service area of each hospital. The map should be similar to Figure 6.18.

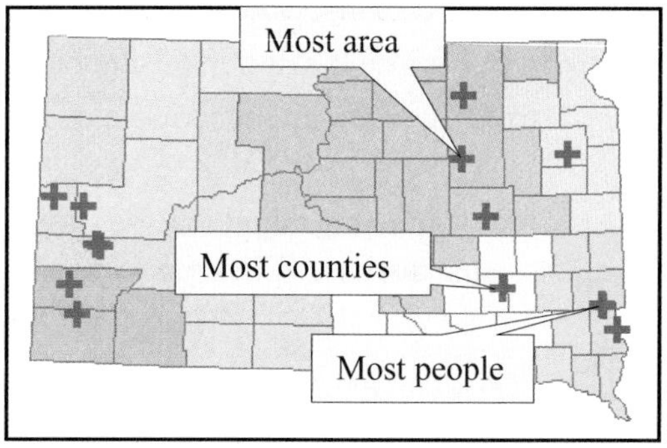

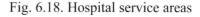

Fig. 6.18. Hospital service areas

We also wish to know the total number of people in each service area. To find this information, we will summarize the hospcount table using the hospital name field and request the sum of the county population as the statistic. It would also be interesting to know the total area served by each hospital, so we will request the sum of the county areas as well.

> 12➔ Open the hospcount attribute table (from either of the layers).
>
> 12➔ Right-click the hospitals_NAME field and choose Summarize.
>
> 12➔ Request the Sum statistic for the POP2000 field and the AREA field.

12➔ Name the output table **hospstat** and click OK. Choose Yes to add the table to the map. Close the hospcount table.

12➔ Open the hospstat table and examine the fields. Sort the table to answer the following questions.

3. Which hospital serves the most counties? _____ Which hospital serves the most people? _____ Which hospital serves the greatest area? _____

Note that, similar to the tourist attraction problem, the second question regarding the number of people and area could have been solved using a summarized join with a reversed cardinality, that is, by joining counties to hospitals with hospitals as the destination layer.

13➔ Before going on, open the hospcount table again and examine the Distance field more carefully. The values are clearly too large to be miles or kilometers.

4. What do you think are the units of this distance field? _____

The units of the Distance field added during a distance join will always match the units defined for the coordinate system of the first input feature class. In this case, those units are meters.

TIP: If the distances seem small (0–180), then the source data are in a geographic coordinate system with units of decimal degrees. The data should be projected before doing any distance spatial joins. Setting the data frame coordinate system is not sufficient—the input data must be projected.

13➔ Open the Properties of the Hospitals layer and check the coordinate system on the Source tab to confirm that the units are meters.

We'll do some point-to-point distance joins to examine the relationships between towns and hospitals. Two different questions might be asked. First, for any given town, what is the closest hospital? Second, for each hospital, how many towns does it serve and what is the total population?

STOP and think about the first question, what is the closest hospital to a given town? In this problem, the solution layer would contain towns with an additional field for the closest hospital. Thus, Towns is the destination layer. Since only one hospital can be closest to a town, the cardinality is one-to-one, and a simple distance join will suffice.

14➔ Click the Display tab. Turn off all layers except Hospitals, Towns, and Counties. Collapse the legends of the joined layers if you wish.

14➔ Right-click the Towns layer and choose Joins and Relates > Join.

14➔ Set Hospitals as the layer to join.

14➔ Choose to give each point the attributes of the closest point to it.

14➔ Name the output feature class **townhosp**. Click OK.

15➔ Open the townhosp attribute table. Find the Distance field created from the join.

15➔ Right-click the NAME field and choose Properties. Give it the alias HOSPITAL.

5. How many towns in South Dakota are more than 200 km from a hospital? _____

16➔ Close the townhosp attribute table.

16➔ Create a unique values map based on the HOSPITAL field to show which towns would go to which hospital.

The second question asks how many towns each hospital serves. **STOP** and work through the logic. The desired outcome is a hospital feature class. The table contains the list of hospitals and the number of towns that are closer to that hospital than to any other. Thus, Hospitals is the destination layer. The cardinality is one-to-many, so a summarized distance join must be used.

17➔ Turn off the townhosp layer and collapse its legend.

17➔ Right-click the Hospitals layer and choose Joins and Relates > Join.

17➔ Set Towns as the layer to be joined.

17➔ Choose the summarized join option and select Sum as the statistic so that we can determine how many townspeople rely on each hospital.

17➔ Name the output feature class **hosptown**. Click OK.

6. Which hospital serves the greatest number of towns? _____
   Which one serves the greatest number of people (POP100 field)? _____

Notice in the last two joins how we used the same two layers, Hospitals and Towns, but asked two different questions, had two different destination layers, and used two different types of joins. This example demonstrates the importance of thinking each problem through before trying to solve it.

18➔ Close all windows and the table.

18➔ Adjust the display so that only the Counties, Major Roads, and Tourist Spots layers are turned on.

Next, we'd like to analyze the number of visitors at different tourist attractions as a function of distance from the interstates. As a first step, we need to determine the distance of each attraction from the closest interstate. **STOP** and think it through.

We want a table showing each tourist site and its distance from the interstate, so clearly Tourist Spots must be the destination layer. A simple distance join will suffice to solve this problem.

However, notice that the Major Roads layer contains both interstates and state highways. Before we can do the join, we need to place the interstates in their own separate layer so that we can join with them alone. Otherwise, many of the tourist sites will be assigned attributes of highways rather than interstates. We need a new layer containing only interstates.

19➔ Open the Select By Attributes window from the main menu bar.

19➔ Choose Major Roads as the layer to query, and enter the expression [FUNC_CLASS] LIKE '*Interstate*'. Click Apply. Close the window.

20➔ Right-click Major Roads and choose Selection > Create Layer from Selected Features.

20➔ Name the new layer **Interstates** and give it an Expressway symbol. Turn off the Major Roads layer.

One of the properties of a layer is the selected features. When we create a layer and then pass it as input to a spatial join or other command, only the selected features will be used. Now do the spatial join.

21➔ Right-click the destination layer, Tourist Spots, and choose Joins and Relates > Join from the context menu.

21➔ Choose to join by spatial location, and make Interstates the source layer. Choose to join each point to the line that it is closest to.

21➔ Store the output in the chap6results geodatabase as tour_dist. Click OK.

22➔ Create a graduated symbol map based on the new Distance field in the new tour_dist layer. Flip the symbols to make the closer tourist spots larger and the more distant ones smaller (Fig. 6.19).

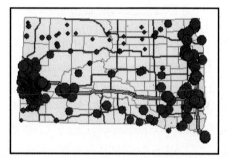

Fig. 6.19. Tourist spots with size indicating closeness to interstates

Examine the distance values in the legend of the map. As before, the values are in projected units because the source data are projected. However, we still may not be sure whether the units are meters or feet. We can find out, though.

23➔ Open the tour_dist layer properties and examine the Source tab. Scroll down the coordinate system description to find the linear unit. Close the window when done.

7. What are the map units of this coordinate system? _____

Now that the distance units are known, we could add a new field and calculate the distances in miles, if desired. We'll just go on.

Some polygon joins

Polygon-polygon joins can be done two ways. The first method assigns the attributes based on whether the source polygon is closest to the destination polygon, a simple inside join. The second method creates a summary for all source polygons that intersect the destination polygon, a summarized inside join. We will perform examples of both.

Consider the problem of determining the county responsible for the water quality in each South Dakota lake. When the state Fish and Game Department receives complaints from citizens about fish kills or water quality in a lake, the department can use such a list to quickly identify the county to be contacted. For the most part, each lake resides in one county, constituting a one-to-one join. Some large lakes may intersect more than one county, so it will be interesting to examine the results in those cases. **STOP** and think it through.

Since we want a list of lakes, with an attribute indicating the responsible county, Lakes is the destination layer. A simple inside join will solve the problem.

      24➔ Turn off and collapse all layers except for Counties and Lakes.

      24➔ Right-click the Lakes layer and choose to Join.

      24➔ Set Counties as the layer to join to.

      24➔ Choose the option to join attributes from the polygon that it falls completely inside.

      24➔ Name the output feature class lakecount and click OK.

      25➔ Open the lakecount attribute table and examine the results.

8.   Which county is responsible for Salt Lake? _____

What happens when a lake does not fall completely within a single county? Lake Oahe, the large dammed reservoir on the Missouri River in the center of the state, provides a good example because it touches several counties and serves as a boundary between them.

      25➔ Find Lake Oahe in the lakecount table and examine its attributes.

9.   How was Lake Oahe treated in the spatial join? _____

      25➔ Close the lakecount table and clear any selected features.

Another case of polygon-polygon joins involves the possibility of many polygons in the source table falling inside one polygon in the destination table. For example, consider the number of urban areas falling within each watershed in South Dakota. Urbanized areas often pose special problems for water quality, so knowing the urban population in each watershed provides important clues for designing monitoring programs. We can use a spatial join to determine the total urban population in each watershed. **STOP** and think it through.

The output table will contain a list of watersheds, each with a value for the total urban population. Thus Watersheds is our destination layer. Since more than one urban area can exist in each watershed, the cardinality is one-to-many, and a summarized inside join is required.

      26➔ Turn on the Watersheds and Urban Areas layers and turn off all other layers except Counties.

      26➔ Right-click the Watersheds layer and choose Joins and Relates > Join.

      26➔ Choose to join spatially, set the join layer to Urban Areas, and fill the top button for the summary option. Check the box next to Sum for the statistic.

      26➔ Enter the name shed_urban for the new feature class. Click OK.

27➔ Open and examine the attribute table of the new shed_urban layer. Close it when finished looking at it.

27➔ Create a graduated color map of the watersheds showing the total urban population. Notice that the watersheds without urban areas inside them received a <Null> in the table and are not drawn in the map (Fig. 6.20).

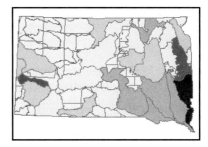

Fig. 6.20. Watershed urban populations

10. Which watershed UNIT has the most urban areas in it?
_____

11. Which watershed UNIT has the most people in urban areas in it? _____

It is even possible to join polygons with different shapes. Consider trying to find the total population living in each watershed instead of just the urban population. Towns and urban areas fall neatly inside watershed boundaries, but since much of South Dakota is rural, the results would be misleading. The county population data includes all the people, but significant overlaps between counties and watersheds occur. Rarely does a watershed occur completely inside a county or vice versa.

28➔ Turn off all layers except the original Watersheds and Counties and examine the relationships between the two.

However, we can still perform a summary spatial join on these two layers. Since we want a total population for each watershed, Watersheds is the destination layer. But which statistic should we request? At first, Sum might seem appropriate because we are trying to find total population. However, consider the watershed labeled in Figure 6.21, containing portions of five different counties. If we sum the populations, we will clearly overestimate the population in the watershed. Instead, an average population for the five counties would provide a better, although not an exact, estimate. An area-weighted average of the county population *densities* would probably give the best estimate, but for now we will use the simpler method of averaging the county populations.

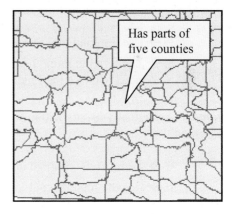

Fig. 6.21. The counties (gray lines) and watersheds (purple lines) have many overlapping areas.

29➔ Right-click the Watersheds layer and choose to Join.

29➔ Enter Counties as the layer to join.

29➔ Choose the summarize option for the join method and request the Average statistic.

29➔ Name the output feature class **watercount** and click OK.

30➔ Open the **watercount** attribute table and examine the fields. Close the table.

31➜ Create a graduated color map from the Avg_POP200 field, as shown in Figure 6.22.

32➜ Create a graduated color map based on the Avg_POP90_SQMI (population density) field and compare to Figure 6.22. Is the result similar or substantially different?

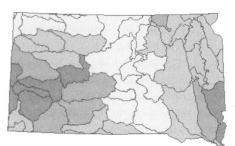

Fig. 6.22. Estimated population of watersheds in South Dakota

Matching issues during joins

Now we will do one more polygon-to-polygon join to demonstrate some issues to consider when joining. We would like to have a field in the Urban Areas layer indicating the state in which each urban area lies. We can do this with a spatial join, but there are a few pitfalls, as we shall see.

33➜ Collapse the South Dakota data frame and activate the USA data frame (right-click the frame name and choose Activate).

33➜ Zoom out, if necessary, to see the lower 48 states.

34➜ Join the states to the urban areas, with urban as the destination layer.

34➜ Choose the second option, for polygons completely inside.

34➜ Name the output layer urban_state and put it in the chap6results geodatabase.

Now let's examine an interesting feature of the output layer.

35➜ Use the Find tool to search for Kansas City in the urban_state layer.

35➜ Zoom to the found feature by right-clicking its name in the Find list and choosing Zoom To. Close the Find box.

Kansas City actually occurs within two states, Kansas and Missouri. To which state was it assigned?

36➜ Turn off the original urban layer.

36➜ Use the Identify tool to display the attributes of Kansas City in the urban_state layer.

The state name and all the other state fields received a <Null> value. Evidently, NO state was assigned to this urban area because it occupied two states rather than one.

36➜ Close the Identify box.

36➜ Open the urban_state attribute table and sort it in ascending order based on the STATE_NAME field.

Notice that many urban areas failed to join to a state. Some of these are not two-state cities, either. Why didn't these join?

37➜ Close the attribute table.

37➜ Use the Find tool to locate and zoom to Galveston, Texas.

37➜ Close the Find box.

Here we begin to get a clue to the problem. Galveston lies on a barrier island off the coast of mainland Texas. The urban area is more detailed than the state outlines, with the result that the urban area boundaries actually fall outside the state boundaries in some areas (Fig. 6.23). Even this small amount of difference is enough to prevent the join from uniting these two records.

Thus, when joining polygons together, it is important to be aware of possible mismatches that can occur because of the misalignment of polygon boundaries.

This is the end of the tutorial.

➜Close ArcMap. You do not need to save your changes.

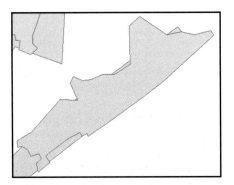

Fig. 6.23. Mismatch of urban areas and state boundaries prevents the features from joining.

Exercises

Open document ex_6b in the MapDocuments folder and use it to answer the following questions.

1. Give the name of the watershed in Rapid City that contains the most gas stations. How many stations does it have? What is the total number of gas nozzles in the watershed?

2. How many gas stations are in a residential land-use zone?

3. Which land-use category in Rapid City contains the most gas stations? How many gas stations are in the floodway?

4. Which gas station is farthest from a restaurant? What is the distance (include the distance units in your answer)? What's the name of the closest restaurant?

5. Which restaurant has the most gas stations that are closer to it than to any other restaurant? How many gas stations?

6. Which watershed in Rapid City has the highest number of septic systems and how many does it have? (**Note:** Each point in the Septic layer represents the center of a one-mile by one-mile section, and the DWELLING field contains the number of septic systems in the section. Assume that, if the center point is in the watershed, all the septic systems are in the watershed.)

7. Using the Streams and West Septic layers, create a map showing the number of septic systems that are closest to each stream. Be sure to use the number of septic *systems,* not the number of septic points, in the map. **Capture** the map.

Use the data in the mgisdata\Oregon\oregon geodatabase to answer the following questions.

8. Which town in Oregon is farthest from an airport? What is the distance in *kilometers?*

9. Assuming that an airport's service area includes all the cities that are closer to it than any other airport, determine how many cities each airport serves. Which airport serves the most cities? How many cities? Which serves the most people? How many people?

10. A boating club would like to know which parks in Oregon have the best access to lakes. Find the number of lakes and total lake area in each park. Which park(s) have the most lakes and how many? Which park has the greatest area of lakes? Examine parks with only one lake. The sum of the area fields for these lakes keeps repeating the same few numbers. Explain why.

Challenge Problem

Design an analysis to estimate which watershed in South Dakota has the most cattle living in it (in 1992) and do it. The best approach is to determine which counties overlap the watershed and average their cattle populations. Create a map showing your answer and **capture** it.

Hint: Create a counties shapefile that contains the cattle data as part of it, not simply joined to it, before you begin. You can do this by doing the join and then using Export to create a new shapefile, which will then contain the information from both tables.

Skills Reference

Performing a spatial join 249

Performing a spatial join

In a spatial join, the attributes of features in the source layer are appended to the features in the destination table. The output layer always contains the same features as the destination table. When choosing the join options in step 4, keep in mind the cardinality of the join to pick the most appropriate one.

1. Right-click the destination layer in the Table of Contents and choose Joins and Relates > Join.

2. Choose to join to another layer based on spatial location (Fig. 6.24).

3. Choose the source layer from the drop-down box, or click the Browse button to choose one from the disk.

4. Choose one of the two join options and specify any summary statistics desired, if applicable.

5. Specify the shapefile or feature class file to contain the new output layer.

6. For more information on joins, click the About joining data button.

7. Click OK to execute the join.

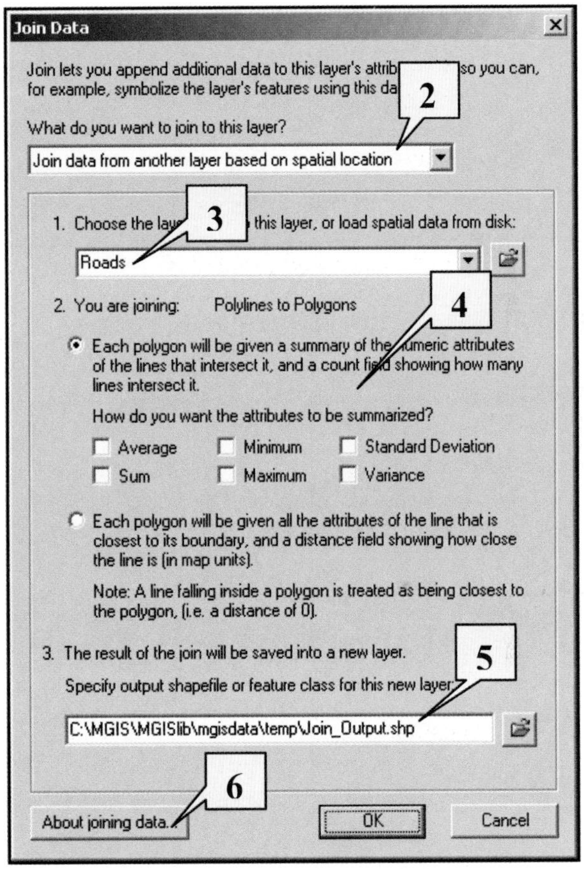

Fig. 6.24. Join Data window for spatial joins

Chapter 7. Geoprocessing

Objectives

> ➢ Learning about spatial analysis functions including overlay, clipping, and buffering

> ➢ Using map overlay to analyze multiple spatial criteria

> ➢ Understanding differences between spatial joins and overlays

> ➢ Geoprocessing with menus, ArcToolbox, the command line, and Model Builder

Mastering the Concepts

GIS Concepts

Over the years many procedures have accumulated for characterizing spatial relationships within and between spatial features. Functions can find areas shared by two or more conditions, evaluate distances, extract or erase areas of interest, merge similar features together, examine distance relationships between points, and more. The entire enterprise of GIS, including gathering the data, putting it in digital form, designing the geodatabase, managing the data, and creating maps, is justified by the ability to apply these tools to extract information that might be difficult to obtain any other way. GIS shares many capabilities with CAD systems and databases software, but spatial analysis gives GIS unique power to get the most out of map and attribute data.

Chapter 6 discussed spatial joins as one way to analyze spatial relationships. In this chapter we'll learn more tools to solve spatial problems. Often two or more functions will be strung together to solve a specific problem, a practice called **geoprocessing**. We'll begin with a general discussion of spatial functions and then investigate how these functions are implemented within ArcGIS.

Map overlay

Map overlay functions combine layers to create a single output. Overlay functions fall into two categories, those that do not combine attributes and those that do. The first set is sometimes called extraction and includes **clip** and **erase**. The second set includes **intersect** and **union**.

Extraction functions

Extraction functions separate features of interest from a larger group based on some criteria. Queries are one form of extraction and can be based upon attribute data as in Select By Attributes or upon spatial criteria as in the Select By Location function. Queries, however, can only extract whole features from a data set. Two new functions, **clip** and **erase**, are able to extract both whole features and portions of features. They have the ability to truncate features where they cross a boundary, for example, a road crossing a state line. This ability differs from the Select by Location function, which must select (or not select) the entire feature.

A **clip** works like a cookie cutter to truncate the features of one file based upon the outline of another. In Figure 7.1, the county roads file (gray lines) has been clipped by the land-use polygon coverage (beige) to produce a shapefile of the roads that fall inside the city planning boundaries (red). The clipping layer must always be a polygon file, but the clipped layers may be points, lines, or polygons. Only the outside boundary is used for clipping; internal boundaries have no effect on the output layer.

TIP: To temporarily clip features for display only in ArcMap, use the Data Frame tab of the Data Frame Properties menu.

Erase works in the opposite sense of a clip, keeping features that fall outside the erase layer and eliminating those inside. Given the same input layers in Figure 7.1 (all roads in the county and the city boundary), an erase would keep the roads outside the city (gray) and eliminate the roads inside the city (red). An erase might be used to eliminate privately owned lands from forest protection areas. Erase is only available to users with an ArcInfo license.

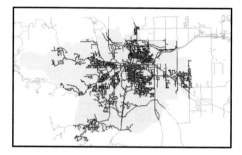

Fig. 7.1. A clip truncates the features of a layer by the boundary of another.

Attributes from both truncated and nontruncated features are preserved intact during a clip or an erase operation. Note, however, that fields related to a feature's area or length are not automatically updated after a clip or an erase. A field containing highway miles or acres, for example, would contain an incorrect value for any truncated feature and would need to be updated manually.

Overlay with attributes

In a clip or an erase, the features that are extracted retain their attributes, but the attributes of the overlay layer are ignored. **Intersect** and **union** combine both areas and attributes.

The intersect and union functions are related to spatial joins in that they correlate features based on their spatial relationships to each other. However, spatial joins fail when spatial features do not overlap exactly. Consider the map of roads and land use shown in Figure 7.2. The state government has requested a report of the total miles of road falling into each land-use category. At first glance, a spatial join might be considered to generate this information; by joining the land-use polygons to the roads, one might expect to get a field showing the land-use type each road crosses.

Fig. 7.2. The selected road inside the square crosses three different land-use zones. An overlay splits the road so that each segment lies within a single polygon.

However, the single selected road circled in Figure 7.2, shown in blue, crosses three different land-use classes. How would a land-use type be assigned in this case? If one used the *completely within* join option, no land-use class would be assigned for the road at all because it does not lie entirely inside a polygon. A *summary* would do no good because a nominal data type such as land-use class can't be averaged.

The ideal method would split the road into three sections and assign the land use to each section, and this is what overlay functions do, thereby enforcing a one-to-one relationship between features and enabling a perfect correspondence when joining the tables. In Figure 7.2, the original single road is now three sections, and each segment in the output table (Fig. 7.3) is assigned the attributes of the land-use polygon containing it.

Fig. 7.3. During an overlay, each new road segment retains its original attributes and also receives the land-use code and other attributes of the polygon in which it falls.

FID	Shape*	LENGTH	ROADNAME	LANDUSE_ID	LU_CODE
2348	Polyline	0.00964	RANGE RD	72	Office/Commercial
2535	Polyline	0.00964	RANGE RD	70	Medium Density Residential
2564	Polyline	0.00579	CANYON LAKE DR	73	Public
3568	Polyline	0.00574	HILLSVIEW DR	354	Low Density Residential
3569	Polyline	0.00574	HILLSVIEW DR	354	Low Density Residential
3570	Polyline	0.00242		354	Low Density Residential
3653	Polyline	0.00482	CANYON LAKE DR	75	General Commercial
4984	Polyline	0.00733	32 ST	40	Floodway

Selected Attributes of road-lu

Map overlay with attributes occurs in two forms. **Union** combines two polygon layers, keeping all the areas and merging the attributes for both layers. **Intersect** also merges the attributes but retains only the areas common to both layers.

A **union** creates all possible polygons from the combination of features in both layers. In a union, both input layers must contain polygons. Figure 7.4 shows a union performed on

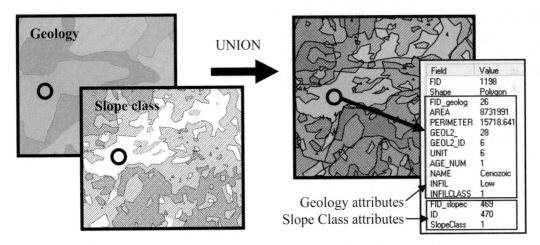

Fig. 7.4. A union creates every possible new polygon from the combined input layers. Each new polygon assumes the attributes from both layers; the new polygon marked with the circle contains attributes both from the geology and slope class layers.

geology and slope class layers to create a landslide hazard map. Landslide risk depends primarily on two factors, the strength of the geologic units (sandstones are less susceptible to sliding than shales) and the slope. The lighter areas in the slope class map have lower slopes, and the dark areas have higher slope. A union produces a new feature class with new polygons, each of which possesses the original attributes of its parents. The new feature class can then be evaluated for various combinations of slope and geology that may constitute a landslide hazard.

Overlaying layers often produces small extraneous polygons called slivers (Fig. 7.5). Sometimes these slivers represent real features. Often they arise when overlaying layers share some boundaries, such as voting districts and counties. In theory the shared boundaries should match exactly, but in practice few data sets have been corrected to this level of integrity. A tolerance can be set to eliminate features smaller than a specified distance, which will eliminate many slivers but not all. However, using this tolerance may also move the boundaries of the other features more than is acceptable and may degrade the spatial accuracy of the data set. Caution should be used in setting tolerances and in using feature classes that are outputs from geoprocessing that uses

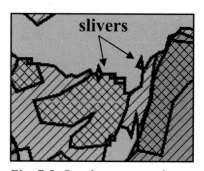

Fig. 7.5. Overlay may produce small sliver polygons.

tolerances. This problem highlights a fundamental rule of GIS—that the output of an analysis can only be as good as the data that are used to generate it.

The **intersect** operation is similar to a union, except that it only keeps the polygon areas that are shared by both input layers. This function provides a way to find out where two or more conditions hold simultaneously. Finding the potential habitat for a species with specific environmental requirements constitutes one example of using intersect.

Imagine that a rare species of snail inhabits the Black Hills. This snail prefers limey soils in cool and dense coniferous forest and is rarely found above an elevation of 1,600 meters or below 1,200 meters. By combining a geologic map, a vegetation map, and an elevation map, one can identify the likely areas of snail habitat. Such a map would help biologists to create sampling strategies for counting populations, to analyze whether the habitats are interconnected or widely separated, and to make decisions about forest management to protect the snails.

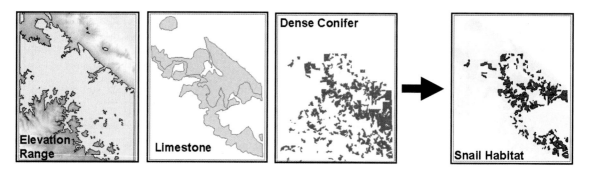

Fig. 7.6. Intersecting elevation, limestone areas, and dense conifer vegetation can help identify areas of potential snail habitat.

The analysis includes a series of steps using the layers shown in Figure 7.6. First, a digital elevation model (DEM) is queried to create a polygon showing the areas between 1,200 and 1,600 meters in elevation. Second, a query is applied to a geology map to extract the limestone units as a separate layer. Third, another query separates the dense conifer areas from the other vegetation. Finally, the intersection is performed in two steps, first intersecting the elevation range and limestone layers and then intersecting the result with the dense conifer map. The final output contains the areas common to all three layers—the snail habitat. These habitat polygons contain the full attributes of all three layers, should that information be needed, such as whether the conifers are ponderosa pine or white spruce.

Figure 7.7 summarizes the different overlay operations that can be performed using two polygon input layers, including the extraction functions clip and erase, which do not combine attributes, and the intersect and union functions, which do. Clip and erase can also accept point and line feature classes for one input, but the boundary input must contain polygons. The output feature classes will have the same geometry as the input classes.

Whereas the union function requires two polygon input layers, the intersect command is more versatile. Both input and overlay layers may contain points, lines, or polygons. The output geometry may vary, but it cannot exceed the dimensionality of the lowest dimension input. In other words, if the inputs are both polygons, the output may be points, lines, or polygons (Fig. 7.8a). If the inputs are both lines, the output may be lines or points (Fig. 7.8b). If the inputs are polygons and lines, the output may be points or lines (Fig. 7.8c). If the input is points, the output must be points.

Other spatial analysis functions

Overlay is only one type of spatial analysis available in GIS systems. The remainder of this chapter presents additional commonly used functions including dissolving, buffering, appending, and merging. All of these analysis functions, and many others, are found in ArcToolbox.

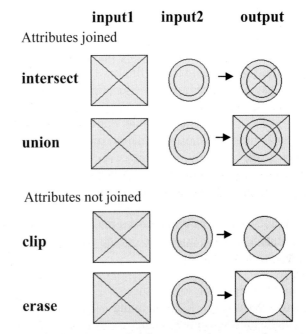

Fig. 7.7. Summary of polygon overlay operations and results

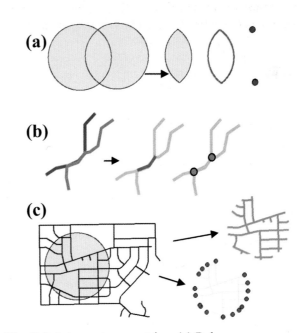

Fig. 7.8. Intersect geometries. (a) Polygons can yield polygons, lines, or points. (b) Lines can yield lines or points. (c) Polygons and lines can yield lines or points.

255

Dissolve

A **dissolve** is used to group features together based on whether they share the same value of an attribute field. For example, the road segments in Figure 7.9 have the same street name (Main St), but they are separate features. A dissolve based on the street name field would yield a new file in which all streets having that name are one feature. A dissolve can also be used to remove lines between polygon features that share the same value. For example, the map in Figure 7.10 shows ponderosa pine stands of different ages. Dissolving on the cover type field removes the age boundaries and yields polygons based on the cover type (ponderosa) only.

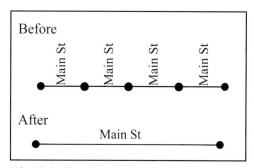

Fig. 7.9. Separate road segments were combined using the dissolve function.

By default, a dissolve will produce multifeatures in which spatially distinct lines or polygons may constitute a single feature. Each of the orange "polygons" in Figure 7.10 is actually part of a single polygon multifeature.

When features are dissolved, the output layer is a new file containing the single attribute on which the dissolve was based. However, the user can specify additional fields to summarize the information from the original features. For example, in the dissolve shown in Figure 7.10, one might request the average tree age of each output polygon, based on averaging the tree ages of the polygons before the dissolve.

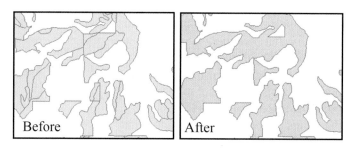

Fig. 7.10. A dissolve removes boundaries between polygons with the same attribute value, in this case, tree species.

Buffer

Buffering is used to identify areas that fall within a certain distance of a set of features. Buffers can be created for points, lines, or polygons (Fig. 7.11). They could be used to find 300-yard drug-free zones around schools or sensitive protected areas within 100 meters of a stream. Negative buffers can be applied to determine setback limits from the edge of a piece of property. Buffers can be created as simple rings or as multiple

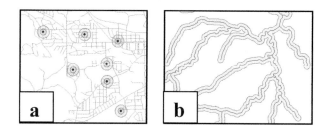

Fig. 7.11. Examples of buffers around (a) wells and (b) roads

rings. An attribute can even supply different sizes of buffers for different features, for example, buffering primary roads by 200 meters and secondary roads by 100 meters. Buffers may be saved as graphics in the map layout, as a shapefile, or as a geodatabase feature class.

Buffers are created for each individual feature and may overlap. This option might be appropriate in analyzing the distribution of soils in buffers around wells. In many cases, however, it is best to dissolve the individual buffers to eliminate overlapping areas. This step can be performed within the buffer operation by setting the Dissolve Type to ALL. Figure 7.12 shows the difference in results obtained using the default NONE option and the ALL option. (Determining areas for the buffers around the roads using the NONE option would result in grossly overestimated areas because the overlapping areas would be counted multiple times. Instead, the ALL option must be used so that no areas overlap.)

Finally, buffering is an intensive process, and the time involved increases quickly as the number of features increases. Any steps to reduce the number of buffers created at one time will facilitate analysis, for example, by performing queries before buffering rather than afterwards.

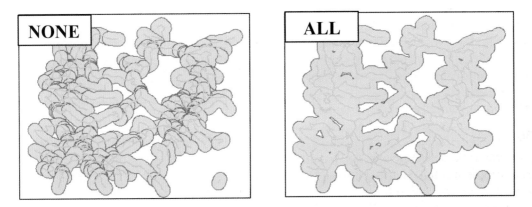

Fig. 7.12. By default the Buffer tool creates undissolved buffers using the NONE option, as shown on the left. In this case the ALL option produces the desired result.

Append and Merge

The **Append** tool is used to combine the features of two or more layers to create a single existing layer (Fig. 7.13). The appended layers must have the same feature type (i.e., both polygons, both lines, or both points), and the output layer will have this same feature type. The two layers must also share the same coordinate system. Overlapping of layers is permitted.

The treatment of attribute tables during an append requires some consideration. If you wish to combine the attribute information from both layers, the attribute fields must have the same definition and must occur in the same order in both tables. If the two tables differ, one can either modify the tables to match (not easy in many cases) or combine the features without bringing the attribute information into the output layer (NOTEST option).

Merge is similar to append, except that it offers more flexible treatment of attribute tables. Instead of insisting that the two tables match, it allows the user to specify the fields to be included in the output feature class and which table they will come from.

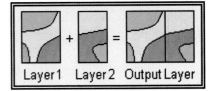

Fig. 7.13. Append combines the features of two adjacent layers.

257

About ArcGIS

About geoprocessing

GIS analysis involves many functions that operate on data objects, such as feature classes or tables. Geoprocessing applies one or more of these functions in sequence to solve a problem or investigate the properties of data sets. Most ArcGIS tools operate on any allowed data type. In performing a road buffer, for example, one may specify a coverage, a shapefile, or a feature class from a geodatabase. The tools make the necessary allowances to execute the function on the various data types. A few tools operate only on certain data types. An ArcInfo license, for example, provides the user with a suite of tools that work only on coverages.

Ways to run tools

Geoprocessing tools may be executed several ways. *Toolbars and menus* provide interactive control of tools. Users may customize menus and toolbars to add frequently used tools that don't appear on the toolbars by default. Users can also create tools and place them on menus, on toolbars, or in ArcToolbox.

ArcToolbox organizes all of the installed tools into one central location. It provides a window with most of the functionality of ArcGIS, containing tools that can be run from ArcMap or from ArcCatalog. Each tool has a set of parameters, or inputs, which must be specified before the tool is run. Some parameters are required, but others are optional, which means that the software supplies a default value that can be changed, if necessary.

ArcMap has a command line that allows the user to type a command and its parameters rather than filling in boxes in a window. Typing requires greater familiarity with the tools, but for experienced users entering commands by typing may be faster and more efficient. Becoming familiar with the command line also paves the way to writing scripts to execute functions. The command line possesses a sophisticated interface to help beginners correctly enter the commands.

Model Builder provides a graphic canvas to string tools together and execute them in sequence. Models have several advantages. First, they record the steps and parameters used to execute an analysis in case questions on methodology arise weeks or months later. Second, models can be used to explore the consequences of different parameters on the final outcome. Consider a model that calculates fire danger from precipitation, temperatures, vegetation, and structure locations through a series of tools. With a model, one could redo the analysis based on the current day's precipitation and temperature conditions more quickly than if all the steps were executed interactively and with less chance of mistakes. Models can be shared with other users, and they permit a less experienced user to reliably repeat an analysis set up by an expert. Models can be incorporated inside other models in order to perform sophisticated chains of analysis functions. Finally, models can be saved as scripts to provide a starting point for geoprocessing programs.

Scripts are programs that may contain conditional statements (if/then), iterative loops, and other control structures that permit sophisticated analysis. Scripts may be written in one of several programming languages, although in each case the geoprocessing commands are identical and only the control statements differ. ESRI, Inc., has adopted the readily available language Python as the recommended scripting platform. A copy of Python is installed with ArcGIS, and all ESRI scripting examples and supplemental tools are written in Python. However, users may elect to use any COM-compliant language such as Jscript, VBScript, or Perl.

Geoprocessing Environments

The operation of tools and commands is impacted by environmental settings specified by the user. For example, the default coordinate system setting is "Same as Input", meaning that the output file has the same coordinate system as the input file. The user could alternatively specify a particular coordinate system, such as UTM Zone 13 NAD 1983. Under this setting, every output feature class would have the UTM coordinate system regardless of its input system.

Environmental settings can be set at several levels and are hierarchical; they carry through to lower levels unless changed. Application settings are set using the main menu bar in ArcMap or in ArcCatalog. Models, scripts, and tools can have their own environment settings, which take precedence over the application settings. Even tools inside models may have settings, which override the model and application settings.

The default settings work fine for most applications. Users may set them when doing so provides more efficient or accurate work. For example, Amy might have to clip and project 10 different feature classes when building a geodatabase for study. Instead of running both the clip and project tools on each layer, she could set the environment coordinate system to the desired projection and then run the clips, reducing the amount of work.

Coordinate systems

Tools will accept any coordinate system or combination of coordinate systems for input, but specific rules of precedence dictate the coordinate system of the output. Outputs will be projected on the fly, if necessary, according to the following rules.

> ➤ If the output is placed in a feature data set, the coordinate system will always match that of the feature dataset.

> ➤ If the coordinate system is set in the Environment settings and the output will not become part of a feature data set, the Environment settings coordinate system is used.

> ➤ If the Environment setting is not set, the default rule applies—that the output will match the coordinate system of the first input to the tool.

When setting the output coordinate system in the Environment settings, you can choose Same as Input (the default) or, Same as Display (the current data frame), or set it to match the coordinate system of a particular data set.

Geoprocessing tools are fundamentally spatial in nature and commonly manipulate areas and distances as part of the processing. When geoprocessing, it is best to always use a projected coordinate system, rather than a geographic coordinate system (GCS). As with spatial joins, it is not enough to set the data frame coordinate system to a projection. The tools operate on the saved feature class on disk, and the saved coordinate system is the one being used.

Often a GCS is employed by data providers because they cannot predict what coordinate systems the user will need. When downloading data, it will often be in a GCS, and the user must choose an appropriate coordinate system and project the data to it. Chapter 11 explains this process. When setting up a database for a mapping and analysis project, a suitable projection should be chosen based on the scale and the extent of the data being analyzed. It must be kept in mind that distortions present in the map projection, if significant, will affect distance and area measurements. It remains important to choose a suitable map projection with minimal distortion of area and distance. Review the coordinate system properties in the inside front cover and the coordinate system selection guidelines in Chapter 3, if necessary.

Areas and lengths of features

Geodatabases and coverages automatically create and update fields containing the area and perimeters of polygons or the lengths of lines. Shapefiles do not maintain this information. Users may create AREA or LENGTH fields for shapefiles and calculate the values manually. However, employ caution when using information from these fields. If polygons in a shapefile are clipped, dissolved, intersected, or undergo any other operation that changes their shape, the AREA fields will not automatically be updated. The user must manually update the fields again to ensure they are correct. Never assume that an AREA or a PERIMETER or a LENGTH field in a shapefile contains the correct values unless you yourself have created them and kept them up to date. Chapter 4 introduced the Calculate Geometry tool for determining areas, lengths, and perimeters in a variety of units.

Using ArcToolbox

ArcToolbox may be used either in ArcCatalog or in ArcMap. When a tool is opened (Fig. 7.14), required inputs are marked with a green dot (a). To enter data sets as inputs or outputs, use the Browse button next to the appropriate box (b). If using the tool in ArcMap, you can also select layers in the map from a drop-down list as inputs. Optional inputs are marked (c) and can usually be left to their defaults. Some tools have input parameters that extend beyond the visible pane, and the scroll bar must be used to view and enter them. To get more information about the tool or an input, click the Show/Hide Help button (d) for a brief description of the tool's function. For more detailed help, especially on the input/output data, click the Tool Help button (e).

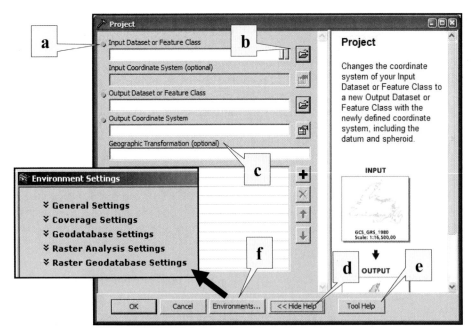

Fig. 7.14. Preparing to run a tool

TIP: When using a tool for the first time, it is strongly recommended that you read the detailed help information. It may contain critical information for making the tool work properly.

The operation of tools may be affected by the Environment settings for the geoprocessing environment. For example, the user can set a default output directory or a default *x-y* extent for the output. These settings can be accessed by clicking on the Environments button at the bottom of the tool (f).

The input pane for a tool posts symbols to help you use the tool (Fig. 7.15). A green dot next to an input parameter indicates that it is required and must be entered before the tool will run. A yellow exclamation point indicates that the user should proceed with caution. Placing the cursor over the yellow sign for a moment will display a pop-up message describing the problem; in Figure 7.15, the user is attempting to define a coordinate system for a data set that already has one. A red X indicates an error that will cause the tool to fail. Again, holding the cursor over the X will display the error message.

When the tool begins running, a dialog box appears. If an error occurs, it will be reported in

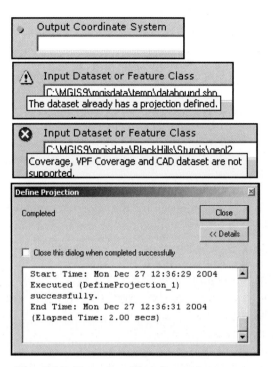

Fig. 7.15. Error handling for tools

the dialog box. By default the dialog will remain open after the tool finishes executing; a check box allows the user to direct it to close automatically upon successful completion of the task. If an error occurs, the dialog will remain open so that the error message can be read.

Users interested in the full capabilities of ArcToolbox and the geoprocessing environment should consult the extensive sections in the ArcGIS Help (search on the index tab for geoprocessing).

Summary

➢ Extraction functions, such as clip and erase, separate features of interest based on a "cookie cutter" provided by a polygon layer and can truncate features at the cutter boundaries. Attributes are carried through without change.

➢ Map overlay resembles a spatial join, but it splits features when they partly overlap. This function enforces a one-to-one relationship between features when their attributes are joined in the output table.

➢ Map overlay comes in two basic types. A union keeps all the features from both layers. An intersect keeps all the features that are common to both input layers. Attributes from both layers are joined together in the output.

➢ A dissolve combines features within a data layer if they share the same attribute. This function can be used to convert many street segments into a single line feature or to remove boundaries between parcels with the same zoning.

➢ Buffers are polygon constructions that enclose all the area within a certain distance of a set of features. Buffers may be created for points, lines, or polygons, and they may be constructed as single or multiple rings.

➢ Append and merge allow feature classes with the same feature type to be combined as a single feature class, such as merging two adjacent quadrangles to make a single data file.

➢ Map overlay and spatial analysis are generally best done using a projected coordinate system, especially if one anticipates determining areas and lengths as part of the analysis.

➢ Geoprocessing involves stringing together a sequence of commands during spatial analysis. Spatial functions may be executed from the menus, ArcToolbox, the command line, ModelBuilder, or scripts.

➢ The Shape_Area, Shape_Length, and Shape_Perimeter fields in geodatabase feature classes are stored and updated automatically. Other area-based or length-based fields, such as Area, Acres, or Road_km, are NOT automatically updated.

➢ Lengths and areas of features stored in shapefiles must be calculated and updated manually.

TIP: To temporarily clip features for display in ArcMap, use the Data Frame tab of the Data Frame Properties menu.

Chapter Review Questions

You may need to consult the Skills Reference section to answer some of these questions.

1. How does the clip function differ from Select By Location?

2. What is the most important difference between a spatial join and a map overlay?

3. Why would it be superfluous to use the summarize capability of spatial joins when doing a map overlay?

4. What is a buffer? What two choices are offered for the output format?

5. What function would you use to create a map of a study area such that all the features in the map stopped at the study area boundary?

6. What attribute fields will be present in a layer resulting from a dissolve?

7. Why is it usually advantageous to use a projected coordinate system when doing map overlay?

8. How can you determine the areas of polygons in a geodatabase? In a shapefile?

9. What determines the coordinate system of the output when overlay is used?

10. What is geoprocessing? In what different ways can commands be executed?

Mastering the Skills

Teaching Tutorial

The following examples provide step-by-step instructions for doing basic tasks and solving basic problems in ArcGIS. The steps you need to do are highlighted with an arrow ➔; follow them carefully. Click on the video number in the VideoIndex to view a demonstration of the steps.

To demonstrate map overlays, we will do the problem described earlier in the chapter concerning the rare Black Hills snail. As described, we will use overlays to define the potential snail habitat. This habitat will be defined by three criteria: on a limestone geology unit, in dense coniferous forest, and between the elevations of 1200 and 1600 meters.

➔ Start ArcMap and open the map document ex_7a.mxd.

➔ Use Save As to rename the map document. Remember to save it often as you work.

As in Chapter 7, we don't want the new feature classes created during the analysis to become part of the Sturgis geodatabase, so we'll create another geodatabase to contain them. We'll also set the geoprocessing Current Workspace as the default location to place output data.

1➔ Open ArcToolbox, if necessary, and open the Data Management > Workspace > Create Personal GDB tool.

1➔ Specify the mgisdata\BlackHills folder as the output location and name the output geodatabase chap7results. Click OK.

1➔ Choose Tools > Options and click the Geoprocessing tab. Click Environments.

1➔ Expand the General settings and set the Current Workspace to chap7results. Click OK and OK.

Preparing to overlay

Our first step is to create layers that contain polygons where each condition holds. We will begin with the Geology layer. The limestone areas include the Madison Formation and the Upper Paleozoic units. We will use an attribute query to place these units in a separate layer.

2➔ Choose Select By Attributes from the main menu bar.

2➔ Set the layer to Geology and enter the expression [NAME] = 'Upper Paleozoic' OR [NAME] = 'Madison Limestone'. Click OK.

2➔ Right-click the Geology layer and choose Selection > Create Layer from Selected features. Name the layer Limestone.

Next, we need to select the dense conifers and create a layer from them. We will do the query in two steps so the expression is less complicated. First we will select the ponderosa pine (TPP) and white spruce (TWS). Then we will select the dense areas (class contains C or 5) from the already selected set.

3➔ Turn on the Vegetation layer and turn off the Limestone layer.

3➔ Clear all selected features.

3➔ Open the Select By Attributes box again and set the layer to Vegetation.

3➜ To select only the conifers, enter the expression [COV_TYPE] = 'TPP' OR [COV_TYPE] = 'TWS'. Click Apply.

4➜ Change the selection method to *select from current selection*.

4➜ Clear the expression and enter another one that says [SSTAGE96] LIKE '*C*' OR [SSTAGE96] = '5'. Click Apply.

4➜ Close the Select By Attributes window.

4➜ Convert the selected features to a new layer and name it **Dense Conifer**. It should look like the polygons shown in Figure 7.16.

4➜ Turn off the Vegetation layer.

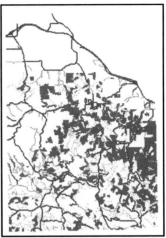

Fig. 7.16. Dense conifer

The elevation range has already been prepared. You are almost ready to begin.

5➜ Turn off all layers except the Elevation Range layer, the Limestone layer, and the Dense Conifer layer.

5➜ Zoom in to the middle right part of the map and examine the Dense Conifer polygons.

Notice that these polygons have many boundaries inside them because they are divided based on age and density as well as species. The amount of time needed to intersect a layer is proportional to the number of features in the layer. To streamline the intersect, we are going to remove the unnecessary boundaries between these polygons using the Dissolve tool.

6➜ Navigate the ArcToolbox tree through Data Management Tools > Generalization > Dissolve and double-click the Dissolve tool to start it (Fig. 7.17).

When using a tool for the first time, it is wise to select Show Help to find out what the tool does and what the parameters mean.

6➜ Click Show Help on the Dissolve tool and read the description.

6➜ Click on the Input Features box. The Help message changes to describe the Input Features parameter.

6➜ Click on and read the descriptions for the other input parameters in the window.

6➜ To bring the tool description back, click on the gray tool area.

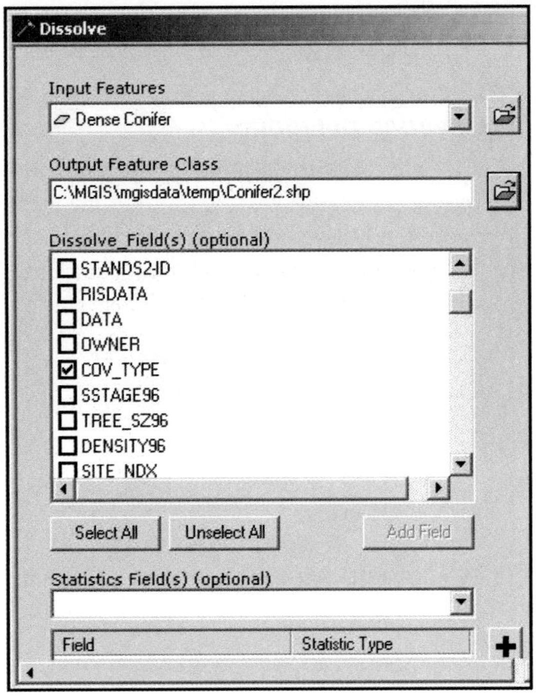

Fig. 7.17 The Dissolve tool

7➔ Click on the Tool Help icon at the bottom of the tool for more detailed information (or the Help icon at the top of the tool for version 9.2). Read about the Dissolve tool.

7➔ Scroll down to the bottom of the description. Note in particular the sections pertaining to the use of the tool from the command line and in scripts.

7➔ Close the ArcGIS Desktop Help window and click Hide Help on the Dissolve tool.

8➔ Click the drop-down arrow in the Input Features box and set it to Dense Conifer.

8➔ Click the Browse button to place the Output Feature class in the BlackHills/chap7results geodatabase and name it conifer2.

8➔ Check the box to dissolve on the COV_TYPE field.

8➔ Notice the other options in the tool. We don't need to change any of them, however.

8➔Click OK. A processing window will inform you of the status of the analysis. If any errors or warnings are encountered, messages will appear in the window.

8➔ When the tool finishes, click the Close button to close the message window. Check the box if you want it to close automatically every time it finishes.

TIP: When using a tool, be aware that some options may be out of sight, and you must scroll down to see them. Often the defaults are fine, but get in the habit of checking before you finish.

9➔ Examine the output file and note that the intervening boundaries have disappeared.

9➔ Right-click the Dense Conifer layer and choose Remove.

9➔ Rename the conifer2 layer Dense Conifer.

Intersecting polygons

Now we are ready to do the overlay. We will use Intersect because we want to find the areas common to all three criterion layers. Because the tool can only intersect two layers at a time, we must use it twice.

10➔ Open the ArcToolbox > Analysis Tools > Overlay > Intersect tool (Fig. 7.18).

10➔ Click Show Help and examine the entries.

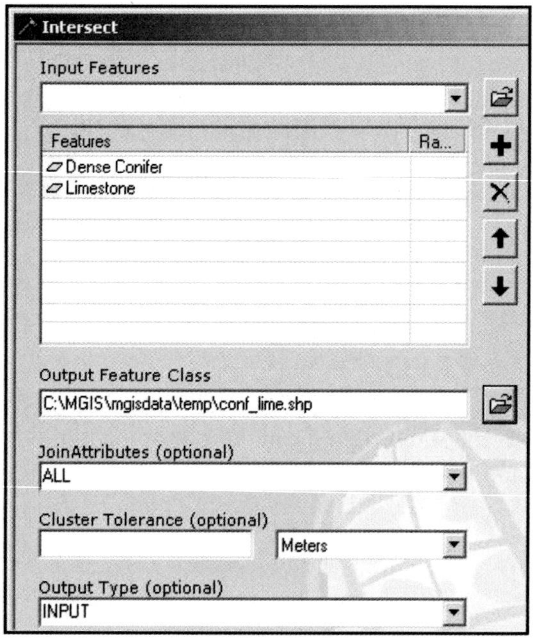

Fig. 7.18. The Intersect tool

1. What are the options for the Join Attributes parameter? Which one do you think is best to use? _____

11➜ Click on the drop-down button under Input Features and choose the Dense Conifer layer. It will be added to the list of Features.

11➜ Click the drop-down button again to add the Limestone layer to the list.

If you have an ArcInfo license, you can intersect multiple layers in one step. Add the Elevation Range layer now, name the output **snailhab**, and then skip to step 13. If you have an ArcView or an ArcEditor license, continue with step 11 to do it in two steps.

11➜ Enter the Output Feature Class, placing it in the chap7results geodatabase and naming it **conf_lime**.

11➜ Set the Join Attributes option to ALL.

11➜ The remaining options can be left with their defaults. Click OK and close the message window after it finishes, if necessary.

11➜ Examine the output. The conifers outside the limestone area have disappeared.

12➜ Use the Intersect tool again to intersect conf_lime and the Elevation Range layers, naming the output **snailhab** and putting it in the chap7results geodatabase.

13➜ Zoom back out to the full extent of the data.

13➜ Rename the snailhab layer **Snail Habitat**.

13➜ Remove the Limestone, Dense Conifer, and conf_lime layers.

13➜ Zoom in to the extent of the Snail Habitat and turn on the Roads layer.

Overlay of lines in polygons

The snails have a three-week breeding season in early June. During this period they seek the open areas offered by roads, and many get crushed. The Forest Service wants to consider closing primitive roads that traverse through snail habitat during the breeding season to lessen the number of crushed snails. They need to assess which roads must be closed. You will make a map with the proposed road closures. This process involves a line-in-polygon intersection.

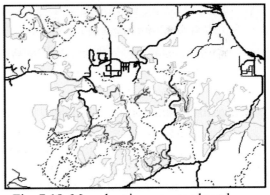

Fig. 7.19. Map showing proposed road closures in snail habitat areas

14➜ Use Select By Attributes to select the Primitive roads [TYPE] = 'PR'. (Make sure you set the method back to Create a new selection.)

15➜ Use the Intersect tool to intersect the Roads layer (with the primitive road selection) and the Snail Habitat layer. Name the output **propclose** and put it in chap7results.

15➜ Rename the propclose layer **Proposed Closures**. Clear the selected features.

TIP: The ArcToolbox tools automatically use the selected features from the input layers, so we could skip creating a separate layer. However, the intervening selection layers can be helpful if we make a mistake and must repeat the analysis.

16➔ Create a map similar to Figure 7.19, with the closure roads highlighted in red and the other roads in black. Turn off the Elevation Range layer. Set the transparency of the Snail Habitat layer to 50% to help bring out the road patterns.

Clipping layers

The vegetation layer does not extend as far north as the rest of the data, and it may give readers the false impression that the analysis is valid there. Clipping the Roads layer at the edge of the vegetation layer will prevent any misunderstanding.

Now you know that you need the Clip tool. Let's practice finding tools when you know the name but not where to look for it in the Toolbox.

17➔ In ArcToolbox, click the Index tab. Type clip in the box.

17➔ Two functions appear, Clip (analysis) and Clip (management). The words in parentheses refer to the area of the Toolbox.

17➔ Open each tool and click the button to Show Help.

17➔ It becomes clear that the Clip (management) tool is for rasters, so Clip (analysis) is the one to use. Close both Clip windows for the moment.

17➔ To learn where the Clip tool resides in the Toolbox, click Clip (analysis) to select it and then click the Locate button.

18➔ Double-click the Clip tool to open it.

18➔ Set the input features to Roads. Set Vegetation as the clip features.

18➔ Store the output in the chap7results as roadclip. Click OK.

You can quickly transfer the symbology of the old Roads layer to the clipped roads.

19➔ Open the properties of the roadclip layer and click the Symbology tab.

19➔ Click the Import button and choose to import symbology from a layer or layer file.

19➔ Click the drop-down box to set the layer to Roads. Click OK.

19➔ The road classification is based on the TYPE field. The clipped roads use the same field, so leave the TYPE name in the box. Click OK and OK.

20➔ Remove the old Roads layer.

20➔ Rename the roadclip layer Roads.

20➔ Click and drag the Roads layer below the Snail Habitat layer to more easily see the proposed road closures.

Working with buffers

The Forest Service biologist wants to do one more analysis. To prevent snail death on the major roads, which cannot be closed, they are considering thinning the tree stands within 200 meters of the roads. This thinning would make the stands less ideal snail habitat, which might keep the snails away from the roads. The biologist wants to prepare a map showing the stands to be cleared and the total percentage of snail habitat that the clearing would eliminate.

21➔ Use Select By Attributes to select the primary and secondary roads from the Roads layer, using the expression [TYPE] = 'P' OR [TYPE] = 'S'.

21➔ Create a layer from the selected roads and name it **Major Roads**.

TIP: Buffers are time consuming to create, so it is helpful to reduce the number of features before buffering. We will use the Elevation Range layer to clip the roads before buffering.

22➔ Open the Clip tool. Set Major Roads as the Input Features and Elevation Range as the Clip Features.

22➔ Save the result in chap7results as **majroadclip**.

22➔ Turn off the other road layers to examine the new file.

Now buffer the clipped roads.

23➔ Locate the Buffer tool in ArcToolbox and open it.

23➔ Set the Input Features to majroadclip and the Output Feature Class to **roadbuf** in chap7results.

23➔ Set the Linear unit to meters and type **200** in the box.

23➔ Scroll down, if necessary, and set the Dissolve Type to ALL. Click OK.

TIP: By default the Buffer tool does not dissolve boundaries between buffers. This is one case where the default NONE is not the desired option, and it must be set to ALL.

Now intersect the road buffers with the snail habitat to determine the potential areas to be cleared.

24➔ Open the Intersect tool.

24➔ Choose Snail Habitat and Roadbuf as the Input features.

24➔ Specify the Output file as **proposedthin** in chap7results. Click OK.

25➔ Clean up the map display to show the results. Turn off the roadbuf layer and make the proposedthin layer a symbol that contrasts well with the snail habitat (Fig. 7.20).

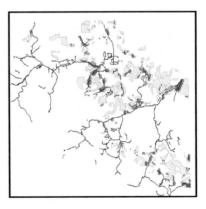

Fig. 7.20. Proposed thinning areas (purple) and road closures (red)

Determining lengths and areas of features

Now the final step is to determine the area of snail habitat that will be eliminated by the thinning. Geodatabases automatically keep track of the areas, lengths, and perimeters of features.

26➔ Open the proposedthin attribute table and examine the Shape_Area and the Shape_Perimeter fields.

These fields were renamed Shape_1_Area and Shape_1_Perimeter during one of the intersections, but they will still contain the correct areas and perimeters for the feature class. The units of the Shape_Area field will be in the same coordinates as the map units of the feature class coordinate system, in this case, square meters.

26➔ Use the Statistics function to determine the total area to be thinned.

2. What is the total area of the proposed thinning in km^2? _____

3. What is the total area of snail habitat in km^2? _____

4. What percentage of the habitat would be eliminated if the proposed thinning is done? _____

TIP: Shapefiles do not automatically maintain area, length, or perimeter fields. Values found in these fields were originally created by users and may not be correct if geoprocessing actions have been performed on the shapefile since the fields were created. These fields may be updated using the Calculate Geometry function described in Chapter 4.

Geoprocessing from the command line (optional)

Some people find that typing commands and arguments is more efficient than running a tool. For those who prefer this method, the command line is available. Let's run the previous buffer and intersect operations from the command line to see how it works.

27➔ Click on the Show/Hide Command Line button on the main menu bar.

27➔ Dock or float the command line window underneath the map or wherever you find it comfortable.

When typing inputs and outputs for commands, it is easier to type only the name of the feature class rather than the entire path. Thus, setting the current workspace is often the first command you will type in the Command Line window. The current workspace is the default location for all input and new output feature classes, tables, and so on that result from commands.

28➔ Click in the box that says Enter Commands Here and start to type the command **workspace**. Notice that a pop-up alphabetical list of commands appears once you type the **W**, allowing you to find a command if you don't know the full name for it.

28➔ After typing **workspace**, press the space bar.

A new pop-up appears above the command line, showing the syntax and arguments for the commands. Required arguments are enclosed in <angle brackets>, and optional arguments are enclosed in {curly brackets}. The argument currently being entered is shown in **boldface**. This command has only one argument, the pathname to the desired workspace.

28➜ Type the pathname to the chap7results geodatabase. The exact pathname will depend on where you installed it, but the command should look similar to this:

workspace c:\mgisdata\blackhills\chap7results.mdb

28➜ Press Enter. Notice that commands and arguments are not case sensitive.

TIP: The command line is supposed to use the current workspace specified in the Geoprocessing Environment settings. However, the author has found that setting it there either does not work or only sometimes works, at least up to version 9.2. The Workspace command produces more reliable results.

29➜ Type buffer in the command window and press the space bar.

The buffer command has many arguments. The first argument is the input layer to be buffered. Note the appearance of yet another pop-up showing the current map layers that could be used for this argument.

29➜ Start typing the letters majro until the majroadclip layer is highlighted in the list. Press the Tab key to enter the rest of the name.

29➜ Press the space bar to enter the next argument, the output feature class. Type roadbuf400. It will be placed in the current workspace.

29➜ Press the space bar and enter 400 for the buffer distance.

The remaining arguments are optional. If you leave them all blank and finish the command here, the defaults will be used. However, you want to use a nondefault value, ALL, for the dissolve option. To skip an argument and its default, enter a pound sign (#) followed by a space. We'll skip the first two optional arguments, enter ALL for the third, and ignore the last.

29➜ Press the space bar and type # for the {line_side} argument.

29➜ Press the space bar and type # for the {line_end_type} argument.

29➜ Press the space bar and type or choose ALL (lowercase is OK).

29➜ The final command should look like this:

Buffer majroadclip roadbuf400 400 # # all

29➜ Press the Enter key to complete the command and execute it.

TIP: The command line is usually most efficient when you type every argument rather than mixing typing with using the mouse. Knowing the syntax of the common commands helps also.

You can recall and edit previously entered commands, another way to increase efficiency.

30➜ Click the Up Arrow key on the keyboard. The previous command appears. Use the Up and Down Arrow keys to scroll through previously executed commands.

30➜ Edit the previous buffer command so that the output feature class is called roadbuf600 and the buffer distance is 600.

30➔ Press Enter to execute the command again with the new arguments. (The cursor may be anywhere in the command when you press Enter, not just at the end.)

Next, we'll execute the Intersect command between the two layers just created, roadbuf400 and roadbuf600, to create a 200-meter-wide strip between them. The intersect command takes two feature class inputs separated by a semicolon.

31➔ Type the command intersect roadbuf400;roadbuf600 roadstrip and press Enter.

Finally, if you want to use a feature class that is not in the map and not in the current workspace, you can type the relative pathname to get there. Suppose that we want to dissolve the vegetation layer veg2006 that is in the Sturgis geodatabase, in the same BlackHills folder as the current workspace. The relative pathname starts with ".." to get us up and out of chap7results and then "\sturgis.mdb" to get down into the other geodatabase.

32➔ Type dissolve and press the space bar.

32➔ Enter the input feature class ..\sturgis.mdb\veg2006 and press the space bar.

32➔ Enter the output feature class covtype06 and press the space bar.

32➔ Enter the dissolve field cover_type from the list.

32➔ The entire expression should read

dissolve ..\sturgis.mdb\veg2006 covtype06 cover_type. Press Enter.

➔ Close the command line window.

TIP: Notes and information on command syntax are found in the Help files. The command line is persnickety about minor details such as spaces, so spend some time learning the command arguments and syntax. Once you become familiar with it, however, the command line can be an efficient way to analyze data and a valuable way to begin learning to create scripts.

Geoprocessing with Model Builder (optional)

This next section shows some basics of working with Model Builder to record and redo sequences of geoprocessing steps. Since the snail habitat problem is already familiar, we'll use that as our example.

➔ Open the original version of ex_7a.mxd without any changes.

➔ Use Save As to save it under a new name and save often as you work.

Models are created inside your own toolbox. Thus, to create a model, you must first create a toolbox to contain it.

33➔ Open ArcToolbox, if necessary, and dock it underneath the Table of Contents.

33➔ Right-click the ArcToolbox entry at the top and choose New Toolbox. Name the toolbox **Chap7**.

33➔ Right-click the Chap7 toolbox and choose New Model. The Model Builder window appears, currently empty.

34➔ Choose Model > Model Properties from the Model Builder menu bar.

34➔ Name the model **SnailHab** and give it a Label and Description (Fig. 7.21.) Click OK.

34➔ Choose Model > Save.

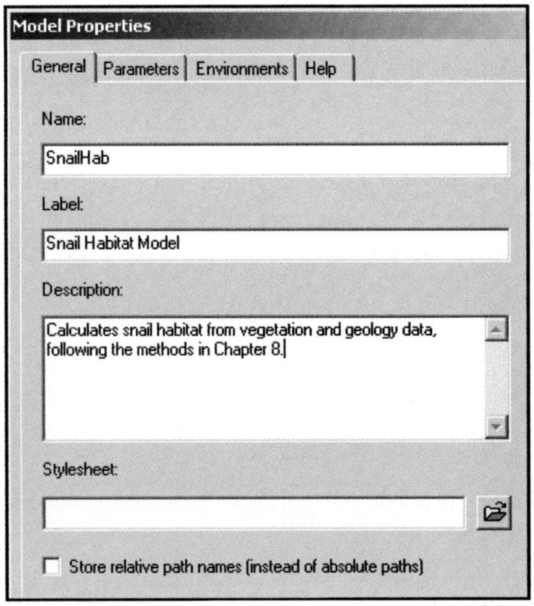

Model Properties

General | Parameters | Environments | Help

Name:
SnailHab

Label:
Snail Habitat Model

Description:
Calculates snail habitat from vegetation and geology data, following the methods in Chapter 8.

Stylesheet:

☐ Store relative path names (instead of absolute paths)

Fig. 7.21. Naming and describing a model

TIP: Storing relative pathnames for models is a good idea if you want to be able to use it on data in multiple folders.

In Model Builder, you paste the tool to be used in the window, set arguments for the tool, and run it. Each step can be run separately as the model is built or all together. For the first time, it is usually better to execute the steps one at a time. In the beginning of this tutorial, we used Select By Attributes from the menu to create layers for input from the geology and vegetation layers. In Model Builder, we use the Make Feature Layer tool.

35➔ Expand the toolbox to find the ArcToolbox > Data Management Tools > Layers and Table Views > Make Feature Layer tool. Click and drag it to the Model Builder window.

35➔ Right-click the Make Feature Layer tool in the Model Builder window and choose Open.

35➔ Set the Input Features to Geology and the Output Layer as **Limestone**, as shown in Figure 7.22.

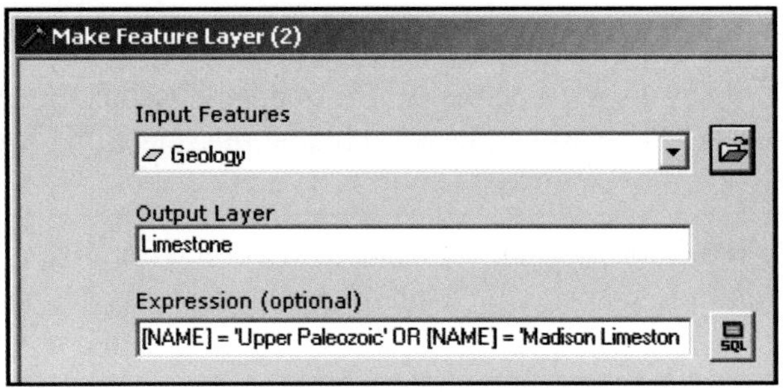

Make Feature Layer (2)

Input Features
▱ Geology

Output Layer
Limestone

Expression (optional)
[NAME] = 'Upper Paleozoic' OR [NAME] = 'Madison Limeston

Fig. 7.22. Setting the arguments for the Make Feature Layer tool to create the Limestone layer

36➔ Click the SQL button to enter the selection expression from Step 1, [NAME] = 'Upper Paleozoic' OR [NAME] = 'Madison Limestone'. Click OK to close the SQL window and OK to close the tool.

36➔ The boxes in the model become colored, indicating the tool is ready to run. Choose Model > Run, or click the Run button to execute this part of the model. A drop shadow behind the model shapes indicates that the tool has been run (Fig. 7.23).

The blue ovals indicate inputs to tools, and the green ovals indicate outputs. Both inputs and outputs have properties that can be set.

36➔ Right-click the green Limestone oval and choose Add to Display. The new layer appears in the Table of Contents and is shown on the map.

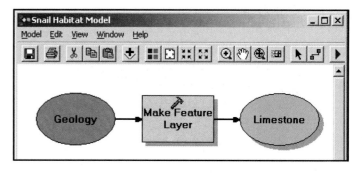

Fig. 7.23. The model after the first step is run

Now we'll use the same tool to select the dense conifers. Because we're using a tool, it is easier to do the selection in one step.

37➔ Drag another copy of the Make Feature Layer tool from ArcToolbox onto the model canvas.

37➔ Right-click the tool to open it. Set the Input Features to Vegetation and the Output Layer to Dense Conifer.

37➔ Click the SQL button and carefully enter the expression (without the line break):

([COV_TYPE] = 'TPP' OR [COV_TYPE] = 'TWS') AND

([SSTAGE96] LIKE '*C*' OR [SSTAGE96] = '5').

37➔ Click the Verify button to ensure that you did not make a mistake in the query.

> **TIP:** Be careful when entering the expression, as it is complicated. Be sure to include the parentheses around the two COV_TYPE conditions and around the SSTAGE96 conditions to ensure the correct order of interpretation.

38➔ Click OK and OK to close the SQL window and close the tool window.

38➔ Click the Run button. Right-click the Dense Conifers output and choose Add to Display. Make sure that the selection looks correct.

> **TIP:** Click on a model shape to select it. Use click and drag boxes or the Shift key to select multiple shapes. Click and drag selected shapes to move them around on the model canvas or to resize them. Use these techniques to arrange the model in neat patterns.

The Auto Layout button can be used to quickly arrange the model in a logical pattern.

We'll skip the dissolve step and proceed now directly to the intersection steps, assuming that we have only an ArcView license and must intersect in two steps.

274

39➜ Locate the Intersect tool in the ArcToolbox > Analysis > Overlay toolset. Click and drag the tool onto the model canvas.

39➜ Right-click the Intersect tool on the model canvas and open it.

39➜ Set the Input Features to Limestone and Dense Conifer. When choosing the layers, be sure to choose the ones with the blue symbol, indicating that they are outputs from the model.

39➜ Name the Output Feature Class conf_lime and store it in the chap7results geodatabase.

39➜ The tool should look like Figure 7.24. Leave the other arguments set to their defaults.

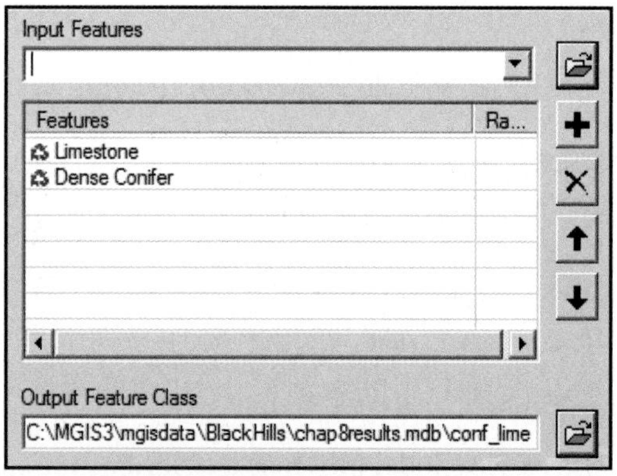

Fig. 7.24. Setting the arguments for the first Intersect tool

Notice the yellow warning sign next to the output feature class. Put the cursor on it to read the message that says that this output already exists. Of course it does, since we've done this analysis before. Since it's just a temporary step along the way, though, we can overwrite it. Check to make sure this option is set.

40➜ Click OK to close the tool.

40➜ Choose Tools > Options from the main ArcGIS menu bar and click on the Geoprocessing tab.

40➜ Make sure the box is checked to Overwrite the results of geoprocessing operations. Click OK.

41➜ Drag another copy of the Intersect tool from ArcToolbox to the model canvas.

41➜ Open the second Intersect tool.

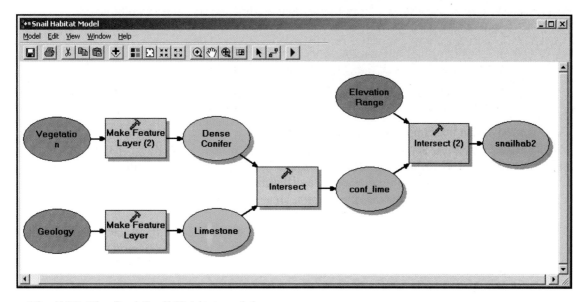

Fig. 7.25. The final Snail Habitat model

42➔ Select the lime_conf model output from the Input Features drop-down box.

42➔ Choose Elevation Range from the Input Features drop-down box.

42➔ Name the output feature class **snailhab2** and place it in chap7results.

42➔ Leave the rest of the arguments to their defaults and click OK.

43➔ Right-click the snailhab2 output and choose Add to Display.

43➔ Choose Model > Save to save the model in its current form. It should look similar to the model in Figure 7.25.

43➔ Run the model. The two intersections you set up will both be executed, and the final output will be added to the map.

The saved model provides a record of the inputs and parameters used to derive the output. It can also be edited and run again to try out different solutions. Imagine that a geologist colleague has commented that the Lower Paleozoic geologic units also contain limestone and should be included with the other geology layers. Edit the model and run it again with the new conditions.

44➔ Right-click the Make Feature Layer input box for the Geology query and open it.

44➔ (If using version 9.2, copy the expression in the box so you can paste it in the query space before editing it.) Click the SQL button and add OR [NAME] = 'Lower Paleozoic' to the expression so that all three units will be selected. Verify the query and then click OK to close the tool.

Notice that the model boxes lose their shadows (except the unchanged Vegetation query), indicating that they need to be run again with the new conditions.

45➔ Remove the Limestone and snailhab2 layers from the Table of Contents, if they are currently there, to make room for the new versions.

45➔ Right-click the Make Feature Layer tool from the Geology query again and choose Run to run only this step of the model to check your query.

45➔ Add the Limestone output to the display, if necessary, and examine it.

46➔ Click the Run button in the Model Builder toolbar to run the rest of the model steps.

46➔ Examine the new snailhab2 output to see how it changed.

46➔ Save the new version of the model by choosing Model > Save.

Finally, models can be set up to run as tools, by turning the inputs into parameters. This approach is helpful when the same series of steps is to be run on many different inputs. Imagine that the model is being used to evaluate the effects of timber treatments on snail habitat. You have 10 different vegetation layers, each representing a different management scenario, and you need to know which scenario maximizes the snail habitat area.

47➔ Right-click the Vegetation model input and choose Model Parameter. A small P appears to the right of the blue oval, indicating that it is now a parameter.

47➔ Right-click on the snailhab2 output and make it a parameter.

47➔ Save the model and close it.

47➔ Find the model in the toolbox and double-click it, or right-click it and choose Open.

276

The model is now opened as a tool, with input boxes for the two parameters you specified. Take a moment to examine the titles above the boxes, Vegetation and snailhab2. These titles are taken from the names assigned to the corresponding shapes in the model. Let's rename them so they make more sense to a user.

48➔ Close the tool. Right-click on the model in the toolbox and choose Edit.

48➔ Right-click the Vegetation oval and choose Rename. Call it **Input Vegetation**. (You can slightly enlarge the box so that the text fits more neatly.)

48➔ Rename the snailhab oval **Output Habitat**.

48➔ Save the model and close the model canvas.

48➔ Double-click on the model to open it as a tool.

TIP: Double-clicking or choosing Open will open the model as a tool to be run. To open the model canvas, right-click the model and choose Edit.

There is a more recent vegetation feature class in the Sturgis geodatabase called veg2006. We'll run the model again using this new data and see if the snail habitat has changed in the last decade.

49➔ Click the browse button and find the veg2006 feature class in the Sturgis geodatabase.

49➔ Set the output feature class to **snailhab06** and put it in the chap7results geodatabase. Click OK to run the tool.

Oh dear. The new output snail habitat layer appears empty. You also may have noticed some green text scrolling through the processing window as the tool ran, a warning about empty output. The tool did not crash, but it certainly didn't work properly. Some troubleshooting is in order.

With a little detective work, you would discover that the veg2006 feature class does not contain the SSTAGE96 field that you used for the vegetation query. The appropriate field is now called FV_COV_SS. Also, the COV_TYPE field is now named COVER_TYPE instead. You could edit the tool to reflect this. However, it might be better to make the tool more flexible by making the SQL statement a variable that can be reset as a parameter each time the tool is run.

50➔ Remove the snailhab06 output from the map.

50➔ Right-click the Snail Habitat tool in the toolbox and choose Edit.

50➔ Right-click on the Make Feature Layer box associated with the vegetation query and choose Make Variable > From Parameter > Expression. A new blue input oval appears for the expression.

50➔ Rename the new oval **Veg Query**. Move or resize as needed.

50➔ Right-click the Veg Query oval and choose Model Parameter.

50➔ Save the model and close it.

Now let's run the model again.

51➔ Open the Snail Habitat model and set the input and output feature classes as before to veg2006 and snailhab06.

51➜ Click the SQL button next to the new Veg Query input box.

51➜ Edit or reenter the query so that it reads (without the line break)

([COVER_TYPE] = 'TPP' OR [COVER_TYPE] = 'TWS') AND

([FV_HAB_SS] LIKE '*C*' OR [FV_HAB_SS] = '5').

51➜ Use the Verify button to ensure that the expression is free of syntax errors before you click OK to close the SQL window.

51➜ Click OK to run the tool again. Watch the processing window. This time no error messages should appear.

Compare the old and new snail habitat outputs to see the changes in habitat.

Finally, models can be used to build scripts, which can be used as is or enhanced with full programming features, such as conditional statements (if/then) or looping, to create advanced and flexible processing tools.

52➜ Right-click on the Snail Habitat model in the toolbox and choose Edit.

52➜ Choose Model > Export > To Script > Python.

52➜ Put the script in your BlackHills folder and name it **snailhab**. It will automatically be given the proper extension, .py.

52➜Close the model.

➜ To view the script, open the Python language program that is installed on your system when ArcGIS is installed. Go to Start > Programs > Python 2.x > IDLE.

➜ Choose File > Open and locate the snailhab.py file just created. Examine the program, looking for familiar ArcGIS command names within the Python code.

This brief introduction shows only a small part of what Model Builder can do. To learn more, read about Model Builder in the ArcGIS Help files. To learn more about geoprocessing in general, including Model Builder, read *Geoprocessing in ArcGIS* from the ESRI digital library.

This is the end of the tutorial.

➜ Close Python and ArcMap. You don't need to save your changes.

Exercises

Open the map document ex_7b in the MapDocuments folder and use it to answer the following questions.

1. Create a map highlighting the residential streets of Rapid City, that is, the streets inside a residential zoning category. Zoom in to the main part of town for better legibility. **Capture** your map.

2. The water resources office is concerned about fertilizer runoff contaminating the Minnekata aquifer. Create a map showing agricultural areas in the Minnekata Limestone. Remember to minimize the number of polygons you are intersecting. **Capture** your map.

3. What is the total length of streets, in kilometers, that fall inside the Rapid Creek floodplain (LU-CODE = 'Floodway')?

4. What is the area in square kilometers of agriculture inside the Minnekata formation?

5. What is the length of streams (include both perennial and intermittent) crossing the Minnelusa Formation?

6. Clip the septic systems and the land-use layers with the watersheds and create a map of Rapid City showing those features. **Capture** your map.

7. Create a map showing the drug-free zones around schools (areas within 300 meters of a school). Show the roads and land use in the map also. **Capture** your map.

Use map document ex_7a to answer the following questions.

8. The Forest Service is concerned about septic systems contaminating the Madison Limestone. Create a map showing private land (OWNER = PVT in the Vegetation layer) on the Madison. **Capture** the map.

9. You wish to investigate whether different soil units influence the net productivity of the forest in terms of canopy cover. Find the mean crown cover percent of *only* the forested stands within each soil unit. Which soil unit supports the highest crown cover? Which one supports the lowest? Interpret your findings. (**Hint:** Use the Dissolve tool. The forested stands have a cover type code that begins with T, e.g., TAA, TPP, etc.)

10. The Forest Service is considering adopting a policy to not cut timber within 200 meters of a primary road (TYPE = P). Assuming that cutting timber includes all [SSTAGE96] LIKE '4%', determine the total area in square kilometers lost to harvesting if this policy were adopted,. **Capture** a map showing the off-limits areas.

Challenge Problem

Imagine that you have always wanted a vacation home in an aspen stand with a little pond. Create a map showing all aspen stands (COV_TYPE = TAA) that lie on low infiltration geology units (INFIL = LOW) and are within 500 meters of a primary or secondary road (TYPE = P or S). **Capture** your map.

Skills Reference

Performing an intersection 280

Creating buffers..281

Clipping a layer 282

Appending layers..282

Dissolving...283

Using tools in ArcToolbox 284

Performing an intersection

1. Choose the Analysis Tools > Overlay > Intersect tool from ArcToolbox (Fig. 7.26).

2. Enter the Input Features by clicking a choice from the drop-down list or by clicking the Browse button to select a file from the disk.

3. Enter a second Input Features layer.

4. Enter the name and location of the Output Feature Class.

5. (optional) Set the JoinAttributes option to NO-FID, the best choice in most cases. ALL joins all attributes from both tables, NO-FID joins all attributes except the FIDs, and FID-ONLY joins only the FIDs.

6. Leave the other optional parameters alone unless you know what you're doing.

7. Click OK.

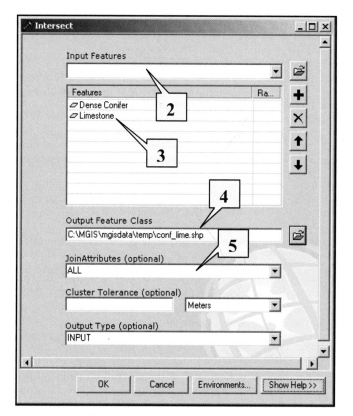

Fig. 7.26. The Intersect tool

TIP: Two input layers must be entered if the user has an ArcView license. With an ArcInfo license, three or more layers may be entered and the intersections will be performed sequentially.

Creating buffers

Note: Buffering is a time-intensive process, especially for lines and polygons.

1. Choose Analysis Tools > Proximity >Buffer from ArcToolbox.

2. Choose the Input Features to buffer (Fig. 7.27).

3. Specify the name and location for the Output Feature Class.

4. Specify the units and value for a fixed buffer distance,

5. OR choose a field from the input layer attribute table containing the buffer distances.

6. (optional) Change the Side and End type settings.

7. Set the Dissolve Type; use ALL to dissolve boundaries between overlapping buffers or NONE to leave them as is.

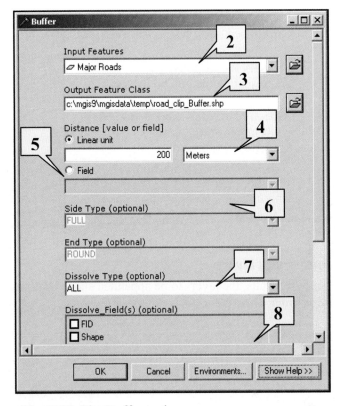

Fig. 7.27. The Buffer tool

8. (optional) Choose one or more Dissolve Field(s). Buffers sharing the same value in the field will be dissolved.

9. Click OK.

Clipping a layer

1. Choose Analysis Tools > Extract > Clip from ArcToolbox (Fig. 7.28).

2. Specify the Input Features layer to be clipped.

3. Specify the Clip Features layer representing the clip boundary.

4. Specify the name and location of the Output Features.

5. Leave the other options with their defaults.

6. Click OK.

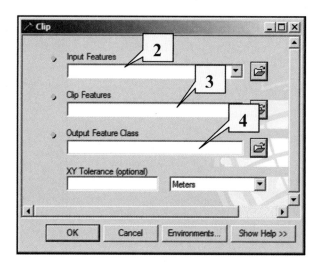

Fig. 7.28. Clipping a layer

Appending layers

For best results, the layers to be appended should have attribute tables with the same fields in the same order.

1. Choose Data Management > General > Append from the ArcToolbox (Fig. 7.29).

2. Specify the layers to be appended together.

3. Specify the *existing* output feature layer to contain the appended features.

4. Choose the Schema type to specify the handling of attributes. Use TEST if you know that the attribute tables match (i.e., they have the same fields in the same order). Use NOTEST if the attribute tables don't match—the features will be appended without the attributes.

5. Click OK.

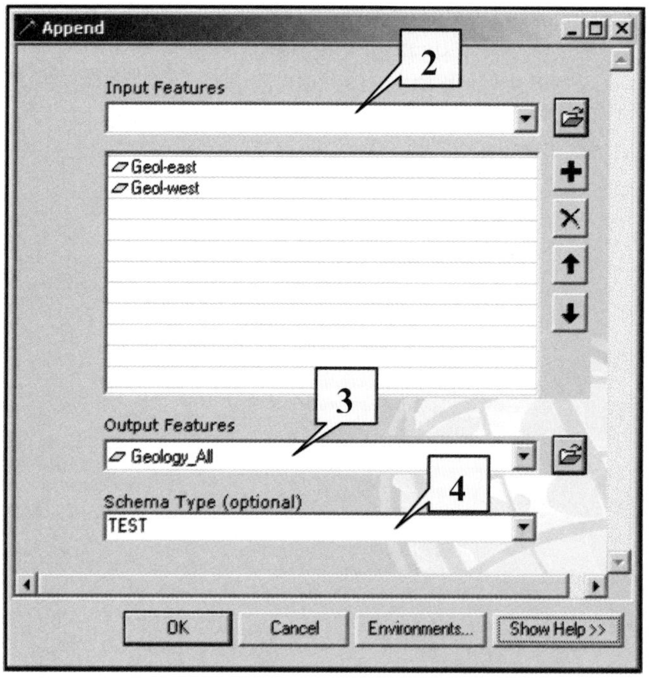

Fig. 7.29. Appending layers

Dissolving

1. Choose Data Management Tools > Generalization > Dissolve from ArcToolbox.

2. Specify the Input Features layer to be dissolved (Fig. 7.30).

3. Specify the name and location of the Output Features to be created.

4. Choose one or more attribute field(s) on which to base the dissolve. These fields will appear in the output layer; all other fields will be dropped.

5. To get summary statistics, check the boxes to choose the fields and the statistics to place in the output layer. Specify as many fields and statistics as desired or none at all.

6. (optional) Scroll down and uncheck the Multipart box if you wish to ensure that all polygons remain separate features in the output, even if they share the same attribute value.

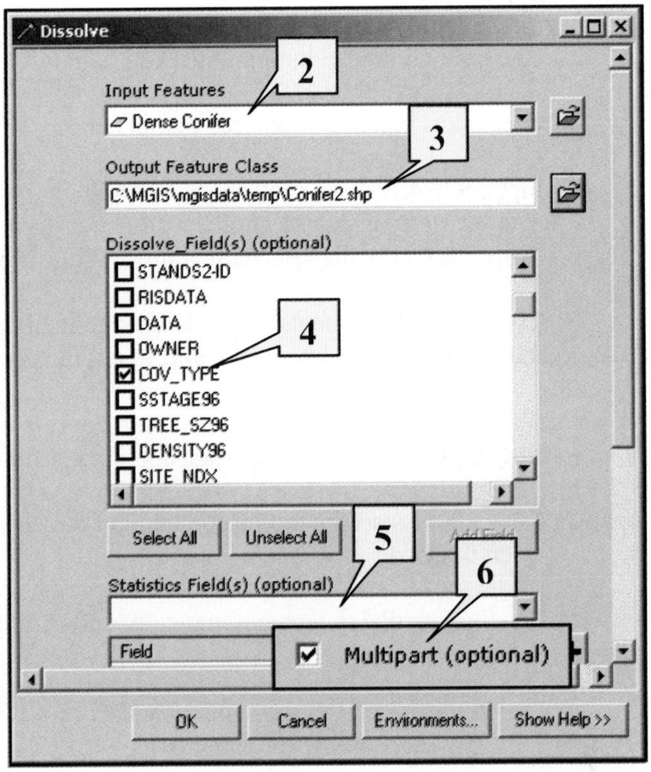

Fig. 7.30. Dissolving a layer

7. Click OK to begin dissolving the layer.

Using tools in ArcToolbox

1. Click the ArcToolbox icon to open it (Fig. 7.31).

2. If using a tool for the first time, click the Show Help button and read about the tool.

3. With the Help showing, click on a parameter box to get more information about it.

4. Click the Tool Help button to get full help information about the tool.

5. Green dots indicate a required parameter.

6. Input/output features may be set by clicking the Browse button and locating the spatial data set, or by making a choice from the drop-down list (ArcMap only). If the drop-down list is used, only the selected features will be used in the tool.

7. Click the Environments button to change the Environment settings (only recommended for advanced users knowledgeable about the geoprocessing environment).

8. If a yellow warning or a red error icon appears after entering a parameter, place the cursor over it to see the message about the error or potential error.

9. Click OK to start the tool. Keep track of the messages in the progress box. Close the box when finished.

10. To close the dialog box automatically upon completion of the task, check the box.

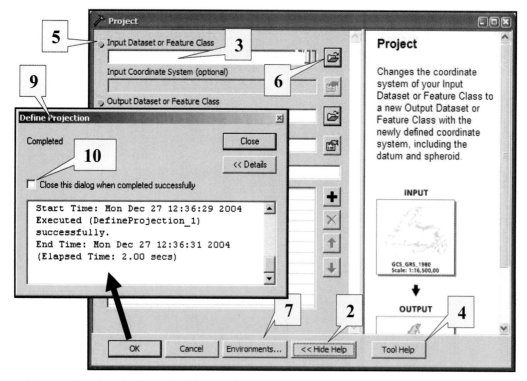

Fig. 7.31. Using tools in ArcToolbox

Chapter 8. Raster Analysis

Objectives

➤ Exploring raster analysis through map algebra and Boolean overlays

➤ Getting familiar with other common raster analysis functions

➤ Understanding the raster data model and the storage of raster grids in ArcGIS

➤ Using Spatial Analyst for raster analysis and geoprocessing

➤ Controlling the analysis environment when using Spatial Analyst

---

NOTE: This chapter requires Spatial Analyst, an optional extension to ArcGIS. If you do not have it installed on your computer, you will not be able to perform the functions described.

Mastering the Concepts

GIS Concepts

Raster data

The availability of two different data models, raster and vector, provides added flexibility to options for data storage and analysis. (Readers may wish to review the section in Chapter 1 on raster and vector data models.) Neither data model is intrinsically superior. They both have areas in which they excel and areas in which they are at a disadvantage. The GIS professional who has a grasp of both tools holds the keys to developing the most efficient and accurate analysis.

The analysis functions available for rasters number in the hundreds and include very advanced and sophisticated applications. This chapter provides an introduction to some of the basic concepts and analysis functions of raster analysis, but the topic spans a much greater breadth of theory and technique than we have time to explore here. The serious student of raster analysis will find plenty of additional material in the software documentation for Spatial Analyst.

Recall that spatial data comes in two forms (Fig. 8.1). Discrete data, such as soil types or land use, take on relatively few values and form distinct regions on the map. Continuous data, such as elevation or precipitation, vary smoothly over a range of values, forming a surface or field. Points, lines, and polygons constitute spatially discrete features, and the vector data model can ONLY model discrete data. With raster data, however, we can store continuous data, which varies across the surface at very small scales, such as elevation and precipitation.

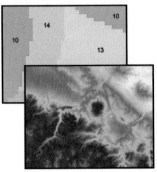

Fig. 8.1. Discrete land-use data and a continuous digital elevation model

Raster coordinate systems

Just like vector layers, rasters have a defined coordinate system with a projection and a datum. The raster is georeferenced by specifying the *x-y* coordinates of the center of the upper-left pixel and the cell size (the length of the sides of the pixels). Usually pixels are square. A 30-meter raster has pixels composed of 30-meter × 30-meter squares. The cell size is always specified in the coordinate system's map units, and meters or feet are typically used.

Although one can store a raster in a GCS, this practice is not common because the inherent distortion gives an unsatisfactory view of the data. Furthermore, many raster analysis functions incorporate distance and angle formulas that will give inaccurate results when applied to a raster in a GCS or in a projection with significant distortion. Therefore, raster data are normally stored in whatever projected coordinate system gives the best and most undistorted view of the data. The main exception occurs for data providers, such as the U.S. Geological Survey, which provide topographic or other data in GCS format. Such data should be projected before it is used.

Raster analysis

Raster processing is fundamentally different from vector processing. Vector analysis analyzes features with *x-y* coordinates. Raster processing proceeds on a cell-by-cell basis and can be very efficient. This section describes some of the most commonly used raster analysis methods, but the reader should be aware that there are many more.

Map algebra

A raster is an array of numeric values arranged in rows and columns. Several rasters representing different attributes of the same area and sharing the same gridded reference system (same extent and cell size) can be analyzed using cell-by-cell formulas based on mathematical, logical, or Boolean operators. This technique is called **map algebra**. Consider the stack of two input grids in Figure 8.2. These rasters can be added together cell by cell to produce a new raster with cells containing the sum of the corresponding cells in the inputs. If the cells of the input grids are not already aligned, then they must be resampled to achieve this.

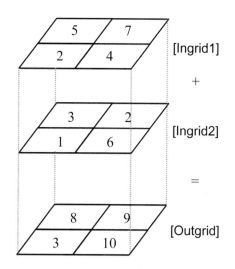

The algebraic expression for the operation in Figure 8.2 would appear as [Ingrid1] + [Ingrid2] with the grid names enclosed in brackets. Imagine a grid of precipitation in inches which must be converted to centimeters. The expression [Precip] * 2.54 would produce an output grid with precipitation in centimeters. The formula expressions can include multiple grids, functions, and values. A more complex expression to predict erosion potential as a function of rainfall, slope, soil erodability, and vegetative cover might read

Fig. 8.2. Map algebra addition

[Precip] * 2 + [Slope] * 4 / ([Erosion] − [Vegcover]).

Map algebra also includes logical functions that can return Boolean grids, which have values of true (1) or false (0). Figure 8.3 shows the result of entering an expression [Elevation] > 1,400; the blue area shows the cells in the elevation grid that meet this criterion.

Boolean overlay

Chapter 7 described a vector process called overlay, used to find areas that meet multiple conditions, as we did for snail habitat. In many cases, raster overlay has advantages over vector overlay. First, it can incorporate conditions from continuous data sets such as elevation, slope, and aspect, which cannot be stored as vectors. Second, the process is faster because corresponding cells are simply added or multiplied together, instead of splicing the existing polygons and creating new ones from them. Third, raster overlay can process more than two layers at once.

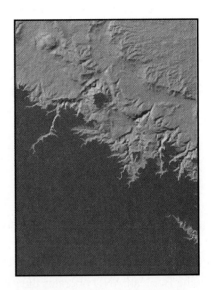

Fig. 8.3. The blue portion of this map shows the pixels where Elevation > 1,400 meters.

Map overlay with rasters is accomplished using map algebra with Boolean grids and operators. Chapter 5 introduced Boolean operators, which evaluate two input conditions A and B to determine a result that is either true or false. In raster analysis, this function is performed on a cell-by-cell basis (Fig. 8.4). A Boolean grid contains two possible values, ones or zeros. A one corresponds to a true result, whereas a zero corresponds to a false result.

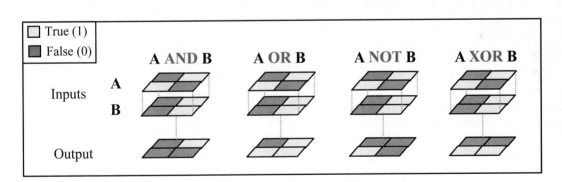

Fig. 8.4. The Boolean operators are applied to rasters. Each input cell is 1 or 0 (true or false). Matching cells in the inputs are evaluated to produce the output result.

Imagine creating a potential habitat map for lodgepole pine trees based on two criteria: lodgepole pine tends to prefer areas above 1500 feet in elevation and with more than 15 cm of precipitation. In Figure 8.5a, the Boolean grid shows where the precipitation conditions for lodgepole are favorable. Areas where the precipitation is greater than 15 cm have the value true (1), and areas labeled false (0) have precipitation < 15 cm. Figure 8.5b shows the favorable conditions based on elevation, where the cells containing 1 have elevations >= 1500 meters.

To find the lodgepole habitat, we apply the Boolean operator AND to the input grids. If both inputs are 1, the output cell is 1 also. If either of the input cells has a 0 or if both of the input cells have a 0, then the output will be 0. Figure 8.5c shows the output raster with the potential lodgepole habitat areas identified as cells with the value 1 (true). The raster AND operation is the equivalent of the vector Intersect tool.

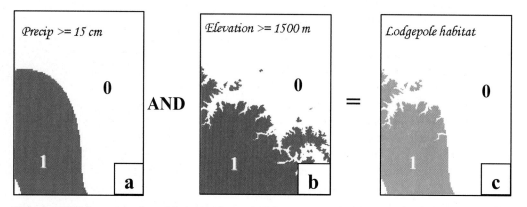

Fig. 8.5. Boolean overlay. (a) A Boolean grid in which the value 1 (true) indicates where precipitation is >= 15 cm. (b) A grid indicating elevations >= 1,500 meters. (c) The result of multiplying the first two grids, showing where both conditions are true.

The Boolean AND operates the same way as multiplication of the two input grids. Figure 8.6 compares the outputs of AND and multiplication on two Boolean inputs, and they are identical. Thus, either AND or multiplication can be used to find the lodgepole habitat. Since both of these operators are symmetrical, meaning that the order of inputs does not matter, they can be applied to three or more grids as easily as two.

1 AND 1 = 1	$1 \times 1 = 1$
1 AND 0 = 0	$1 \times 0 = 0$
0 AND 1 = 0	$0 \times 1 = 0$
0 AND 0 = 0	$0 \times 0 = 0$

Fig. 8.6. Equivalence of the AND and multiplication operators

Map *addition* provides another way to model the habitat—to *rank* the possibility of finding lodgepole. In this model, if both elevation and precipitation are ideal, then lodgepole is most likely to occur. If only one is ideal, then it can occur but is not as likely. If neither condition is ideal, then lodgepole is very unlikely. To implement this model we add the two input maps instead of multiplying them. The final map in Figure 8.7 has three possible values: 0 if neither condition is true, 1 if either condition is true, or 2 if both conditions are true. Thus, 2 indicates the highest probability of finding lodgepole, 1 a moderate probability, and 0 a low probability.

Fig. 8.7. A Boolean addition overlay

The best model to use will depend upon which most accurately predicts the actual habitat areas. GIS modeling is always tied to knowledge of the real world. If this project were real, we would proceed to test the model predictions by comparing them to the actual distribution of lodgepole pine. If the model proved accurate, then we might use it to make predictions in other areas.

Distance functions

Distance functions calculate distances across grids or find the most suitable path across a surface. Figure 8.8a shows a straight-line distance grid in which each cell represents the distance from the closest stream. In a cost-distance analysis (Fig. 8.8b), weights or costs can be assigned to affect travel; for example, high slopes or dense cover increase the cost of travel. Several steps are needed to use these functions. In the example, a digital elevation model (DEM) was used to calculate a slope grid. The slope grid was used as input to the cost-weighted distance function to create a grid representing the cost-weighted distance from the first point, as well as a grid representing the best direction of travel from any given cell. Finally, the shortest-path function was used to create the shapefile showing the best path.

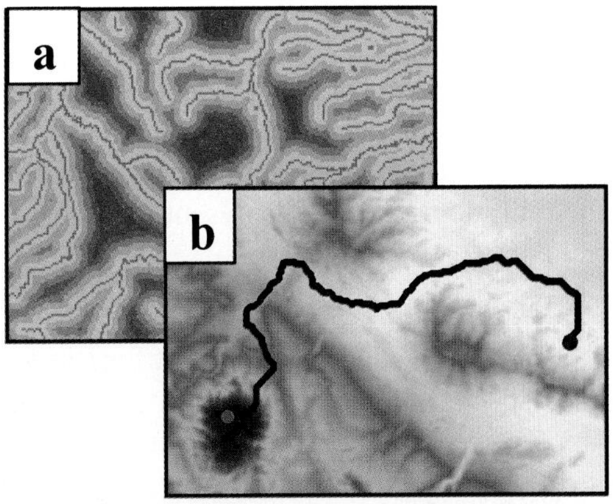

Fig. 8.8. (a) Straight-line distance; (b) least-cost path between two points

Density

The density function is used to create grids showing the density of features within a given radius. Figure 8.9 shows a density map of population in the United States. The darker areas represent a higher density of people and the lighter areas a lower density of people. The raster was calculated using a point file of cities (black dots) with a population attribute and a search radius of 500 miles. Density functions have a moving window that counts the number of occurrences (people) inside the search radius and divides by the area. A large radius produces a smooth map with higher density values (in this map, 0–90 persons/km^2); a smaller radius produces more detail and smaller density magnitudes.

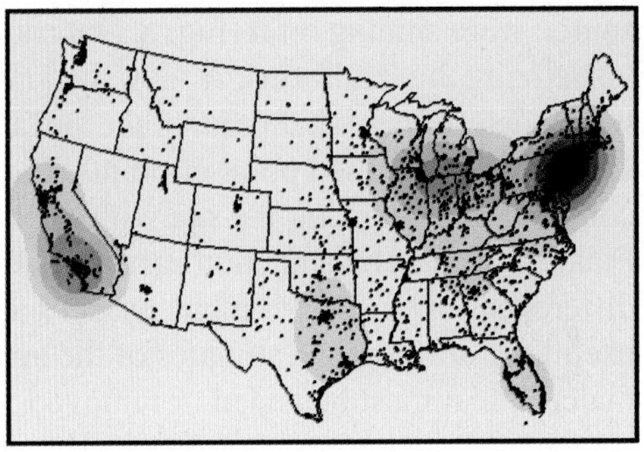

Fig. 8.9. Density of population in the United States

Interpolate to raster

Interpolation takes values from points and distributes them across a grid, estimating the values at points in between the measurements. A common example involves taking precipitation measurements at weather stations and interpolating the precipitation values over an area. Figure 8.10 shows the interpolation of the total precipitation measured at the stations over the Black Hills region, with the watersheds superimposed on the image.

Three different interpolation methods are offered: inverse distance weighted (IDW), splining, and kriging. Each method uses different assumptions and different means to obtain the gridded result. More information about these three interpolators

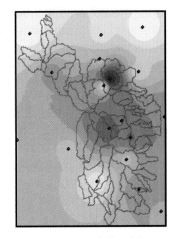

Fig. 8.10. Rainfall grid interpolated from weather station measurements

can be found in the Help documentation for Spatial Analyst. Interpolation is a complex topic and the subject of entire textbooks. GIS professionals should become familiar with the different methods in order to understand the benefits, drawbacks, and assumptions of each one.

Although interpolated and density maps appear similar, the functions work differently. Density functions count occurrences within a radius and divide by area, yielding a density, such as people per square kilometer. Interpolation estimates a value for each grid cell based on its neighbors, and the rainfall grid values represent inches of rain.

Surface analysis

Surface analysis includes a set of functions for calculating properties of a surface, such as slope and aspect. Although most commonly applied to elevation data, these functions can be used on any type of continuous grid. Using these functions on a DEM could help you find an ideal location for a cabin, for example. Contouring the grid would produce a set of contour lines representing the surface. A slope map, showing the steepness of the terrain, could help identify sites flat enough to build on. An aspect map, showing the compass direction of the slope, could locate south-facing hillsides to give the cabin a warm southern exposure. A hillshade map, which represents the surface as though an illumination source were shining on it (Fig. 8.11), can help in visualizing the landforms. A viewshed analysis, which determines what parts of the surface can be seen from a specified point or points (Fig. 8.11), could help in choosing the site with the best view. Finally, the Cut/Fill function calculates the differences between two input surfaces and is useful for developing estimates of the amount of material that must be added or removed.

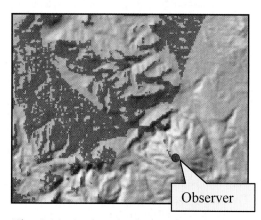

Fig. 8.11. A viewshed map shows the areas visible from a point on a hilltop in red. The background map is a hillshade.

Neighborhood statistics

Neighborhood functions, also known as **focal functions**, operate on an area of specific size and shape around a cell, known as a neighborhood. A moving window passes over each cell in turn (the target cell), calculates a statistic from all the cells in the neighborhood, and places the result in the target cell. Neighborhoods may have several shapes, including squares, circles, rectangles, and annuli. The statistical functions include minimum, maximum, range, sum, mean, standard deviation, variety, majority, minority, and median.

Consider a 3-cell × 3-cell averaging window, for example (Fig. 8.12). For each target cell in the grid, the algorithm finds the average value of the nine cells within the window and puts the average value in the target cell of the output grid. The window then moves to the next cell. An edge cell is still evaluated but will have fewer cells in the neighborhood, such as the cell in the upper-left corner of Figure 8.12. Only four values are averaged, the target and its three neighbors. The "1" cell to its right would have the average of six cells.

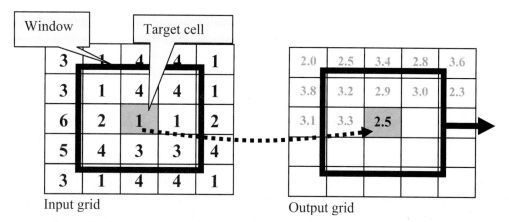

Fig. 8.12. A 3 × 3 averaging window averages the nine cells in the window and places the result in the target cell of the output grid.

Zonal statistics

The zonal statistics function resembles the neighborhood function with two important differences. First, the neighborhoods are not a fixed size or shape but instead are defined by another layer, called the zone layer. The zones are regions in a raster, which share the same attribute. A zone might be a watershed, a parcel, or a soil type. Second, instead of creating an output grid, the zonal statistics function creates a table containing statistics for each zone.

Figure 8.13 shows a zonal function used to find the average slope within the watersheds. In Figure 8.13a, the watersheds are displayed on top of a hillshade map, and you can clearly see which watersheds have more variable topography and a higher average slope. When the zone function is executed, the zone grid contains the watersheds, and a slope grid provides the values to be averaged. The analysis produces a table of statistics, which is then joined to the original watershed polygons using the zone field. From this joined table, a map showing the average slope of each watershed can be created (Fig. 8.13b). In this map, the darker colors indicate higher slopes, and the average slope values are also labeled in each watershed.

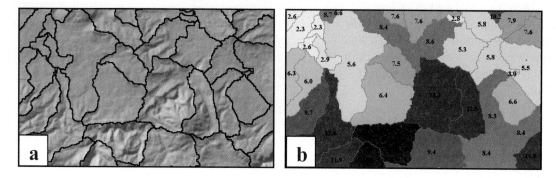

Fig. 8.13. Using zonal statistics to find the average slope of watersheds. (a) The zone grid shown over a hillshade. (b) The watersheds colored by and labeled with the average slope.

Zones need not be contiguous like the watershed zones, however. The geology map in Figure 8.14 contains many unconnected areas with the same geologic unit. The map only has six zones, one for each color, and the two orange regions comprise a single zone.

The zonal statistics function can also assign values from a surface to line or point data, such as finding the average slope along a stream section or attaching an elevation value from a DEM to a layer of wells. Each line or point should have a unique ID value that identifies it as its own zone.

Fig. 8.14. This geology map has six zones.

Reclassify

Map reclassification changes the values of a raster according to a scheme designed by the user, such as classifying a slope map into three regions of low, medium, or high slope. For a map with only a few values, such as land-use codes, the user might specify a new value for each code individually. Perhaps a town has annexed a nearby planning unit that used different land-use codes and must convert one set of codes to another; the code 143 becomes 347, and so on. With continuous data covering a range of values, such as slope, the values are typically classified into a discrete number of ranges. Classification uses a tool similar to the one described in Chapter 2 for classifying attribute data.

Resample

Often grids must have their values converted to a grid with a different cell size. This process is called resampling. Users may resample purposely in order to align several grids with each other or to decrease the resolution of a data set that is too large to work with effectively. Resampling must also occur any time two grids with different cell sizes are analyzed together. Even if the grids have the same cell size, the cell grids might not be aligned with each other, requiring that resampling occur.

Figure 8.15 shows a set of cells being resampled from one cell grid (X1-Y1) to another (X2-Y2) shown by the black lines. Resampling uses one of three methods. In **nearest neighbor** resampling, the new cell is given the value of the old cell that falls at or closest to the center of the new cell. In **bilinear** resampling, a distance-weighted average is taken from the four nearest cell centers. The **cubic convolution** resampling method determines a new value by a curve fit through the nearest 16 cell centers. Nearest neighbor is the fastest method and should always be used with categorical raster data because the values do not change. The bilinear method is usually better for continuous rasters, such as elevation.

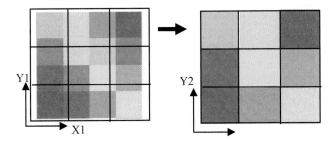

Fig. 8.15. Resampling raster cells using the nearest neighbor method

Resampling can degrade the quality of a data set. Best practices dictate that the user retain full awareness of the coordinate systems and cell sizes of all input layers being used for analysis and endeavor to standardize them before processing occurs. Some GIS systems automatically resample without warning. Users must be especially careful with these systems if they are attempting to retain a high degree of spatial integrity in their rasters.

Convert

Users can easily convert between rasters and vectors, such as turning a polygon shapefile of geology into a raster. The user must choose *one* attribute field to populate the values of the raster, such as the geology unit name. Rasters can only store numbers. Numeric fields are converted as is, but text values are assigned arbitrary numbers for storage in the raster. Some raster formats place the original text values in a raster table for reference.

You may also convert discrete grids to polygons as long as the number of polygons does not become excessive. Continuous data such as DEMs should never be converted to polygons since each pixel is different from its neighbors and will become a separate polygon. If too many polygons will result from a conversion, the software may be unable to complete the task.

About ArcGIS

Storing rasters

ArcGIS can read many different raster data formats, but Spatial Analyst only operates on a format unique to ArcGIS, called a grid. Other rasters, such as TIFF files, SID files, and JPEG files, must be converted to grid format before they can be analyzed. All raster output generated by Spatial Analyst takes the form of grids.

Grids come in two forms. Integer grids store whole numbers in binary format. Floating-point grids can store decimal values, but they require four times as much space as integer grids. Thus, it is foolish to use floating-point grids except when decimal values are required. For example, DEMs store land elevations and typically have a vertical accuracy of 7 to 15 meters. It makes no sense to store a decimal fraction for these data.

When creating output, the grid type will usually be the same as the input. If you add or multiply two integer grids, the result will also be an integer grid. If one of them is floating-point, the output will be floating-point. Some functions always create floating-point grids. Dividing or averaging a grid will result in floating-point output. (Floating-point grids may be converted to integer grids using the INT function.)

TIP: Grids have stricter naming conventions than other ArcGIS data files. Grid names must have no more than 13 characters, and they should contain only letters, numbers, or the underscore character. **Pathnames to the grid directory should contain no spaces.**

Grid attribute tables

Grids can have attribute tables, just as feature layers do. However, since a grid contains only cell values, the attribute tables look different. Each record contains one unique value present in the grid. The table has three fields that are always present: the ObjectID field containing a unique ID for the table rows, the Value field showing each unique cell value in the grid, and the Count field indicating how many cells contain that value (Fig. 8.16). This table indicates that the Precambrian rock unit has the largest area in the grid because more cells (64,597) have that value than any other. The set of all cells sharing the same value and having one record in the grid attribute table is called a **zone**.

The attribute table in Figure 8.16 comes from a raster that was converted from a polygon shapefile of geology using the Name field of the geologic units. As a result, the table contains the Name field with the original text values. However, the numbers in the Value field are the actual values stored in the grid because all grid values must be numeric. The numbers were arbitrarily assigned to the different names. Thus the

ObjectID	Value	Count	Name
0	1	22651	Cenozoic
1	2	8602	Upper Mesozoic
2	3	38657	Lower Mesozoic
3	4	32924	Upper Paleozoic
4	5	1078	
5	6	27805	Madison Limestone
6	7	64597	Precambrian
7	8	24948	Lower Paleozoic

Fig. 8.16. The attribute table of a geology grid

Precambrian unit is represented in the grid by the number 7. Additional fields may be added to grid attribute tables by the user.

Only integer grids can have attribute tables, but not all should. A grid of elevation could have nearly as many different values as cells. In such a case, storing a table in addition to the binary data itself would only increase the storage space without conferring any particular benefit. Users may control the threshold value above which attribute tables will not be created for integer grids.

NoData values

Rasters have a special type of value called NoData to indicate a null value for a cell. Because rasters are rectangular in shape, the NoData value can mask areas outside a nonrectangular study area. NoData values are typically displayed with black or transparent color. In Figure 8.17, some cells at the top of the DEM are missing data and are stored as NoData values. NoData values are far more convenient than using an arbitrary number, such as 0 or –9999, to represent a missing value (although some grids may have these). Arbitrary numbers will be used in calculations and can cause strange problems, whereas NoData values can be ignored for more consistent results.

Fig. 8.17. NoData values at the top of a DEM

Using Spatial Analyst

Spatial Analyst is an extension to the ArcGIS software system and must be purchased separately. It consists of a large set of tools loaded into ArcToolbox under the heading Spatial Analyst Tools (Fig. 8.18a) and organized into a series of tool sets by function type. In ArcMap, the most commonly used tools can be accessed from the Spatial Analyst toolbar and menus (Fig. 8.18b). However, many tools are not accessible from the menu, so to get your money's worth you'll need to investigate the options in the toolbox.

As with other tools, Spatial Analyst functions are fully integrated into the geoprocessing environment. They can be run from the toolbox or the command line, used to build models in Model Builder, and incorporated into scripts.

The tools are well documented in the Help files, in many cases including complete discussions of the concepts behind each tool. Users interested in interpolation, for example, will find extensive information on, and references to, the various interpolation methods (IDW, kriging, and splining). Reading about unfamiliar tools before implementing them is highly recommended.

Raster Calculator

The Raster Calculator is another method to execute functions from the toolbox. The ordinary usage of the Calculator is to perform map algebra operations, as the interface makes clear with its buttons for mathematical, logical, and Boolean operators (Fig. 8.19). However, any Spatial Analyst function has a command-line usage, and these commands can be typed directly into the Raster Calculator and executed. As an example, the INT function can be used to truncate floating-point grid values to integers, and the expression INT([Ingrid] + 0.5) will round the values of a grid to the closest integer. The SETNULL function uses an expression, such as Elevation > 1400, and changes those cells to NODATA while retaining the original values elsewhere. Other operators, such as bitwise operators, can also be accessed from the Calculator. The ability to enter any function or operator in this interface makes the Calculator a powerful analysis tool.

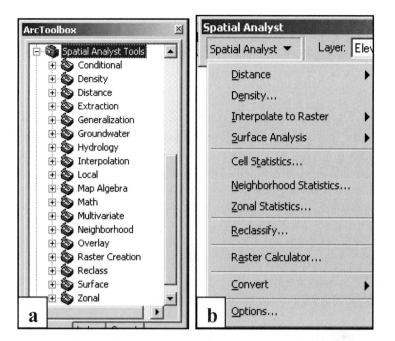

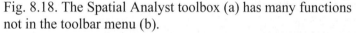

Fig. 8.18. The Spatial Analyst toolbox (a) has many functions not in the toolbar menu (b).

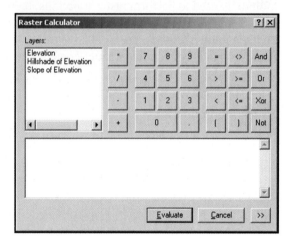

Fig. 8.19. The Raster Calculator

Coordinate system management

Raster functions treat layers as arrays of numbers, which can be processed as a stack. To perform raster analysis, all of the input grids must have the same cell size and extent so that the array values in each stack (the cells) line up. If a command receives inputs with different cell sizes or extents, the grids will automatically be resampled to match each other. By default all grids are resampled to the most coarse cell size present in the input grids. When input rasters have different coordinate systems, Spatial Analyst will automatically project the grids to match.

The resampling and projection process is invisible to the user and seems convenient; however, it does increase the processing time. Resampling or projecting very large grids may overwhelm the computer and cause a system failure. Since resampling or projecting can also degrade the quality of a data set, it is unsettling to realize that it can happen without the user's knowledge. Storing all grids to be analyzed together with the same cell size and extent can reduce unintentional resampling.

If multiple coordinate systems are input for analysis, then the same rules discussed in Chapter 7 for geoprocessing are applied, and the output will be given a coordinate system based on this order of precedence: (1) the feature data set coordinate system if the output is being placed in one, (2) the coordinate system specified by the environment settings, and (3) the coordinate system of the first input to the tool. An additional rule applies to (3): if one of the inputs is a vector data set, it will use the coordinate system of the first vector feature class. The user can override the default rules and choose to have all new rasters adopt the same coordinate system as the data frame.

Analysis settings

Spatial Analyst uses several options to influence the results of an analysis: the working directory, the output extent, the output cell size, and the analysis mask. Settings may be established in the Spatial Analyst menu and remain in effect throughout an analysis session unless changed.

Advanced users who utilize the geoprocessing environment for building models, tools, and scripts have learned that certain environment settings are applied to geoprocessing operations. Some of these settings are identical to the Spatial Analyst settings. Users should be aware that if the toolbox is used to perform raster functions, then the geoprocessing environment settings will take precedence over the settings specified in the Spatial Analyst menu. For example, if Betty sets the Spatial Analyst menu to use a cell size of 50 meters, then that setting will be employed with any function launched from the menu. However, if she executes a tool from ArcToolbox, the cell size setting will not be used.

The working directory

Spatial Analyst produces many new grid files. To help keep the files tidy and in one place, the user should specify a working directory. Unless you request otherwise, all grids created during an analysis session are placed by default in the user's system temp directory, such as C:\Documents and Settings\Maribeth Price\Local Settings\Temp\, but this location may not be desirable.

Two types of output grids are created by Spatial Analyst. Often it produces a temporary grid, saved in the working directory. Temporary grids have generic, automatically assigned names like calc1, calc2, hillshade1, and so on. Frequently a temporary grid is a mere stepping-stone to the next analysis step. A user may choose to convert a temporary grid to a permanent grid, if necessary. Temporary grids are automatically deleted from the disk when the map document is closed and will not be available next time it is opened, so it is best to remove temporary grids from the map document before closing it. Permanent grids have specific names and locations chosen by the user, with the implication that the grid is a final result that must be kept for future use. Most functions ask the user to specify whether a permanent or temporary grid should be created.

The choice of working directory will depend on whether primarily temporary or permanent grids are being created. If you are working on a project creating mostly permanent grids, then the project directory is a good choice for the working directory. If most of the output is expendable, the directory C:\temp on the computer works well. Often a special folder for grids inside the current project directory provides the best alternative; it allows permanent grids to be saved in the same area as the rest of the project data but provides a safety measure against inadvertent corruption of important data files.

> **TIP:** Grids should NOT be stored in folders that have pathnames containing spaces because some functions will not work properly. GIS users are well advised to avoid spaces in ALL folder names, no matter what. An experienced GIS user cringes whenever she sees a space in a folder or a file name (even when it's allowed).

> **TIP: NEVER copy, move, rename, or delete grids using Windows Explorer.** Grids have complex storage formats, and Windows cannot manage them properly. Always use ArcCatalog to manage grids.

The analysis extent

Grids do not need to have the same area in order to be analyzed together. By default, the program will automatically generate an output grid with an extent equal to the intersection of the input grids. For example, if the two overlapping grids in Figure 8.20, shown with black outlines, are multiplied together, the output grid will include only the areas common to both (the small yellow square).

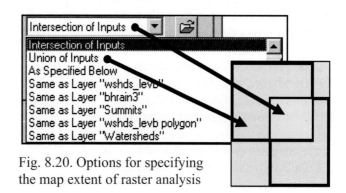

Fig. 8.20. Options for specifying the map extent of raster analysis

However, the user could alternatively specify the Union of Inputs option, which results in an output grid that covers the areas of both input grids (the large blue square). The areas outside both grids, needed to complete the rectangle, would receive NoData values.

A user could also elect to copy the area of a specific grid as the output region. Choosing Same as Layer "bhrain3" would require that all the output grids match this layer's extent. Another choice is to use the area of the current data frame as the output grid size or to type in explicit *x-y* coordinates for the extent. Whatever choice is made remains in effect until changed.

The analysis cell size

The cell size of the output grid is determined according to the current cell size rule. By default, the output size equals the largest of the input cells. If you are adding a 50-meter soil grid to a 30-meter elevation grid, the output cell size would equal the larger of the two, 50 meters. The default rule can be overridden by other options, however, as shown in Figure 8.21. The user can choose to

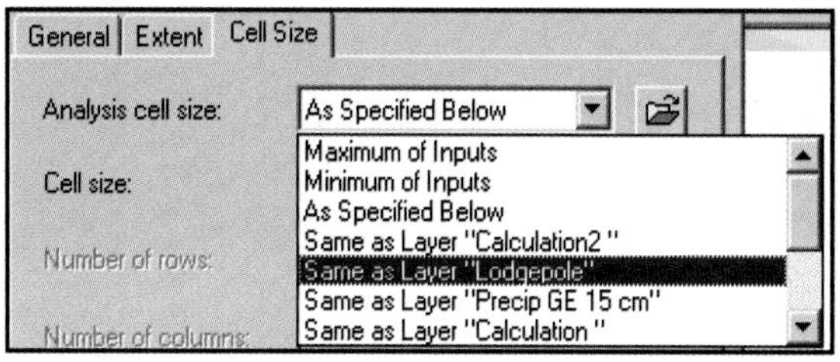

Fig. 8.21. Setting the cell size option for output grids

use the smallest or the largest cell size of all the inputs. The output can be matched to the cell size of a particular grid, or you can simply specify the cell size explicitly, such as 30 meters. The cell size is always measured in map units, which depend on the coordinate system of the raster layer.

Most often the units are meters, but they can be feet. If the grid is stored in a GCS, then the units will be in decimal degrees.

Some analysis functions ask for a cell size. In this case, the cell size box will be filled with a suggested value based on the map units and the area of the analysis extent. The suggested size will usually create a grid with a sufficient, but not excessive, number of cells. The suggested size can be changed, but making a large change in size, such as 200,000 to 50, runs the risk of creating a grid with too many or too few pixels. If the suggested size is far different from what you expect or want, then double-check the extent settings and the map units before proceeding.

The analysis mask

An analysis **mask** is used to screen out areas of unwanted data when new grids are being created. The mask can be a feature class or a grid. If it is a grid, then it must contain NoData in the cells to be excluded. Any cell with a NoData value in the mask will receive a NoData value in the output grid. If a feature class is specified, then any cells outside the feature class boundary will receive NoData values. In Figure 8.22, areas outside the watershed have no elevation after the mask is applied. The mask will be applied to any grid created during analysis and remains in effect until it is changed or turned off.

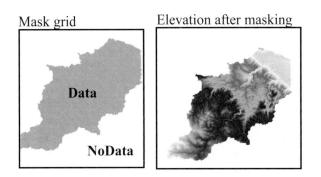

Fig. 8.22. NoData cells in a mask grid are forced to be NoData cells in the output grid.

Summary

➢ Rasters consist of spatial data stored as individual cells in an array and can represent discrete features or continuous fields of information.

➢ Rasters have coordinate systems just like other spatial data. Many raster calculations are based on distance and area functions, so it is preferable to store rasters in a projected coordinate system with minimal distortion.

➢ Map algebra treats grids as arrays of numbers that can be added, multiplied, and so on on a cell-by-cell basis. Arithmetic, logical, Boolean, and bitwise operators and functions offer a wide array of algebra options for analyzing multiple grids and scalars.

➢ Distance functions calculate costs and paths associated with traveling over a surface.

➢ Density functions measure the density of objects over an area.

➢ Raster interpolation estimates a surface based on measurements at isolated locations.

➢ Statistical measures can be calculated for rasters based on single cells, within a specified neighborhood, or within large, irregular zones.

➢ Grids can be reclassified and given new values as needed.

➢ Resampling occurs automatically when rasters with different cell grids are analyzed together but can degrade the quality of a data set. Users should attempt to minimize resampling when possible by aligning raster data before analysis.

➢ Raster analysis is performed in Spatial Analyst using a data format called a grid. Spatial Analyst tools can be executed from menus, toolbars, the command line, ModelBuilder, and scripts. Spatial analyst functions can also be executed by the Raster Calculator along with map algebra expressions.

➢ Output grids are temporary or permanent. Temporary grids are deleted when removed from the map document or when the document is closed.

➢ The analysis environment controls certain settings for all new grids created during a session. The settings include the working directory, the extent and cell size of output grids, and the analysis mask.

TIP: Grids have stricter naming conventions than other ArcGIS files. Grid names must have no more than 13 characters and should contain only letters, numbers, or the underscore character.

TIP: Grids should NOT be stored in folders that have pathnames containing spaces because some functions will not work.

TIP: ALWAYS use ArcCatalog to copy, move, rename, or delete grids. Grids have complex storage formats, and Windows cannot manage them properly.

Chapter Review Questions

1. Describe the difference between the terms *raster* and *grid*.

2. Why is it best to store rasters in the same coordinate system in which you plan to use them?

3. You are adding two grids together in the Raster Calculator using the formula grid1 + grid2. Grid1 is in UTM projection, grid2 is in State Plane, and the data frame coordinate system is USA Equidistant Conic. By default, what coordinate system will the output grid have?

4. In Question 3, you could set an option to change the coordinate system of the output grid. Describe how you would change it and what the new coordinate system of the output would be.

5. List and describe the other four analysis options that influence the output results during raster analysis.

6. Imagine that a parcels map is used as the zone layer for a zonal statistics function. Explain how the results would differ if the zone attribute used was the (1) parcel-ID number or (2) the land-use designation.

7. You are adding three grids with cell resolutions of 30 meters, 50 meters, and 90 meters. List the four available options that would determine the cell size of the output grid and state what the output resolution would be in each case.

8. Using Figure 8.4 as a guide, state the results of the four possible combinations of 0 and 1 for the Boolean operators OR, NOT, and XOR. (The results for the AND operator are already shown in Figure 8.6.)

9. Imagine you have Boolean grids for 10 different criteria regarding the siting of a landfill. If you add the grids together, what is the potential range of values in the output grid?

10. What is the potential range of values in Question 9 if you multiplied the grids?

Mastering the Skills

Teaching Tutorial

The following examples provide step-by-step instructions for doing basic tasks and solving basic problems in ArcGIS. The steps you need to do are highlighted with an arrow ➔; follow them carefully. Click on the video number in the VideoIndex to view a demonstration of the steps.

To become acquainted with raster analysis, we will use rasters to solve the snail habitat problem we did in Chapter 7 (Fig. 8.23). Before using the Spatial Analyst extension, it must be turned on.

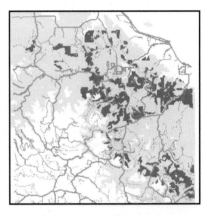

➔ Start ArcMap and open ex_8a.mxd in the MapDocuments folder.

➔ Use Save As to rename the document and remember to save frequently as you work.

1➔ Choose Tools > Extensions from the main menu bar.

1➔ Check the box to activate the Spatial Analyst extension, if it is not already checked.

1➔ Close the Extensions window.

Fig. 8.23. Snail habitat determined by vector analysis

Now that you have activated the extension, open the Spatial Analyst toolbar.

1➔ Right-click one of the menus/toolbars and choose the Spatial Analyst toolbar.

Before starting any analysis, make sure the working directory and default cell size and extent are set. To prevent new grids from getting mixed up with the Sturgis data, create a temp data folder.

➔ Use Windows Explorer or ArcCatalog to create a new, empty folder named **temp** inside the BlackHills folder.

2➔ Choose Spatial Analyst > Options and click the General tab.

2➔ Click the Browse button to set the working directory to the new mgisdata\BlackHills\temp folder. Click OK.

2➔ Click the Cell Size tab and set the analysis cell size to Same as Layer Elevation. Note that this value is 50 (meters).

2➔ Click the Extent tab and make sure the default extent is set to Intersection of Inputs. Click OK.

The map document is the same one we used in Chapter 7 to predict snail habitat from geology, vegetation, and elevation factors using intersection. In this chapter we will perform the same analysis using three Boolean rasters, each one representing the ideal conditions in one of the three layers. We will then multiply them together to find the habitat. We'll begin by creating the elevation range from the DEM.

Performing a Boolean overlay analysis

We can use reclassification to choose cells with elevation from 1,200 to 1,600 feet. This step is analogous to classifying an attribute field for a graduated color map, except that the output is a new raster and the output class values are single numbers rather than ranges. Because we are planning a Boolean overlay, we wish to create a Boolean raster in which the cells meeting the elevation criterion have a value of 1 and the rest have a value of 0.

3➜ On the Spatial Analyst toolbar, set the layer to Elevation, if necessary.

3➜ Choose Spatial Analyst > Reclassify.

You can edit the entries here directly (be sure to put spaces around the hyphens if you do). However, it is often easier to use Classify to set up the desired class breaks first, either manually or using one of the classification schemes described in Chapter 4.

3➜ Click the Classify button and set the number of classes to three.

3➜ In the Break Values box, click on the last entry and change it to 2000. The maximum break value can be set to any value above the maximum grid value seen in the Statistics section in this window.

3➜ Click the middle entry and change it to 1600.

3➜ Click the first entry and change it to 1200. Click OK.

Now the classes are correctly defined, but the new desired values must still be set.

3➜ Click in the first new values box and type 0.

3➜ Click in the second new values box and type 1.

3➜ Click in the third new values box and type 0.

> **TIP:** Be sure to hit the Tab key after the last new value is typed, to make sure it is actually entered. If you neglect to do this, then that last category may be absent in the output raster.

To save this grid permanently, you could click the Browse button by the output raster to specify a folder and file name. This time a temporary grid is fine.

3➜ Make sure the output file is temporary and click OK.

4➜ Find the output, named Reclass of Elevation. Rename it Snail Elev Range.

4➜ Right-click the symbol for 0 in the new raster and change its color to No Color.

Notice that the new grid has only one value, 1, and the rest of the cells are 0. Their extent matches the snail elevation habitat range used in Chapter 7.

A faster and easier way to create a Boolean grid uses the logical operators in the Raster Calculator. Let's repeat the previous steps using the alternative method.

4➜ Choose Spatial Analyst > Raster Calculator.

4➔ Enter the expression [Elevation] >= 1200 And [Elevation] <= 1600 and click Evaluate.

4➔ Examine the new Calculation grid.

4➔ When finished, right-click the new Calculation grid and remove it.

Next, we will select the desired polygons from the geology and vegetation shapefiles and convert them to grids. A single field in the shapefile provides the values for the grid.

5➔ Turn off the Elevation grid and the Roads layer and turn on the Geology layer.

5➔ Use Select By Attributes to select the Geology polygons where [NAME] = 'Madison Limestone' OR [NAME] = 'Upper Paleozoic'.

6➔ Choose Spatial Analyst > Convert > Features to Raster.

6➔ Select Geology as the input features and set the Field to NAME (Fig. 8.24).

6➔ Click the Browse button and enter the folder and file name for the output grid. Call it **limegrid**. Click Save and OK.

6➔ Turn off Geology and turn on limegrid.

6➔ Right-click Geology and choose Remove from the context menu.

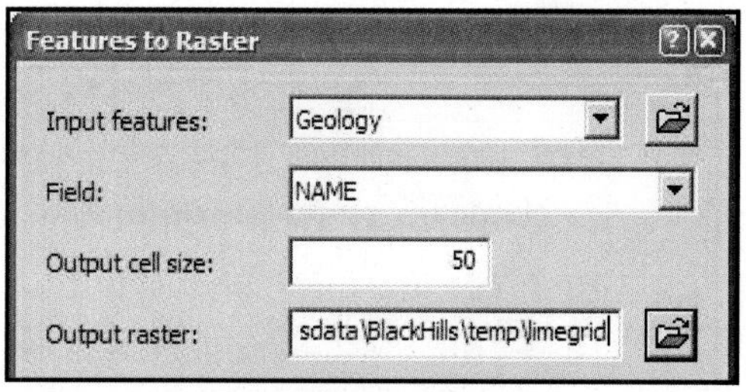

Fig. 8.24. Converting geology polygons to a raster

Notice that limegrid has two colors for units 1 and 2. Because a grid must be numeric, the conversion automatically assigned a numeric value to each unique text instance as the polygons were converted. A discrete grid like geology will have an attribute table created for it. If you want to see the grid values associated with each geologic name, open the attribute table.

7➔ Right-click the limegrid layer and choose Open Attribute Table.

The attribute table contains two entries, one for each unique value present in limegrid (1 and 2, stored in the Value field). The Count field indicates the number of cells for each value (Fig. 8.25).

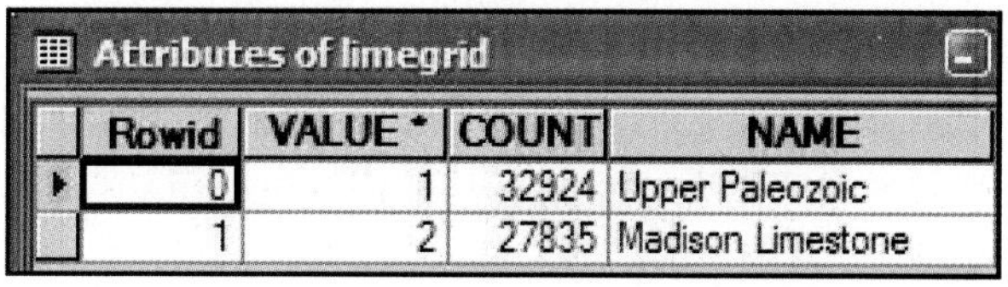

Rowid	VALUE *	COUNT	NAME
0	1	32924	Upper Paleozoic
1	2	27835	Madison Limestone

Fig. 8.25. The limegrid attribute table

1. What percentage of the limestone in this grid consists of Madison Limestone? _____

Next, we must create a Boolean grid containing only 1s and NoData using either Reclassify or the Raster Calculator. We will use the second method.

7➔ Close the limegrid attribute table.

8➔ Choose Spatial Analyst > Raster Calculator.

8➔ Enter the expression [limegrid] > 0 and click Evaluate.

8➔ The new grid is currently called Calculation. Name it **Limestone**.

8➔ Right-click and remove the limegrid layer.

Notice that we are being careful to name the output grids. Otherwise we can end up with dozens of grids with confusing names, like NbrMajority of Reclass of Calculation, which make it easy to use the wrong grid for a later step. If keeping a grid, name it something recognizable.

Next, we will repeat the procedure to create a vegetation Boolean raster.

9➔ Turn on the Vegetation layer.

9➔ Use Select By Attributes on the Vegetation layer to select the polygons where [COV_TYPE] = 'TPP' AND [DENSITY96] = 'C'.

9➔ Close the Select By Attributes window.

10➔ Convert these features to a grid just as for Geology, using the COV_TYPE field. Name the raster file **conifers**.

10➔ Rename the output layer **Conifers**.

10➔ Turn off the Vegetation layer to see the raster.

10➔ Open the attribute table for the Conifers layer.

This time the only grid value is 1 because there was only one value of cover type, TPP. So this time we don't need to reclassify the grid.

Now we are ready to do the overlay. We have three grids, each containing 1s where the desired condition holds and 0 or NoData where the condition is absent. If we multiply these three grids together, any cell that has a 1 in each input grid will have a 1 in the output grid, signifying that it meets all three conditions (elevation, limestone, conifers). Use the Map Calculator to perform this simple map algebra expression.

11➔ Close the Conifers attribute table.

11➔ Choose Spatial Analyst > Raster Calculator.

11➔ Double-click the grid names and click the operator buttons to enter the expression [Snail Elev Range] * [Limestone] * [Conifers] and click Evaluate.

11➔ Turn off all layers except for Calculation.

11➔ Zoom to the extent of the Calculation layer and examine it carefully.

TIP: Although *file names* of grids must contain only 13 characters and no spaces, this rule does not apply to naming a grid layer in the Table of Contents. Give it any name you like.

Notice that the Calculation output has few zero values (pink areas in Fig. 8.26) as a consequence of the inputs. The Geology and Vegetation grids contained NoData values where no polygons were present, whereas the Snail Elev Range layer contained either 1 or 0 everywhere. In the Raster Calculator multiplication, any cell with a NoData input was assigned NoData in the output rather than a 0. NoData and 0 values are different and are treated differently by functions.

In Chapter 7, the next step used overlay to identify which primitive road sections crossed snail habitat. This analysis is not well suited to raster data. Although you could convert the roads to raster, the 50-meter cell size would not represent the roads well, and a smaller cell size would take longer to calculate. Instead, we will convert the snail habitat raster to polygons and overlay them with the vector roads.

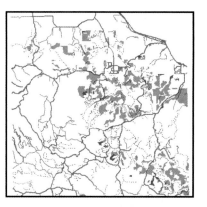

However, converting the current Calculation layer will yield polygons with values of both 1 and 0. We should eliminate the 0 values before doing the conversion to polygons.

Fig. 8.26. Snail habitat raster

12➔ Choose Spatial Analyst > Reclassify and change the 0 values of Calculation to NoData, leaving the 1 values as 1.

12➔ Rename the result **Snail Habitat**, and remove Calculation.

12➔ Choose Spatial Analyst > Convert > Raster to Features.

12➔ Set the layer to Snail Habitat, set the field to Value, and name the output layer **snailhab2.shp**. Make sure the output type is Polygon.

12➔ Make sure the Generalize box is checked. This option removes extraneous vertices from the output polygons for a smoother result that uses less disk space. Click OK.

12➔ Examine the snailhab2 polygons.

At this point we could use the Intersect tool in ArcToolbox to overlay the snailhab2 polygons with the roads to get the proposed road closures just as we did in Chapter 9. Instead of repeating that step, though, we'll just go on with more raster analysis.

The last step in the snail habitat analysis included finding the habitat within 200 meters of primary and secondary roads in order to propose thinning the trees. Buffering with rasters is based on a distance function and requires two steps, but it is still faster than buffering vectors.

13➔ Turn off all the layers, and turn on the Roads layer.

13➔ Select the primary and secondary roads using the expression [TYPE] = 'P' OR [TYPE] = 'S'.

13➔ Create a layer from the selected roads and name it **Major Roads**.

13➔ Turn off the Roads layer.

14➔ Choose Spatial Analyst > Distance > Straight Line.

14➔ Set the distance layer to Major Roads (Fig. 8.27).

14➔ Do NOT check the boxes to create direction or allocation grids.

14➔ Let it create a temporary raster. Click OK.

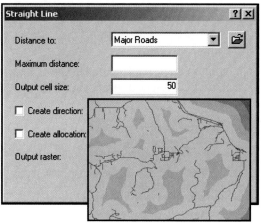

Fig. 8.27. Creating a distance grid using the Straight Line distance function

14➜ Name the output grid **Road Distance**.

Each cell of the new grid represents the distance of that cell from the nearest road. To complete the buffering process, we will use the Raster Calculator to create a Boolean grid, with a 1 representing the area within 200 meters of the roads and a 0 representing the areas elsewhere.

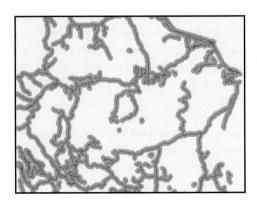

15➜ Choose Spatial Analyst > Raster Calculator.

15➜ Enter the expression [Road Distance] <= 200 and click Evaluate.

15➜ Name the output grid **Road Buffers** (Fig. 8.28).

Fig. 8.28. Raster road buffers

Finally, a Boolean overlay will yield the snail habitat inside the road buffers. This time we'll use AND instead of multiplying.

16➜ Remove the Road Distance grid.

16➜ Turn on the Snail Habitat layer and move it above the Road Buffers layer in the Table of Contents.

16➜ Turn on the Roads layer, clear the selection, and change the colors of all three road types to medium gray.

17➜ Choose Spatial Analyst > Raster Calculator and enter the expression [Snail Habitat] AND [Road Buffers]. Click Evaluate.

17➜ Name the output **Proposed Thin** and change its 0 value symbol to No Color.

Fig. 8.29. Your map showing proposed areas to thin timber

17➜ Turn off all layers except Proposed Thin, Snail Habitat, and Roads.

18➜ Change the symbols so that the map shows the roads as gray, the snail habitat as green, and the proposed thinning areas as purple (Fig. 8.29).

2. Using the cell size (50 meters) and attribute table of Proposed Thin, calculate the total area in square kilometers of proposed thinning. _____

Using topographic functions

➜ Begin a new, empty map. Choose to save your work from the previous section.

19➜ Add the raster mgisdata\BlackHills\rasters\dem2 to the map.

19➜ Set the dem2 symbology to a stretched color elevation ramp.

Note: For performance reasons, the videos will display the elevation grid using the classified method instead of the stretched method. It will look slightly different from yours.

20➔ Set the Analysis Options cell size to 50 meters so that all grids default to this value. The previous Options settings were discarded when the new map was opened.

20➔ Be sure that the working directory is still set to your temp folder.

21➔ Choose Spatial Analyst > Surface Analysis > Slope.

21➔ Set the input surface to dem2, the output measurement to Degree, the Z factor (vertical exaggeration) to 1, and make it a temporary grid. Click OK (Fig. 8.30).

21➔ Turn off the dem2 grid. Rename the Slope of dem2 grid to **Slope**. Examine the range of slopes present.

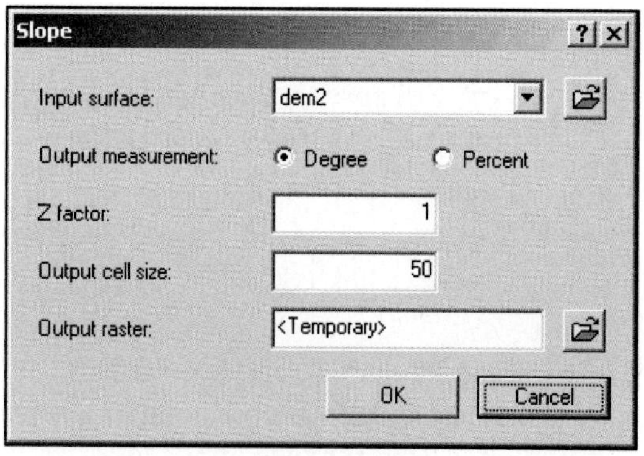

Fig. 8.30. Calculating the slope of a surface

3. What is the maximum slope of this grid? _____

Next, we'll calculate an aspect grid. The aspect is the direction that a slope faces—in other words, if you are standing on a hill looking down the steepest direction and facing south, then the hill has a southern aspect. Aspect is measured in degrees from north, with true north being 0 (and 360) degrees, east 90 degrees, south 180 degrees, and west 270 degrees.

22➔ Turn off the Slope grid.

22➔ Choose Spatial Analyst > Surface Analysis > Aspect.

22➔ Make sure the input grid is dem2, the cell size is 50, and the output is temporary.

22➔ Click OK. Rename the new result **Aspect** and examine it. Notice that the aspect map is automatically colored and labeled according to the compass directions.

4. Which color indicates a northern aspect? _____ A southern aspect? _____

A hillshade function creates a grid that mimics the illumination of the surface from a light source at a specified direction (**azimuth**) and altitude (**zenith angle**). The result looks similar to what a person might see when flying over the surface. Note that both azimuth and altitude are measured in degrees.

23➔ Turn off the Aspect raster.

23➔ Choose Spatial Analyst > Surface Analysis > Hillshade.

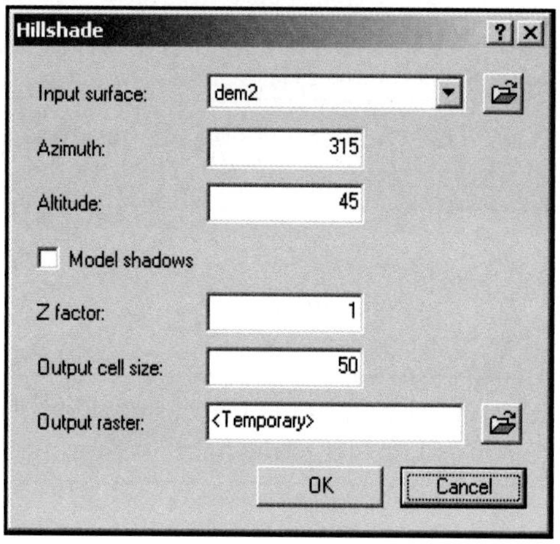

Fig. 8.31. Calculating a hillshade map

23➔ Make sure the input grid is dem2, the cell size is 50 meters, the Z factor is 1, and the output is temporary. Accept the default azimuth and altitude. Click OK (Fig. 8.31).

23➔ Rename the output **Hillshade** and examine it.

Compare both the elevation and hillshade grids and notice how the hillshade brings out fine topographic details that are difficult to see in the elevation grid. Setting a transparent layer over a hillshade looks even better.

24➔ Collapse the legends for the Aspect and Slope rasters.

24➔ Drag the dem2 layer above the Hillshade layer in the Table of Contents and turn it on.

24➔ Open the dem2 layer properties, click the Display tab, and set the transparency to 60%. Click OK.

The Viewshed function determines what can and cannot be seen from a point or a set of points. It is used for such applications as placing fire towers for maximum viewing area or determining whether proposed timber clear-cut areas are visible from the main roads. We will use it to determine the visibility offered by three proposed fire towers (Fig. 8.32).

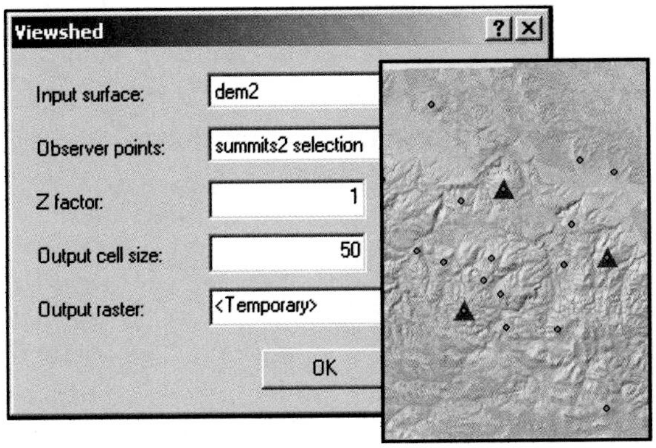

Fig. 8.32. Calculating the view from three proposed fire towers

25➔ Add the **summits** feature class from the Sturgis geodatabase.

25➔ Use the Select Features tool to select the three summits shown by pink triangles in Figure 8.32.

25➔ Create a layer from the selected summit features and name it **Towers**.

25➔ Change the symbology of Towers to pink triangles, and turn off the **summits** layer.

26➔ Choose Spatial Analyst > Surface Analysis > Viewshed.

26➔ Make sure the input surface is dem2, the observer points are set to Towers, and the Z factor is 1. Click OK.

TIP: The Toolbox version of Viewshed allows the user to specify tower heights for more accurate results.

27➔ Remove the **summits** layer, collapse the legend of the Hillshade layer, and turn off all layers.

TIP: Notice that as you accumulate grids, you have to wait while each draws. For best performance, get in the habit of turning off or removing any grids not currently in use.

Using neighborhood functions

A neighborhood function, or **filter**, examines the area around each cell in the grid, one at a time, and calculates statistics. For example, a square 3-cell-by-3-cell neighborhood with an average statistic would calculate the average of the nine cells in the neighborhood and set the value of the center cell to the average. A majority statistic determines the value that occurs most frequently and provides a valuable tool when converting from raster to features. Imagine creating polygons showing areas with slope less than 10 degrees as an initial step for finding land on which to build a cabin. First, reclassify the slope grid already created.

28➜ Turn on the Slope layer.

28➜ Choose Spatial Analyst > Reclassify.

28➜ Choose the Slope grid as the input and create two classes, 1 for slopes less than 10 and 0 for slopes greater than 10.

29➜ Rename the temporary output grid Reclass of Slope to *Slope Class*.

29➜ Zoom in for a closer look (Fig. 8.33).

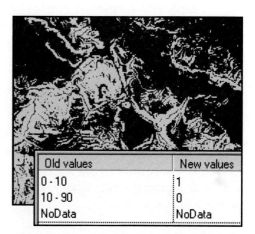

Fig. 8.33. Reclassifying a slope grid into classes for slope < 10 and slope > 10

Notice that the Slope Class raster has many small specks and narrow lines. Converting this raster to features would produce an excessive number of polygons. Furthermore, a tiny flat area surrounded by steep slopes is not a suitable building site, and a small bit of steepness would not disqualify an otherwise flat area. You can use the majority filter to eliminate small areas inside their opposites. If your goal is an area of mostly flat slope that is at least 250 meters square, a 5-cell-by-5-cell majority filter will locate the areas in which the slope is predominantly flat or steep and will simplify the raster.

30➜ Choose Spatial Analyst > Neighborhood Statistics.

30➜ Make sure that Slope Class is the input data. Choose the Majority type and set the neighborhood to a 5-cell rectangle. Make sure the cell size is 50 and the output is temporary. Click OK.

31➜ Change the symbology of the output to unique values to represent 0 and 1.

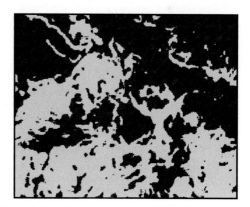

Fig. 8.34. Neighborhood majority filtering simplifies the slope classes before creating polygons.

Notice the much simpler appearance of the new map (Fig. 8.34). However, there are still a few small island areas. Repeating the majority filter on the output file will clean it up more.

32➜ Repeat the 5 × 5 majority filter, this time using NbrMajority of Slope Class as the input and keeping the other parameters the same.

33➜ Rename the output map *SlopeClass2*.

33➜ Change the symbology of the output as you did last time to display only two values.

33➜ Remove the NbrMajority of Slope Class layer.

We are almost ready to create the polygons. However, recall that the raster contains zeros and ones. To prevent the zeros from becoming polygons, we must convert them to NoData.

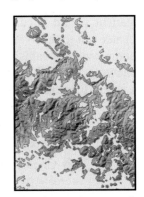

34➜ Use the Reclassify function to set the 0 values in SlopeClass2 to NoData. Set the other values to 1. Be sure to press the Tab key after entering the final value.

34➜ Name the result *SlopeClass3*, and remove SlopeClass2.

34➜ Turn off all of the rasters except for SlopeClass3.

Fig. 8.35. Polygons of low-slope areas

35➜ Zoom to the extent of SlopeClass3.

35➜ Use the Convert > Raster to Features function to create polygons of the high-slope areas from SlopeClass3. Name the output shapefile *lowslope*.

35➜ Turn off SlopeClass3 and examine the new shapefile, which shows potential locations for building a cabin based on slope (Fig. 8.35).

Using zonal functions

A common application of zonal functions occurs when hydrologists are finding watershed parameters for hydrologic runoff models, such as the average slope of a watershed. In this case, the watershed features constitute the zone layer, and the slope map is the values layer.

➜ Click the Open button and open ex_8b.mxd in the MapDocuments folder. Save the changes.

➜ Use Save As to rename the document and save frequently as you work.

36➜ Use Spatial Analyst > Options menu and make sure the working directory is set to your temp folder.

36➜ Make sure the default extent is Intersection of Inputs and the default cell size is 50 meters. Click OK.

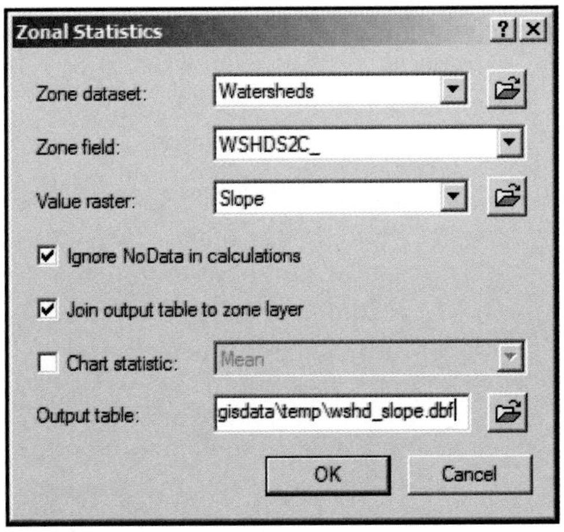

37➜ Choose Spatial Analyst > Zonal Statistics.

37➜ Fill out the dialog box as shown in Figure 8.36, and name the output table **wshd_slope.dbf**. Click OK.

Fig. 8.36. Calculating zonal statistics from slope in watersheds

Recall that an ArcInfo coverage has a special field called the *cover#* (renamed to *cover_* in the geodatabase), which contains a unique number for each feature. We use this field to ensure that each watershed is its own zone.

The joined attribute table is opened automatically. Note that each watershed now has several statistics fields, MIN, MEAN, and so on, indicating the slope values for that watershed. Let's create a map showing the mean slope of each watershed (Fig. 8.37).

38➔ Close the watershed_slope table.

38➔ Open the properties for the Watersheds layer and click the Symbology tab. Create a graduated color map based on the MEAN field.

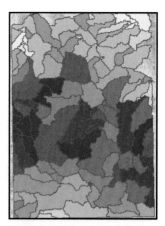

Notice how the high slopes in the center of the grid generate the watersheds with the highest mean slope (Fig. 8.37). Exporting the attribute table would now provide the average slope of each watershed for hydrologic modeling.

Fig. 8.37. Mean slope of watersheds after step 39

5. How many watersheds have an average slope greater than 15 degrees? _____

Zone grids need not be polygons; they could also be points or lines. You could use the same routine with streams to determine the mean slope along a stream line—another important parameter for hydrologic models. When a point grid is used, the result essentially assigns a value from the grid to each point.

39➔ Collapse the Watersheds legend. Turn on the Summits and Elevation layers, and turn off the others.

39➔ Open the Summits attribute table and examine the fields. Each summit has a name but no elevation.

39➔ Close the attribute table.

40➔ Choose Spatial Analyst > Zonal Statistics and fill out the dialog box as shown in Figure 8.38. Name the output **summit_elev.dbf**. Click OK.

40➔ Examine the summit_elev table.

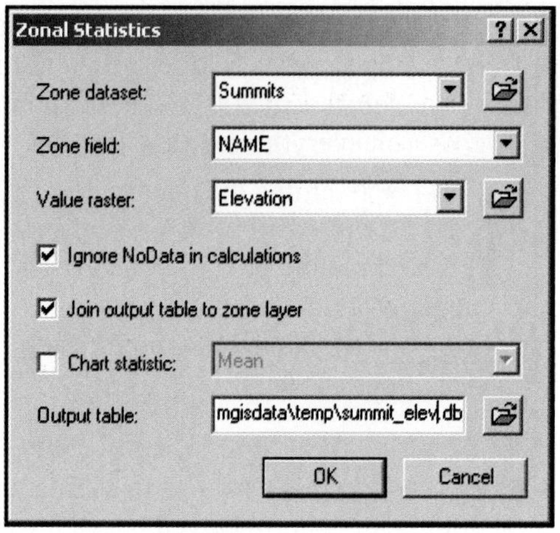

Fig. 8.38. Finding elevations of summits using Zonal Statistics

Notice that all of the statistics have the same value because the mean, min, max, sum, and so on of a single point is the same number. This number, however, is the elevation of the summit. We can now create a map with the summit elevations.

41➔ Close the summit_elev table.

41➔ Open the Summits layer properties and click the Labels tab. Give the layer white labels in 12-point italic bold Arial font. Use any of the zone fields, such as MIN.

6. What is the elevation of Elkhorn Peak? _____

Using interpolation functions

The Black Hills has weather stations that record temperatures and precipitation. To know the precipitation at a point at some distance from a weather station requires interpolating between measurements. We will create a precipitation grid from the stations.

42➔ Activate the Black Hills data frame.

42➔ Open the Climate Stations attribute table and examine the fields. Notice that this data set contains monthly precipitation values in centimeters for the year 1997.

42➔ Close the attribute table.

This data set covers a larger area than the others, and keeping the default cell size at 50 will require an excessively large storage space for the grids. Thus, set the default cell size to a larger value for this part of the analysis.

43➔ Choose Spatial Analyst > Options and set the default cell size to 200 meters.

44➔ Choose Spatial Analyst > Interpolate to Raster > Inverse Distance Weighted.

44➔ Fill out the dialog box as shown in Figure 8.39 to create a precipitation grid for July. Set the input points to Climate Stations, the Z value field to JUL, and the cell size to 200. Keep the defaults on everything else. Click OK.

44➔ Examine the new grid.

7. Do you think the precipitation grid is an integer or a floating-point grid? _____ Examine its properties to check your answer.

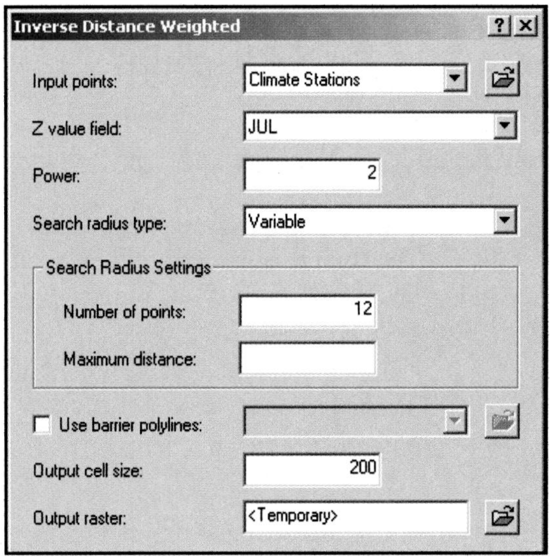

Fig. 8.39. Interpolating rainfall using IDW

Now we can use this new grid to estimate the total volume of water received by each watershed during the month of July. The first step is to calculate the total precipitation for each cell in cubic meters. First, we convert the precipitation measurements to meters.

45➔ Rename the IDW of Climate Stations grid Precip-cm.

45➔ Choose Spatial Analyst > Raster Calculator and enter the expression [Precip-cm] / 100. Click Evaluate.

45➔ Rename the result Precip-m.

Next, we calculate the total volume of water for each cell. The area of each pixel is 200 m × 200 m = 40,000 m^2. Hence, the volume of water V in each pixel is V = Precip-m * 40,000, in cubic meters.

46➜ Choose Spatial Analyst > Raster Calculator and enter the expression [Precip-m] * 40000. Click Evaluate.

46➜ Rename the result Precip-V.

Finally, we use zonal statistics to sum the water volume of cells over the entire watershed. The field SHED_ contains the unique ID number for each watershed, so we use that as the zone field.

47➜ Choose Spatial Analyst > Zonal Statistics.

47➜ Fill out the dialog box as shown in Figure 8.40, naming the output file shedvolume.dbf. Click OK.

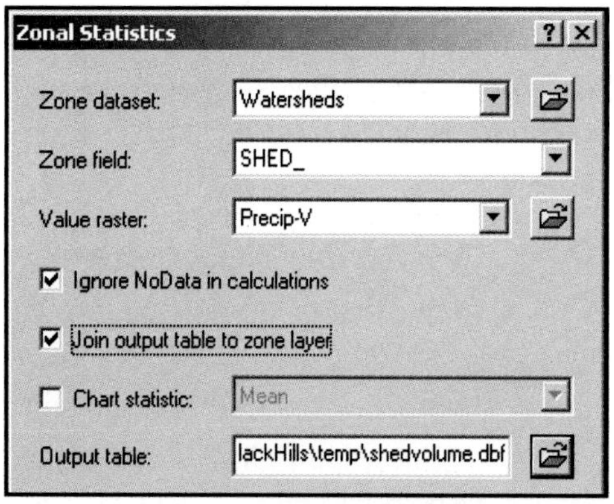

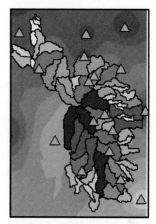

Fig. 8.40. Summing precipitation volume over the watersheds

The output table now lists the total volume of water received in July by each watershed. The values range from about 1 to 46 million cubic meters. Create a map, if you wish (Fig. 8.41).

8. What is the name of the watershed that has the largest volume of water? What is the volume?

_____

Using the analysis options

The analysis options provide detailed control of raster processing. Imagine that you want only information inside a watershed to appear in maps and that the project parameters call for a consistent cell size of 50 meters. The analysis options can enforce both of these conditions.

Fig. 8.41. Map showing climate stations and volume of water in each watershed

48➜ Close the shed-precip table.

48➜ Set the selectable layers to Watersheds, and use the Select Features tool to select the watershed study area (Fig. 8.42).

48➜ Export the selected watershed into a shapefile named studyarea.shp. Add it to the map.

When using features to mask, the software uses the extent of the entire feature class rather than the extent of only the selected feature(s). So we had to put the study area watershed in its own shapefile.

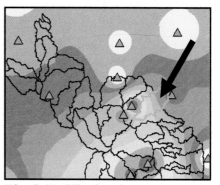

Fig. 8.42. Clipping data to a single watershed

49➜ Copy the studyarea shapefile and paste it into the Sturgis data frame.

49➜ Activate the Sturgis data frame.

49➜ Click the minus sign to collapse the Black Hills data frame out of the way.

49➜ Turn off the Summits and studyarea layers.

Now for the study that we wish to do in the watershed, every time we create an output grid we want it to have the same cell size as the Elevation raster and be masked by the watershed boundary.

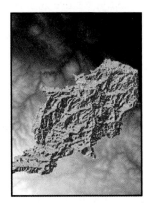

50➜ Choose Spatial Analyst > Options.

50➜ Click the Extent tab. Set the extent to Same As Elevation.

50➜ Click the Cell Size tab and set the cell size to As Specified Below. Enter **50** for the cell size.

50➜ Click the General tab and set the Analysis mask to studyarea. Click OK to accept the options.

The mask option uses a polygon layer or a raster to specify a "cookie cutter" to eliminate unwanted areas in a grid.

Fig. 8.43. Aspect masked to a watershed

51➜ Choose Spatial Analyst > Surface Analysis > Aspect and create an aspect grid from Elevation.

Notice how the study area mask has eliminated the aspect values outside the watershed (Fig. 8.43). The mask grid will remain in effect until you set it back to <None>, and all new grids will be clipped to the mask boundary. All new grids will also have the 50-meter cell size and the extent of the elevation grid, as indicated in the settings.

TIP: Masks may be either grids or feature classes. To mask a single grid, use the Spatial Analyst Tools > Extraction > Extract by Mask tool.

This is the end of the tutorial.

➜ Close ArcMap. You do not need to save your changes.

More skills

Consult the Skills Reference section of this chapter to learn to do the following:

➢ Calculating cell-by-cell statistics for a stack of multiple grids

➢ Finding a least-cost path

Exercises

Use the data layers in the BlackHills\rasters folder and the Sturgis geodatabase to answer the following questions. Be sure to set the cell size option to 50 meters.

TIP: Be sure to turn off any mask grids before going on to the next problem.

---

1. What percentage of the Sturgis area DEM lies above 1500 meters of elevation?

2. You speculate that the chance of finding new cave entrances is highest for areas of Upper Paleozoic rocks with slopes of greater than 20 degrees. Create a map showing the limestone outcrop area and the areas that meet both these criteria. **Capture** your map.

3. Create an elevation map of the Sturgis area, such that the only portions shown are inside the polygons of wshd2b. **Capture** your map.

4. Within the watersheds in wshds2b, what is the farthest you can be from a stream?

5. Which of the summits is farthest from a stream? How far away is it? Which is closest and how far away is it?

6. Create a map showing the areas that are closer to a stream than they are to a road. Make the areas closer to a stream blue and the areas closer to a road gray. Show only the areas inside the watersheds. **Capture** your map.

7. Is there any Forest Service land (grid PVT2 = 0) more than 1 km from the roads in usfsrds? How many square kilometers?

8. Which geologic unit has the highest average slope? Which has the lowest? What is the average slope for each one?

9. The geology feature class has a field called INFILCLASS. Assume that 10% of precipitation infiltrates for class 1, 20% infiltrates for class 2, and 30% infiltrates for class 3. Assume that all the remaining precipitation runs off. Create a raster that portrays the total runoff in centimeters for the Sturgis area in May 1997. **Capture** your map and include a legend.

10. Create an erosion index map for the Sturgis area based on the following assumptions. The basic erosion potential equals the slope divided by the infiltration class times the vegetation factor. The vegetation factor is determined from the DENSITY96 field in the vegetation feature class in the Sturgis.mdb geodatabase, such that 0 = 0.9, A = 0.8, B = 0.6, and C = 0.4. Finally, areas within 200 meters of a stream have their erosion index doubled. Use a Natural Breaks classification with nine classes and a yellow-brown color ramp. **Capture** your map and include a legend. (**Hint:** Use integers when classifying the vegetation and divide by 10 later.)

Challenge Problem

Imagine that you would like to build a house somewhere in the Sturgis area. Develop a map showing potential land on which to build based on the following criteria. Be sure to turn off any masks before you begin.

➢ Must be on private land (grid pvt2 = 1).

➢ Must have a slope less than 20 degrees.

➢ Must have a southerly aspect (between 90 and 270 degrees).

➢ Must be within 500 meters of an existing road.

➢ Must be more than 100 meters from a stream.

➢ The land-use category in lulc must be Evergreen Forest or Mixed Forest Land.

Create and capture a map showing your potential sites with the roads shown in light gray. Also show the private/forest service land in pastels behind the sites.

What is the total area in square kilometers of potential land you found?_____

Skills Reference

Turning on Spatial Analyst..317

Managing grids...317

Setting the analysis environment...318

Converting between grids and features 319

Reclassifying a grid..320

Calculating cell statistics 321

Calculating neighborhood statistics..321

Calculating zonal statistics 322

Using surface functions...323

Calculating a viewshed..324

Creating a density map 324

Interpolating between points 325

Finding a least-cost path..326

Turning on Spatial Analyst

1. Open ArcMap, if necessary.

2. Choose Tools > Extensions from the main menu bar (Fig. 8.44).

3. Check the box to turn on Spatial Analyst.

4. To open the Spatial Analyst toolbar, right-click on the menu and choose the Spatial Analyst menu.

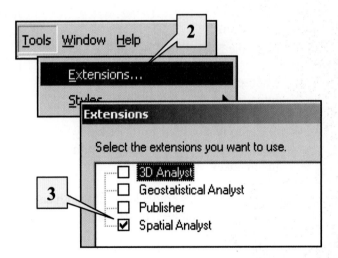

Fig. 8.44. Turning on Spatial Analyst

Managing grids

Spatial Analyst creates many grids, both temporary and permanent. Grids are a complex data format and should ONLY be copied, moved, renamed, or deleted using ArcCatalog. Using Windows Explorer may damage the grids and/or the directories containing the grids.

Copy, move, and delete grids just as you do other data sets in ArcCatalog.

TIP: Grids have much stricter naming conventions than other ArcGIS data files. Grid names must have no more than 13 characters and should contain only letters, numbers, or the underscore character. Pathnames to the grid directory should contain no spaces.

Setting the analysis environment

All settings in the analysis environment remain in effect until changed or until the session ends.

1. Choose Spatial Analyst > Options from the Spatial Analyst toolbar.

Setting the file creation options

2. Click the General tab (Fig. 8.45a).

3. Set the working directory.

4. Choose a grid as an analysis mask, if desired.

5. Set the coordinate system option desired.

6. Check the box if you want to be warned when projecting grids during analysis.

Setting the extent

7. Click the Extent tab (Fig. 8.45b).

8. Choose the analysis extent as the Intersection or Union of Inputs,

9. OR specify a specific grid or display as the extent, or enter the extent coordinates in the boxes.

10. To make the cells coincide exactly with an existing grid, choose that grid in the Snap Extent To box.

Setting the cell size

11. Click the Cell Size tab (Fig. 8.45c).

12. Click the drop-down box to set the cell size method: the maximum or minimum of input grids, to a specific grid, or as specified.

13. To specify a size, type the value in map units in the box. Press the Tab key to see the number of rows and columns that will result,

14. OR type in the number of rows and columns desired, and press the Tab key to see the cell size that results.

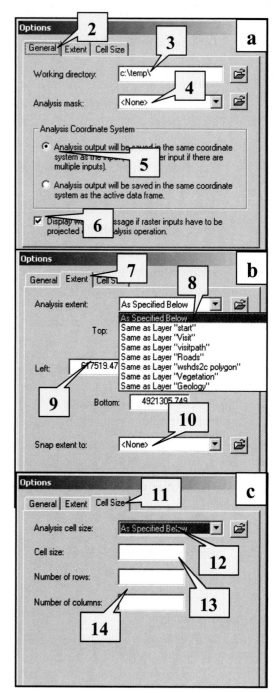

Fig. 8.45. Setting the analysis environment

Converting between grids and features

Converting features to a raster

1. Choose Spatial Analyst > Convert > Convert Features to Raster from the Spatial Analyst toolbar.

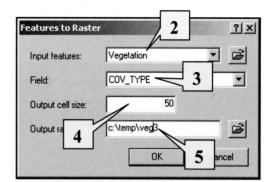

2. Choose the layer to convert from the list, or click the Browse button to locate a file on the disk (Fig. 8.46).

3. Choose the field that will become the values in the output grid.

4. Specify the cell size.

Fig. 8.46. Converting features to a raster

5. Specify a location and name for the output grid. Click OK.

Converting a raster to features

1. Make sure that the raster to convert contains discrete data that will convert well to features.

2. Choose Spatial Analyst > Convert > Raster to Features from the Spatial Analyst toolbar.

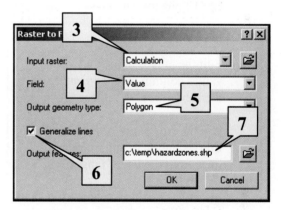

3. Select the raster to convert (Fig. 8.47).

4. Choose the field of the raster (usually the Value field) to become the defining attribute for the polygons.

Fig. 8.47. Converting a raster to features

5. Choose the output feature type. Be sure it is appropriate for the type of raster being converted.

6. Check the box to generalize lines. You will usually want this option, which reduces the number of vertices in the output features to a manageable number.

7. Specify the location and name of the output shapefile or geodatabase feature class.

TIP: Grids have much stricter naming conventions than other ArcGIS data files. Grid names must have no more than 13 characters and should contain only letters, numbers, or the underscore character. Pathnames to the grid directory should contain no spaces.

Reclassifying a grid

1. Choose Spatial Analyst > Reclassify.

2. Choose the grid to reclassify (Fig. 8.48).

3. Choose the field in the grid to reclassify (usually Value).

If the grid has only a few values

4. Click the Unique button to see each value individually.

5. Edit the right-hand column to specify the output values.

If the grid has many values

6. Edit the ranges in the Old values column, and then enter the New values to replace them.

7. Use the Add Entry or the Delete Entries button to add or remove lines from the list of ranges.

8. Instead of entering ranges by hand, click the Classify button to open the Classify Values window, exactly like the one described in Chapter 4. Modify the classification until it is satisfactory.

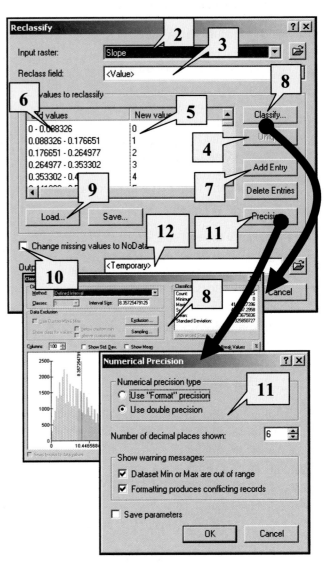

Fig. 8.48. Reclassifying a grid

9. Save or Load previously created classifications, if desired.

10. Check the box to convert missing values to NoData, if desired.

11. If the grid contains floating-point data, click the Precision key to set the format and the precision of the output.

12. Specify the name and location of the output grid, or leave it as is to create a temporary grid.

TIP: Grid names must have no more than 13 characters and should contain only letters, numbers, or the underscore character. Pathnames to the grid directory should contain no spaces.

Calculating cell statistics

Cell Statistics calculates a specified statistic for each cell in a stack of grids.

1. Choose Spatial Analyst > Cell Statistics from the Spatial Analyst toolbar.

2. Click the layers in the left column, and click Add to add them to the list of layers to be included in the stack (Fig. 8.49).

3. To remove a grid, click on it in the right-hand column and click Remove.

4. To add a grid from the disk, click the Browse button.

5. Specify the statistic to be calculated for the stack.

6. Specify a name and location for the output grid or leave it to create a temporary grid.

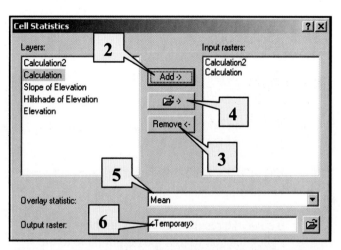

Fig. 8.49. Calculating statistics from a stack of grids

Calculating neighborhood statistics

Neighborhood Statistics utilizes a roving window centered on a target cell. It calculates a statistic for all cells in the window and assigns the value to the target cell in the output grid.

1. Choose Spatial Analyst > Neighborhood Statistics from the Spatial Analyst toolbar.

2. Choose the grid to use (Fig. 8.50).

3. Choose the grid field (usually Value).

4. Choose the statistic type.

5. Choose the neighborhood shape.

6. Specify whether the size is in cells or map units.

7. Specify the neighborhood size. The settings will differ for different shapes.

8. Specify the cell size of the output grid.

9. Specify the name and the location of the output grid or leave it to create a temporary grid.

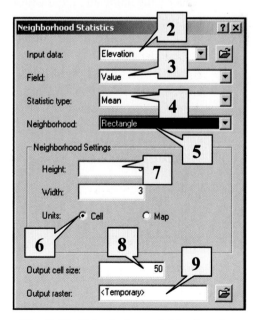

Fig. 8.50. Calculating neighborhood statistics

Calculating zonal statistics

Zonal Statistics calculates statistical measures for cells within zones. Two input grids are required, one defining the zones and another containing the values to use in calculating the statistics. The output is a table containing a record for each zone and a field for each statistic.

1. Choose Spatial Analyst > Zonal Statistics from the Spatial Analyst toolbar.

2. Select the grid or the feature layer that defines the zones (Fig. 8.51).

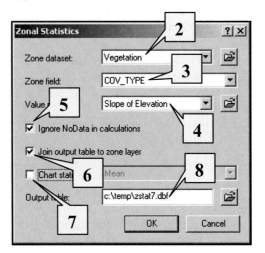

3. Select the field from the zone layer that defines the zones.

4. Select the grid containing the values to be calculated for the statistics.

5. Check the box to ignore NoData cells when calculating (usually the best option).

6. To use the results right away in the map, check the box to join the output table to the zone layer. You will usually want this option.

Fig. 8.51. Calculating zonal statistics

7. Check the box to create a chart showing one of the statistics.

8. Specify the name and the location of the output table.

Using surface functions

The surface functions contour, slope, aspect, and hillshade all produce output grids based on an input surface and a few arguments. The general procedure is as follows:

1. Choose Spatial Analyst > Surface Analysis > *function*, where *function* is the name of the desired operation.

2. Choose the surface to use as input.

3. Specify the parameters for each function as described in the following paragraphs.

4. Enter the cell size for the output grid (except for contouring).

5. Specify the name and the location of the output grid or leave it as is to create a temporary grid. (If you are contouring, the output will be a shapefile.)

To **Contour**, specify the base contour value and the contour interval. For example, a base of 10 and an interval of 20 will give the contours 10, 30, 50, etc. You can also specify a Z factor to multiply with the surface before contouring, for example, to create foot contours on a meter surface.

For **Slope**, enter a Z factor, if desired, and specify degrees or percent for the output units.

For **Aspect**, there are no additional parameters. The resulting grid contains values between 0 and 360 degrees representing the aspect, with 0 = 360 = North. A value of –1 indicates a flat area with no aspect.

For **Hillshade**, change the default values for the azimuth and altitude of the illumination source, if desired. The azimuth is given in degrees between 0 and 360 with 0 = 360 = North. The altitude is given in degrees between 0 (horizontal) and 90 (vertical). A Z factor > 1 may be specified, which will exaggerate the topographic relief. A Z factor < 1 will subdue the topographic relief. You may also check the box to model shadows. If you choose this option, cells in the shadow of another will be set to 0 so that later you can extract these to create a map of shadows. The default option calculates the illumination of all cells, including shadowed ones. The visual difference between the two options is negligible.

For **Cut/Fill**, specify the before and after surfaces. Adjust the Z factor, the output cell size, and the output raster, if necessary. The output grid will have negative values for areas that have been removed and positive values where material has been added.

Calculating a viewshed

1. Create a feature layer that contains the observation points. If desired, this layer can contain attributes describing additional viewing factors, such as the height of the observer and the angle of view. For more information on these parameters, consult the Help document for Spatial Analyst.

2. Choose Spatial Analyst > Surface Analysis > Viewshed from the Spatial Analyst menu.

3. Choose the elevation raster (Fig. 8.52).

4. Specify the point layer containing the observation points.

5. Check the box to use the curvature of the earth in calculating the viewshed, if desired.

6. Enter a Z factor, if desired.

7. Enter the cell size of the output grid.

8. Enter the name and the location of the output grid or leave it as is to create a temporary grid.

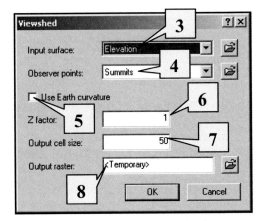

Fig. 8.52. Creating a viewshed grid

Creating a density map

1. Choose Spatial Analyst > Density from the Spatial Analyst toolbar.

2. Specify the point layer (Fig. 8.53).

3. Choose an attribute of the point layer, such as population, on which the density will be based. To simply count the number of points, choose <None>.

4. Choose Simple to count values in the search radius, or choose Kernel to weight the center values more. Kernel gives a smoother result.

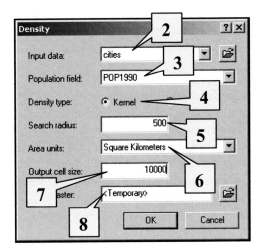

Fig. 8.53. Creating a density map

5. Enter the search radius in map units (maximum radius to include data points).

6. Enter the area units in which the output values will be reported, such as people/square kilometer.

7. Enter the cell size of the output grid in map units.

8. Specify the name and the location of the output grid or leave it as is to create a temporary grid.

Interpolating between points

1. Choose Spatial Analyst > Interpolate to Raster > *Method* from the Spatial Analyst menu, where *Method* is Inverse Distance Weighted, Kriging, or Spline.

2. For Inverse Distance Weighted, choose the point layer containing the data to interpolate (Fig. 8.54).

3. Enter the field containing the values to be interpolated.

4. Enter the power law for the distance weighting. Entering 1 will give a linear weighting in which the influence of each point decreases linearly with distance. Entering 2 causes the influence to decrease as a square of the distance and so on. The larger the number, the smaller the influence of each point on its surroundings.

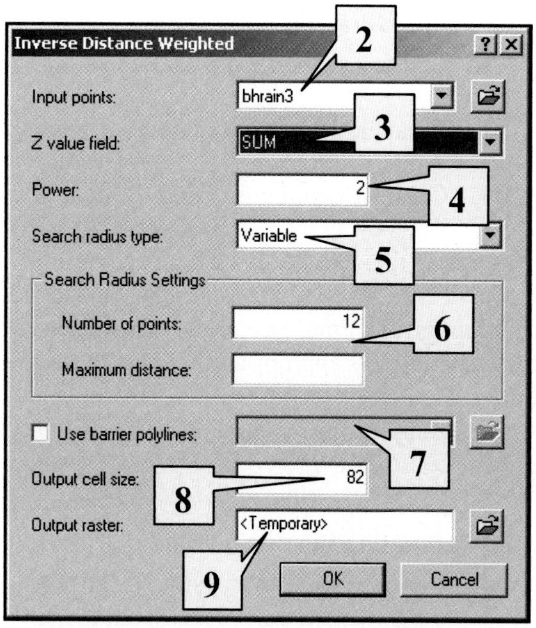

5. Enter the type of search radius.

Fig. 8.54. Interpolation using IDW

A variable radius searches for a specified number of closest points, with an optional requirement to stop searching outside a specified maximum distance. A fixed search radius uses all points within a specified distance, and it has an optional requirement to limit the points used to a specified maximum number.

6. Enter the values for the search distance and the number of points.

7. You can specify a line layer that places linear barriers between points to keep them from being used, such as using watershed boundaries to prevent interpolation from points outside the watershed.

8. Specify the cell size of the output grid.

9. Specify the name and the location of the output grid or leave it as is to create a temporary grid.

Finding a least-cost path

Step 1: Preparing the input layers

1. Create a shapefile containing a single point representing the start location.

2. Create a shapefile containing a single point representing the destination.

3. Create a grid representing the cost of moving across the surface.

This cost grid could be a single layer, such as slope, or it could combine rankings from several different grids. If multiple grids are used, ensure that each input is appropriately scaled to the others. You can also apply different weights when adding the grids. The final result is the cost grid.

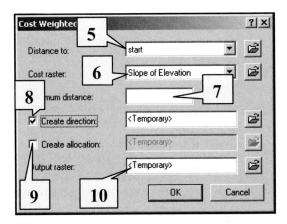

Fig. 8.55a. Calculating the cost distance and direction grids

Step 2: Creating the cost distance grids

4. Choose Spatial Analyst > Distance > Cost Weighted from the Spatial Analyst toolbar.

5. Enter the layer containing the start location (Fig. 8.55a).

6. Specify the name of the cost grid.

7. Specify a maximum distance, if desired.

8. Check the box to create a direction grid, which is required to create the path.

9. An allocation grid is not required, but check the box if one is desired.

10. Enter the name and the location of the output cost distance grid or leave it as is for a temporary grid.

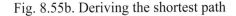

Fig. 8.55b. Deriving the shortest path

Step 3: Finding the least-cost path

11. Choose Spatial Analyst > Distance > Shortest Path from the Spatial Analyst toolbar.

12. Specify the point layer containing the destination (Fig. 8.55b).

13. Specify the cost distance and cost direction grids created in step 2.

14. Choose to find the Best Single path.

15. Specify the name and the location of the output shapefile.

Chapter 9. Network Analysis

Objectives

➢ Learning the function and terminology of networks

➢ Understanding the different properties of transportation and utility networks

➢ Performing tracing analysis on networks

➢ Understanding how networks are constructed

Mastering the Concepts

GIS Concepts

About networks

Networks consist of a system of paths traveled by a variety of things, such as traffic, water, sewage, or electricity. Common examples of networks include roads, utility lines, airline routes, and streams. Modeling the behavior of a "commodity" flowing through the network can help answer questions such as these: How long would it take a chemical spill in this watershed to reach the city water supply? If this transformer blows, which parts of the city will be out of power? If a problem develops with a sewer line, which shutoff valve can we use to halt flow during repairs while affecting the fewest possible people?

All feature classes participating in a network must take the role of either an **edge** or a **junction** (Fig. 9.1). An edge feature represents a path along the network. Examples of edges include roads, pipelines, and electric cables. Edges must come from a line feature class. A junction represents a point in a network where edges meet or where an object such as a valve exists. A street intersection, for example, is represented by a junction. T-valves or straight-line connectors would constitute junctions in a pipeline network. Junctions can come from two sources: they can be automatically constructed during network building wherever edges meet, or they can be specific objects that come from a point feature class, such as valves.

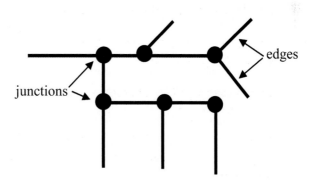

Fig. 9.1. A network is composed of edges and junctions.

Network features can take one of two states, enabled or disabled. An enabled edge or junction allows material to flow through it. A disabled edge or junction blocks travel, whether of water, electricity, or cars. The feature state is initially assigned during construction of the network; however, it may be temporarily modified during analysis, such as disabling a street edge to represent road construction that blocks travel. The states of the network features are stored in the attribute table of the feature class in a field named Enable.

Network storage in a data model has two aspects (Fig. 9.2). The **geometric network** consists of the actual points and lines in the feature classes that participate in the network. These features are stored as the familiar feature classes containing points and lines. Certain fields are added to the attribute tables of the feature classes to help track the participation of each feature in the network. The **logical network** includes a set of tables that store information about the construction and operation of the network elements. Users do not usually have direct access to the logical network. The files are created at the time the network is built from the feature classes. The geometric network and the logical network are interconnected, and changes in the geometric network require changes to the logical network. During editing, the relationships between the geometric and the logical networks are maintained automatically.

Types of networks

Networks fall into one of two basic categories, transportation networks or utility networks. The material in a transportation network (cars) has few constraints on travel; the drivers go where they choose (Fig. 9.3). The vehicles can usually travel in either direction along an edge and through intersections, represented by junctions.

Utility networks have established directions of flow determined by the topology of the connections and the location of sources or sinks (Fig. 9.4). A **source** point provides material to the network and "pushes" the material away from itself. A power plant would be a source for an electric grid; a water treatment plant might act as a source for a water line network. A **sink** represents a location where material is used or leaves the network. Sinks draw the material toward themselves. Electric meters or water meters on a building might be considered sinks, as would a sewage treatment facility discharging to a lake.

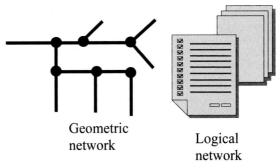

Geometric network

Logical network

Fig. 9.2. The geometric network consists of features in a feature class. The logical network contains relationships, connections, and behavior as tabular data.

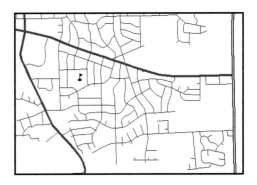

Fig. 9.3. In a transportation network, the drivers choose the way to travel.

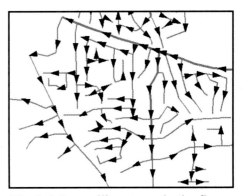

Fig. 9.4. In a utility network, the flow of the material is determined by the network configuration.

Transportation networks and utility networks have different types of analysis functions available for use in modeling flow. A transportation network could be used to solve problems such as the following:

> ➤ What is the best path to travel to 16 delivery locations?

> ➤ What is the likely service area of a fire station based on travel time?

> ➤ What is the shortest path from point A to point B?

Utility networks might be used to address other kinds of problems:

> ➤ If this valve fails, which customers will be affected?

> ➤ If I have to close this pipe for repairs, can I reroute water through another path to minimize service disruption?

> ➤ How will contamination at one location propagate through the network?

> ➤ Which sewer lines serve only residential customers?

Utility network flow must be established before the network can be analyzed. You establish flow by specifying the sources or the sinks in the network and solving for the flow direction. A utility network edge may have three possible states: it may be unassigned if network flow has not yet been established; it may have a determinate flow, meaning that a single flow direction was established; or it may have an indeterminate flow, meaning that the flow direction is ambiguous and cannot be uniquely determined. The third case generally results from a topological error in the network or from an error in specifying the sources and the sinks.

Network analysis

Network analysis problems are tackled using programs called **solvers**. Many types of solvers are possible. Figure 9.5 shows the set of solvers that come with ArcMap. Some general features of using solvers are discussed in this section.

Flags pinpoint locations of interest, such as the starting and the endpoints of a journey, the delivery stops on a route, or a location from which to trace up or downstream. Flags may be placed along edges or at junctions (Fig. 9.6).

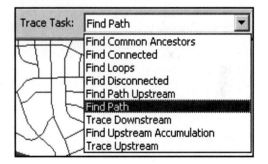

Fig. 9.5. Tracing solvers in ArcMap

Barriers represent temporary blockages or outages in the network, such as a damaged bridge, a closed valve, or a broken power line. As with flags, barriers may be placed on either edges or junctions (Fig. 9.6). When placed on a feature, barriers temporarily override the normal "Enabled" state of the feature, as stored in the layer's attribute file.

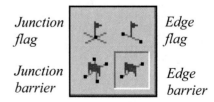
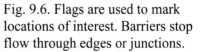

Fig. 9.6. Flags are used to mark locations of interest. Barriers stop flow through edges or junctions.

Network weights can be stored for each edge or junction as attributes and specify the "cost" of traversing the network. Common edge weights include distance along the edge or travel time as dictated by the street length and speed limit. Common junction weights include a loss of pressure at a pipe T-junction or a waiting time imposed at a traffic light. Solvers can take these weights into account when finding solutions, such as finding the path with the least wait time at intersections rather than finding the shortest path between two points.

The output of a network tracer usually takes the form of a set of edges and junctions along the traced path. Network analysis offers several options concerning the form of this output. The path solution may be returned either as a graphic or as a set of selected features. If returning a selection, one may choose to return only edges, only junctions, or both.

Generic trace solvers

The solvers in this section may be applied to either utility or transportation networks. They carry out their work without regard to the direction of flow through the edges and junctions.

Finding paths

This solver traces the least-cost path between two or more flags placed on the network. It can find a way to travel from one location to another or to plan a path to most efficiently visit a series of locations such as delivery stops. When finding paths, the solver keeps track of the "cost" of traveling the path. By default the cost is the number of edges traversed. Figure 9.7a shows the lowest-cost path, in terms of the number of edges traversed, for driving from a student's house to a school. The flags are shown as green boxes at the locations of interest; in this case edge flags are used.

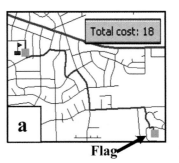

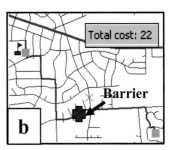

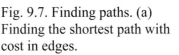

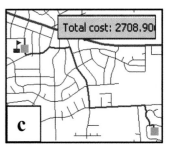

Fig. 9.7. Finding paths. (a) Finding the shortest path with cost in edges.
(b) Finding the shortest path with a barrier.
(c) Finding the shortest path using a distance weight (meters).

The user can modify solver behavior by specifying barriers to travel, such as road construction. Figure 9.7b shows the new path found when a barrier is placed along the original one. The cost has increased to 22 edges, but the new path is still the least-cost path, given the presence of the barrier.

Users can also specify a weight, such as distance or travel time. Without weights, the solver simply counts the number of edges traversed. However, the lowest number of edges does not always mean the shortest path; in some cases the shortest path could have more edges. Finding the shortest path requires a distance weight. In Figure 9.7c, the distance along the edges has been chosen as the weight. A slightly different path results than when simply counting edges. The cost shows up in the distance units—in this case, meters.

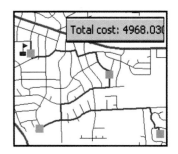

Fig. 9.8. Finding the best path to pick up a child's friends on the way to school

Finally, the user can specify multiple flags in order to find the least-cost route for making several stops in sequence. Figure 9.8 shows a path to pick up a child's friends and take them to school. The Find Path tracer simply visits the stops in the order given. Other solvers must be used to solve the classic "traveling salesman" problem: given a set of locations to visit, devise the most efficient sequence and path to follow.

Find Connected/Find Disconnected

The Find Connected solver traces along the features and highlights those that are connected to the flag(s) placed on the network. Such a solver might be used to find all of the water lines that are supplied by a single water intake gallery along a river (Fig. 9.9). Such information might be useful in gauging the impact of a chemical spill near the intake and determining which customers must be warned to purchase drinking water until the contamination is removed.

Barriers placed at specific locations affect the results of this solver. Any feature with a barrier between it and the flag location will be considered disconnected from the network (Fig. 9.9).

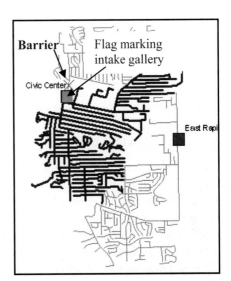

Fig. 9.9. Find Connected returns all parts of the network connected to one or more points of interest.

Find Disconnected performs the opposite function of Find Connected; it locates portions of the network that can't be reached from the flagged locations. This solver is particularly useful for diagnosing problems when constructing networks. Sometimes a failure in snapping two line ends together or neglecting to split a line at an intersection will result in part of the network being disconnected. The solver finds these problem areas so that they can be fixed.

The Find Connected and Find Disconnected solvers are most useful for utility networks, but they can also be applied to transportation networks. A road network is designed to facilitate connecting all locations for easy travel. Thus, a Find Connected usually results in the selection of the entire network (not a very useful result most of the time). However, one useful application might include finding which parts of the city might be seriously affected by construction placed at a critical location. Find Disconnected also helps locate network errors, as previously noted.

Finding loops

A loop is a section of a network that forms a continuous loop or ring. In a transportation network, loops are generally the rule, as most streets eventually connect to other streets and can be traversed in a circular fashion. This solver finds little utility in transportation networks. In utility networks, loops indicate a topological error rather than a legitimate situation. A water pipe that forms a loop would cause problems when the emerging flow reentered the loop. The Find Loops tracer, then, has its primary use as a tool for detecting and fixing loops. A flag must be specified as the place to begin tracing, and the solver finds all loops that lie downstream of the flag.

Figure 9.10 shows an example of a loop in a utility network. The flag has been placed at the upper right of the picture. The arrows indicate the direction of flow along the network. The loop found by the tracer is marked with small circles, which indicate that flow in the loop is indeterminate; because of the loop, the flow direction cannot be established from the two possible solutions. To fix the loop, one end must be disconnected from the circle or an edge or a junction must be disabled.

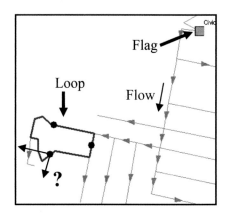

Fig. 9.10. A loop in a utility network cannot flow.

Utility trace solvers

The solvers in this section are used only for utility networks. They rely on an established flow direction in order to carry out the analysis.

Find Path Upstream

This solver works similarly to the Find Path solver previously discussed, except that movement along the edges is always constrained to be in the upstream direction. The tool can locate a path between a single location and its source (Fig. 9.11). If more than one flag has been specified, the paths of each flag are traced upwards.

Find Upstream Accumulation

This solver is a variation of the Find Path Upstream tool, but it keeps track of the weights on each edge and junction and applies them to determine a total accumulation of resources along the path.

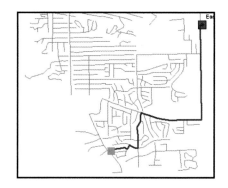

Fig. 9.11. Find Path Upstream traces the flow from a location to its source.

Trace Upstream/Downstream

Trace Upstream and Trace Downstream provide a way to follow the flow from a specified location. Trace Upstream is essentially the same solver as Find Path Upstream. The downstream trace finds all the edges that lie downstream of a flagged location or locations. Such a solver might be used to find the service areas that would be disrupted by a downed power line or a water line break (Fig. 9.12).

Find Common Ancestors

This solver takes a group of flags and finds any upstream edges and junctions that are shared by all the individuals in the group. The tool can be used by utility companies to help locate failures in the system based on reports by customers. Suppose that a utility company receives 50 calls in one evening from customers complaining about a loss of water pressure. By entering the locations of these reports as flags, the company can isolate possible locations of the suspected

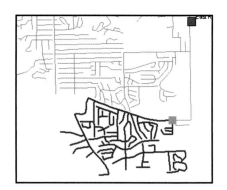

Fig. 9.12. Trace Downstream could find the service area disrupted by a broken pipe.

break by finding the set of pipes that are common to all the calls. The break will most likely be found near the lower end of the common ancestors (Fig. 9.13).

About ArcGIS

Geodatabases are used in ArcGIS to create and store networks. Networks are topological data structures that can be built from existing simple feature classes. A feature class containing roads, for example, can be used to create a transportation network. Multiple layers can be combined to form more complex networks. A water supply network might include feature classes representing pumping stations, pipes, valves, and meters.

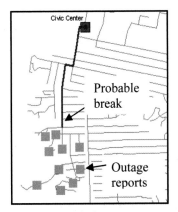

Fig. 9.13. The Find Common Ancestors solver can locate service failures.

Networks may only be created in feature datasets. All feature classes participating in a network must belong to the same feature dataset; however, the dataset can contain other feature classes that do not participate in the network. A Transportation feature dataset containing roads and rail lines could have a roads network that did not include the rails.

Figure 9.14 shows a geodatabase containing two networks. The Transportation feature dataset contains a single line feature class, roads, plus two feature classes that comprise the network, Road_Net and Road_Net_Junctions. The Utilities feature dataset contains four feature classes plus the Water_Net and Water_Net_Junctions. It is not obvious from looking at the geodatabase which feature classes in the feature dataset participate in the network; however, this information can be obtained by examining the properties of the network in ArcCatalog.

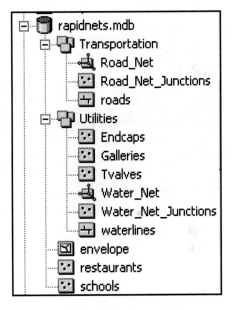

Fig. 9.14. Networks are built from feature classes in feature datasets.

The Utility Network Analyst toolbar

ArcMap comes with simple solvers called **traces**, which follow network connections (Fig. 9.5). More advanced and complex network analysis tools are also available for users who purchase the Network Analyst extension. Knowledgeable users may employ the special customization languages provided with ArcGIS to write new solvers. ArcMap solvers require input in four forms: the network itself, the weights of elements along the network, the locations of interest, and any barriers to flow.

ArcMap contains a special toolbar used to direct and control the analysis of networks (Fig. 9.15). The main features of this toolbar are as follows:

> ➤ The Network drop-down list controls which network is being analyzed. One network may be analyzed at a time.

> ➤ The Flow menu controls the display of symbols representing flow directions in a utility network. The user can turn the symbols on and off and change the symbols being used.

> ➢ The Establish Flow button next to the Flow menu is used during editing to establish the initial flow direction in a network, and it requires an ArcEdit or ArcInfo license. ArcView can perform analysis on networks for which this task is already complete.

> ➢ The Analysis menu manages the analysis by placing flags or barriers, clearing both flags and barriers, or clearing previous results in preparation for a new analysis. The Trace Task menu specifies which kind of analysis will be performed.

> ➢ The drop-down flag tool controls which type of flag or barrier to add to the map.

> ➢ The Solve button must be clicked to perform the trace. All of the trace tasks require that at least one flag be added to the map before any analysis can occur. If no flags are present, the Solve button will be dimmed.

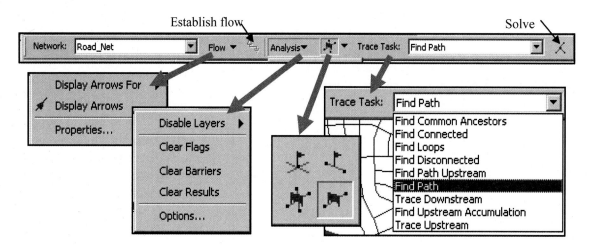

Fig. 9.15. The Utility Network Analyst toolbar contains all the functions needed to analyze networks in ArcMap.

The overall process of analyzing a network includes specifying the network to be used, choosing the tracing task, adding flags or barriers, and solving. Once a solution is obtained, it may be cleared before going on.

The Analysis > Options menu choice opens a window with four tabs. These options control the analysis by specifying which features will be traced, whether the features will be weighted, and the format of the results. The output of a trace may take the form of a drawing or a set of selected features (Fig. 9.16). The features returned can include all features matching the criteria of the trace task or simply those stopping the trace. The user can also choose whether to return edges or junctions or both.

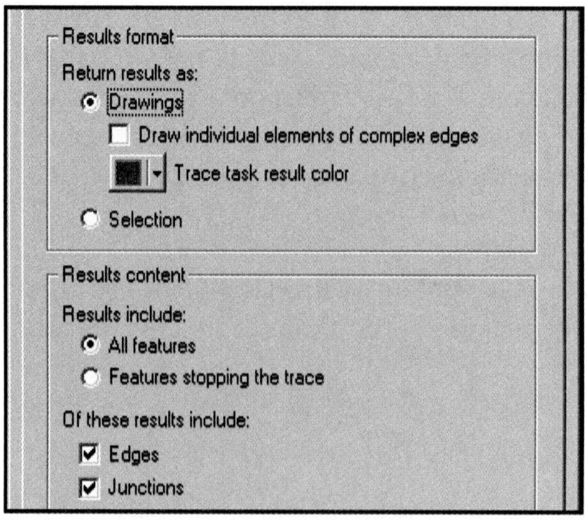

Fig. 9.16. Choosing the format of the output results

Building networks

Building valid networks requires ingenuity and attention to detail, especially if the network is to be used for sophisticated modeling. The planning stage is very important, and the builder must consider a number of issues in designing the network. These issues include using simple or complex edges, assigning weights, using domains or subtypes, and establishing connectivity rules.

Although ArcView can perform analysis on existing networks, an ArcEditor or an ArcInfo license is required in order to create networks. We will not further address building networks in this chapter, but the following paragraphs introduce basic design issues that can affect network analysis. If you have the license and want to try building a network, consult the entry in the Skills Reference.

Simple edges or complex edges?

Network edges occur in two different styles, simple edges and complex edges (Fig. 9.17). A simple edge can only have junctions at its endpoints. A complex edge can have one or more junctions in its middle.

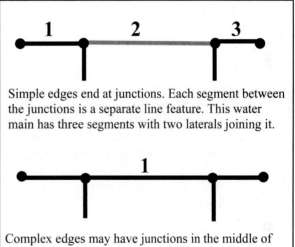

Simple edges end at junctions. Each segment between the junctions is a separate line feature. This water main has three segments with two laterals joining it.

Complex edges may have junctions in the middle of the line feature. This water main is a single feature with two laterals joining it in the middle.

Complex edges offer advantages when keeping track of long features. For example, a water main along a street has junctions composed of T-valves that branch off to laterals carrying water to each individual house. Using simple edges, the main would have to be broken into multiple pieces between each T-valve. However, for purposes of queries and maintenance, the water main is more logically and easily treated as a single feature. When clicking on the water main, it would be better to select the entire feature, not just one tiny segment between two laterals. By using a complex edge network, the water main can remain a single feature while allowing the junctions of the laterals to connect to it.

Fig. 9.17. Networks with simple edges versus networks with complex edges

Junctions also can be implemented as simple or complex features. A complex junction can be built from a set of simple junctions and simple edges. For example, a dam in a stream network might be represented as a complex junction, which itself is composed of edges and junctions representing flow gates, generators, turbines, and more. The geodatabase model comes equipped with simple junctions, simple edges, and complex edges. Complex junctions must be created using advanced modeling tools, and their construction is beyond the scope of this book.

Network weights

When creating a network, the user can include certain attributes that establish weights for each edge and junction. Such weights allow for more realistic modeling of flow. For example, in analyzing possible routes from a fire station to a fire, the shortest route is not necessarily the fastest. A longer route with less traffic, fewer traffic lights, and a higher speed limit might serve better than the shortest route. A more realistic travel network could include a travel time attribute for the road features and point junctions representing traffic lights with different average delays.

When creating the network, these attributes would be specified as weights, to take into account costs associated with traversing the edges or passing through the junctions. Weights can only be established during network creation.

Summary

> Networks are used to model the flow of objects or materials along an interconnected system of edges and junctions.

> Transportation networks are distinguished by the fact that the material can independently determine how it will flow. In a utility network, the direction of flow is determined by the network topology and the location of sources and sinks.

> A variety of programs called solvers may be applied to analyzing flow through a network. Solvers typically use three types of input: the network itself, flags indicating points of interest, such as delivery locations, and barriers that prevent flow.

> Network weights are stored as attributes of edges. Weights can be used during analysis to determine, for example, the shortest path or the shortest travel time. They can also be used to accumulate the costs of moving along the network.

> Networks are stored in feature datasets, and they may include some or all of the feature classes in the dataset.

> Network topology, connectivity, and relationships are stored in data tables and collectively represent the logical network. The actual line and point features represent the geometric network.

> Network topology is created in ArcCatalog. Once established, it is updated and maintained during editing. Building networks requires an ArcEditor or ArcInfo license.

> Networks may be composed of simple edges that always end at junctions, complex edges that can contain junctions in the middle of the feature.

TIP: Weights are set up at the time the network is created. If no weights are listed in the drop-down boxes, then the network contains no weights, and none can be used.

TIP: Some tracing tools will not work if the flags or the barriers are of different types. For an analysis, try to always use edge flags and edge barriers together or junction flags and junction barriers together.

Chapter Review Questions

You may need to consult the Skills Reference section to answer some of these questions.

1. What kinds of feature classes can be used to create networks?

2. Explain the difference between a geometric network and a logical network.

3. What characteristic distinguishes a transportation network from a utility network?

4. Networks are composed of what two spatial elements? What is the role of each?

5. What are sources and sinks? In what type of network are they found?

6. What purposes do flags and barriers serve?

7. What are loops? Are they desirable?

8. By default, does the Find Path solver always find the shortest path? Explain.

9. What is the purpose of using network weights?

10. How do simple edges and complex edges differ?

Mastering the Skills

Teaching Tutorial

The following examples provide step-by-step instructions for doing basic tasks and solving basic problems in ArcGIS. The steps you need to do are highlighted with an arrow ➔; follow them carefully. Click on the video number in the VideoIndex to view a demonstration of the steps.

    1➔ Start ArcCatalog and navigate to the mgisdata\Rapidcity folder.

    1➔ Expand the rapidnets geodatabase.

    1➔ Expand the Transportation feature dataset.

    1➔ Right-click the Road_Net network and choose Properties.

1.   What, if any, weights have been created for this network?_____

    1➔ Close the Transportation network properties.

    1➔ Open the properties for the Utilities network.

2.   Which feature classes participate in the network, and what roles do they play?

_____

3.   What weights does the Utilities network have?

_____

    1➔ Close the Utilities network properties.

    ➔ Close ArcCatalog and start ArcMap with the ex_14.mxd map document in mgisdata\MapDocuments.

    ➔ Use Save As to rename the document and remember to save frequently as you work.

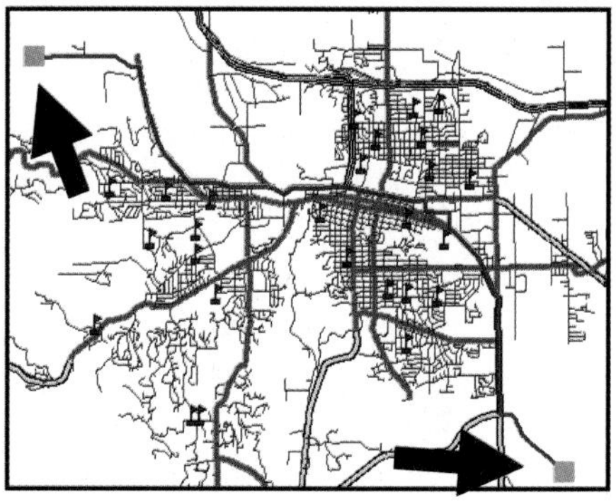

Finding paths

We will begin by finding some simple paths through this transportation network.

Fig. 9.18. Finding a path between two flags

TIP: Be sure to use the *Utility* Network Analyst toolbar, not the Network Analyst toolbar. Network Analyst is an extension that must be purchased.

    2➔ If the Utility Network Analyst toolbar is not open, right-click on the menu area on top of the window and turn it on.

    2➔ Make sure that the Network is set to Road_Net in the Utility Network Analyst toolbar.

    2➔ Set the trace task to Find Path.

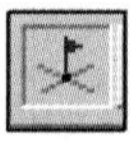

    2➔ Click on the Junction Flag tool.

2➜ Click on the ends of two roads on opposite ends of town as shown in Figure 9.18.

2➜ Click the Solve button.

TIP: Flags and barriers should match types. For best results, always use junction flags and junction barriers together or edge flags and edge barriers together.

The path solution appears as a bold red line connecting the flags. This path meets the criterion that it follow the minimum number of edges through the network. Since this is a transportation network, traffic can travel in either direction. Now let's try another one.

3➜ Choose Analysis > Clear Flags from the Utility Network Analyst menu.

3➜ Choose Analysis > Clear Results from the Utility Network Analyst menu.

3➜ Click the Junction Flag tool and add two flags at any locations you choose.

TIP: If the flag does not snap to the right location, clear the flags and start again. Zooming in will make it easier to snap to the right spot.

3➜ Click Solve.

4➜ Clear the flags and the results again using the Analysis menu.

4➜ Choose the Edge Flag tool, try adding a pair of edge flags, and then solve.

4➜ Clear the flags and the results.

TIP: Flags and barriers will be snapped to the closest feature within the snap tolerance set from the General tab in the Analysis > Options menu.

Now let's explore the use of barriers to help find the best way to take a child to school. Imagine that you live on Harney Dr. and that your child goes to Southwest Middle School.

5➜ Choose Bookmarks > School Trips 1 from the main menu bar.

5➜ Click the Find tool.

5➜ Type **Harney** as the text for which to search ,and set the layer to Roads. Click Find.

5➜ Hold down the Ctrl-key and click on each Harney Dr. entry in turn to select them all. Don't select Harney Pl. or the others.

5➜ Right-click on one of the entries and choose Select. Harney Dr. will be selected.

5➜ Use the Find tool to locate and to select the text **Southwest** in the Schools layer.

5➜ Close the Find window.

6➜ Zoom in to the extent of Harney Dr. to more easily mark the right spot with the flag.

6➜ Click the Edge Flag tool.

6➜ Place a flag near the middle of Harney Dr., and use the Zoom to Previous Extent button to return to the original extent.

7➜ Place another edge flag on Park Dr. next to the school, zooming as necessary.

7➜ Zoom back to see both flags, and make sure the trace task is set to Find Path.

7➜ Click Solve.

Well that was probably the answer you expected. Perhaps you are wondering, however, how far it is from home to school.

8➜ Click the Solve button again without moving the mouse off it and this time keep an eye on the lower-left corner of the map window to see the cost.

4. What is the cost of this trip and what are the units of the cost? _____

It would be more interesting to learn the distance. To find distances, you must use a weight.

8➜ Choose Analysis > Options from the Utility Network Analyst menu bar.

8➜ Click the Weights tab.

8➜ Set both the along and the against weights to Distance. Click OK.

8➜ Clear the results (but not the flags).

8➜ Click Solve again and watch for the cost.

5. What is the cost and what are the units? _____

> **TIP:** When specifying weights in a Transportation network, it is critical to specify a weight both FOR and AGAINST the direction of travel. If only one weight is chosen, then cost will accumulate only when travel is in the same direction as the line, and the total will be too small.

On the first day of school, you discover that an enormous traffic jam builds up at the intersection south of the school. You wonder if there is an alternate route through the tangle of residential streets north of the school and what the street names are. First, though, let's turn on Map Tips.

9➜ Right-click the Roads layer and choose Properties.

9➜ Click the Display tab. Check the Show Map Tips box.

9➜ Click the Fields tab and set the primary display field to ROADNAME. Click OK.

With Map Tips on, use the cursor to hover over a street to find out its name. Also, the streets will be labeled if you are zoomed in to scales greater than 1:20,000.

10➜ Clear the results (but not the flags).

10➜ Zoom in to the middle school to see it better and to see the road labels appear.

10➜ Click the Edge Barrier tool and place a barrier on Corral Dr. between Park Dr. and Sheridan Lake Rd.

10➜ Zoom to the previous extent when done.

10➜ Click Solve.

6. What is the distance of this alternate route? _____

To make it easier to find the way tomorrow morning, make a list of the streets involved in this path. One way to get a list is to return the path as a selection instead of as a drawing.

11➔ Choose Selection > Interactive Selection Method from the main toolbar, and make sure that it is set to Create New Selection.

11➔ Choose Selection > Set Selectable Layers and make sure all layers are selectable.

12➔ Choose Analysis > Options from the Utility Network Analyst toolbar.

12➔ Click the Results tab.

12➔ Change the results format to Selection, and uncheck the box for returning junctions. The edges will contain the street names. Click OK.

12➔ Clear the results (but not the flags or the barriers).

12➔ Solve again.

> **TIP:** If no features are selected when solving a trace, check that the selection method is set to Create New Selection and that the network is one of the selectable layers.

13➔ Open the Roads attribute table.

13➔ Click the button to show only the selected records.

7. How many (different) street names are included on this path? _____

13➔ Set the results format back to Drawings in the Analysis > Options menu and check the box to return junctions once more. Click OK to close the window.

13➔ Close the attribute table.

Finally, your other child attends Meadowbrook Elementary School. Normally she walks to school, but you want to plan a route to take both children on rainy days. In addition, you have agreed to pick up her friend who lives on Player Dr. Find a new path including all of these elements. The Middle School starts earlier than the others, so you plan to stop there first, pick up the child on Player Dr. next, and then arrive at Meadowbrook. You need to enter the flags in the order of your planned stops (Fig. 9.19).

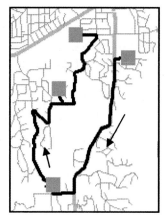

14➔ Use Clear Selected Features from the Selection menu to clear the trace.

14➔ Clear the barriers and the flags.

14➔ Place an edge flag on Harney Dr.

14➔ Place an edge flag next to Southwest Middle School.

Fig. 9.19. Traveling a sequence of stops

15➔ Locate Player Dr. using the Find button or Map Tips, and place an edge flag on it.

15➔ Place an edge flag next to Meadowbrook Elementary School (west and slightly north of Harney Dr.).

15➜ Solve for the path.

15➜ When finished, clear the results and the flags.

Advanced path problems

Combinations of selections, barriers, and other techniques can tackle more advanced path problems. To start, assume that you live on Player Dr. and work at the South Dakota School of Mines and Technology (SDSM&T). You decide to map a route to work.

16➜ Find Player Dr. again and place an edge flag on it.

16➜ Zoom to the extent of the Roads layer.

16➜ Find SDSM&T (search for Mines), zoom in, and place an edge flag on E. St. Joseph St. next to it.

16➜ Zoom to the extent of the Roads layer again.

16➜ Solve for the path.

Now imagine that a doctor has placed you on a strict low-cholesterol diet. Driving home from work in the evening, you find it nearly impossible to drive by a restaurant without stopping for something wonderfully greasy to eat. Try to find a route to work that avoids all the restaurants in town. To begin, select all the road segments that are close to restaurants.

17➜ Choose Selection > Select by Location from the main menu bar.

17➜ Select roads that are within 100 meters of a restaurant.

17➜ Close the window after completing the selection.

Next, we will set the trace analysis options so that the selected roads act as barriers to travel.

18➜ Choose Analysis > Options from the Utility Network Analyst toolbar.

18➜ Click the General tab and fill in the button to trace only on unselected features.

18➜ Click OK.

18➜ Clear the previous trace results.

18➜ Solve for the path (Fig. 9.20).

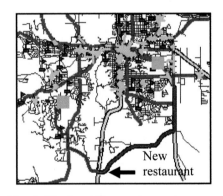

Fig. 9.20. Avoiding restaurants on the way home from work

Unfortunately, you discovered that last week a new restaurant with juicy burgers and crisp fries opened at the intersection of Catron Blvd. and Highway 16. This new restaurant is not in the database, so enter barriers by hand.

19➜ Zoom in to the intersection of Catron Blvd. and Highway 16 (see Figure 9.21).

19➜ Add two edge barriers to block off the intersection.

19➜ Zoom back to the previous extent.

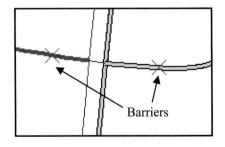

Fig. 9.21. Placing barriers on Catron Blvd.

19➔ Clear the previous result and solve for the path again.

Looks like you can still get to work, although the path is a little tortuous!

20➔ Clear the flags, the barriers, and the results.

20➔ Clear the selected set of roads.

20➔ Turn off the Restaurants and Schools layers.

Finding transportation network errors

Errors in topology are often created during the process of digitizing and creating networks. Once the network is built, the tracing tools can help you locate and fix these problems. The most useful tool is the Find Disconnected solver. In a road network, it should theoretically be possible to get to any road. A disconnected road is usually the result of a failure to snap the end of a new road to the existing network or to split a road at a junction.

21➔ Zoom to the extent of the Road Network group layer.

21➔ Choose the Junction Flag tool and click to add a flag anywhere on the network.

21➔ Set the trace task to Find Disconnected and click the Solve button.

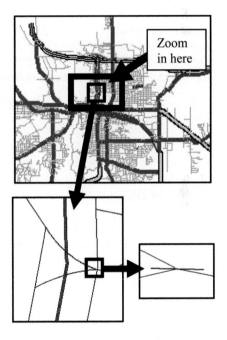

At first glance, no trace result appears. This network is in good shape. However, look closely at the north-south extension of the yellow interstate, I-190, for a small red glint.

22➔ Zoom in to this area for a closer look (Fig. 9.22).

22➔ Zoom in to the triangle with the two small red spots.

22➔ Zoom in to one of the spots until the red street fragment is clearly visible.

Here's the problem. A tiny sliver of road, which is probably a digitizing error, lies across the intersection. With ArcEditor or ArcInfo, the slivers could be deleted to fix the network. For now, we will simply go on; these small extra segments will not affect the behavior of the network.

Fig. 9.22. Finding a disconnected road segment

22➔ Zoom back to the extent of the Road Network group layer.

We can also use the Find Loops on transportation networks.

23➔ Change the trace task to Find Loops.

23➔ Click the Solve button.

This time nearly the entire network is returned as the result. We can expect this because typically road networks DO let people drive in loops. (How many times have you driven around the block?) However, the roads that are not part of the result do have a common feature—they are all "dead end" streets. Of course some of them end merely because they go outside the city limits. Generally, finding connections and loops is not very useful with transportation networks.

Tracing on utility networks

Although finding driving routes is amusing, utility networks also have interesting tracers. Now we will take a look at a portion of the water utility system for east Rapid City.

24➡ Clear the flags and the results.

24➡ Turn off the Road Network group layer and collapse it using the minus box.

24➡ Turn on the Water Network group layer.

24➡ Zoom to the Water Network group layer.

24➡ Change the network to Water_Net in the Utility Network Analyst toolbar.

24➡ Examine the network.

This network contains more elements than the roads network. The water mains themselves form the edges. In addition, there are galleries, end caps on the pipes, and T-valves at the pipe junctions. All of these features participate in the network.

25➡ Right-click the Galleries layer and choose Open Attribute Table.

The attribute table contains three fields. The Name field simply indicates the name of the gallery. The Enabled field is a part of the network setup. The value True indicates that the galleries are turned on and functioning as part of the network. The AncillaryRole field says Source, indicating that the galleries are water sources for the network. They are underground chambers that pull water in from Rapid Creek, filtering and treating it before it enters the water lines.

25➡ Close the attribute table.

26➡ Choose Bookmarks > Pipes from the main menu bar.

26➡ Choose Flow > Display Arrows from the Utility Network Analyst toolbar.

Arrows appear at the center of each edge, indicating the direction of flow. Here's how to edit the symbols used for portraying flow.

27➡ Choose Flow > Properties from the Utility Network Analyst toolbar.

27➡ Click the Arrow Symbol tab.

There are three different conditions of flow possible in the network. Uninitialized flow means that the flow direction has not yet been established for the network. Indeterminate flow means that flow has been initialized but cannot be determined for that section. Determinate flow indicates that flow has been established.

27➡ Click on each flow type to examine the symbol used to display it.

8. What is the symbol for Indeterminate flow? _____

> 28➜ Click the symbol for Determinate flow. The Symbol Selector appears.
>
> 28➜ Change the symbol color to blue. Click Apply. The map arrows change color.
>
> 28➜ Change the symbol for Indeterminate flow to a 12-point pink square. Click Apply.

Like other features in ArcMap, a display scale can be set so that the networks are not cluttered with arrows at small scales.

> 29➜ Click the Scale tab (still in the Flow Display Properties).
>
> 29➜ Change the settings to not display the arrows when zoomed out past 1:30,000. Click OK.
>
> 29➜ Zoom to the extent of the Water Network group layer to check the scaling. The arrows should disappear.
>
> 29➜ Return to the previous extent.

Notice that the arrows take a few moments to draw. Turn them off for now.

> 30➜ Choose Flow > Display Arrows again to turn the symbols off.

Let's explore this network more by using some of the tracing tools. Let's start by seeing which parts of the network are connected to each gallery.

> 30➜ Zoom to the extent of the Water Network group layer to show the entire network.
>
> 30➜ Choose the Junction Flag tool and click to add a flag at the Civic Center gallery.
>
> 30➜ Set the trace task to Find Connected. Click the Solve button.

Now it is easy to see which areas of town are served by the Civic Center gallery. We can use Find Disconnected to highlight the other pipes that are not connected to this gallery.

> 31➜ Clear the results and change the trace task to Find Disconnected.
>
> 31➜ Click the Solve button.

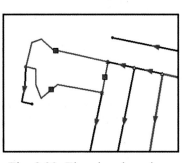

Next, let's use the Find Loops tracer to see if there are any loops in this network.

> 32➜ Clear the results.
>
> 32➜ Add another junction flag to the other gallery so that both subnets are examined for loops.
>
> 32➜ Change the trace task to Find Loops and Solve.
>
> 32➜ Zoom in to the single red loop result that appears (Fig. 9.23).
>
> 32➜ Choose Flow > Display Arrows from the Utility Network Analyst toolbar.

Fig. 9.23. Flow in a loop is indeterminate.

Note that all of the lines except the loop have blue determinate flow arrows. The loop has the pink square indeterminate flow symbols. The water could go around the loop in either direction. Fixing this would require editing the loop to break the connection at one point.

33➔ Clear the flags and the results.

33➔ Zoom to the extent of the Water Network group layer.

One useful benefit of using network models is the ability to predict what areas of town will experience service disruptions due to repairs and outages. The city has been planning to repair a leaking pipe on St. Patrick St. between 6th and 7th Ave. They will need to turn off the flow to the pipe for several hours, and they want to notify the customers who will be affected.

34➔ Turn on the Road Network group.

34➔ Use Select by Attributes to select the roads where [STREET] = 'ST PATRICK'. Click Apply rather than OK to keep the selection window open.

34➔ Change the Selection method to Add to Current Selection.

34➔ Select the roads where [STREET] = '6' OR [STREET] = '7'.

34➔ Close the Selection window.

35➔ Zoom in and add an edge flag to the segment of the water line that follows St. Patrick St. between 6th and 7th. Then return to the previous extent.

35➔ Change the trace task to Trace Downstream.

35➔ Click the Solve button.

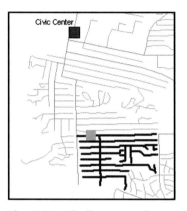

Fig. 9.24. Finding a service disruption due to repairs on St. Patrick St.

Now you know where to notify customers (Fig. 9.24). If this network were complete down to the building connections for each parcel, then the utilities company would be able to automatically compile a mailing list to send to the affected customers.

The Trace Upstream tool performs the opposite task, finding the path that water has traveled to get to a particular location.

36➔ Clear the flags and the results.

36➔ Clear the selected features.

36➔ Use the Find tool to locate and select the streets N. and S. Grand Vista Ct. They are on the middle of the west side of the network (Fig. 9.25).

Tracing can also be done with accumulation, for example, to find the total distance water travels to get to a particular location.

37➔ Zoom in to the location of these two streets.

37➔ Place a junction flag on the end of the water main serving these streets, and return to the previous extent.

37➔ Change the task to Find Upstream Accumulation.

37➔ Click Solve and watch for the cost in the lower-left corner of the window.

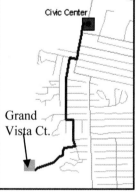

Fig. 9.25. How water gets to Grand Vista Ct.

9.   What is the cost in edges to get from the Civic Center to Grand Vista Ct.? _____

38➔ Choose Analysis > Options from the Utility Network Analyst toolbar.

38➔ Click the Weights tab and set both the along and the against weights to Distance. Click OK.

38➔ Click the Solve button again.

10. How far does the water travel to get there? _____

There is no downstream equivalent to tracing with accumulation. However, we can use the selection result option to select the downstream features from a point and determine the total length of pipe beyond that point.

39➔ Clear the flags and results.

39➔ Place a junction flag at the East Rapid gallery.

39➔ Choose Analysis > Options from the Utility Network Analyst menu.

39➔ Click the Results tab and change the result format to Selection. Click OK.

39➔ Change the trace task to Trace Downstream and Solve.

Now that the water lines are selected, it is easy to get their total length.

40➔ Open the attribute table for the Waterlines layer.

40➔ Right-click the Shape_Length field and choose Statistics.

11. What is the total length in kilometers of pipe served by this gallery? _____

40➔ Close the Statistics window and the attribute table.

40➔ Clear the selected set.

This is the end of the tutorial.

➔ Close ArcMap. You do not need to save your changes.

More skills

Consult the Skills Reference section of this chapter to learn to do the following:

➢ Build a simple network (requires an ArcEditor or an ArcInfo license).

Exercises

1. Find the shortest path from Hidden Timbers Dr. to Stevens High School. **Capture** a map showing the route.

2. What is the distance covered by the route in Exercise 1?

3. What is the shortest alternate route if you want to avoid the before-school congestion on Park Dr. by Southwest Middle School? **Capture** the map. How long is the new route?

4. The cafeteria at Stevens High School makes daily lunches for several elementary schools in town. They are delivered in this order: Pinedale, South Canyon, Upper Rapid, Canyon Lake, Meadowbrook, and Cleghorn. Find the most efficient route to make the deliveries. **Capture** the map.

5. You are trucking a load of Canadian hemp from a distributorship on Olde Orchard Road to a cowboy rope-making factory on Stacy St. Because of a bizarre technicality in the drug laws, you can be arrested if the truck is stopped within 250 meters of a school. Find a path to the destination that avoids this risk. **Capture** the map.

6. A construction worker accidentally breaks a water line at the intersection of E. St. Patrick St. and Ivy Ave. Create a map showing the area of disrupted service. **Capture** the map.

7. Of the two water mains leaving the Civic Center gallery, which serves a greater length of pipes? What is the total pipe length for each main (north and south)?

8. What is the total length of water *mains* served by the East Rapid gallery versus the Civic Center gallery? (Do not include laterals or other water line types.)

9. The water main that ends at 5th St. and Cathedral Dr./Fairmont Blvd. passes through how many T-valves?

10. Joe did Exercise 9 with Selection as the result format and got 26 *selected* features. Mary did the same problem and got 51 *selected* features. Explain why they got different numbers. What is the cost in each case?

Challenge Problem

Find a route that goes from Alta Vista Dr. to the east end of E. Knollwood St. and that travels ONLY on local streets. (**Hint:** You cannot ignore the role of junctions in this problem.) **Capture** your map.

Skills Reference

Examining network properties 349

Performing a trace 350

Setting general tracing options 350

Changing the results format 351

Using weights when tracing 351

Viewing flow directions 352

Adding and clearing flags and barriers...353

Finding a path...354

Finding the shortest path 354

Finding connections and loops 355

Tracing upstream/downstream 355

Tracing with accumulation...355

Building a simple network 356

Examining network properties

1. In ArcCatalog, right-click a
    network in the catalog list and
    choose Properties.

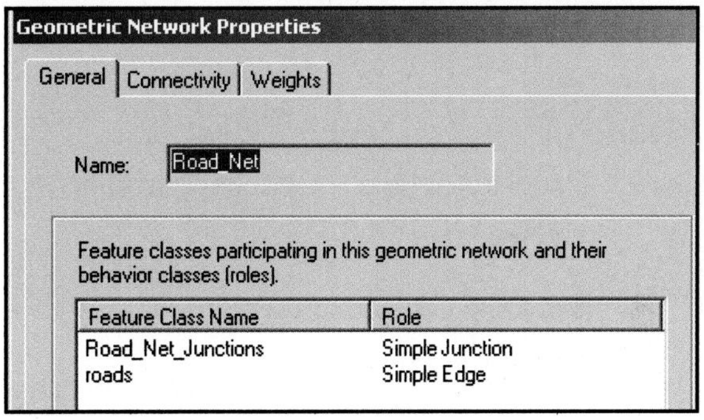

The General tab reports which feature
classes participate in the network (Fig.
9.26).

The Connectivity tab shows any rules
that have been established regarding
how the network elements can connect
(e.g., a three-inch pipe can connect to
a six-inch pipe only through a coupler
junction feature). Connectivity rules
are not required but are provided to
help maintain network integrity.

Fig. 9.26. General properties of a network

The Weights tab lists weights that have been set up for the network. Weights are created at
the time the network is built, but they may be used during an analysis, if desired.

Performing a trace

1. In ArcMap, click the Add Data button and choose a *Name*_Net layer, such as Road_Net, in the geodatabase containing the network. The network and its participating feature classes will be added to the map.

2. Right-click in the menu area and turn on the Utility Network Analyst toolbar (Fig. 9.27).

3. Select the network to be analyzed, if more than one is available.

4. Choose the Trace Task to be performed.

5. Click the Flag/Barrier tool and add desired flags and barriers at edges or junctions. For reliable results, do not mix edge and junction markers; use either but not both.

6. Click the Solve button to perform the trace.

Fig. 9.27. Performing a simple tracing task

Setting general tracing options

1. Choose Analysis > Options from the Utility Network Analyst toolbar.

2. Click the General tab (Fig. 9.28).

3. Choose which features to trace. You can choose to trace only on selected features, only on unselected features, or on all features.

4. Choose whether to include edges with indeterminate or uninitialized flow when tracing.

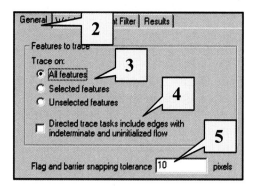

Fig. 9.28. Setting general options

5. Choose the snapping tolerance to be applied when adding flags and barriers. If the tool clicks a location within this distance from an edge or a junction feature, the flag or barrier will automatically snap to the feature.

6. Click OK to apply the options.

Changing the results format

1. Choose Analysis > Options from the Utility Network Analyst toolbar.

2. Click the Results tab (Fig. 9.29).

3. Choose whether the trace result will take the form of a drawing or a selected set of features.

4. If choosing to return a drawing, change the color of the trace, if desired.

5. Choose whether the result will include all features included in the trace or only the features that are stopping the trace.

6. Choose whether the trace result will include edges, junctions, or both. By default, both are returned.

7. Click OK to apply the changes.

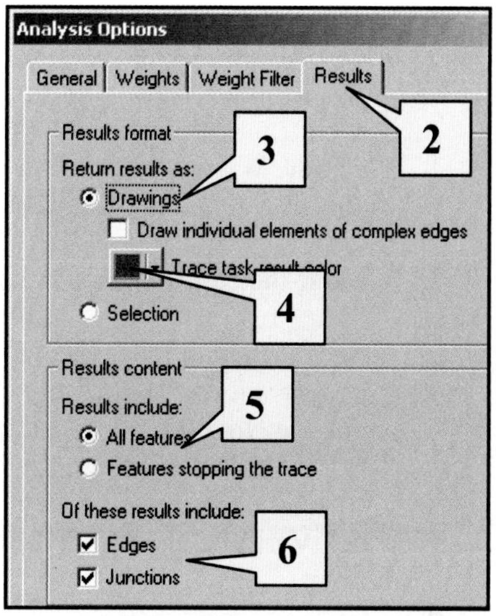

Fig. 9.29. Changing the output format

Using weights when tracing

1. Choose Analysis > Options from the Utility Network Analyst toolbar.

2. Click the Weights tab (Fig. 9.30).

3. To use weights on junctions, specify the name of the weight desired.

4. To use weights on edges, specify the name of the weight to be applied in the same direction that the line was digitized.

5. If desired, specify a weight to be applied against the digitized direction.

6. Click OK to apply the changes.

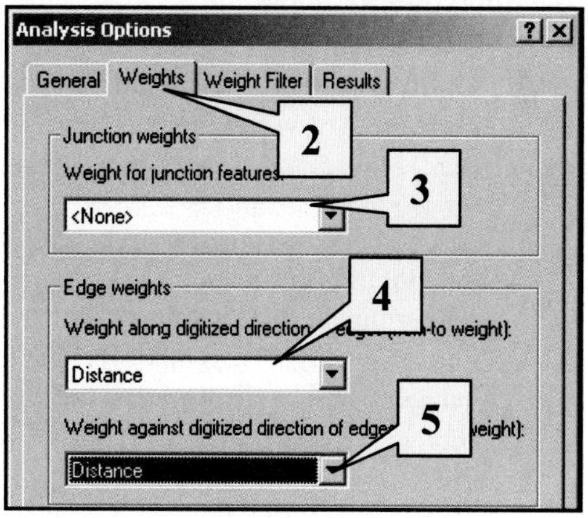

Fig. 9.30. Choosing weights to use when tracing

TIP: Weights are set up at the time the network is created. If no weights are listed in the drop-down boxes, then the network contains no weights, and none can be used.

TIP: To accumulate distances or other weights regardless of the digitized direction, specify a weight in both directions (along and against).

Using weight filters

Filtering weights means to apply only weights that fall within a specified range. For example, a junction weight called StopTime might specify an average waiting time at intersections with stop lights or stop signs. You may only be interested in adding up the total time spent at traffic lights and ignoring the wait at stop signs, which is usually 10 seconds or less. A weight filter could screen out waits of less than 10 seconds.

1. Click the Weight Filter tab in the Analysis > Options window (Fig. 9.31).

2. Choose the edge or junction weights.

3. Enter the weight range as min – max.

4. Click the Not button to use the weights NOT in the specified range.

5. Click Verify to check that the syntax is correct.

6. Click OK or Apply.

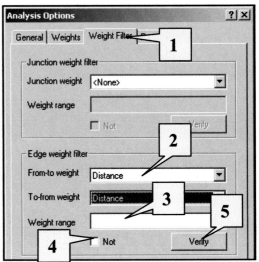

Fig. 9.31. Using weight filters

Viewing flow directions

1. Choose Flow > Display Arrows from the Utility Network Analyst toolbar.

Changing flow symbols

2. Choose Flow > Properties from the Utility Network Analyst toolbar and click the Arrow Symbol tab (Fig. 9.32).

3. Select the flow category to see the symbol used for that category (Determinate, Indeterminate, or Uninitialized). The symbol for that category will be displayed to the right.

4. To change the symbol, click on it. The Symbol Selector appears. Choose the new symbol, set its size and color properties, and click OK.

5. Click OK or Apply.

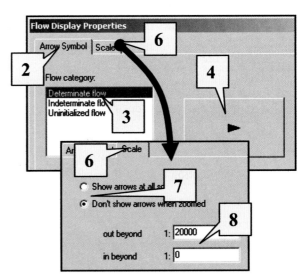

Fig. 9.32. Setting the flow symbols and display scale

Setting the display scale

Use this feature to avoid clutter in the map by making symbols appear only at certain scales.

6. Click the Scale tab (Fig. 9.32).

7. Choose to always show arrows, or to show arrows when zoomed in or out beyond a specified scale.

8. Enter the scale.

9. Click OK or Apply.

Adding and clearing flags and barriers

1. Click the Flag tool menu drop-down button on the Utility Network Analyst toolbar (Fig. 9.33) and select the type of flag or barrier to add (edge or junction).

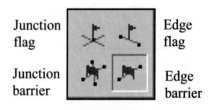

Fig. 9.33. Types of flags and barriers

2. Click on a junction to add a junction flag or barrier. Click along an edge to add an edge flag or barrier.

3. To clear flags or barriers, choose Analysis > Clear Flags or Analysis > Clear Barriers from the Utility Network Analyst toolbar. All flags or barriers are cleared together.

> **TIP:** Flags and barriers will be snapped to the closest feature within the snap tolerance set from the General tab in the Analysis > Options menu.

> **TIP:** Some tracing tools will not work if the flags or the barriers are of different types. For an analysis, try to always use edge flags and edge barriers together or junction flags and junction barriers together.

Finding a path

By default, the trace will return the path that traverses the fewest number of edges. This path will not necessarily be the shortest path.

1. Choose the network to trace on.

2. Set the trace task to Find Path.

3. Choose the Junction Flag or Edge Flag tool.

4. Click on the starting location to enter a flag.

5. Click on the end location to enter another flag.

6. Click the Solve button.

Total cost: 18

TIP: To see the cost (number of edges in the path), look in the lower-left corner of the ArcMap window before the mouse moves off the Solve button. The cost will vanish once the mouse is moved. Click Solve again to bring it back.

TIP: Enter more than two flags to visit a series of locations. The stops will be visited in the order in which they are entered.

Finding the shortest path

To find the shortest path, the network must have a Distance weight based on the length of the edges. (It may be called something other than Distance.)

1. Choose the network on which to trace.

2. Set the trace task to Find Path.

3. Choose the Junction Flag tool or the Edge Flag tool. Enter the starting, end, and any intermediate flag locations.

4. Choose Analysis > Options from the Utility Network Analyst toolbar and click the Weights tab.

5. Set both the along and the against edge weights to Distance (or whatever the weight name is). Click OK.

6. Click the Solve button.

Total cost: 4968.

TIP: The distance in map units along the path will be displayed in the lower-left corner of the ArcMap window. It will vanish when the mouse is moved. Click Solve again to make it reappear.

Finding connections and loops

These trace tools can be used to find features connected to a certain location, those features disconnected from a certain location, and those features that form loops downstream of a certain location.

1. Choose the network on which to trace.

2. Set the trace task to Find Connected, Find Disconnected, or Find Loops.

3. Click the Edge or Junction Flag tool and enter the location from which to start tracing. More than one location flag may be entered, in which case tracing will commence from each location separately.

4. Click the Solve button.

Tracing upstream/downstream

These tracers work only on utility networks.

1. Choose the network on which to trace.

2. Set the trace task to Trace Upstream or Trace Downstream.

3. Click the Edge or Junction Flag tool and enter a flag at the location to begin the trace. If more than one flag is entered, tracing will begin at each location separately.

4. Click the Solve button.

Tracing with accumulation

1. Choose the network to trace on.

2. Set the trace task to Find Upstream Accumulation.

3. Choose Analysis > Options and click the Weights tab to specify the weights to be accumulated during tracing.

4. Click the Edge or Junction Flag tool and enter a flag at the location to begin the trace. If more than one flag is entered, tracing will begin at each location separately.

5. Click the Solve button.

6. Read the accumulation value in the weights' units in the lower right corner of the ArcMap window.

> Total cost: 4968.

TIP: The accumulation value will vanish when the mouse is moved off the Solve button. To bring it back, click Solve again.

Building a simple network

Building a network requires an ArcEditor or an ArcInfo license. If you have one and want to try it, follow this example using the road network from the usdata geodatabase.

1. The features forming the network must be feature classes in a feature dataset in a geodatabase.

2. Right-click the Transportation feature dataset in the mgisdata\Usa\usdata.mdb geodatabase and choose New Geometric Network. A wizard will appear. Read the information and click Next.

3. Choose to build a geometric network from existing features and click Next.

4. Check the box to include majroads in the network. Do NOT include interstates. Leave the default name for the network as Transportation_Net. Click Next.

5. Choose No, indicating that you don't want complex edges in the network.

6. Choose No, indicating that the features do not need to be snapped. See the tip about snapping on the next page.

TIP: To set weights, you must know the data type of the field used for the weight. Examine the fields in ArcCatalog before creating the network to obtain this information.

---

7. Click Yes to assign weights to the network. We will assign two distance weights based on both miles and kilometers (Fig. 9.34).

8. Click in the #1 Weight Name box and enter Distance-km. Click in the Type box and choose Double. Leave the Bitgate Size box blank.

9. Click the Blank page button to add a new weight. Name it Distance-mi. Set the Type to Double and leave the Bitgate Size blank. Click Next.

10. The Available network weights box should currently list Distance-km::Double. Click in the Associated field box and choose the attribute field KM.

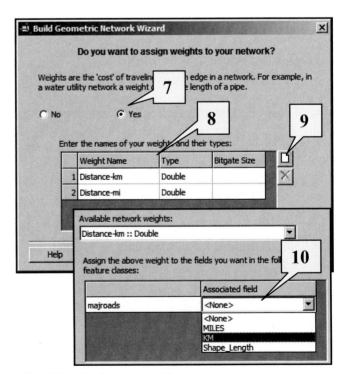

Fig. 9.34. Assigning weights

11. Change the Available network weights box to Distance-mi::Double using the drop-down arrow and; then choose the attribute field MILES. Click Next.

12. Review the summary and choose Finish.

TIP: For the network to operate correctly, all edges must meet each other at junctions with no gaps or dangles. This adjustment may have already been done using planar topology, in which case you don't need to snap when creating the network. If not, you may be able to automatically fix the dangles by indicating a snap tolerance in map units. It must be large enough to fix the errors but not so large that features are incorrectly snapped together. Setting a good tolerance can be tricky, which is why snapping is often done interactively using topology. Choosing Yes may make changes to your data, so be careful and have a backup copy just in case.

TIP: If you made a mistake or need to change the network, the easiest way is to delete it and start over. Right-click the Transportation_Net entry in the feature dataset and choose Delete. It will remove the logical network but leave the majroads feature class as is.

Chapter 10. Geocoding

Objectives

➢ Understanding the geocoding process

➢ Knowing the different styles of geocoding

➢ Setting up an address locator service

➢ Converting addresses to locations

➢ Converting *x-y* coordinates in a table into point locations

Mastering the Concepts

GIS Concepts

What is geocoding?

As everyone knows, a street address tells a postal worker where to deliver mail. It contains a type of spatial information. However, the postal worker requires additional knowledge to get the letter to the house. Unlike a latitude-longitude point, which uniquely locates a house on the globe, a street address requires information about the city and state, the location of the street, and the sequence of house numbers. A newly hired postal worker delivering a piece of mail in a strange city would need a good map in addition to the address.

Likewise, geocoding combines map information with street addresses in order to locate a point uniquely. It enables someone to convert a list of addresses into points on a map (Fig. 10.1). Geocoding has many uses.

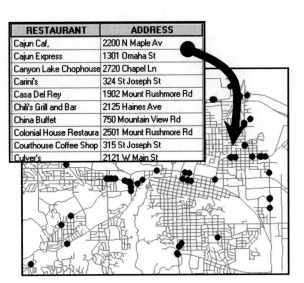

RESTAURANT	ADDRESS
Cajun Caf,	2200 N Maple Av
Cajun Express	1301 Omaha St
Canyon Lake Chophouse	2720 Chapel Ln
Carini's	324 St Joseph St
Casa Del Rey	1902 Mount Rushmore Rd
Chili's Grill and Bar	2125 Haines Ave
China Buffet	750 Mountain View Rd
Colonial House Restaura	2501 Mount Rushmore Rd
Courthouse Coffee Shop	315 St Joseph St
Culver's	2121 W Main St

Fig. 10.1. Geocoding can convert restaurant addresses in a table to points on a map.

911 Addressing

One critical use being developed by many government agencies concerns finding street addresses given over the phone during an emergency call in order to quickly route response vehicles to the location. Routing of vehicles is accomplished by providing a map or a Global Positioning System (GPS) location in vehicles equipped with GPS units. Because lives are at stake, this application requires highly accurate address information, and it is a challenging application to implement.

Driving directions

Many people these days have gone to one of the popular Web sites, such as MapQuest, to find directions from a starting address to a destination. The results include both step-by-step driving directions and a map showing the route. Other people may have tried the GPS systems installed in cars, such as OnStar or Magellan, which guide people to their destination from their current position as identified by GPS. Both of these applications are built upon geocoding technology.

Crime Analysis

Police records of crime incidents include the address, and these locations can be converted to points on a map through geocoding. Analysts could use these data to help identify "hot spots" in order to help plan patrol routes for officers or to target efforts to develop neighborhood watch programs. Police departments can glean information about spatial patterns for different types of crime or target the time of day that particular crimes tend to occur. Such information can be critical in developing strategies to reduce crime and to protect residents and businesses.

Marketing

Many companies collect address information about their customers as a regular part of conducting business. Converting addresses to locations and analyzing their distribution might help a company provide better service by opening new stores in underserved areas. Mail-order businesses can locate areas with low sales for targeted marketing efforts. Sales and service companies can better plan where to place agents to serve their clients. Recently a study helped a university in South Dakota to understand the national distribution of its students, and it is helping them to plan marketing strategies to increase national enrollment (Fig. 10.2). A whole host of business applications is available once information about customer locations is known, and many companies now routinely use GIS in their operations.

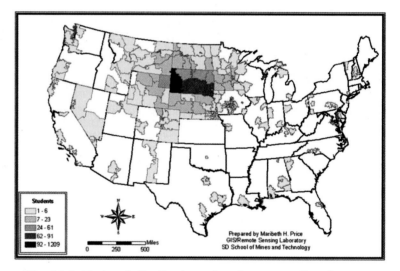

Fig. 10.2. National distribution of students attending the South Dakota School of Mines and Technology in 2004

How does geocoding work?

Just as a stranger requires a map to find an address in a city, the computer requires map data in order to turn an address into a location. This spatial data is called the **reference layer**. Geocoding typically uses a street layer composed of polylines with specific attributes indicating each street's name and the address ranges present on that street. However, other types of geocoding can use point or polygon reference data, such as matching store franchises to points representing cities or matching a list of owner addresses to parcel polygons.

Although humans automatically recognize a variety of address styles, a computer requires specialized software to interpret addresses. An address might be written 123 Maple Street, or 123 Maple St., or 123 N. Maple St., or even 123 Maple, and the postmaster would have no difficulty finding the house. To computers, however, each of these addresses is different. Geocoding programs must interpret different styles of addresses and manage typographical and other errors, such as someone entering Maple Rd instead of Maple St. They accomplish this task by dividing the address into different components and then scoring how well they match the reference components. A high score (>80) indicates a good match, a medium score (60–80) indicates a tentative match, and a low score (<60) suggests a poor match.

To interpret an address, a geocoder begins by parsing the address into individual fragments of information. The address "235 W. Main Str." could be divided into four pieces, including the address number (235), the direction prefix (W.), the street name (Main), and the street type designation (Str.). After separating the components, each fragment is standardized so that it has the best chance to match the information in the reference layer. All letters may be capitalized (MAIN), periods removed (W), and common abbreviations reduced to a single standard (ST). This process is called **address standardization**. This step is necessary because a computer, confronted with the problem of comparing (Str.) in the address table to (ST) in the reference layer, will conclude that the two do not match. A human would hardly notice the difference, but computers can be tiresomely literal. An address may include these components:

Prefix direction. A direction indicator that comes before the street name, such as *E* Maple St. or *North* Main St. Some towns and cities use prefixes for directions; others use suffix directions (e.g., Main St. North). This part may have any of the standard directions, N, E, S, W, NE, SW, SE, and NW. The complete spellings North, East, South, and West are also recognized.

Prefix type. An indicator of the road type which comes before the street name, such as *Highway* 85. In this case, 85 is the street "name" and Highway, or Hwy, is the prefix type.

Number. The address number, such as *123* Maple St.

Street name. The name of the street, such as 123 *Maple* St. or *Mount Rushmore* Road.

Street type. A type indicator that comes after the street name, such as Main *St.* or Maple *Ave.* The software recognizes a variety of street types (Street, Lane, Place, Road, etc.) and understands most common abbreviations, such as St, St., Rd, Av, and Ave.

Suffix direction. A direction indicator that comes after the street name, such as Florman St. *E* or 87th St. *North*. This part may have the same values as the prefix direction.

Zip code. The standard U.S. ZIP code or, optionally, the ZIP+4 nine-digit codes.

Table 10.1 shows some examples of addresses and their standardizations. Note that in addition to separating the address into components, the text is converted to capital letters to facilitate comparisons in case-sensitive databases. The complete words in the directions and types are also converted to standard abbreviations, so that Road becomes RD and the periods are omitted.

Table 10.1. Examples of address standardization

Address	Prefix Dir	Prefix Type	Number	Street	Type	Suffix Dir
123 Maple Road			123	MAPLE	RD	
15 Center St. East			15	CENTER	ST	E
314 Hwy 85 N		HWY	314	85		N
234 E. St. Patrick St.	E		234	SAINT PATRICK	ST	

Once the address is standardized, the components are compared to attributes in a spatial data layer, such as a line shapefile containing the city streets. The shapefile must contain certain attribute fields to compare with the standardized components. The score for each component match is averaged to give an overall match score for the address, in terms of percentage. A score of 100 indicates a perfect match. A zero indicates that none of the components match. Usually a score of 80 or better is considered a good match, and a point will be created for the address. When the score is lower than 60, the geocoding software looks for other possible candidates. Typically candidates have scores above 30. The user can examine the candidates individually to find out if any of the candidates might provide a reasonable match.

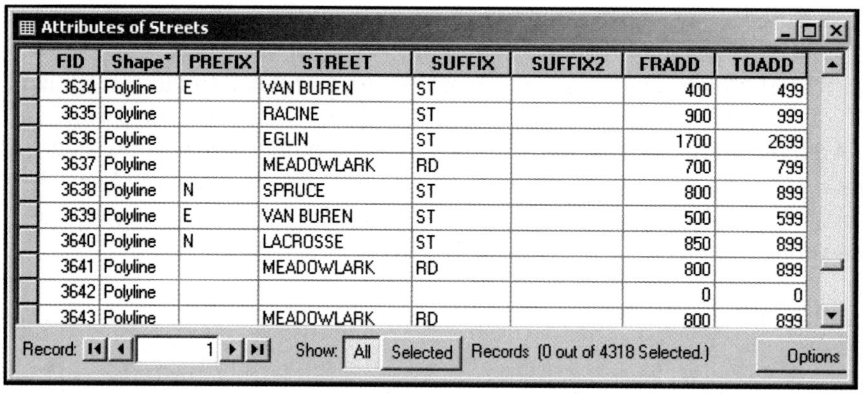
Attributes of Streets

FID	Shape*	PREFIX	STREET	SUFFIX	SUFFIX2	FRADD	TOADD
3634	Polyline	E	VAN BUREN	ST		400	499
3635	Polyline		RACINE	ST		900	999
3636	Polyline		EGLIN	ST		1700	2699
3637	Polyline		MEADOWLARK	RD		700	799
3638	Polyline	N	SPRUCE	ST		800	899
3639	Polyline	E	VAN BUREN	ST		500	599
3640	Polyline	N	LACROSSE	ST		850	899
3641	Polyline		MEADOWLARK	RD		800	899
3642	Polyline					0	0
3643	Polyline		MEADOWLARK	RD		800	899

Record: 1 Show: All Selected Records (0 out of 4318 Selected.) Options

Fig. 10.3. The attribute table of a shapefile that could be used as the basis for geocoding

Figure 10.3 shows an example of a street feature class used for geocoding. Notice that most of the fields described are present although the names are not the same. Also notice that instead of a single address, the street has two fields (FRADD and TOADD) that describe the range of addresses that fall along the street. The FRADD field is the From address and contains the lower range value; the TOADD field is the To address and contains the higher range value.

To create a point feature based on an address and street shapefile, the following logic is used. The number of the address is tested to see if it falls within the range of addresses on that street section. If it does, the point locating the address is placed along the street in proportion to its value between the range endpoints. In Figure 10.4, for example, Meadowlark St. has an address range of 700 to 799. If matching the address 750 Meadowlark St., the point feature would be placed halfway between the street ends because 750 is halfway between 700 and 799. Likewise, 725 Meadowlark St. would be placed a quarter of the distance along the street. This method of placing points is called the *One Range* method because a single address range is used for each street.

For more advanced geocoding, a *Two Range* method can be used. In this method each street has four attributes defining its address ranges, a Left From and Left To address range for the left side of the street and a Right From and Right To address range for the right side. When using the Two Range method, the addresses will be located on the left or the right side of the street, as dictated by the address ranges. In addition, the user can specify an offset distance such that the actual point is placed slightly to the side of the street instead of directly on top.

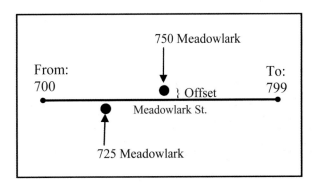

Fig. 10.4. Placing a located point along the street based on its proportional distance from the ends

The assumption that a house or other structure is located along a street segment at a distance proportional to its address number does not always work as well as it should. Parcel sizes may vary along the street, or some address numbers may be skipped entirely. This problem occurs most severely in rural areas where the spacing of addresses may be very irregular. If Meadowlark St. in Figure 10.4 represents a mile-long section line and all the houses are clumped together at one end, then the actual location and geocoded location may be quite different. Geocoded locations should always be considered approximations.

One also encounters great differences in the accuracy of the reference data used for geocoding. Large national data sets, such as TIGER, are generally less accurate and up to date than locally produced data intended for city governments and 911 applications. High-growth areas may experience difficulty keeping reference layers up to date. Users need to be aware of these potential pitfalls when geocoding. Creating and maintaining reference layers is a time-consuming and expensive process, and free data are rarely high quality.

One Range and Two Range are examples of different geocoding styles. The style chosen will depend on the way in which the **streets** layer has been set up for geocoding. Other styles are also available as described in the next section. The user must ensure that the attributes in the reference layer support the desired geocoding style.

Although geocoding is most often associated with placing addresses, in fact it applies to many types of data. Points can be located by city and state, by ZIP code alone, or even by township/range information. These different styles of geocoding have two common themes: they convert attributes to point locations, and they allow the user to link data that are similar but are not necessarily an exact match.

Available geocoding styles

Geocoding takes a variety of approaches, from finding a specific house location on one side of the street to simply finding a city and country on a world map. The available options depend on the attributes of the reference layer, the setup in the address table, and the type of matching to be done. ArcMap offers several predefined styles for geocoding. Each style requires a different set of attributes in the reference layer. To choose an appropriate style, the user must be familiar with the organization of the reference data.

Single Field

A single field locator uses one field to match to the reference layer, which typically contains points or polygons although lines could also be used if desired. Any single field, such as the ZIP code, a state name, or a place name, can be used as the key to match the locations. This method expands geocoding from addresses to *any* type of information. For example, a table might contain information about petroleum wells keyed to specific well IDs such as "TenRun Hole No. 6." Such IDs may change slightly from database to database, depending on who typed them in (Ten Run Hole #6 or Ten Run Num 6). Performing an attribute join to link the table with a point layer of well locations would yield many missing values because the matches are not exact for each ID. Geocoding would work better because it can link close matches, not just exact matches.

US Alphanumeric Ranges

Many people are familiar with using grid zones to locate streets on a map by consulting a list of streets, finding its zone (such as H7), locating the zone by means of letters and numbers on the map margin, and scanning the streets in the zone to find the one in question. This geocoding style uses this approach to reduce the search time while geocoding. Each address is attributed with its zone identification, and the geocoding algorithm need only search inside that zone. The required fields are the grid zone, the street name, and the From and To ranges for both sides of the street.

US Cities with State

This style accommodates national scale matching of location based on the city and state name only. Cities may be matched to points or polygon features. This style could create a map showing the locations of customers across the United States. The required fields are the city name and a state name or state abbreviation.

Fig. 10.5. The US One Address style codes to point or polygon layers, such as these parcels.

US One Address

This style uses points or polygon reference data set up so that each address corresponds to one point or polygon (Fig. 10.5). For example, it could be used with points representing buildings or polygons representing parcels. The required fields include street number and street name. The user may also include prefix directions, suffix directions, and other address components.

US One Range

This style uses a single address range along a street rather than having ranges for both the left and right sides of the street. The required fields are the street name, the From address, and the To address. Prefixes, suffixes, and street types may also be used if desired.

US Streets

This style represents the Two Range option discussed previously and relies on a reference layer with two address ranges, one for each side of the street. The required fields include the street name, a From and To address for the left side of the street, and a From and To address for the right side. Prefixes, suffixes, and street types may also be used if desired.

US Hyphenated

This style follows conventions in locations, such as Queens, New York, where housing locations are indicated by a cross-street designation followed by the number and the street of the house. For example, the address 72-40 38th Ave specifies a location at 40 38th Ave near the cross street 72nd.

World Cities with Country

This style allows matching of a city and a country on a world map. The reference layer may contain cities as points or as polygons. The required fields include the city name and the country name or abbreviation.

ZIP 5-Digit and ZIP+4

The ZIP style matches ZIP codes to points or polygons in a ZIP code reference layer. The only field required is the 5-digit ZIP code. The ZIP+4 style works the same way but has an additional separate field containing the 4-digit extension.

With Zone

Many of the above styles offer a zone option. A zone is an additional piece of information in the reference layer, such as state, city, or ZIP code, which is added to the other required fields for that style. The addition of the zone allows matching over larger areas. For example, the US Streets style can only find addresses within a single city because only the street address is included; there is no city name to distinguish between Main St. in Rapid City and Main St. in Sioux Falls. To match street addresses across all of South Dakota requires a city name or a ZIP code as a zone, provided that such a field exists in the reference layer and the address table.

The reference data

The reference layer provides the key link that allows the interpretation of an address to a more universal coordinate system—the Cartesian x-y coordinates of a map projection. It does this by linking the specific attributes of a spatial feature stored in a table, such as the street name or the address range, to the x-y coordinates stored for the feature. As such, the validity of the attributes is of prime importance in making this translation.

The requirements for reference data vary widely with the type of geocoding being done. A reference layer might be as simple as a point layer of cities with states or ZIP codes in the attribute table. Street address geocoding probably has the most stringent requirements because it needs address ranges for one or both sides of a street. Such information is tedious to gather and requires significant preparation and maintenance time to ensure that it remains up to date and has minimal errors. U.S. Census data, or TIGER data, is the most common starting point for assembling reference layers although it requires a fair amount of conversion, assembly, and error checking before it is ready to use. Reference data are also available from a number of commercial vendors, with considerable variability in completeness, quality, and price.

Adding x-y coordinates

Another useful GIS function involves taking x-y coordinates from a table and converting them to locations on a map. The x-y coordinates must be given in a real-world coordinate system, such as latitude-longitude or UTM meters. The coordinate system details must also be known, including the geographic coordinate system and datum and the projection. Global Positioning Systems

(GPS) frequently provide this type of data. A GPS receives signals from four or more special satellites and uses the information to calculate its position. Other examples of *x-y* data sources include benchmarks, surveyed points, or locations measured on a map. In any event, the procedure for adding the points is identical.

The layer added from *x-y* points is called an **event layer**. This terminology comes from considering each entry in the table to be an event, such as an earthquake or a traffic accident. Not all tables contain events, of course, but the terminology remains. An event layer does not constitute an actual shapefile or feature class; it is merely the combination of the table and the located points on the map.

Event layers appear similar to point feature classes, but they do not constitute a feature class in the truest sense for they remain a table of objects represented by spatial points. Some geoprocessing functions will not accept event layers as input. In many cases users will export the event layer to a shapefile or a geodatabase features class for use in a tool or for permanent storage. When exported, all the attribute fields in the original table become attributes of the feature class.

About ArcGIS

The geocoding process

Before starting, the user must first verify that an address locator has been established. Setting up an address locator service is covered in the next section. The user must also have a table of addresses to be matched.

To begin geocoding, start an ArcMap session and load the address table and the reference layer. An address locator is loaded from ArcCatalog and then geocoding can begin.

The first geocoding attempt is usually done in batch mode; that is, the locator tries to match all of the addresses it can automatically. When finished, it issues a report of how many addresses were matched and how many remain unmatched. You then have the option of examining the unmatched addresses interactively to find more matches.

Interactive mode examines each unmatched address and suggests one or more match candidates. A candidate has a lower score than a match but scores high enough to provide a possible match. If no candidates appear, you can try lowering the candidate cutoff score or relaxing some of the geocoding options, such as the spelling sensitivity.

Once you have matched all the addresses that can be matched, the software will create the output shapefile or the feature class of points. The attribute table will contain information, such as the standardized address, the match score, and the status of the record (matched, unmatched, or tentative).

Occasionally the address table or the geocoding layer contains errors or inconsistencies. In this case the user may decide to halt geocoding, fix the problems, re-create the address locator, and try again to match the addresses. For example, the original reference layer prepared for this textbook had many streets named for saints, such as Saint Patrick St. In the reference layer as well as the address table, the street names were abbreviated to St. Patrick, St. Anne, and so on. However, during geocoding the software automatically converted *St* to *Saint* during the address standardization. As a result, dozens of addresses remained unmatched. To streamline matching, the author edited the streets layer to replace all the St. abbreviations with Saint.

In another example, students were analyzing crime patterns using street addresses and the StreetMap USA product. The police reports used the standard address prefixes for Rapid City (*West* Main St.), but StreetMap uses suffixes (Main St. *West*). In this case the easiest solution involved editing the spreadsheet containing the addresses so as to place them in the address file with suffixes instead of prefixes.

These examples demonstrate that geocoding is not a simple push-button operation but requires some investigation and ingenuity on the part of the user. A thorough understanding of the geocoding process, address formats, and reference data being used helps in detecting and solving barriers to matching.

Setting up an address locator

Before geocoding can be performed, the user must create an address locator that specifies the reference layer and the geocoding options. It takes a snapshot of the reference data and stores it within the locator. Locators may be copied and moved to new locations without losing the source data. To create a locator you must decide which geocoding style will be used, assign the correct fields in the feature class to the required matching fields, and set the options that control how the locator performs. Once an address locator is defined, it can be used over and over again.

Address locators are created using ArcCatalog. The first step chooses the style of the locator and specifies whether the geocoding layer is a shapefile or a geodatabase feature class. The second defines the parameters of the locator. Figure 10.6 shows part of the menu for defining a locator based on the US One Range for shapefiles style. You must enter the name and location of the feature class that contains the streets and the attributes. Next, you must assign the correct attribute from the streets table to the fields needed for geocoding. In this case, for example, the FRADD field in the shapefile is used as the House From field. The fields shown in boldface are required; the others are optional.

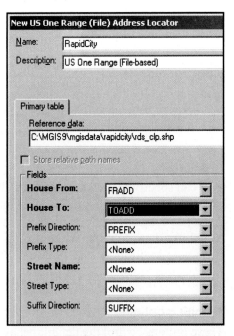

Fig. 10.6. Defining the reference layer for a new address locator

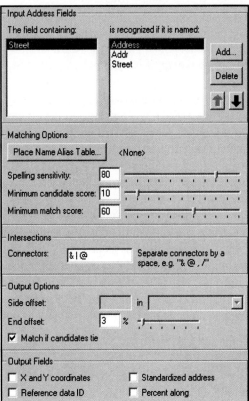

Fig. 10.7. Address locator options

The fields shown in the setup will vary depending on the style of locator requested. A ZIP code locator, for example, would need fewer fields, as would a city-state style of geocoding.

The next part of the menu (Fig. 10.7) contains various options for how the locator will work. You can accept these default options, modify them immediately, or change them during geocoding.

Input address fields

You can specify the names of fields in the address table, which are assumed to contain the addresses to be matched. It saves time if the locator automatically recognizes the address field in the address table being matched.

Alias table

You can specify an alias table, which allows certain names to substitute for addresses (Fig. 10.8). For example, the address table might only list Rushmore Mall as the "address" for a series of stores, or the place name "City Hall" might often appear instead of the actual address. An alias table links these common name listings with the actual addresses at which they occur; that is, Rushmore Mall sits at 2100 N. Maple Av. and City Hall sits at 300 Sixth St. Then if the address file contains "Rushmore Mall" as the location of a store, the location can be found.

Matching options

These options control the default behavior of the locator although they can be adjusted during geocoding. The **spelling sensitivity** specifies how closely a street name must match to be a valid candidate. With this option, small misspellings or typing mistakes do not prevent a match from being made. Thus, Pierce St would still be recognized even if it was spelled Peirce St.

OID	ALIAS	ADDRESS
0	Rushmore Mall	2100 N Maple Av
1	City Hall	300 6th St
2	County Courthouse	315 St Joseph St
3	Public Library	610 Qunicy St
4	Rushmore Plaza	444 Mt Rushmore Rd

Fig. 10.8. An alias table contains common place names with their addresses.

The **minimum candidate score** is used to find possible candidates when a good match can't be found. During interactive matching, the user examines each unmatched address individually. During this process, the computer suggests possible candidates with lower scores than a good match. If no candidates are displayed, lowering the minimum score sometimes helps. The minimum score can also be increased if too many candidates are showing up.

The **minimum match score** is the lowest score that will constitute a match. In applications in which an occasional mismatch is not important, it is acceptable to lower the default in order to match as many records as possible. To a company analyzing customer locations, a few wrong addresses are not likely to significantly affect the analysis. In other cases, the minimum score might be raised. To an emergency management team, a poorly matched address may be a matter of life and death. In this case, poor matches should be identified and individually considered by the dispatcher to avoid sending the team out to a wrong location.

Intersections

Geocoding can also match intersections, such as the corner of 5th St. and St. Joseph St. This option specifies the symbols that are considered to be connectors so that the address locator can properly interpret the address as an intersection rather than a regular address. The default values are **&**, |, and @ characters, but additional characters may be used. However, the separation character should not be one that might ordinarily appear in an address.

Output options

This section controls parameters related to the output file. First, for a Two Range style, the user can set an offset distance to place the locations slightly off the street to the left or to the right. A setback from the street endpoints may also be specified so that no point ends up at the end of a street. Finally, when an address matches more than one street with the same score, marking the check box dictates that the address will be matched to the first street found.

Output fields

The user can request additional fields in the output file for more information about the locations. These fields include the *x-y* coordinates of the point, the reference data ID (the feature-ID of the street it was matched to), the standardized address, and the percentage distance along the street where the point was placed.

Summary

➤ Geocoding converts addresses in a table to points on a map, using reference map layers that contain information about street names and address ranges.

➤ The process breaks down addresses into standard components, compares these to the attributes in the reference layer, and scores how closely the components match possible candidates in the reference layer.

➤ Once a match is found, the address location is placed along the street at a distance proportional to the address value between the street address ranges.

➤ Geocoding styles vary on a number of factors, such as whether ZIP codes are used and whether the address ranges are given for the street as a whole (One Range) or for both the left and the right sides of the street (Two Range).

➤ Before geocoding, you must create an address locator by choosing a reference layer and setting the desired coding options. Locators store a snapshot copy of the reference layer. If changes are made to the reference layer, a new locator should be created.

➤ The process begins by matching as many addresses as possible in batch mode. Then the user may interactively match single addresses or change the geocoding options and try for another batch match. The process may be repeated until as many addresses as possible have been matched.

➤ Event themes are sets of point locations created from tables containing *x-y* coordinates. They may be exported to shapefiles or geodatabases for permanent storage.

TIP: Geocoding occasionally has some odd effects with editing routines. If editing problems occur after a geocoding session, save the work, exit ArcMap, and start the program again.

Chapter Review Questions

1. What two data objects are required for geocoding?

2. Describe the difference between One Range and Two Range geocoding.

3. If the 1200 block of Maple Street is 1000 ft long, how many feet from the 1200 corner would the address 1275 Maple be located during geocoding?

4. What is a street type? Give examples.

5. Which components are present in the address, 1298 Bear Valley Road NW?

6. How does batch mode matching differ from interactive matching?

7. What is an address locator and what does it include?

8. What is an alias table and what is its purpose?

9. True or False? Once a geocoding result is completed, you can no longer change the locations or rematch additional addresses.

10. Discuss the factors that would impact the geometric accuracy of a feature class derived from geocoding.

Mastering the Skills

Teaching Tutorial

The following examples provide step-by-step instructions for doing basic tasks and solving basic problems in ArcGIS. The steps you need to do are highlighted with an arrow ➔; follow them carefully. Click on the video number in the VideoIndex to view a demonstration of the steps.

> ➔ Start ArcMap and open ex_10.mxd in the MapDocuments folder.
>
> ➔ Use Save As to rename the document and remember to save frequently as you work.

Note: The rematching process changed between version 9.2 and version 9.3. The windows are quite different although the overall process has not changed. Users with version 9.2 might find it slightly difficult to follow this tutorial. Experimenting or consulting the Help files might be needed.

Setting up an address locator

For our first example, we will geocode restaurants in Rapid City. The addresses were obtained from the telephone book and typed into a .dbf file for geocoding. However, before we can start, we need to decide what style of geocoding we will use, and we need to create an address locator.

> 1➔ Open the Streets attribute table and examine the fields.

1. For each of the following address components, write down the name of the field in the table, which contains that information. If there is no field with that information, write None.

    Prefix direction   _____

    Prefix type   _____

    Street numbers   _____ and _____

    Street name   _____

    Street type   _____

    Street direction   _____

2. Based on the fields just listed, which geocoding style is best to use? _____

    > 1➔ Close the Streets attribute table.
    >
    > 1➔ Open the restaurants table and examine the fields and records. (If the table is not in the Table of Contents, click the Source tab to show it.)

3. Which field contains the restaurant addresses? _____

4. Are there any intersections for addresses? _____ What connector symbol is used? _____

    > 1➔ Close the restaurants table.
    >
    > 1➔ Open the Properties of the Streets layer and examine the Source tab contents.

5. What is the source layer for Streets and where is it located? _____

6. Is the reference layer a coverage, shapefile, or geodatabase feature class? _____

Now we have all the information we need to set up an address locator using the Streets as the reference layer.

➜ Open ArcCatalog.

2➜ Right-click the mgisdata\Rapidcity folder and choose New > Address Locator.

2➜ Choose US One Range for the style. Click OK.

2➜ Name the address locator RapidCity.

2➜ Click the Reference data Browse button and navigate to the mgisdata\Rapidcity folder. Select the rc_roads.shp shapefile and click Add.

Next, we specify the fields to be used for the address components. Note that the required fields are in boldface type.

3➜ Choose the FRADD field for the House From component and TOADD for the House To component.

3➜ For the Prefix Direction, choose PREFIX. For the Prefix Type, choose <None>.

3➜ For the Street Name, choose STREET. For the Street Type, choose SUFFIX. For the Suffix Direction, choose SUFFIX2.

3➜ Examine the geocoding options on the right side of the window.

The default geocoding options are fine for our purposes, so we are finished setting up the locator.

3➜ Click OK to create the locator.

➜ Close ArcCatalog and return to the ex_10.mxd document in ArcMap.

Geocoding

Now we can geocode the restaurants.

4➜ Choose Tools > Geocoding > Geocode Addresses from the main menu.

4➜ Click the Add button because the locator you created is not listed yet.

4➜ Navigate to the mgisdata\Rapidcity folder, if necessary, and click on the RapidCity locator.

4➜ Click Add to finish adding the locator.

4➜ Click the newly added locator to highlight it and click OK.

The Geocode Addresses window appears.

5➜ Choose restaurants as the address table.

5➜ Select ADDRESS as the address field.

5➜ Click the Browse button to specify the output file. Make sure you set the Save as Type to Shapefile. Put it in the Rapidcity folder and name it rceats.shp.

5➔ Click OK to begin geocoding.

5➔ Examine the progress bar. When it is finished, you have some unmatched addresses. Click Rematch to do them interactively.

7.   How many restaurants were matched or tied? _____

8.   How many restaurants are unmatched? _____

At this point you have several ways to proceed. You could change the geocoding options and try to rematch. However, it is usually instructive to examine the unmatched addresses in interactive mode to find out why they won't match.

6➔ Change the Show Results box from All Addresses to Unmatched Addresses.

6➔ Scroll the table to the right until you can see the restaurant names and addresses.

FID	Shape	Status	Score	Side	ARC_Street
1	Point	U	0		I-90 & Haines Ave
3	Point	U	0		Rushmore Mall
	Point	U	0		720 LaCrossta St
			0		Rushmore Mall

Click here to select

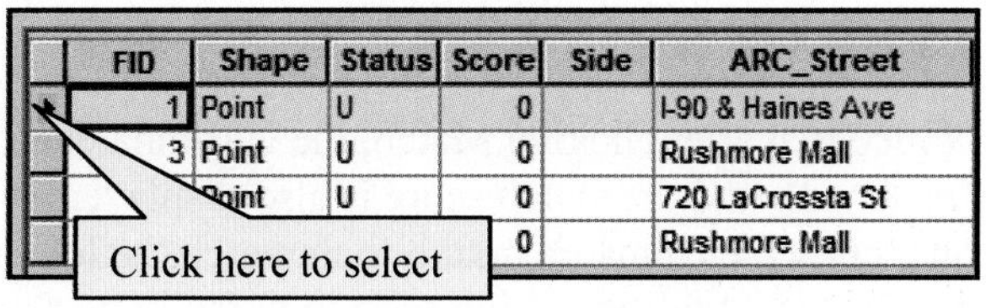

Fig. 10.9. The unmatched restaurants

Some of these fields are wide and cumbersome. Let's rearrange this window a little.

6➔ Right-click the Match_addr field heading and choose Turn Field Off.

6➔ Resize the ARC_Street field so that it is not so wide (click and drag right edge).

6➔ Examine the four unmatched records (Fig. 10.9).

Now you can see the restaurant names and addresses. Notice that two of these entries contain no addresses, only the place name "Rushmore Mall." An alias table can help match place names.

7➔ Click the Geocoding Options button at the bottom left of the window.

7➔ Click the Place Name Alias Table button. The Alias Table window appears.

7➔ Click the Browse button and navigate to the mgisdata\Rapidcity folder. Select the file named rc_alias.dbf and add it.

7➔ Choose ALIAS as the Alias field and ADDRESS as the Address field.

7➔ Click OK in the Alias Table window and again in the Geocoding Options window.

When you have made a change that you expect will match many records, it is helpful to rematch them all together instead of individually.

8➔ Click the Rematch Automatically button. It will try to rematch the four currently unmatched addresses. Accept the warning and click OK.

8➔ It matched two of the four, as reported in the window. Click Close.

8➔ Turn off the Match_addr and resize the Arc_Street fields again.

9➔Click just to the left of the address record that reads 720 LaCrossta St so that the record is highlighted.

No candidates appear for this address. However, we can try relaxing the spelling sensitivity to see if we can find any candidates.

9➜ Click the Geocoding Options button in the lower-left corner.

9➜ Use the slider bar to set the spelling sensitivity to 50. Click OK.

Now some candidates appear, with the top score of 55 going to the 700 block of N LaCrosse St. With a little consideration, you decide that LaCrossta is probably a misspelling of LaCrosse and that these two addresses match. Of course, familiarity with the city can help a great deal when deciding about matches.

9➜ Click the LaCrosse St candidate to select it (Fig. 10.9).

9➜ Click the Match button.

Notice that after clicking Match, the U (unmatched) in the address listing changes to an M (matched), and the match score is also displayed. If you decide later that this match is not a good one, click the Unmatch button to unmatch the location. We will leave the spelling sensitivity at 50 for the rest of this matching session to facilitate finding candidates.

Let's try matching another one.

10➜ Click the intersection address I-90 & Haines Ave. No candidates appear.

10➜ Click on the Standardized Address text to expand it.

The standardized address shows how the address locator parsed the address into different elements. Occasionally, it makes a mistake, so it's worth examining the standardization. Haines Ave appears correctly, but the I of I-90 ended up in the PreType box instead of being interpreted as part of the street name, Interstate 90.

10➜ Delete the I from the PreType box.

10➜ Enter I-90 in the box for StreetName (Fig. 10.10).

10➜ Click the Search button. No candidates appear.

10➜ Change I-90 to INTERSTATE 90.

Still no candidates appear. This address will have to remain unmatched for now.

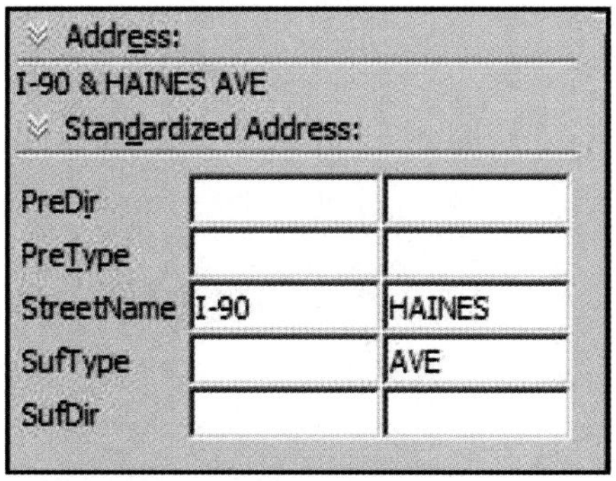

Fig. 10.10. Editing the standardized address

Reviewing tentative matches

Now the window reads 67 matched, 28 tied, and 1 unmatched. However, there are degrees of matching that you can investigate. Let's look at the matched addresses with the lower scores to make sure that the matches look reasonable and to see what kinds of issues lower the match score.

11➜ Change the Show Results box to Matched Addresses with score below 80.

11➜ Turn off the Match_addr and resize the Arc_Street fields again.

11➜ Click the first address, 535 Mountain View Dr, to examine its candidates.

Note that the Status field has a T for tentative instead of an M for matched. Notice that the reference layer contains Mountain View *Rd* instead of Mountain View *Dr*. This is a reasonable candidate, as people often mix up Rd and Dr in addresses. However, be careful about assuming this in every case. Harney Blvd and Harney Ct probably don't match because a boulevard is generally larger and busier than a quiet residential court, and a user is less likely to have made this mistake. In such a case, you might want to unmatch the record.

The second address was previously matched by hand, the LACROSSTA street typing error. The third address, 1803 LaCrosse St., has the prefix direction missing.

11➔ Examine the remaining addresses and candidates one by one.

9. What is wrong with each of the following addresses?

   1301 Omaha St   _____

   710 Meridian Ln   _____

   2250 Haines   _____

None of the tentative matches seems to have any major issues, so finish matching.

11➔ Click Close in the Interactive Rematch window, and examine the rceats shapefile of restaurants.

➔ Save your map document.

Updating the reference layer

Let's return to the problem we had in matching the Applebee's restaurant at I-90 and Haines Ave. Sometimes the problem lies not in the address table or in the geocoding engine but in the reference layer itself.

12➔ Use Find to search for the text **90** in the Street field of the Streets layer.

12➔ Scroll down the extensive list until you see the entries for I-90.

Now we begin to get a clue to the problem. The creator of this roads layer used a nonstandard abbreviation, INTSTE, for Interstate. The geocoding engine knows many abbreviations, but not this one. This reference layer table needs editing to change INTSTE to INTERSTATE.

13➔ Close the Find window.

13➔ Use Select By Attributes to select the records from the Streets layer, using the expression "STREET" LIKE '%INTSTE%'. Notice that the Interstate is highlighted on the map when it is selected, as a visual confirmation of your selection.

13➔ Open the attribute table for streets, and click Selected to view only the selected records to make sure you will not change anything that shouldn't be changed.

Only one of these records appears as though you might not want to edit it. The N INTSTE 90 SVC road might no longer fit in the field if you changed its name, so let's leave this one out.

14➔ Right-click to the left of the record for the N INTSTE 90 SVC RD and choose Unselect.

14➔ Right-click the Street field and choose Field Calculator.

14➜ Enter "INTERSTATE 90" in the box, including the double quotes. Click OK.

15➜ Repeat the same calculation for the ROADNAME field.

15➜ Close the table and clear the selected features.

Because the address locator stores a snapshot of the reference layer at the time the locator is created, we need to create a new locator in order to use the updated information. This time we'll also include the alias table as part of the locator.

16➜ Repeat steps 2 and 3 to create a new address locator named RapidCityFix. Use the same settings but remember to specify the alias table in the geocoding options (see step 7 if you need a reminder).

17➜ Choose Tools > Geocoding > Geocode Addresses and add your new locator.

17➜Choose to geocode restaurants again on the Address field. Specify a new shapefile named rceats2.shp as the output. Click OK to begin.

When the progress window appears, notice that only one address is unmatched. Fixing the INTSTE problem and including the alias table resulted in better performance by the locator. Only the misspelled LaCrossta address remains unmatched.

18➜ Choose to Rematch. Change the Show Results to Unmatched Addresses.

18➜ Fix the LaCrossta match as before and finish geocoding. The table is now entirely matched. Click Close in the Interactive Rematch window.

➜ Save the map document.

Geocoding U.S. cities

Next we will practice our geocoding skills by matching a table of precipitation data for U.S. cities to a shapefile of U.S. cities. When finished, we'll make a map of monthly normal precipitation totals for the United States. As before, we must create the address locator as our first step.

19➜ Start ArcCatalog, if necessary, and navigate to the mgisdata\Usa\usdata geodatabase.

19➜ Use the Preview tab to examine the attributes of the precipitation table, normal_precip.

10. Which style of geocoding will we use this time? _____

20➜ Also examine the attribute table of the cities feature class in the usdata geodatabase. This feature class will be the reference layer.

11. Which field will we use for the city name? _____ Which field do we use for the state abbreviation? _____ (**Hint:** Look at the far right of the table.)

21➜ Right-click the usdata geodatabase and choose New > Address Locator.

21➜ Choose the US Cities with State style. Click OK.

21➜ Name the locator US_City_State.

21➔ Click the Browse button and specify the cities feature class in the mgisdata\Usa \usdata geodatabase as the reference layer.

21➔ Specify the CITY_NAME field for the Cities entry and the ST_ABBRV field (at the bottom of the fields list) for the State.

21➔ Click OK to create the locator.

➔ Close ArcCatalog.

Now let's prepare to do our geocoding and examine the table we'll be using for the "addresses."

22➔ Return to ArcMap and your geocoding map document with the restaurants.

22➔ Insert a new data frame and name it **US Weather**.

22➔ Add the cities and states feature classes from the mgisdata\Usa\usdata geodatabase to the new data frame. Also add the table normal_precip.

23➔ Zoom in to the conterminous United States.

23➔ Open the normal_precip table and examine the fields. The precipitation values are given in inches for each month, and a yearly total is also given.

23➔ Close the normal_precip table.

All right, let's give it a try.

24➔ Right-click the normal_precip table and choose Geocode Addresses.

24➔ In the Choose an Address Locator window, click the Add button and add the US_City_State locator just created. Click Add and then OK.

24➔ In the Geocode Addresses window, specify CITY for the City field and STATE for the State abbreviation field.

24➔ Click the Browse button and specify the location and name of the output shapefile. Put it in the mgisdata\Usa folder and call it **cityprecip.shp**.

24➔ Click OK to start geocoding.

24➔ Examine the results window.

12. How many matched records are there? _____ How many unmatched? _____

That is not a bad effort. Let's take a look at the unmatched records to see if we can get more cities matched.

25➔ Click Rematch and change the Show Results to Unmatched Addresses.

25➔ Turn off the Match_addr and resize the Arc_City and Arc_State fields.

25➔ Choose the first city, ANNETTE, AK, and see if any candidates appear. No!

25➔ Click the Geocoding Options button and reduce the spelling sensitivity to 10. Still no matches appear.

Perhaps some of these cities do not appear in the reference layer. Write down the names of the first five cities: _____

26➔ Click Close to stop interactive matching.

26➔ Use the Find button to search the CITY_NAME attribute in the Cities layer for the first five cities you wrote down. Close the Find window when finished.

Apparently our guess that these cities are missing from the reference layer was correct. However, this is not true of all the cities. Let's go back and examine the geocoding result again, making use of our ability to review and rematch a finished geocoding result.

27➔ Choose Tools > Geocoding > Review/Rematch Addresses > Geocoding Result: cityprecip from the main menu bar.

27➔ Change Show Results to Unmatched Addresses.

27➔ Turn off the Match_addr and resize the Arc_City and Arc_State fields.

28➔ Click on Geocoding Options and set the spelling sensitivity to 20.

TIP: You can actually use Find while the rematch window is open.

Many of the cities are unmatchable because they are not present in the reference layer. The only solution would be to find a reference layer with more cities in it, but there are a few more we can match. It is helpful to know that some cities have double names. For example, the next matchable address is FID 102 Lexington, KY, which can be matched to Lexington-Fayette, KY. Some have multiple entries; for example, you can match three different stations to New York City (two airports and Central Park).

29➔ Locate FID 102, Lexington, KY, and click on it. The candidate Lexington-Fayette appears. Click Match.

29➔ Keep moving through the table, examining each address in turn. Look for candidates to appear.

29➔ At FID 126, a candidate should appear for Minneapolis-St Paul. Match it.

29➔ At FID 128, a candidate should appear for Saint Cloud. Match it.

30➔ FID 168 is in Central Park, NY, but the name is so far off it does not find a candidate. However, you know it matches New York.

30➔ Change the Address City on the left to New York (Fig. 10.11). When the candidate appears, match the record.

Fig. 10.11. Retype the address.

➔ Keep moving through the table, examining each address in turn. If it is a major city, edit the address to find a match. If you find candidates that look reasonable, match them. If not, leave them unmatched.

When finished, you should have 222 matched addresses and 51 unmatched addresses. Once matched, the data can be used to analyze the distribution of rainfall across the United States.

31➔ Click Close to close the Interactive Rematch window.

31➔ Turn off the Cities layer to see the precipitation locations better.

31➔ Create a graduated symbol map showing the annual precipitation (field ANN) in symbols that increase in size with precipitation (Fig. 10.12).

13. Can you think of two reasons why the geocoding method works better to accomplish this task, as opposed to joining the precipitation data to the Cities layer based on a common CITY_NAME attribute?

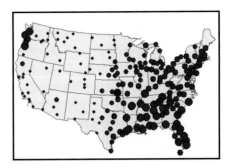

Fig. 10.12. Annual precipitation of the matched cities

Recall that during the interactive matching, you found three entries for New York City based on the three different stations. How were the points created for these three cities?

32➔ Zoom in to the southernmost tip of New York state and find the point on the western side of Long Island (Fig. 10.13).

32➔ Use the Identify button to click on the point.

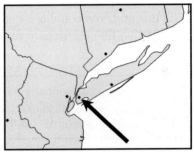

Notice that three entries come up in the Identify window. Click on each entry in turn and note that each of the three stations in the precipitation table was geocoded as its own point. The three points lie directly on top of each other, however, and appear as one point.

Fig. 10.13. Click the geocoded point for New York City.

32➔ Close the Identify window, and zoom back to the previous extent of the conterminous United States.

We have accomplished our task, but let us briefly see some applications of the data we have created. If you have Spatial Analyst, do the following steps. If not, skip down to step 36.

33➔ Choose Tools > Extensions from the main menu bar.

33➔ Turn on the Spatial Analyst extension and open the Spatial Analyst toolbar.

34➔ Choose Spatial Analyst > Options.

34➔ Under the General tab, set the Mask layer to the States feature class, and choose to save in the same coordinate system as the data frame.

34➔ Under the Extent tab, choose Same As Display.

34➔ Under the Cell Size tab, choose As Specified Below and enter **10000** (meters) in the cell size box. Press the tab key to enter the value and see if this gives a reasonably sized grid. If the number of rows or columns is greater than 1000, increase the cell size.

34➔ Click OK to close the Options window and keep these settings.

35➔ Choose Spatial Analyst > Interpolate to Raster > Inverse Distance Weighted from the Spatial Analyst menu.

35➔ Specify the Geocoding result as the input points and ANN as the Z value field. Leave the defaults on everything else. Click OK.

35➔ Give the States layer a hollow symbol to view the raster result.

36➔ If you don't have Spatial Analyst, load the grid ann_precip_in from the folder mgisdata\MapDocuments\Results.

36➔ Create a stretched color map of the grid (Fig. 10.14).

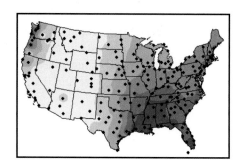

Fig. 10.14. Rainfall grid interpolated from city precipitation data

With this grid, you could use a zonal statistics function in Spatial Analyst to find the average precipitation for each state or for each county.

TIP: Geocoding occasionally has some odd effects with editing routines. If you have been geocoding and you start to experience problems with editing, save your work, exit ArcMap, and start the program again. We will do this now to ensure that the rest of the tutorial session works.

➔ Save the map document.

➔ Exit ArcMap. If you are asked whether to save your edits, do so.

Adding x-y data

As our final example, we will use the Add XY Data function to create points from latitude and longitude values in a table. In the Sdakota folder, we have a table containing climate stations for the state. These stations were downloaded from a Web site of the National Climatic Data Center. They come from a cooperative network of stations and so are called the COOP stations. Each climate station has a latitude and a longitude. The coordinates are degrees latitude/longitude in datum NAD 1983, like the rest of our South Dakota data.

➔ Start ArcMap again with a new empty map.

37➔ Name the data frame **South Dakota**.

37➔ Add the counties from the mgisdata\Sdakota\southdakota geodatabase.

37➔ Add the table coop_sta.dbf from the mgisdata\Sdakota folder.

37➔ Open the coop_sta table and examine its fields.

Notice that there are two fields with latitude and longitude values. These data originally came from text files downloaded from the Climatic Data Center, imported into Microsoft Excel, formatted, and then saved as a .dbf file so that it could be opened in ArcMap. We are ready to create a point map of the stations.

38➔ Close the coop_sta table.

38➔ Choose Tools > Add XY Data from the main menu bar.

38➔ Fill out the fields as shown in Figure 10.15. Set the X field to LONDD and the Y field to LATDD.

38➔ Click on Edit, and use the Select button to set the coordinate system to GCS NAD 1983.

38➔ Click OK to add the points.

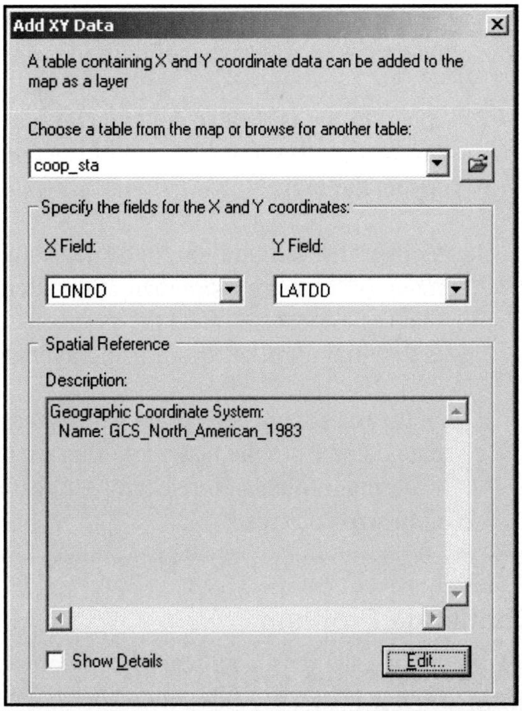

Add XY Data

A table containing X and Y coordinate data can be added to the map as a layer

Choose a table from the map or browse for another table:

coop_sta

Specify the fields for the X and Y coordinates:

X Field: Y Field:

LONDD   LATDD

Spatial Reference

Description:

Geographic Coordinate System:
  Name: GCS_North_American_1983

☐ Show Details   Edit...

When *x-y* data are added to a map, they take the form of an event layer. Each pair of *x-y* coordinates is referred to as an event. An event layer only exists inside the map document in which it is created. Notice that the stations are being displayed in a projected coordinate system, even though they came from a geographic coordinate system. This happens because they are being projected on the fly to the data frame coordinate system, South Dakota State Plane North, just like any other data set in the data frame. To save it permanently and make it available to other map documents, export the event layer as a shapefile or a geodatabase feature class.

Fig. 10.15. Adding the coop stations from *x-y* data

39➔ Right-click the coop_sta Events layer and choose Data > Export Data.

39➔ Choose to use the same coordinate system as the data frame.

39➔ Click the Browse button and set the Save As Type to File and Personal Geodatabase feature class.

39➔ Navigate to the southdakota geodatabase and name the new feature class coopstat.

39➔ Click OK to start exporting.

39➔ Choose Yes to add the new feature class to the document.

39➔ Right-click the original event layer and Remove it from the map document.

Now the coop stations feature class is ready for use to analyze South Dakota's climate (Fig. 10.16).

This is the end of the tutorial.

➔ Save your map document.

➔ Exit ArcMap.

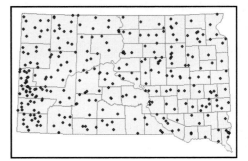

Fig. 10.16. South Dakota climate stations

Exercises

1. In the mgisdata\Rapidcity folder, locate the file gas_loc.dbf. Geocode these gas stations using the rc_roads shapefile in the same folder. **Capture** a map showing the gas stations.

2. The gas_loc table also contains latitude-longitude coordinates in NAD27. Create another layer of gas stations based on these coordinates. **Capture** this map also. Are the two layers the same?

3. Convert the latitude-longitude gas stations to a shapefile in UTM 13 and determine how many of them are more than 100 meters from the geocoded positions. (You can assume that the closest station is the same one although this might not in fact be the case.) What is the minimum, maximum, and average distance between points in the two data sets?

4. In the mgisdata\Usa\usdata geodatabase, you will find a table called customers. These data come from the test marketing of a catalog for natural landscaping to minimize watering of plants (xeriscaping). Create a map showing the location of the customers. **Capture** your map.

Challenge Problem

Search the Internet for a data set which interests you, is geocodable, and contains at least 50 data points. Then geocode it. With your answer, include the source of the data, the style of locator used, and the reference layer used.

Suggestion: Search **www.GeographyNetwork.com** for the Dynamap product, which contains downloadable street addresses by ZIP code. Locate the sample data for Concord, New Hampshire. (There is a charge for the actual product in other ZIP codes, but the sample data are free. Be sure to read the License Agreement for this data set.) Then search for Concord addresses to geocode. An online telephone directory could help you locate restaurants, for example, which could be entered into an Excel spreadsheet, converted to a .dbf file, and geocoded.

Skills Reference

Creating an address locator 383

Geocoding addresses 384

Rematching a geocoding result 384

Interactive matching 385

Adding *x-y* data 386

Creating an address locator

1. Examine the reference layer to ensure that all the fields required for the geocoding style are present, and identify the names of the fields.

2. In ArcCatalog, right-click the name of the folder in which you want to store the locator and choose New > Address Locator.

3. Choose the address locator style (Fig. 10.17) and click OK.

4. Name the address locator.

5. Specify the name and the location of the reference layer.

6. Set the required fields (bold) and the optional fields to the appropriate fields in the attribute file of the reference layer.

7. If the field name containing the addresses in the address table is different from the ones shown, add it to the list.

8. Choose a place name alias table, if one is available.

9. Change the matching options, if desired.

10. Enter a different set of connectors, if necessary.

11. Set the side offset, if using Two Range, and the end offset.

12. Check the optional fields to add to the output file.

13. Click OK to create the locator. It will be saved with its name appended to your user name (e.g., John Smith.AddressLocator) and will appear in ArcCatalog as an address locator.

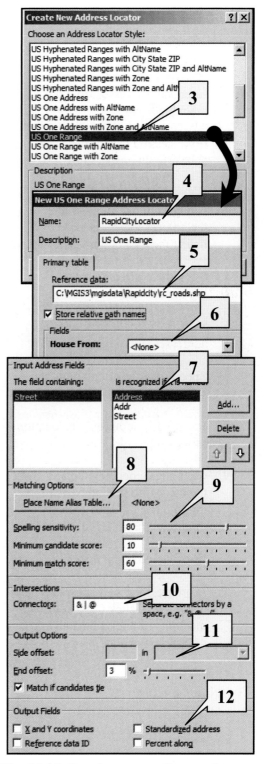

Fig. 10.17. Creating a geocoding service

Geocoding addresses

1. Add the address table and the reference layer (optional) to ArcMap.

2. Choose Tools > Geocoding > Geocode Addresses from the main menu bar.

3. If the desired locator does not appear in the menu, click Add (Fig. 10.18).

4. Navigate to the folder where the address locator is stored. Select the locator and click Add to close the menu and add the locator to the list.

5. Click the desired locator to highlight it and click OK.

6. Specify the table containing the addresses (Fig. 10.19).

7. Specify the field containing the addresses to be matched.

8. Specify the name and the location of the output file.

9. Click the Geocoding Options button to change the spelling sensitivity, minimum candidate score, or minimum match score.

10. Click OK to begin matching.

11. A progress bar will appear. When it is complete, click Close to finish matching or click Rematch to try matching more addresses.

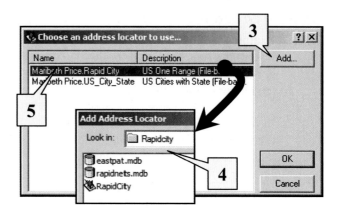

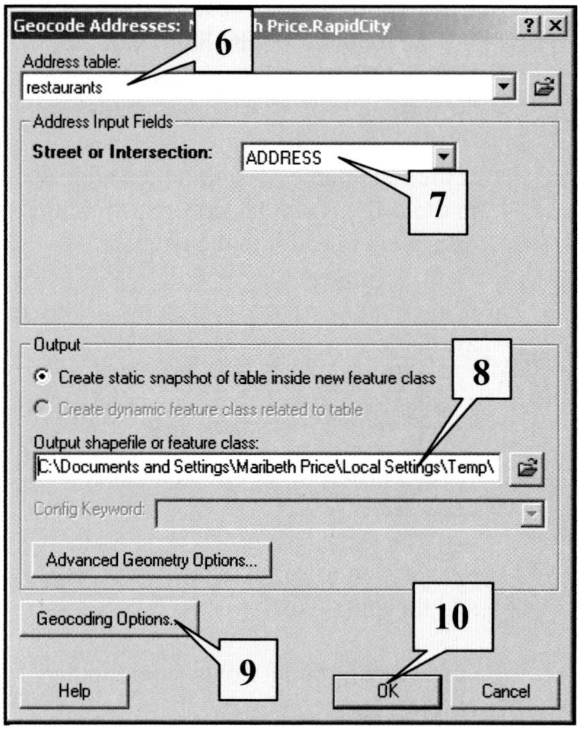

Fig. 10.18. Adding an address locator

Fig. 10.19. Preparing to match in batch mode

Rematching a geocoding result

You can reopen a completed matching session to rematch more addresses.

1. It is best to terminate any editing before rematching addresses. Choose Editing > Stop Editing from the Editor menu and save any changes, if necessary.

2. Choose Tools > Geocoding > Review/Rematch Address > and choose the geocoding result to rematch. Click the Rematch button to proceed with Interactive Rematching.

Interactive matching

1. Change the Show Results box to Unmatched Addresses (or another group) (Fig. 10.20).

2. Click on one of the unmatched addresses to highlight it.

3. If a list of candidates appears in the bottom box, click the best match to choose it, and click the Match button. (The score for each candidate is shown to help identify the closest match.) The address will be matched, and the U in the upper list will change to an M. The match score will also be shown.

4. If no candidates appear, try clicking the Geocoding Options button to modify the spelling sensitivity or the minimum candidate score.

5. If still no candidates appear, expand the Standardized Address section to see if the address is being interpreted correctly.

6. Change the address standardization, if necessary. Press Enter after typing each change to make it take effect and allow new candidates to appear.

7. If still no candidates appear, the address is probably unmatchable. It may be incorrect, or the reference layer may be incomplete. Click another address and begin again.

8. If you are trying to decide between several candidates, click the Zoom to Candidates to view them on the map. Click Original Extent to go back to the previous extent.

9. You can use the Manage Result sets to save groups of records to work on later.

10. When you are finished rematching, click Close.

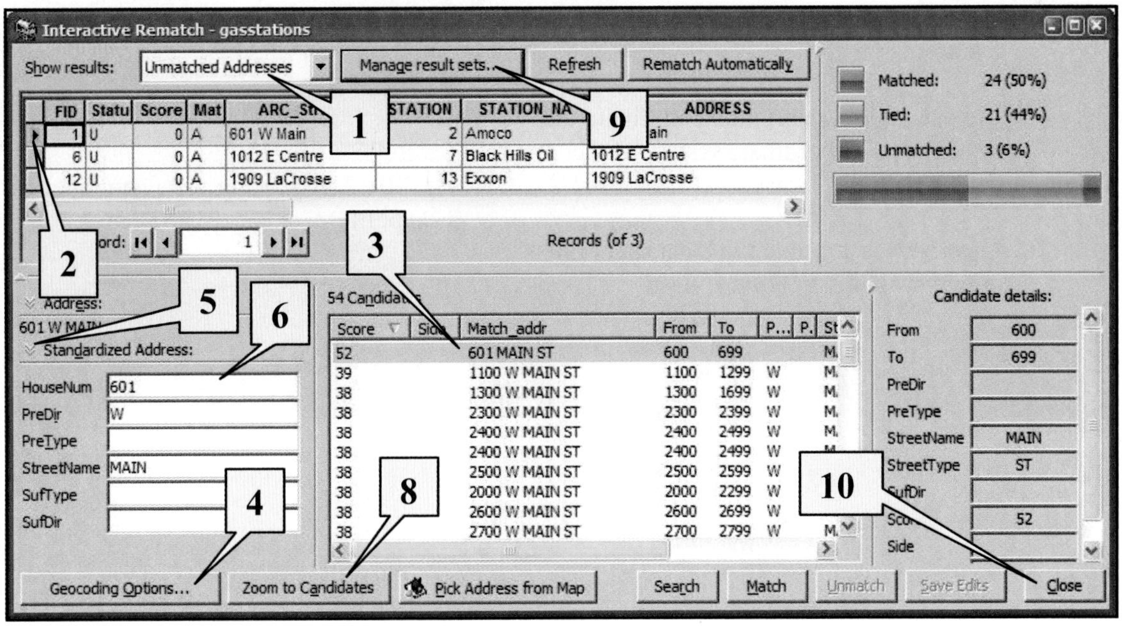

Fig. 10.20. Rematching addresses interactively

Adding x-y data

1. Make sure the table has fields for *x* and *y* coordinates and that you know the coordinate system of the coordinates.

2. In ArcMap, choose Tools > Add XY Data from the main menu bar.

3. Select the table containing the *x-y* coordinates (Fig. 10.21).

4. Specify which fields contain the *x-y* coordinates. Recall that longitude is the *x* coordinate and latitude is the *y* coordinate.

5. Click the Edit button to specify the coordinate system of the points in the table.

6. Click OK to create the points.

To save the event layer permanently as a shapefile or as a feature class in a geodatabase, do the following steps:

7. Right-click the new event layer and choose Data > Export Data from the context menu.

8. Choose to export All features or Selected features (Fig. 10.22).

9. Choose to use the original coordinate system or the coordinate system of the data frame.

10. Specify the name and the location of the output shapefile or feature class.

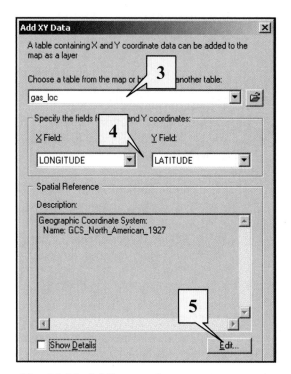

Fig. 10.21. Adding *x-y* data to a map

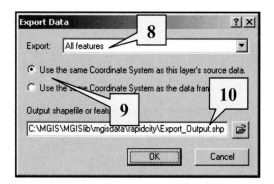

Fig. 10.22. Exporting the event layer as a shapefile

Chapter 11. Coordinate Systems

Objectives

- ➤ Learning the basic properties and uses of coordinate systems

- ➤ Understanding the difference between geographic coordinates and projected coordinates

- ➤ Getting familiar with different types of map projections

- ➤ Understanding how different projections cause distortions

- ➤ Managing and troubleshooting coordinate systems of feature classes and images

Mastering the Concepts

GIS Concepts

About coordinate systems

A successful GIS system depends in large part on managing coordinate systems correctly, and a person's skill (or lack of it) can make the difference between usable and trashed data. Unfortunately, coordinate systems are also a vast and complex topic. Experience in working with coordinate systems and regular review of these concepts will ultimately provide users a sound basis for effectively working with coordinates.

In order to display and analyze maps on the screen, a GIS system uses **coordinate pairs**, which specify the location and shape of a particular feature. A point is described by a single x-y coordinate pair located in space; a line is a series of x-y coordinates. A coordinate system is an agreed-upon range of coordinates used to portray the features.

Imagine that a surveyor is mapping an industrial site using typical surveying instruments that measure angles, directions, and distances. He probably begins at a fence corner at the lower right of the complex and assigns that point the coordinates (0, 0). This is the **origin** of the coordinate system (Fig. 11.1). The surveyor then determines that the southwest corner of building A is 175 meters east and 200 meters north of the origin, and he assigns that corner the coordinates (175, 200). The surveyor uses meters to record the distances and determine the coordinates, thereby establishing meters as the **map units**. The combination of origin and map units

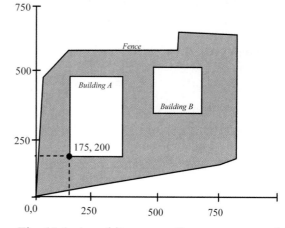

Fig. 11.1. An arbitrary coordinate system used for surveying a site

becomes the **coordinate space** of this map. In this case the coordinate space is arbitrary and chosen for convenience.

Now the surveyor can use this coordinate space to tell other people to go to a particular location. He might tell his assistant to walk to (225, 398) and place a flag to mark the location for a new monitoring well. The assistant can do this as long as he knows the coordinate space in use. Without knowing the coordinate space, however, the directions would be useless.

Now imagine that the surveyor's colleague is surveying another site across town. She also chose the corner of her site as the origin, but she is measuring her distances in feet. The two surveyors take their drawings back to the office and decide to create a map showing the location of both sites within the city, so they each transfer the coordinates into a GIS. When they plot the sites, however, the two different sites show up in exactly the same spot, even though in real life they are several miles apart. This problem occurs because both surveyors used an arbitrary (0, 0) origin unrelated to the other site. Moreover, the woman's buildings look about three times as large as the man's buildings because she is measuring in feet instead of meters. To plot these together the two surveyors would need to define a single common origin and convert both sites to the same map units.

To avoid problems such as this, geographers and map creators usually employ standard **coordinate systems** whose characteristics are agreed upon beforehand. The surveyors could have used a Global Positioning System (GPS) instrument to establish the *x-y* coordinates of their origins in a common coordinate system, such as Universal Transverse Mercator (UTM) or State Plane, and then entered their surveyed points relative to that base point. They then would have been spared the extra effort of converting the drawings into the same system.

Geographic coordinate systems

The most commonly used coordinate system is based on spherical units because the earth approximates a sphere (Fig. 11.2). The units are called **degrees** and measure angles. The earth is divided into 360 degrees around the equator, which are called degrees of **longitude**. Usually longitude is expressed as degrees east or west of the **Prime Meridian**, which is the line along which the longitude equals 0 and passes through Greenwich, England. Longitudes thus fall in the range –180 to +180. **Latitudes** measure angles from the earth's equator, which has latitude equal to 0, and latitudes range from –90 at the south pole to +90 at the north pole. This system of measurement is called a **geographic coordinate system**, or sometimes a **GCS**.

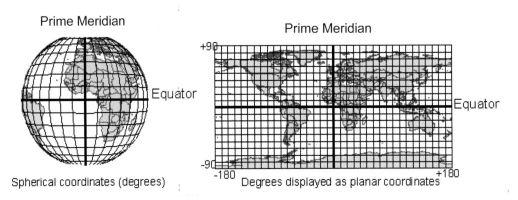

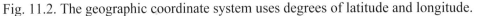

Fig. 11.2. The geographic coordinate system uses degrees of latitude and longitude.

Although strictly speaking a GCS records locations on a sphere, in a GIS the coordinates are drawn as though they represent a planar coordinate system (Fig. 11.2). However, distortions are introduced by this practice. Notice on the sphere that a latitude line at the equator has a distance equal to the circumference of the earth, that a latitude line at the poles has a distance of zero, and that latitude lines in between have intermediate values. Yet in the planar display, these lines are all represented as the same length as the equator. The further north or south an area is from the equator, the more it gets distorted.

Spheroids and datums

A geographic coordinate system is not a perfect system for measuring locations on the earth, however, because the earth is not a perfect sphere. It bulges at the equator because of its rotation. The bulge can be modeled mathematically using a **spheroid**, an oblong sphere with a major and minor axis. (The bulge in Figure 11.3 is highly exaggerated.) Latitude-longitude locations defined on a spheroid are more accurate than those defined on a sphere. The spheroid becomes a part of the GCS definition. (Some people use the term **ellipsoid** rather than spheroid.)

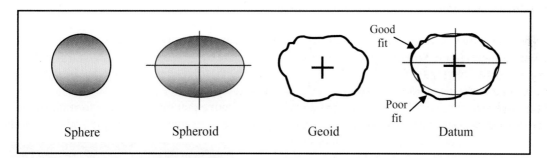

Fig. 11.3. Relationship between the spheroid, geoid, and datum

Many different spheroids have been used in the past because estimates of the earth's shape have varied over time. Clarke 1886 was a common spheroid used in North America, but it is gradually being replaced by more accurate satellite-measured spheroids, such as WGS84. When one uses a GCS to record locations, it is important to specify which spheroid is being used. The same location might have a slightly different latitude and longitude in GCS systems based on different spheroids, and a misalignment of features on the map could occur.

Even when a spheroid is used, however, the map accuracy will still not be perfect because the actual shape of the earth is best represented by the **geoid**, which is the measurement of the earth's radius with respect to a datum, which can be defined either locally or globally (Fig. 11.3). This datum is often sea level, but sea level itself varies from place to place according to differences in gravitational force, which arise from rotational, topographic, and compositional differences in the earth's mantle. The geoid is often described as what the shape of the earth would be if there were only seas and no continents. The geoid can only be approximated mathematically because of local anomalies at various scales.

To achieve a more accurate map, the spheroid is moved until the geoid at the area being mapped (arrow) corresponds best with the spheroid surface (Fig. 11.3). The combination of the spheroid and the location where its center is placed is called a **datum**. The purpose of the datum is to create the best possible fit between the actual earth geoid and the spheroid used for mapping. In this example, the center of the spheroid is slightly offset from the center of the geoid in order to achieve the best fit. Some datums have this offset; others do not. A datum in which these two centers coincide is called an earth-centered datum.

Notice, however, that the improved fit at the arrow means that other areas of the globe, such as the southern hemisphere in this example, have a worse fit. Thus, the best datum will change depending on one's location at the earth's surface. One should not use a North American Datum to map Europe and vice versa. For world maps, one would use a best-fit datum for the whole earth rather than a specialized datum for one part.

Several datums are commonly used in North America. The North American Datum of 1927, often abbreviated NAD 1927 or even NAD27, is based on the Clarke 1866 spheroid. The North American Datum of 1983, or NAD 1983, is based on the GRS_1980 spheroid. These two datums include a network of surveyed points to help define the fit and provide accurately located positions. Another datum commonly used for satellite data is the World Geodetic Survey of 1984, or WGS 1984. Because NAD 1983 is relatively recent, one finds many maps that still use NAD27. However, the coordinates of a particular location can differ by up to several hundred meters between NAD27 and NAD83. Thus, one usually converts all the layers within a particular database into a common datum to avoid alignment problems.

Figure 11.4 shows a typical datum offset. Notice that the roads are displaced about 200 meters to the south relative to the image because the image coordinate system is based on NAD83, whereas the roads are in NAD27.

Datum conversions are called **transformations**. They are not exact mathematical formulas but require localized estimates and fitting. Several different methods may be used, and some methods are unique to particular datums and cannot be used for others. Thus, it is not desirable to convert map layers back and forth repeatedly from datum to datum. Usually one picks a datum to be used for a particular project and sticks with it, converting layers as needed. The choice of datum is sometimes a matter of convenience. If most layers to be placed in a database already use NAD27, it is easier to stick with the older datum. Alternatively, an organization may decide that all layers must use the NAD83 datum as a standard policy, in which case nonmatching layers would require conversion. Most GPS and other data collection systems can be set to work with a variety of datums.

Fig. 11.4. Offset of roads due to different datums

Map projections

Because the shape of the earth is close to a sphere, the most accurate way to map it is to use a spheroid. However, globes and spheroids are inconvenient objects for putting in a report or mounting on a wall. Therefore, a major effort of cartography includes converting the locations of features in three dimensions on a globe to locations in two dimensions on a piece of paper. This process is called **projection**. One starts with a defined GCS based on a spheroid and a datum and applies a set of mathematical formulas that convert degrees of latitude and longitude into planar *x-y* coordinates on the paper. During this conversion the three-dimensional GCS becomes a two-dimensional representation, and degrees are converted into feet or meters. The geographic coordinate system thereby becomes a projected coordinate system and inherits the datum from its underlying GCS.

You can imagine the projection process visually as a clear globe with the graticules of the earth printed in black. If you place a lightbulb at the center of the earth and wrap a piece of paper around the equator in the shape of a cylinder (Fig. 11.5), the graticules will be cast as shadows on the paper. Tracing them gives you a map, which you can unroll and lay flat.

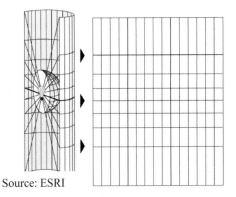

Source: ESRI

Fig. 11.5. Projecting a map from 3D to planar coordinates

Projections are grouped into three major classes, depending on how the paper is situated relative to the globe. A **cylindrical projection** uses a paper cylinder that lays tangent to (touches) the earth at the equator along a great circle (Fig. 11.6). Rotating the paper sideways and making it tangent to the earth along a line of longitude produces a **transverse cylindrical projection**. An **oblique cylindrical projection** places the tangent at an angle. In each case, distortion is absent along the tangent and increases away from it. Cylindrical projections typically preserve shape or direction at the expense of area and distance.

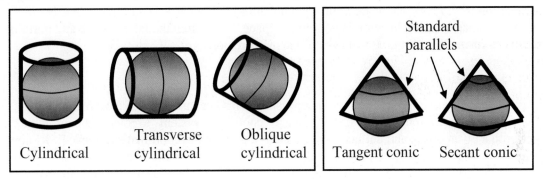

Cylindrical   Transverse cylindrical   Oblique cylindrical

Fig. 11.6. Cylindrical projections

Standard parallels

Tangent conic   Secant conic

Fig. 11.7. Conic projections

A **conic projection** is based on setting a paper cone over the sphere (Fig. 11.7). If the cone intersects the globe along one line of latitude at which the paper is tangent to the globe, then the projection is a **tangent conic** projection. The line of tangency is called the standard parallel. One can also place the cone *through* the sphere so that it touches in two places—a **secant conic projection**. Such projections have two standard parallels. (Cylindrical projections may also be tangent or secant.) Distortion increases away from the parallels. Conic projections typically preserve area or distance at the expense of direction and shape. One can also place the paper flat against the globe at a single point, producing an **azimuthal** projection (Fig. 11.8). Other names for this method include **stereographic** or **orthographic** projection. Typically these projections are used for displaying the earth's poles, and for that reason they are sometimes called polar projections. Distortion is minimal at the point of tangency and increases away from it. Three-dimensional Web services that display the globe often use an azimuthal coordinate system. Distortions

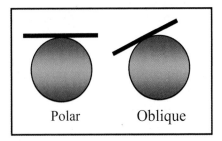

Polar   Oblique

Fig. 11.8. Azimuthal projections

associated with azimuthal projections vary with the projection method, but they generally preserve area or distance rather than direction or shape.

Map projections have certain properties that are described using *projection parameters* (Fig. 11.9). One parameter is the line of longitude, which constitutes the **central meridian** or $x = 0$ line on the map. The central meridian is chosen based on its position near the center of the area being mapped. A map of England would have a central meridian close to 0 degrees. A map of the United States would have a central meridian of about −100 degrees because that meridian is close to the center of the continental United States. The **latitude of origin**, also sometimes called the **reference latitude**, is the $y = 0$ line. The equator is often used as the reference latitude. A conic

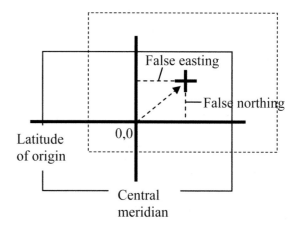

Fig. 11.9. Some parameters of map projections

map projection will also have one or more **standard parallels** in its parameters—the latitudes that define the lines of tangency or secancy for the cone. Maps may also have a **false northing** or **false easting** (Fig. 11.9). These are arbitrary numbers added to the *x-y* coordinates in order to translate the map to a new location in coordinate space. Most often the false easting and northing are used to ensure that all the coordinates of the map are positive.

A note on terminology

The preceding discussion uses potentially confusing terms because they have been applied in different ways by different people. ESRI may use "coordinate system" to refer to a coordinate space with a given origin and defined units (such as a geographic coordinate system). ESRI now also uses the term to indicate the definition of a data layer's coordinate reference, including the GCS, the spheroid, the datum, and the projection. "Projection" is understood as a mathematical transformation from a spherical to a planar coordinate system. However, some people still use the term "projection" to refer to the spatial reference (the complete description of the coordinate system). Users need to be aware of both of these uses and be able to ascertain the correct meaning in a particular instance, which is usually clear in context.

Raster coordinate systems

Rasters have coordinate systems and projections just as vector feature classes do. Imagine scanning both a Mercator and Lambert Conformal Conic map of the United States and creating raster images. The shapes of the states will be very different in the two rasters, just as they are in the two paper maps. Correctly using a raster requires knowledge of the coordinate system upon which it is based. Like vectors, rasters can be projected from one coordinate system to another.

Assigning coordinate systems to rasters

Many rasters are provided with a coordinate system at the outset. However, sometimes a user must assign the coordinate system. A raster that has been assigned a real-world coordinate system is said to be **georeferenced**. Two cases commonly occur.

Case I involves a raster for which the projection, datum, cell size, and *x-y* coordinate of the upper left corner is known, but the information has not been stored in accessible form. Some image formats store georeferencing information inside a header file or in the image itself. Other formats use a **world file**, which is a six-line text file containing the parameters needed to georeference the raster. If the georeferencing information is missing, the user can create a world file with the appropriate parameters. A description of the contents of the world file can be found in the ArcGIS Help.

> **TIP:** A world file has the first and third characters of the image file extension plus a "w" on the end. A .tif file would have a .tfw world file. A .bil file would have a .blw world file.

Case II involves a raster for which the coordinate system parameters are completely unknown. Such files are commonly generated by downloading an image from the Internet or by scanning a map. The coordinate system is in pixel units from the screen or the scanner rather than real-world coordinates in feet or meters. To georeference this raster, the user must establish a list of **ground control points** that link the pixel *x-y* values to real-world *x-y* values in a coordinate system. From this list, a set of equations can be developed to convert the pixel coordinates to real-world coordinates. The final step is to **rectify** the image by using the equations to create a new, georeferenced image.

Several different methods of rectification are used. A first order (**affine**) transformation can translate, rotate, and skew the image to match the points. A second-order or third-order transformation allows the image to stretch or contract in some places relative to others in order to fit the points. This process is often called **rubbersheeting**, likening the image to a sheet of rubber that can be stretched and pulled by different amounts in different areas to get a good fit. After rectification there may still be offsets in the two sets of ground control points. The algorithm calculates a **root mean square** (RMS) error between the two sets of points, which serves as a measure of the accuracy of the transformation. The RMS error is usually reported in map units of the new coordinate system. The transformation RMS error should be included with the geometric accuracy report of the metadata for the image.

The ground control points are established by selecting points on the image that can be matched to another image or feature class that already has a real-world coordinate system. Road intersections, building corners, or other distinctive features are typically used. At least three pairs of points must be chosen for an affine transformation, but more points generally result in a more accurate transformation (lower RMS). Rubbersheeting requires additional control points. Once a set of points is defined, a geometric transformation (rectification) is applied to convert the image to the same coordinate system as the map.

During rectification, the raster is being converted from one grid of cells to another with potentially different sizes and spacing. The old cell grid in many cases will not match the new cell grid, and so the cells must be resampled. This procedure defines cell values for the new grid by interpolating values from the old grid.

Figure 11.10 shows the resampling that occurs during rectification. The cell centers on the left are being rubbersheeted from coordinate system X1-Y1 to X2-Y2, introducing gaps between the cells. The cell grid for the new coordinate system is shown in the output by the black lines. The old cells are resampled to the new cells using one of three methods. In **nearest neighbor** resampling, the new cell is given the value of the old cell that falls at or closest to the center of the new cell. In **bilinear** resampling, a distance-weighted average is taken from the four nearest cell centers. The **cubic convolution** resampling method determines a new value by a curve fit through the nearest 16 cell centers. Nearest neighbor is the fastest method and should always be used with categorical raster data because the values do not change. The bilinear method is usually better for continuous rasters such as elevation.

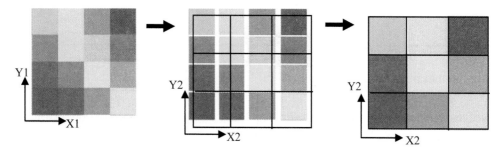

Fig. 11.10. Resampling raster cells after rectifying them using the nearest neighbor method

Projecting rasters

In general, projecting rasters is a more complex and computer intensive process than projecting vector features. When projecting vector features, each *x-y* coordinate is simply converted to its new location through the projection equations. For rasters, each cell center must be relocated, but in the new coordinate system the cells may overlap each other or have gaps between them.

Figure 11.11 shows the same elevation raster in a GCS (Fig. 11a) and a State Plane (Fig. 11b) coordinate system. The GCS raster was obtained from a data provider. (Just looking at the shapes in Figure 11.11 suggests that the GCS pixels will get squeezed together in order to create the State Plane map.) The map units are in degrees, and the cell size is 30 arc seconds or 0.008333 degrees, with 964 columns and 422 rows. A single degree of latitude at the equator is 111.3 km. The height of the cell is

0.00833 degrees × 111.3 km/degree = 0.927 km

or about 900 meters. The cell is not square because 30 arc seconds of latitude is not equal to 30 arc seconds longitude except at the equator—degrees of longitude become smaller with increasing latitude. (At this latitude the cells are about 600 × 900 meters.)

Fig. 11.11. An elevation raster in (a) GCS and (b) State Plane coordinate systems

To project the GCS raster to State Plane, the user must specify a new cell size in the State Plane map units (meters). The old cell grid in degrees will be converted to a new cell grid in meters. Choosing the right size cell is important. Since the resolution in ground units is no better than 900 meters, 900 should be the minimum cell size specified for the new raster. In this case a cell size of 1000 meters was chosen; many people prefer to have nicely rounded cell sizes.

Since we are converting from one grid system to another, the pixels will need to be resampled, just as described for rectifying images. Even if projecting from meters to meters with the same cell size, the projection equations can cause contraction or stretching of the cell grid, causing a change in the shape of the map. (Think of the difference in map appearance between a conic and Mercator projection of the United States.) Thus, the user must also choose a resampling method when projecting rasters. As discussed in Chapter 8, resampling can degrade the quality and characteristics of an image and introduce artifacts. Projecting rasters should be performed only as necessary, and repeated projection or resampling of the same image should be avoided if at all possible.

Common projection systems

Two coordinate systems deserve special description because they are commonly used in the United States: UTM and State Plane. Neither of these is a single coordinate system. Instead the terms UTM and State Plane refer to families of coordinate systems with similar characteristics. The systems divide areas into zones and define the best projection for each zone. As long as the map stays within the specified zone, the distortions in all four map properties are negligible.

The UTM system

The UTM system is based on, as the name suggests, a secant transverse cylindrical projection. Two lines of tangency of the cylinder to the sphere fall about 180 km to each side of the central meridian (Fig. 11.12a). A zone is defined to include 3 degrees on each side of the central meridian between the 80S and 84N latitude lines. Since distortion is zero along the line of tangency and increases with distance from it, the distortion inside the zone is minimal. To map different locations, the cylinder is rotated about the sphere in 6-degree intervals, producing a total of 60 zones around the world. Each zone is indicated by a number and the designation N for the northern hemisphere or S for the southern hemisphere, for example, UTM Zone 14N. Figure 11.12b shows the UTM zones of the conterminous United States.

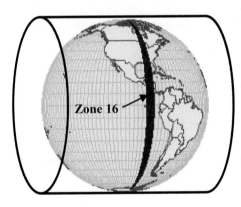

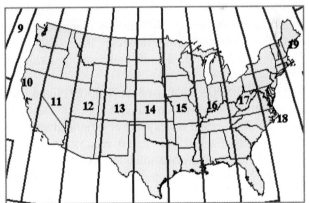

Fig. 11.12 (a) The UTM projection system has 60 north-south zones, each 6 degrees wide.

Fig. 11.12 (b) UTM zones of the conterminous United States

Each UTM zone has its own central meridian and splits at the equator into a north and south component. In order to eliminate negative *x-y* coordinates, southern zones have a false northing of 10 million meters applied, and both northern and southern zones have a false easting of 500,000 meters. UTM is very convenient because users need only know the zone number and hemisphere. Once the zone is specified, the other map parameters are set according to the standards for that zone.

UTM works best for small areas that lie completely within a single zone since distortion increases away from the zone center. The zone used for a particular map depends on the location being mapped. The UTM system is used extensively for U.S. Geological Survey 1:100,000 and 1:24,000 scale publications, for county maps, for maps of states within single zones, and for project maps covering small areas. If the region of interest lies on the boundary between two zones, the user can choose the zone containing the larger area to represent the map. This choice may be acceptable if only a small area extends out of the zone. If not, State Plane zones may provide a better alternative.

The State Plane Coordinate System

The State Plane Coordinate System (SPCS) includes an assortment of coordinate systems. It was developed in the 1930s by the U.S. Coast and Geodetic Survey for large-scale mapping in the United States. To maintain the desired level of accuracy, large states like California and Texas are broken into zones with different parameters used to minimize distortion in each zone (Fig. 11.13). Like the UTM system, an SPCS is identified by its name and zone.

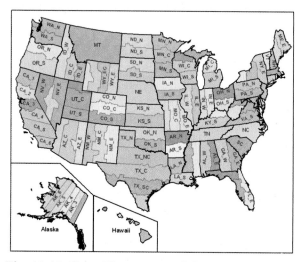

Fig. 11.13. State Plane zones of the United States

The SPCS uses three different projections: Lambert Conformal Conic, Transverse Mercator, and Oblique Mercator. Zones oriented north to south use a Transverse Mercator projection, as it provides the least distortion. Zones oriented east to west use a Lambert Conformal Conic projection. A state may use more than one projection depending on the size and shape of its zones. Alaska has 10 zones and uses all three projections. Zones are formally identified using a FIPS code composed of a two-digit state code and a two-digit zone code (e.g., FIPS 0407 for Zone 7 of California). Users may refer to them informally by state name and zone, for example, California State Plane Zone 7.

The original SPCS was based on NAD 1927 and used units of feet. Improvements in surveying necessitated the revision of the SPCS in 1983. At that time it adopted the NAD 1983 datum and metric units. Certain zones were also revised and in some cases eliminated. Even though the official units changed from feet to meters, users may specify either unit for both versions of SPCS. Additional information about both the State Plane and UTM systems can be found in the ArcMap Help (index headings State Plane Coordinate System and UTM).

State and National Grids

Some large states define a special coordinate system for statewide maps because a single State Plane zone cannot represent it accurately. The Oregon Statewide Lambert coordinate system is one example. Many countries define their own coordinate systems for national maps or their own systems of projections (similar to State Plane) for local and regional maps. Examples include the New Zealand National Grid and the Canada CSRS98 system. A variety of different projections, datums and spheroids are employed. ArcMap maintains a collection of these coordinate system definitions for use with state and country data.

Choosing projections

At the beginning of every new GIS project, one of the first tasks is to establish the coordinate system that will be used. Best management practices dictate that every data set incorporated in the project should be placed in this same coordinate system. Here are some guidelines for choosing it. Preference should be given to one of the standard coordinate systems for the sake of interoperability with others, but if none of the existing systems provide the right properties, a user can define his or her own.

Maps of the world. Distortions cannot be avoided in small scale maps, so they must be managed. Consider the applications. If distances and areas are an important part of the project, then an equidistant or equal area map should be chosen. If navigation or wind/water currents are important, then preserving direction is critical. If general feature locations and properties are the focus, then a compromise projection might be best.

Maps of countries or continents. Many countries or regions have one or more projections defined for general mapping purposes, such as USA Contiguous Equidistant Conic or Europe Albers Equal Area Conic. Some countries have their own defined mapping systems. Like world maps, the application determines which properties of the map are important to preserve.

Maps of states. Smaller states within a single UTM or State Plane zone can use them. Mapping a large state with several zones is not as accurate as using a single zone map, and some states have a statewide projection defined, such as Oregon Statewide Lambert. If none exists, you can modify the central meridian and/or parallels of a State Plane zone to work better for the entire state. A good rule of thumb is that the central meridian should bisect the map extent, and the parallels for secant conic maps should divide the north-south extent into thirds.

Local and regional maps. Any region that fits inside a single UTM or State Plane zone can use either. The preferred system would place the region to be mapped closer to the center of the zone where the distortion is least. However, other factors may take precedence, such as the existing coordinate system of your data sources, the decision of the agency that oversees the project, or your preference for using feet or meters as the map units.

Regional maps in other countries. Many countries such as Canada and Australia have defined their own mapping systems similar to the State Plane system in the United States. Some research can be used to determine the typical projection systems and datums that are commonly used in other countries.

About ArcGIS

Labeling coordinate systems

One benefit of using ArcMap includes being able to display data in a variety of projections. This capability relies on every data set having a coordinate system label, which records the characteristics of the coordinate system including the datum and projection (Fig. 11.14). This label serves two purposes: it documents this information about the coordinate system for the users, and it also helps ArcMap and ArcCatalog to properly display and manage the data. Making sure that all data sets have correct coordinate system definitions is a critical step in building a GIS database. Shapefiles store the labels in a separate file with a .prj extension. Geodatabases store it within the feature class. Rasters must also have coordinate system labels, storing them in various places depending on the raster format.

Data sets may contain unprojected or projected data. If unprojected, the coordinates are stored in decimal degrees of latitude and longitude. The feature class coordinate system label would include simply a GCS and a datum, such as GCS North American 1983. The stored x-y coordinates of a data set on the disk would fall in the range –180 to +180 degrees for x and –90 to +90 degrees for y. The map units would be degrees.

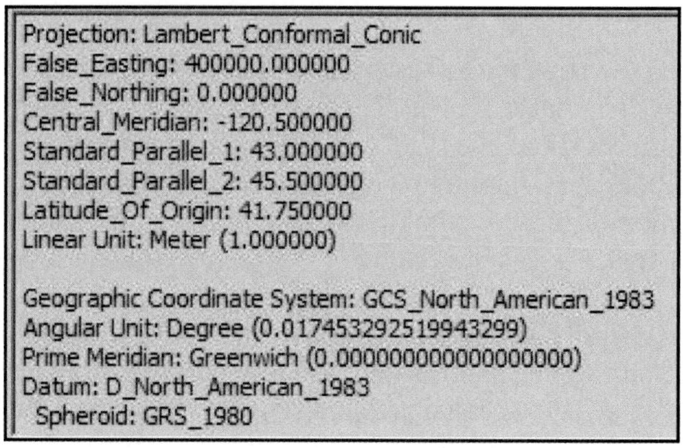

```
Projection: Lambert_Conformal_Conic
False_Easting: 400000.000000
False_Northing: 0.000000
Central_Meridian: -120.500000
Standard_Parallel_1: 43.000000
Standard_Parallel_2: 45.500000
Latitude_Of_Origin: 41.750000
Linear Unit: Meter (1.000000)

Geographic Coordinate System: GCS_North_American_1983
Angular Unit: Degree (0.017453292519943299)
Prime Meridian: Greenwich (0.000000000000000000)
Datum: D_North_American_1983
   Spheroid: GRS_1980
```

Fig. 11.14. A coordinate system label for Oregon Statewide Lambert NAD 1983

Projected data have had a map projection applied to them, converting the GCS lat-long values to feet or meters in a planar coordinate system. The map units become meters or feet. Typically the x-y values in a projected coordinate system encompass millions or billions of meters. The feature class coordinate system definition includes the original GCS and datum plus a description of the applied projection and its parameters, such as the central meridian, reference latitude, standard parallels, and false easting and northing.

The coordinate system is part of a larger description called the **spatial reference**, which includes the coordinate system, the X/Y domain, and the resolution. The **X/Y domain** is the range of allowable x-y values that can be stored in a feature class. Storing a world map in degrees requires a small domain since degree values range at most between –180 and +180. Storing a world map in feet requires a large domain because there are more than 115 million feet in the circumference of the earth. The **resolution** represents the underlying accuracy of the values; a resolution of 0.001 meters means that coordinate values are stored to the nearest thousandth of a meter.

The **extent** of the data layer is the range of x-y coordinates of the features actually in the feature class. You can view the extent of a data layer in ArcMap, and this information can help you decide if the layer is projected or unprojected. Values between +180 and –180 indicate an unprojected GCS coordinate system with map units in degrees, whereas large values indicate a projected coordinate system with units of feet or meters. Figure 11.15 shows the extent and coordinate system label viewed in ArcMap. Notice the large x-y values in the extent.

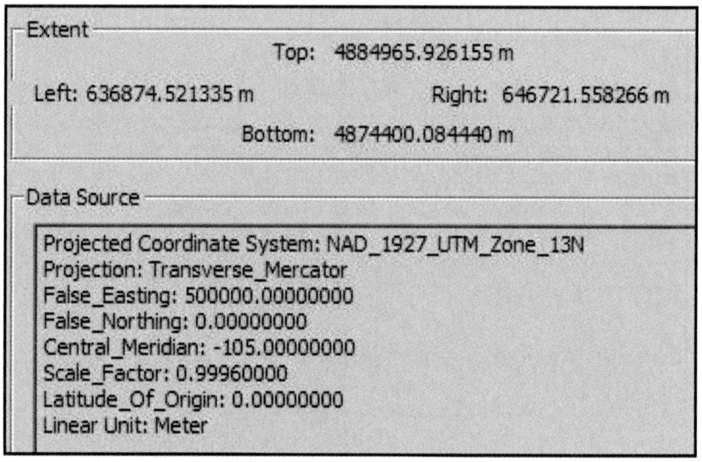

```
Extent
                         Top:   4884965.926155 m
  Left: 636874.521335 m                     Right:  646721.558266 m
                      Bottom:  4874400.084440 m

Data Source
  Projected Coordinate System: NAD_1927_UTM_Zone_13N
  Projection: Transverse_Mercator
  False_Easting: 500000.00000000
  False_Northing: 0.00000000
  Central_Meridian: -105.00000000
  Scale_Factor: 0.99960000
  Latitude_Of_Origin: 0.00000000
  Linear Unit: Meter
```

Fig. 11.15. Coordinate system extent and label

Missing coordinate system labels

Occasionally a user will encounter a GIS feature class or raster without a coordinate system label, and the coordinate system appears as "Unknown" when viewed in ArcMap or ArcCatalog. To use the data effectively, the user must figure out the real coordinate system of the data and create a label for it. Finding the coordinate system can be a challenge because it usually cannot be determined by looking at the data set itself. The user must find documentation supplied by the data provider, such as a Web page or document at the site where the data were downloaded or a separate text or document supplied with the data. The user might have to contact the data supplier in person to determine this information. As a last resort, the user might have to guess one of the more commonly used coordinate systems and test whether it is correct.

Once the coordinate system is known, the user may proceed to create the coordinate system label by using the **Define Projection tool** in ArcToolbox or by using the XY Coordinate System tab from the feature class properties in ArcCatalog. The Define Projection tool labels a data set with a coordinate system and allows the user to specify all the coordinate system properties, including the GCS, datum, central meridian, and so on. Once defined, the data set is ready for use in ArcMap. The same tool is used to define a label for rasters.

On-the-fly reprojection

A data frame may be assigned any desired coordinate system, and then ArcMap will reproject all data in the frame to match. Rasters can also be projected on the fly, although this can take extra time because of the greater complexity and computational requirements. In order to project data to match the frame, ArcMap must know the coordinate system of the data. If the data set coordinate system definition is Unknown, then ArcMap will issue a warning message, and it will display the data with whatever coordinates it has. Sometimes it will fortuitously match the other data in the frame, but other times it will appear totally out of place. If it matches, then you will know that the Unknown layer has the same, or at least a very similar, coordinate system to the data frame, and you can create a coordinate system label for it.

The importance of having a correct coordinate system label on a data set cannot be overstated. Imagine that someone gave you the number 37 and asked you to convert it to meters. You would naturally want to know the units of the number 37 because otherwise you would not be able to choose the correct equation. If the units were millimeters, the conversion would be 37 mm × 0.001 m/mm = 0.037 m. If the units were feet, the conversion would be 37 in × 0.3048 m/ft = 11.3 m. Likewise, ArcMap needs to know the coordinate system of a data set in order to choose the appropriate equations to project it to match the data frame. If you give a feature class the wrong coordinate system label, then ArcMap will choose the wrong equations, and the features will end up in the wrong place.

ArcMap will also adjust differences in datum with on-the-fly projection, but the task of converting datums is more complex. A datum transformation cannot be performed using a set of exact equations but instead relies on various fitting techniques. The accuracy of a particular transformation can range from centimeters to meters. When ArcMap encounters a datum that does not match the datum of the data frame, it converts it using a default conversion method. Because the transformation is not exact and may introduce positional inaccuracies, ArcMap issues a warning to the user when it needs to convert a datum. The user must then decide whether to proceed with the default transformation.

The impact of the transformation on the feature class coordinates depends on the positional accuracy of the data set. If it comes from a state or world map, the data set accuracy will range from hundreds to thousands of meters, and a shift of a few meters due to the datum transfer will not be noticeable. If the data are highly accurate surveying results, then shifting locations during a transformation becomes a potential problem. In most cases a user can safely accept the default on-the-fly transformation but should be aware of the potential positional errors that can result.

Projecting data

Usually GIS projects choose a single coordinate system and convert all data to match it. Many organizations establish an official projection for their area of interest and use it consistently. Outside data from different coordinate systems are always converted to the official one at the outset, using the Project tool.

The Project tool is found under the Data Management section of the Toolbox (Fig. 11.16). It converts the x-y coordinate values in a feature class to a different coordinate system and saves them in a new feature class with a new coordinate system label. The original data remain unchanged. The original data set must have a correctly labeled coordinate system before it is projected. If the label is Unknown, the Project tool cannot work. The Project Raster tool is used to project rasters.

Fig. 11.16. The Projection tools are found in ArcToolbox.

Troubleshooting coordinate system problems

A coordinate system problem occurs when two map layers represent the same region and should map on top of each other—but don't. They might be off by a few hundred meters or by thousands of miles. Most of these problems result from the mislabeling of a data set with the wrong coordinate system. The usual remedy is to determine the correct coordinate system and correct the label.

ArcMap can be used to troubleshoot coordinate system problems. In Figure 11.17, the user is examining the properties of a feature class using the Source tab. The extent shows x-y values in the range −103.375 to −103.250 for x and 44.000 to 44.125 for y. Recalling what you know of coordinate systems, these values should be degrees. However, the coordinate system label shown underneath the extent lists NAD 1927 UTM Zone 13N as the coordinate system label, and the x-y values are listed with meters as the map units. Something is wrong here—the label does not match the x-y extent values. So which is correct?

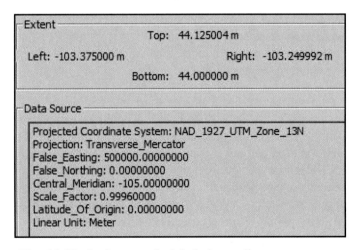

Fig. 11.17. An incorrectly labeled coordinate system

The *x-y* values are always correct because they are obtained from the values stored in the file. This data set really is stored in degrees. Therefore, the label must be incorrect and needs to be changed. Imagine adding this feature class to a data frame set to the UTM coordinate system. ArcMap reads the label and concludes that the *x-y* values are already in UTM meters and don't need to be projected on the fly, even though they are in degrees and do need projecting. So the degrees are not projected to meters, and the data set does not appear with the other feature classes.

To fix the problem, you must know the correct coordinate system and use Define Projection to modify the label. You will need another data set in the same area that is known to be correct so that you can compare the result and test the new label. If the data set aligns with the known one, you set the label correctly. If it is still misaligned, then a different label is needed.

Troubleshooting Steps

➔ Add a data set known to be correct to ArcMap first to set the data frame coordinate system. Record the coordinate system of the reference data.

➔ Add the data set causing the problems. Examine its coordinate system and extent and make sure that the label and extent appear to match. If they don't, or if the data set has an Unknown or Undefined coordinate system, you must try to figure out what it is.

➔ Examine the extent and consider what you know about the source of the data. If the coordinates are in degrees, they are in a GCS. If not, make your best guess as to the coordinate system and datum using any clues or resources you have available.

➔ Use the Define Projection tool to assign your guess to the problem data set. Rasters may need to be removed and added again afterwards to see the difference.

➔ If your guess is correct, the data sets should align. If not, try a different guess.

Data in a GCS are relatively easy to correct because the choice of possible datums is limited. Projected data are virtually impossible to guess unless they are in one of the common projections or you have some outside information to narrow down the possibilities.

Define Projection versus Project

Why would a data set be mislabeled with the wrong coordinate system? The usual culprit is a user who does not understand coordinate systems and the tools used to manage them. Many beginners make the mistake of using Define Projection when they should use the Project tool.

Imagine Frank has a shapefile of cities with GCS coordinates in latitude and longitude and wants to convert it to a UTM projection to match the rest of his data. He carelessly selects the Define Projection tool from ArcToolbox and sets the coordinate system to UTM. However, the tool merely puts a UTM label on the shapefile without actually changing the coordinates inside. So now the shapefile is incorrectly labeled UTM while the coordinates inside are in a GCS. (This situation is analogous to placing an albacore tuna label on a can of cat food, with disastrous results to your dinner recipe.) When the data are added to ArcMap, it will choose the wrong equations to convert the data to the data frame of the coordinate system, and the feature class will no longer appear where it should.

WARNING: Many people get confused about the functions of the Define Projection tool and the Project tool. This confusion can ruin data sets, so **pay attention** to the next two paragraphs and then be careful to select the correct tool for the task at hand.

The **Project tool** acts on the *x-y* coordinates of a layer and converts them to a different coordinate system, producing a new feature class and leaving the original feature class unchanged. The new file has a new coordinate system and a different coordinate system label. You use the Project tool if you have a layer in one coordinate system and want to permanently convert it to a different one. The Project tool should be used only on layers with correct coordinate system definitions.

The **Define Projection** tool only changes the coordinate system label of the feature class, without affecting the coordinates inside. It should only be used on a data set that has an Unknown coordinate system or on a data set that was previously mislabeled and needs to be fixed.

Summary

➢ All GIS data has a coordinate system, which defines the units and axes used to represent map features as *x-y* coordinates.

➢ A geographic coordinate system, or GCS, is based on spherical coordinates of latitude and longitude with units of degrees. A spheroid may be used with the GCS to represent the earth, which is not a perfect sphere.

➢ A datum includes a spheroid and its location relative to the earth's surface and is used to reduce map errors introduced by differences between the spheroid and the earth's actual surface. Common datums include the North American Datums of 1927 and 1983, which are based on different spheroids.

➢ Map projections are mathematical equations that convert degrees from a spherical GCS into planar *x-y* coordinates of meters and feet so that the map may be portrayed on a flat sheet of paper. All map projections introduce distortions of area, distance, direction, and shape.

➢ Rasters also have coordinate systems. Assigning coordinate systems to rasters may involve creating world files for them or georeferencing them based on control points and a rectification process.

➢ Rectifying or projecting rasters involves resampling the cells to a different cell grid using nearest neighbor, bilinear, or cubic convolution methods.

➢ UTM and State Plane are two of the most commonly used projection systems in the United States. Each consists of a family of projections designed to minimize distortion over the area covered.

➢ Every data set used in ArcGIS should have the correct coordinate system defined and stored with the map features. ArcGIS uses these definitions to help manage and display data correctly.

➢ Most coordinate system problems result from missing or incorrect labels on data sets. The Define Projection tool is used to update a data set with an undefined or incorrect label.

➢ Data may be permanently projected from one coordinate system to another using the Project tool or Project Raster tool in ArcToolbox.

TIP: The ArcGIS online documentation includes a good reference book on map projections and coordinate systems, titled *Understanding Map Projections*. The ArcMap Help section also has some good material about projections.

TIP: Take care to understand the difference between the Define Projection tool and the Project tool and be careful to always choose the right one for the task.

Chapter Review Questions

1. If a data set's features have x coordinates between -180 and $+180$, what is the coordinate system likely to be? In what units are the coordinates?

2. What are the x-y coordinates of a map's origin? _____ What is the x coordinate along the central meridian?_____

3. What is the difference between a spheroid and a geoid?

4. Examine Figures 11.5 through 11.7 and then explain why conic projections usually conserve area and distance but cylindrical projections typically preserve direction.

5. What extra step is performed when projecting rasters that is not needed when projecting vector data? What happens during this step?

6. What is the difference between a central meridian and the Prime Meridian?

7. You have a shapefile with an Unknown coordinate system, but a file on the Web site says that the coordinate system is UTM Zone 13 NAD 1983. What is your next step?

8. True or False: A shapefile of the United States with a GCS coordinate system would have an x-y extent that contains entirely positive values. _____ Explain your answer.

9. You have a shapefile with a UTM Zone 13 NAD 1983 coordinate system, and you want to bring it into your city database, which uses the Oregon Statewide Lambert coordinate system. What is your next step?

10. Explain why an X/Y Domain of -1070 to $+1070$ would be inappropriate for a data set with a UTM coordinate system.

Mastering the Skills

Teaching Tutorial

The following examples provide step-by-step instructions for doing basic tasks and solving basic problems in ArcGIS. The steps you need to do are highlighted with an arrow ➜; follow them carefully. Click on the video number in the VideoIndex to view a demonstration of the steps.

Displaying coordinate systems

➜ Open ArcMap and choose to start with a new, empty map.

➜ Use Save As to save the document with the name of your choice.

1➜ Use the Add Data button to add the country and latlong shapefiles from the mgisdata\World directory.

1➜ Change the data frame name from Layers to World.

1➜ Right-click the country layer, choose Properties, and then click the Source tab.

1. What is the coordinate system for this feature class? _____

Observe the values in the top of the window, which show the extent of *x-y* values present in the shapefile. The extent is the range of *x-y* values of features *actually* in a data set. Compare this to the domain, which is the range of *x-y* values that *could be stored* in a data set.

2➜ Close the Layer Properties window.

2➜ Observe the coordinates shown in the lower part of the window and confirm that they are in degrees. The data frame units defaulted to the units of the first data set loaded.

2➜ Zoom in to the tip of Florida in the United States.

2➜ Use the cursor to hover over the most southeastern tip of Florida and observe the *x-y* coordinates (Fig. 11.18).

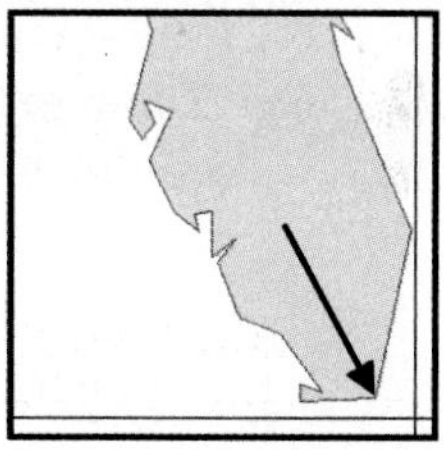

Fig. 11.18. Hover here

2. What are the coordinates of the SE tip of Florida? _____

3➜ Right-click the World data frame name and choose Properties.

3➜ Click the General tab.

3. What are the map units of this frame? _____What are the display units? _____

The map units are dimmed because you can't set them—they are based on the current coordinate system of the data frame. The display units appear in the location display in ArcMap and can be set to any units desired by the user. They defaulted to the data frame map units.

4➜ Set the display units to miles and click OK to close the Properties sheet.

4. What are the coordinates for Florida's tip now? _____

The *x* value tells you the distance Florida lies from the central meridian of the map (in this case the Prime Meridian that runs through Greenwich), and the *y* value tells you its distance from the reference latitude (the equator).

5➔ Click the Full Extent button to see the whole world again.

5➔ Open the World data frame properties.

5➔ Click the Coordinate System tab and notice what it says.

The data frame coordinate system defaulted to match the first data layer added to it. However, the data frame can display maps in any coordinate system, either geographic or projected.

5➔ In the folder tree, expand the Predefined folder by clicking on its plus sign.

5➔ Navigate through the folders to Projected Coordinate Systems > World > Mercator.

5➔ Click on the Mercator (world) projection and click OK.

> **TIP:** As you work with the *data frame* coordinate systems in ArcMap, keep in mind that they can be different from the coordinate systems of the *feature classes* you are working with.

6➔ Zoom in to the tip of Florida again.

5. What are the coordinates and units of the SE corner of Florida now? _____
   (Notice that the display units, previously miles, were reset to the map units when you applied the projection.)

   6➔ Zoom to the previous extent and use the cursor and coordinate box to locate the coordinate origin, where both *x* and *y* equal zero (approximately—look for where the coordinates change from positive to negative).

Mercator is a cylindrical projection. In this example, the central meridian of the map where *x* = 0 happens to correspond with the Prime Meridian that runs through Greenwich, England. A map of North America would have a different central meridian at about –100 degrees longitude. The *y* = 0 line is the equator and is called the latitude of origin. The equator is often the latitude or origin for a projection, but it need not be. Coordinates west of the central meridian and south of the reference latitude will have negative values. Because Florida is west of the central meridian of this projection, its *x* coordinate is negative.

6. Which continent would have primarily negative *x* AND negative *y* coordinates in this projection? _____ Which one would have primarily positive *x* and *y* coordinates? _____

   7➔ Use the Insert menu to add a new data frame to the map. Name it USA.

   7➔ Add the states feature class from the mgisdata\Usa\usdata geodatabase.

   7➔ Zoom in to the contiguous 48 states.

This feature class is stored in a projected coordinate system, North America Equidistant Conic, and the data frame defaults to it when the feature class is loaded.

> 8➔ Open the Properties for the USA data frame and click the Coordinate System tab. Read the properties of the current coordinate system.

7. What longitude is the central meridian of this projection? _____ What latitude is the latitude of origin (or reference latitude)? _____

The projection has several other properties. False easting and false northing are values applied to translate the coordinates in the x-y direction to ensure that all values are positive. This map has no easting or northing applied. The standard parallels are lines where the paper is tangent to the globe and the projection is free of distortion.

8. Is the Equidistant Conic projection shown here a tangent or secant projection? _____ How can you tell? _____

Changing the projection parameters changes the appearance of the map. Let's try an example.

> 9➔ Close the USA Data Frame Properties window when finished.
> 9➔ Right-click the World frame and choose Activate from the menu.
> 9➔ Turn off the latlong layer.
> 9➔ Open the Coordinate System tab in the World data frame Properties sheet.
> 9➔ Note the current coordinate system and then click the Modify button.
> 9➔ Change the central meridian from 0 to 100 degrees. Click OK in both menus.

Notice how the map shifts to put Asia in the center and splits North America in two. The center of the map is now at 100 degrees instead of 0.

ArcMap can also apply datum transformations when it encounters layers that are in a different datum than the data frame. Datum transformations are not exact and may shift the coordinates by a few centimeters to a few meters. Therefore, ArcMap warns the user when a datum shift is being applied.

> 10➔ Zoom in to the countries better, eliminating the huge portrayal of Antarctica (an example of area distortion).
> 10➔ Reset the central meridian to zero.
> 10➔ Add the states feature class from the mgisdata\Usa\usdata geodatabase.
> 10➔ Read the warning carefully (Fig. 11.19).

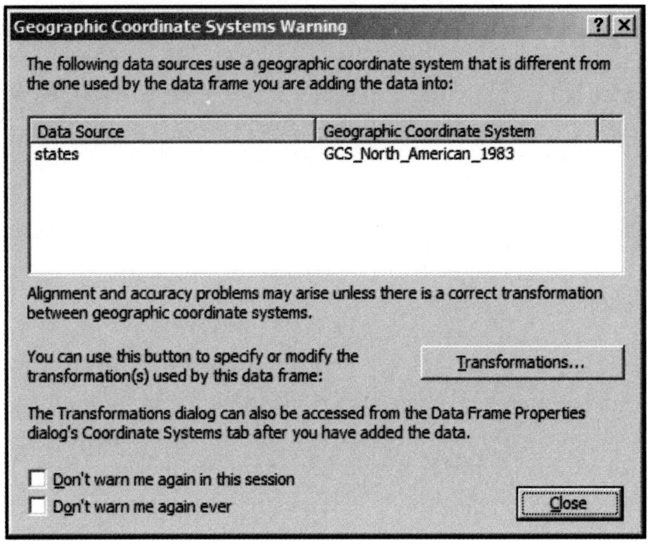

Fig. 11.19. Datum warning

10➔ Click Close.

Both the states and countries feature classes are in a GCS, but the countries are based on the WGS 1984 datum, and the states are based on the NAD 1983 datum. ArcMap uses a default transformation to convert the states from NAD83 to WGS84. World scale data have large geometric uncertainties and are not significantly affected by a shift of several meters, so the user can accept the default. If the data frame contained high-accuracy survey data, an experienced user might wish to choose a different transformation.

Understanding map distortion

The circumference of the world at the equator is about 25,000 miles. The circumference at the poles is zero. Latitudes in between have intermediate lengths.

11➔ Set the World frame coordinate system to GCS_WGS_1984 (look in Predefined > Geographic Coordinate Systems > World).

11➔ Check the box in the Warning window so that it won't notify you again this session.

11➔ Turn the latlong layer back on.

11➔ Zoom in to the United States, noting its elongation in the east-west direction, and then return to the previous extent.

When geographic coordinates are displayed on the screen, they are mapped as though they were a planar *x-y* coordinate system from $x = -180$ to $+180$ and $y = -90$ to $+90$. Although the circumference of the earth shrinks to zero at the poles, it is shown here with the same length as the equator. Any map projection that does not show shortening of the parallels with increasing latitude distorts distance and also area.

12➔ Right-click the states layer in the World data frame and remove it.

12➔ Change the World frame coordinate system back to Mercator.

12➔ Add the circles shapefile from the mgisdata\World folder.

12➔ Drag the latlong layer below the country layer to see the outlines better (switch to the display tab if necessary to rearrange the layer order).

The circles each have radius of 5 degrees and would appear the same size on a globe, but the distortion in this Mercator projection causes them to increase in size toward the poles. Also notice that the latitude lines become further apart toward the poles. In fact, the north and south poles are an infinite distance from the equator in this projection. (Examine the cylindrical projection in Figure 11.5 to figure out why.) Area and distance are distorted in this projection, except at the equator where the sphere touches the paper. However, Mercator does preserve direction and shape. The circles remain circles, and the longitude lines point north.

13➔ Change the coordinate system of the World frame to Sinusoidal (world). (Look in Predefined > Projected > World.)

Sinusoidal is an equal-area projection, meaning that it preserves area and, to some extent, distance. However, it does so at the expense of shape and direction. Notice the squashed look of the United States and that the longitude lines converge toward the poles. The circles have equivalent areas now, but some are no longer circles. The distortion gets worse the further you are from the equator and the central meridian.

14➜ Change the coordinate system of the frame to Robinson (world).

Robinson is called a compromise projection because it is designed to minimize distortion in all four properties but preserves none. Distortion is unavoidable when using small-scale world and national maps. There are excellent projection systems for large-scale maps. Two common systems in the United States are the Universal Transverse Mercator (UTM) system and the State Plane Coordinate System (SPCS).

15➜ Change the coordinate system of the data frame to Predefined > Projected > UTM > WGS84 > WGS 1984 Zone 35N. Zoom in to the map area shown.

16➜ Add the shapefile utmzone35 from the mgisdata\World folder. Make the symbol hollow with an orange outline.

16➜ Zoom in to display the line of circles running north-south across Africa.

The narrow strip of UTM Zone 35 runs along the tangent of an oblique cylindrical projection. Observe that the circles remain circles and that the areas remain constant except in Antarctica. All four properties of area, distance, shape, and direction are preserved within the zone.

17➜ Activate the USA frame.

17➜ Add the shapefile utmzone.shp from the mgisdata\World folder.

17➜ Click the utmzone symbol to change it to a hollow shade with an orange outline.

17➜ Use the Labels tab in the utmzone Properties to turn on the labels and display the ZONE field.

9. Which UTM zone should be used for the state of Nevada? _____

18➜ Turn off the utmzone layer.

18➜ Add the feature class spcszones from the mgisdata\Usa\usdata geodatabase.

18➜ Label the zones using the ZONENAME83 field.

The State Plane coordinate system divides states into one or more zones with its own coordinate system. North-south zones use a Transverse Mercator coordinate system with a central meridian customized for that zone. East-west zones can be more accurately represented in a conic coordinate system that preserves properties along parallels rather than along meridians, and they use the Lambert Conformal Conic projection.

19➜ Open the data frame properties and click the coordinate system tab.

19➜ Select the Predefined > Projected > State Plane > NAD 1983 > StatePlane New York East coordinate system. Don't click Apply or OK, but just read the projection parameters in the upper part of the window.

19➜ Click the StatePlane Texas Central coordinate system and read its parameters.

19➜ Close the Data Frame properties window.

Working with map projections

ArcMap relies on correct coordinate system labels to perform on-the-fly projection.

20➔ Click the New Map File button to start a new empty map. You don't need to save your changes.

20➔ Add the usfsrds feature class from mgisdata\BlackHills\Sturgis.mdb.

20➔ Open the Properties for usfsrds, click the Source tab, and write the coordinate system and datum of the layer here: _____

20➔ Also examine the Extent information, which shows the range of *x-y* coordinates as stored in the file. Close the Properties window.

20➔ Add the majrds feature class from mgisdata\Sdakota\southdakota.mdb and write its coordinate system here: _____ Close the window.

Even though the two coordinate systems are different, the maps overlay each other correctly because each has an accurate label. There are some slight offsets, but these are to be expected for two sets of roads derived from different scale maps (originally 1:24,000 and 1:250,000).

10. State the data frame coordinate system WITHOUT looking at its properties. _____ Why could you know this information without looking?

21➔ Add the precip_ann shapefile from the mgisdata\Sdakota folder. Read the warning carefully.

21➔ Open the Properties for the precip_ann layer and check the Source tab.

This data set has an Undefined coordinate system. Before ArcMap can project these data on the fly, an appropriate coordinate system must be assigned, which means you must figure out what is really stored in the file. Examine the extent range and note that the *x-y* values fall in the range $x =$ –180 to +180 and $y =$ –90 to +90, indicating that it has a geographic coordinate system with units of degrees. The datum is still unknown, but you can guess and then check to see if the guess worked by comparing it with another data set.

22➔ Close the window. Open ArcToolbox by clicking on its button on the main toolbar.

22➔ Locate the Define Projection tool in the Data Management > Projections and Transformations toolset. Double-click the tool to open it.

22➔ Click and drag the precip_ann layer into the Input Dataset box on the tool.

22➔ Click on the button to specify the coordinate system and click Select.

22➔ Navigate to Geographic Coordinate Systems > North America > North American Datum 1983, select it, and click Add and OK. Click OK to start the tool.

Note that you have changed only the LABEL for the precip_ann coordinate system. That's what Define Projection does, and it should only be used for a feature class with an unknown or incorrect coordinate system label, as in this case.

23➔ Add the state feature class from mgisdata\Sdakota\southdakota.mdb and give it a hollow symbol with a contrasting outline color.

23➔ Zoom to the full extent, then Zoom in to the boundaries and compare the state and precip_ann layers. When finished, return to the full extent of the state.

No major difference appears, so this choice of datum appears correct. Notice the upward tilt of the eastern part of the state, which occurs because the data frame is still set to NAD 1927 UTM Zone 13. This zone is only accurate for the western third of the state, and the direction distortion for the eastern part is clearly visible.

Projecting data

We wish to place the precip_ann shapefile in the southdakota geodatabase, and we want its coordinate system to match the other feature classes. We use the Project tool, which creates a new feature class with recalculated *x-y* coordinates.

24➔ In ArcToolbox, locate the Project tool in the Data Management > Projections and Transformations > Feature toolset. *Do not select the Define Projection tool by mistake.*

24➔ Click and drag the precip_ann layer to the Input Dataset box on the tool.

24➔ Click the Browse button for the Output Dataset. Navigate to the southdakota geodatabase and name the new feature class **annprecip**.

24➔ Click the button to specify the Output Coordinate System. Click Import and navigate to southdakota.mdb.

24➔ Choose city as the feature class from which to import the coordinate system (to make sure our new file matches the others in the geodatabase). Click Add and verify that the window reads NAD 1983 State Plane South Dakota South FIPS 4002. Click OK.

24➔ No transformation is needed because the datum is the same for both input and output, that is, NAD 1983. Click OK to start projecting.

Now imagine that we also want to put the usfsrds feature class into the southdakota geodatabase. In this case we must also specify a datum transformation because the roads are in NAD 1927.

25➔ Open the Project tool again. Drag the usfsrds layer to the Input Dataset box (Fig. 11.20).

25➔ Click the Browse icon for the Output Dataset, navigate to the southdakota geodatabase, and enter the name of the new feature class as **sturgisrds**. Click Save.

25➔ Click the button next to the Output Coordinate System box. The Spatial Reference Properties menu appears. Click Import and select city from the southdakota geodatabase. Click Add and then OK.

25➔ Notice that a green dot has appeared next to the Geographic Transformation box, indicating that you need to enter information.

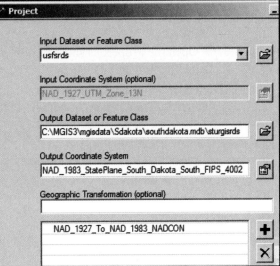

Fig. 11.20. The Project tool

411

25➜ Click the drop-down box and choose NAD_1927_to_NAD_1983_NADCON for the Geographic Transformation. This is a reliable general transformation for NAD 1927 to NAD 1983.

25➜ Click OK to start projecting.

Troubleshooting map projections

In an ideal world, every data set would be perfectly documented with the correct coordinate system and datum as well as all the other metadata information. Unfortunately, we live in a less-than-perfect world and must deal with human error and inefficiency. In this world, it is not uncommon to obtain a data set and be less than certain about its coordinate system.

Imagine that you are a new employee of the Black Hills National Forest, charged with building a GIS for a section of the forest. Your predecessor, Bob, left a directory of data sets with less than ideal documentation, you discover.

26➜ Click the New Map File button to open an empty map document in ArcMap. You don't need to save your changes to the current map document.

26➜ Click Add Data and add the feature class usfsrds from the BlackHills\Sturgis geodatabase.

26➜ Also add the raster file L7_Aug20_W84.bil from the mgisdata\BlackHills\rasters folder, a Landsat 7 satellite image. You should receive a warning that the coordinate system is Undefined.

26➜ Right-click the usfsrds symbol and choose a nice contrasting color that shows up well against the image.

26➜ Zoom in closer and examine the match between the roads in the feature class and the roads in the image.

The roads are offset to the south by about 300 meters. Because the two data sets are this close, you suspect that this is a datum mismatch rather than two different coordinate systems. Now the detective work starts. Begin by investigating the coordinate systems of the two layers.

27➜ Open the Properties sheet for the usfsrds layer and examine the Source tab. Write the projection here: _____ Close the box.

27➜ Open the Properties sheet for the Landsat image, scroll down the information in its Source tab, and locate the spatial reference information.

The Landsat image coordinate system has not been defined by the person who created the file. This does not mean that it has no coordinate system—every spatial data set has an intrinsic coordinate system defined by the units used to store its *x-y* coordinates. It means that the label is missing. You need to determine the coordinate system and create the label.

Since ArcMap cannot reproject a data set with an undefined coordinate system, it shows the raw *x-y* values in the file. The fact that the raw values nearly coincide with the roads suggests that they share the same projection, UTM Zone 13. The fact that they are offset by a few hundred meters suggests that the datum is different. Moreover, the "W84" in the image name suggests that perhaps this image is based on the WGS84 datum. You'll proceed on this assumption. To check your guess, define the image coordinate system as WGS84.

28➔ Close the Properties window. Open the Data Management Tools > Projections and Transformations > Define Projection tool.

11. Why are we using Define Projection in this case and not Project?

_____

28➔ Click and drag the L7_Aug20_W84.bil layer to the Input Dataset box.

28➔ Click the button to set the coordinate system. Click Select and locate Projected Coordinate Systems > UTM > WGS 1984 > WGS 1984 UTM Zone 13N.prj. Select it and click Add. Click OK and OK.

29➔ Remove the L7_Aug20_W84 image and add it again.

> **TIP:** The initial release of ArcGIS version 9.3 had a bug and did not actually change the coordinate system for rasters. By the time you try it, hopefully it will be fixed. If the roads don't look any different, just go on.

The layers match much better now. Once the coordinate systems of both data layers are correct, ArcMap is able to correctly project each layer to the data frame coordinate system. The small offsets still present are probably due to slight errors in digitizing the roads or errors involved in georeferencing the satellite imagery. Data sets from different sources rarely match perfectly.

Creating a new coordinate system

Sometimes you don't have a predefined coordinate system for the area you are working and need to define your own. Imagine creating a geodatabase for Turkey to display the general features with no need for equal areas or equal distances. A search of the available coordinate systems yields no coordinate systems specific to Turkey, so begin by choosing a suitable projection.

30➔ Click on the New Map File document. You don't need to save changes.

30➔ Add the country shapefile from the mgisdata\World folder.

30➔ Use Select by Attributes to select Turkey and create a layer from the selected features. Name it **Turkey**.

31➔ Zoom in and find the approximate central longitude and range of latitudes of Turkey using the cursor and reading the values from the location display.

Central longitude _____ Southern latitude _____ Northern latitude _____

Turkey is oriented east-west and too large for a single UTM zone. A conic projection such as Lambert Conformal Conic would probably be best. The central meridian should go in the center of Turkey, and the parallels should cut it into approximate thirds.

32➔ Open the data frame properties coordinate system tab and choose New > Projected Coordinate System (Fig. 11.21).

32➔ Enter **Turkey_Lambert_Conformal_Conic_WGS84** as the name.

32➔ Choose Lambert_Conformal_Conic from the drop-down list for the projection name.

32➔ Carefully change the Central_Meridian to **35** degrees.

32➔ Set Standard_Parallel_1 to **37.5** degrees. Set Standard Parallel_2 to **39.5** degrees.

32➔ Choose Select to set the Geographic Coordinate System.

32➔ Choose World > WGS 1984. Click Add and Finish and OK.

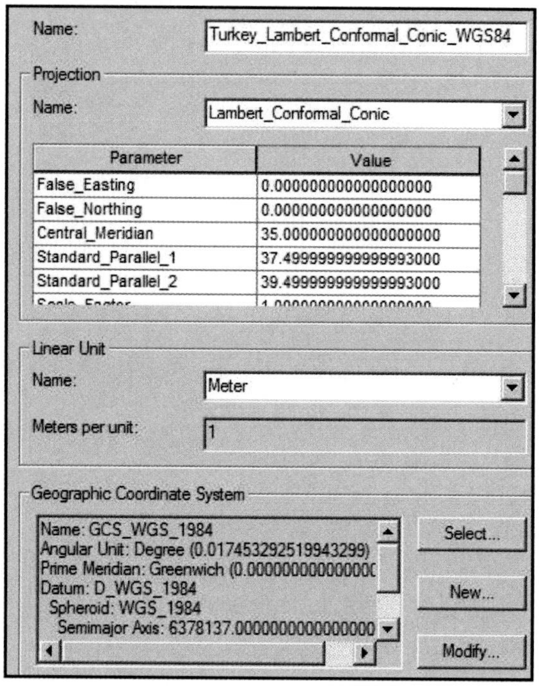

This choice appears to work well. There are no obvious distortions, and it should serve nicely as a regional display map for Turkey. It would be nice to have all positive map coordinates.

33➔ Use the cursor to examine the *x* values at the western edge of Turkey. They are about −800,000. The false easting should be larger than this value.

33➔ Open the data frame properties again to the Coordinate System tab.

Fig. 11.21. Creating a new coordinate system

33➔ Click the Modify button. Set the False Easting to **1000000** (1 million) meters and click OK and OK.

33➔ Examine the *x-y* values again and note that the *x* values are all positive for Turkey.

34➔ Export the Turkey layer to a shapefile named **Turkey.shp** in the mgisdata\World folder. Choose to use the same coordinate system as the data frame. Add the result to the map.

You would now use this shapefile as the template for assigning all your other feature classes to the Turkey Lambert Conformal Conic WGS84 coordinate system.

TIP: Use the New Coordinate System window to define the coordinate system for a data set that is Undefined or Unknown but has the projection information recorded in a separate location, such as a Web page or readme file.

Georeferencing a raster

In this section, you will download an image from the Internet and georeference it. If you don't have Internet access or if you have trouble obtaining an image, use the nwsradar_sample.gif image stored in the mgisdata\Usa folder. Your Web browser instructions might be slightly different if you are not using the Internet Explorer Web browser.

35➜ Open your Web browser and go to http://www.nws.noaa.gov/radar_tab.php.

35➜ Click on Full resolution version.

When georeferencing, you want to use a coordinate system that is as close as possible to the coordinate system of the map in the image. Examine the state outlines and try to determine what kind of coordinate system was used to portray this map. Use the inside front cover of the book to make comparisons, if you wish.

12. On what type of projection does the map appear to be based? _____

35➜ Right-click the enlarged image and choose Save Picture As. Navigate to your mgisdata\Usa folder and save the image as a GIF file. Name it **nwsradar.gif**.

36➜ Open a new map document and add the states feature class from the mgisdata\Usa\usdata geodatabase.

36➜ Examine the coordinate system of the states feature class. Write it here:

_____

To georeference the image, we need to set the data frame coordinate system to whatever system the image appears to be stored in, in this case, a GCS. NAD 1983 is the most common GCS for United States data, so we will assume it for the image.

37➜ Change the data frame coordinate system to Predefined > Geographic Coordinate Systems > North America > North American Datum 1983.

37➜ Zoom to the full extent to see the states again, if necessary.

37➜ Switch back to the browser for a moment and compare the shapes of the states in ArcMap and in the image. They appear similar. We will try to georeference using this GCS.

37➜ Close the browser window and return to ArcMap.

38➜ Zoom in to the conterminous states so that the image and the map window are showing a similar map extent.

38➜ Change the symbol for the states so that it is hollow with a high-contrast outline color, such as red or hot pink. It needs to show up well against the image.

38➜ Right-click the gray area of the menu bar and open the Georeferencing toolbar.

38➜ Click on Add Data and add the nwsradar.gif image to the map.

The image does not appear yet because it is in a different coordinate system from the map (pixels). We can examine it, if we wish.

39➜ Right-click the image name in the Table of Contents and choose Zoom to Layer.

39➜ Move the cursor over the image and watch the *x-y* location update. Notice the values. Although they are reported as degrees in the location display because that is what it was set to, the values are far too large to be degrees. They are in pixels.

39➜ When finished, click the Previous Extent button to go back to the states map.

Now we are ready to start georeferencing and adding control points. Begin by bringing the image into the same map extent as the states. This step will make it easier to match control points.

40➜ On the Georeferencing toolbar, choose Georeferencing > Fit to Display. Your ArcMap session should now look similar to Figure 11.22.

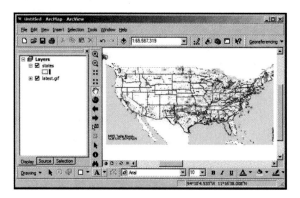

The points work best if they are spread around the map. We will first add two at diagonal corners and then two near the other corners. If the state corners or boundaries are obscured by weather formations on your map, choose another point that is close by and clearly visible.

Fig. 11.22. Ready to begin georeferencing

40➜ Locate the southwestern corner of California and *mentally* match the corner of the state outline on the image to the corner of the state outline on the map.

40➜ Use the Zoom In tool to zoom in to an area containing the two matching points.

40➜ Switch to the Add Control Points tool on the Georeferencing toolbar.

40➜ Carefully click the corner on the image first and then click the corresponding corner on the states map. This is your first control point.

TIP: The pairs of control points must always be added in the same order. Decide if you will click the image point first and then the map or vice versa. Then stick with your choice! For the rest of this exercise, we will always click the image first since we did the first one that way.

40➜ Notice that the map updated after you added the control point so that the corner of California is aligned.

41➜ Zoom to the previous extent, and then zoom in to the New England region.

41➜ Mentally locate a matching point in the northeast, such as the southeast corner of Maine.

41➜ Switch back to the Add Control Points tool.

41➜ Click the location on the image first and then click the corresponding location on the states map.

The match will already look very good, indicating that you did a good job of choosing the coordinate system. Continue switching between the Zoom and Control Point tools to add the next points, even if they are almost identical. You need at least three points for an affine transformation.

42➜Add a control point on the Florida peninsula. Click the image first!

42➜ Add a control point in the Pacific Northwest. Click the image first!

Examine the map and locate any areas where the state boundaries are still poorly aligned. These areas should receive additional control points. If you don't see any at the national scale, zoom in to a couple of locations in the middle of the country to check.

43➔ Zoom to the state corner between Idaho, Wyoming, and Utah. Add a point, if necessary (image first)!

43➔ Zoom to the western corner of Kentucky. Add a point, if necessary (image first)!

43➔ For many maps, you would continue to add more control points in misaligned areas until you no longer see any improvement when new points are added.

44➔ Click on the View Link Table button to see the points you've added.

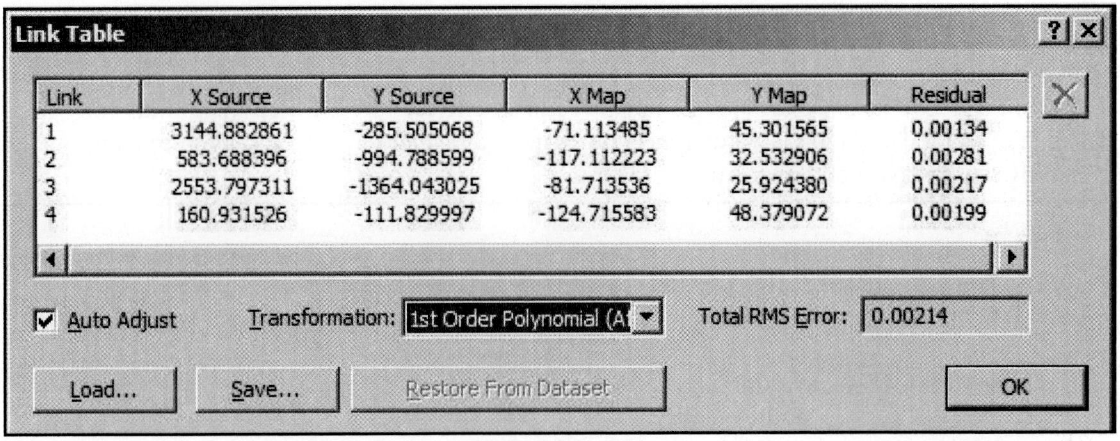

Link	X Source	Y Source	X Map	Y Map	Residual
1	3144.882861	-285.505068	-71.113485	45.301565	0.00134
2	583.688396	-994.788599	-117.112223	32.532906	0.00281
3	2553.797311	-1364.043025	-81.713536	25.924380	0.00217
4	160.931526	-111.829997	-124.715583	48.379072	0.00199

☑ Auto Adjust   Transformation: 1st Order Polynomial (A ▾)   Total RMS Error: 0.00214

Load...   Save...   Restore From Dataset   OK

Fig. 11.23. The Link Table shows the ground control points that have been added.

The points are listed in the order that you added them (Fig. 11.23), and yours will differ slightly from the list in the figure. The X and Y Source fields show the coordinates of each control point in the original image units (pixels). The X Map and Y Map columns show the point coordinates in map units (degrees). These points are being input to the transformation equation indicated in the Transformation box, currently the 1st Order Polynomial (Affine) Transformation.

Note the residual for each point. This value is the distance by which the points are still offset (since a perfect match between the pairs of points is impossible). The root mean square of all these residuals is reported as the RMS error. Both the residuals and RMS error are reported in map units. In Figure 11.23, the residuals range between 0.001 and 0.003 degrees, and the RMS error is about 0.002 degrees. At the equator that error would be about 220 meters, a reasonable value for a national map. Yours should be in about the same range if you entered the control points carefully and always in the same order.

44➔ Look for any points with a much larger residual than the others. You might have entered it incorrectly.

44➔ If you find one, highlight it and click the X button to remove it from the list.

44➔ Replace it with another, better point. You can temporarily close the Link Table to add the points and then open it again.

45➜ On the Link Table, click the Transformation button. The first-order transformation is the only one available right now because you don't have enough points for a second or third order.

TIP: If the RMS error of a map is large and does not improve when more points are added, the best strategy is to try a second-order transformation. This will often reduce the residuals and the RMS error. Third-order transformations tend to introduce more distortion than they fix and should be avoided unless necessary. Second- and third-order transformations require more control points, especially around the edges.

45➜ Examine the map and make sure that all areas are still well adjusted after removing a link. If not, zoom in to any misaligned areas and add another point.

45➜ Once you are happy with the registration of the image to the map, close the Link Table. If you achieve an RMS error of 0.002 degrees, then you have done well.

TIP: You can save these control points in a text file by clicking the Save button in the Link Table. You can Load them to georeference the same image later. However, georeferencing generally works best if completed in one session.

The final step is to permanently save the image in the new coordinate system.

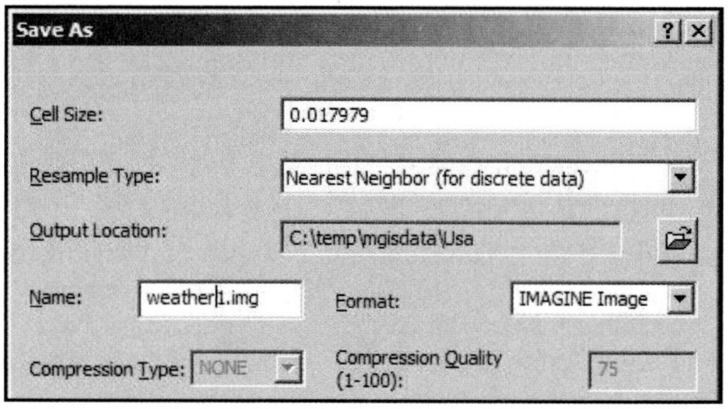

46➜ Choose Rectify from the Geoprocessing toolbar (Fig. 11.24).

46➜ The suggested cell size is usually a good choice. If this were to be the final image, you might choose to round it off. However, you plan to project this image when you are done, so leave it to the calculated value for minimal resampling.

Fig. 11.24. The Rectify window

46➜ This image has discrete color values, so choose the Nearest Neighbor resample type.

46➜ Change the output location of the image, if necessary.

46➜ Name the output **nwsradar1.img**, as it suggests, and keep the IMAGINE image format. Click Save.

46➜ Add the rectified image to the map. The colors may have changed slightly because of the rectification process (Fig. 11.25).

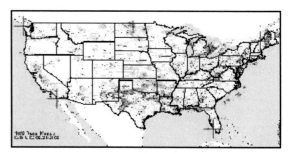

Fig. 11.25. Rectified radar map

Projecting a raster

We want to place the rectified image in the same coordinate system as our other United States data, so we must project it. First, however, estimate the expected resolution of the raster once it is projected. The cell size is 0.017979 degrees. One degree of latitude is about 111 km, so the cell size in meters should be about 0.017979 × 111,000 = 1995.7, or about 2000 meters.

47➔ Open ArcToolbox and find the Data Management > Projections and Transformations > Raster > Project Raster tool. Open it.

47➔ Set the Input raster to nwsradar1.img.

47➔ Name the output raster usradar and place it in the mgisdata\Usa\usdata geodatabase.

48➔ Click the button to set the output coordinate system. Click the Import button and choose one of the other feature classes in usdata to import from. This ensures that the new raster coordinate system matches the coordinate system of the other feature classes in the geodatabase. Click Add and OK, but don't start the tool yet!

48➔ No green dot appears by the Geographic Transformation, so no change of datum is required. Leave it blank.

48➔ Set the resampling technique to NEAREST since the image is discrete data.

48➔ You are pleased to see that the suggested cell size is close to your estimate. Round it to 2000 meters.

TIP: The optional Registration Point is used to ensure that a set of grids always starts at the same place so that the grids are perfectly aligned for analysis in Spatial Analyst.

48➔ Click OK to start projecting.

The new raster appears, but because the data frame coordinate system is still set to GCS NAD 1983, the raster is being reprojected on the fly back to the old system.

49➔ Click the New Map File button to clear the layers and reset the coordinate system of the data frame. Don't save the changes.

49➔ Add the usradar raster and the rivers feature class. The final product should look similar to Figure 11.26.

This is the end of the tutorial.

➔ Close ArcMap.

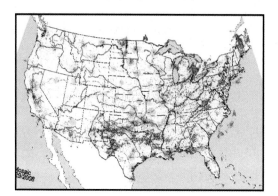

Fig. 11.26. The projected radar image

More skills

Consult the Skills Reference section of this chapter to learn to do the following:

➢ Defining a coordinate system using ArcCatalog

419

Exercises

1. Examine the coordinate systems for the files rc_roads.shp and geologywest.shp. For each one, answer the following questions: What is the name of the coordinate system? Is it projected or unprojected? What are the map units? What is the *x-y* extent?

2. The shapefile rcw_oops in the mgisdata\Rapidcity folder should overlay with the rc_roads shapefile in the same folder. Examine them both and then explain what is wrong with rcw_oops and why it won't overlay with rc_roads. How could you fix it?

3. Open the map document ex_11.mxd in the mgisdata\MapDocuments folder. What is the coordinate system of the data frame in this map? _____ What are the map units of the frame? _____ What are the display units of the frame? _____ What are the map units of the feature class coordinate systems? _____

4. Measure the distance in miles from Olympia, Washington, to Tallahassee, Florida. What is it? Change the data frame coordinate system to North America Lambert Conformal Conic and measure it again. Then measure it once more in Mercator projection centered at 103°W using NAD83. Discuss the differences in the measurements.

5. Choose the best coordinate system from the predefined coordinate systems in ArcGIS for maps of the following areas.

    Humboldt County, California

    Grafton County, New Hampshire

    State of Nevada

    England

    Antarctica, equal distances needed

6. You are tasked with creating a statewide map for Illinois. Choose a UTM or State Plane zone and modify it to best represent the entire state. Explain your approach and **Capture** the coordinate system description you created. (Hint: Don't modify the existing zone! Start by setting the data frame to the chosen zone. Then create a new coordinate system and enter the properties from the chosen zone, changing the properties as needed.)

7. You are tasked with creating a statewide map for Colorado. Choose a UTM or State Plane zone and modify it to best represent the entire state. Explain your approach and **Capture** the coordinate system description you created.

8. Find a map image on the Internet, save it as a JPEG, and georeference it. Display it with at least one other data set from the mgisdata folder.

Challenge Problem

Pick one state, besides South Dakota or Oregon, and choose the best coordinate system to use for it. Create a geodatabase for that state containing the same feature classes as the usdata geodatabase and in the projection you've chosen. (**Hint:** Use the Clip tool to crop linear features at the state boundary.) Find an image of the state on the Internet and include that in the geodatabase also, correctly referenced and projected to match the other feature classes.

Skills Reference

Defining coordinate systems of data sets 421

Projecting feature classes 422

Projecting rasters 423

Defining coordinate systems of data sets

Defining a coordinate system for a data set allows you to display it with other data. It is important to set the coordinate system to match the actual *x-y* coordinates contained in the file. This tool should only be used if the data set coordinate system is undefined or incorrectly defined.

Using the Define Projection tool

1. Open ArcToolbox, if necessary.

2. Choose the Define Projection tool from the section Data Management Tools > Projections and Transformations.

3. Click the Browse button and select the data to define (Fig. 11.27a).

4. Click the icon to open the Spatial Reference Properties window.

5. Use the XY Coordinate System tab (Fig. 11.27b) to select a coordinate system from the list, import one from another data layer, modify an existing one, or create a new one using the buttons in the window.

6. Click OK and OK.

Using ArcCatalog

1. Right-click the feature class and choose Properties from the menu.

2. Click the XY Coordinate System tab.

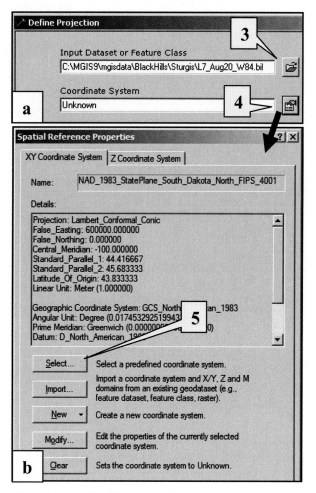

Fig. 11.27. The Define Projection tool

3. Select an existing coordinate system from the list provided, import one from another data layer, modify an existing one, or create a new one using the buttons in the menu as shown in Figure 11.27b.

TIP: If a layer is open in ArcMap, then a file lock is placed on it to prevent it from being modified. To define the coordinate system in ArcCatalog, the data set cannot be open in ArcMap.

Projecting feature classes

Projecting data changes both the coordinate system definition AND the *x-y* coordinate values of the features, creating a new file in the process and leaving the original unchanged.

1. Start ArcToolbox and choose the Project tool from the Data Management Tools > Projections and Transformations > Feature section (Fig. 11.28).

2. Click the first Browse button to select the data set to be projected.

3. Click the second Browse button to specify a folder, a geodatabase, or a feature dataset to contain the output and give the output shapefile or feature class a name. Click Save to enter the new name. Click Next.

4. Click the button to open the Spatial Reference Properties window to select a coordinate system to which to project the data, using Select, Import, or Modify as in Figure 11.27b.

5. A green dot will appear by the Geographic Transformation box if one is required to convert from one datum to another. Select a transformation from the drop-down box.

6. Click OK to project the data.

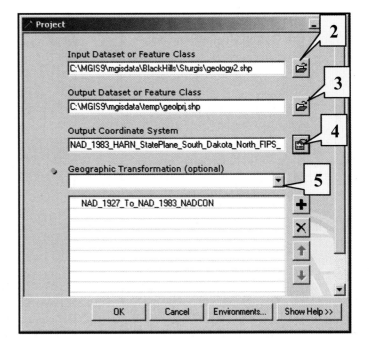

Fig. 11.28. The Project tool

TIP: Projecting coverages and grids requires an ArcEditor or ArcInfo license. If you only have an ArcView license, a red X button will appear by the Input Dataset box.

Projecting rasters

1. If converting from a GCS to a projected coordinate system in meters or feet, calculate the target output cell size in meters and convert to feet, if necessary. Round the result to an even number. You can also usually take the cell size suggested by the tool and round it.

   Cell size in degrees × 111.3 km/degree × 1000 = output cell size in meters

2. Find the Data Management Tools > Projections and Transformations > Rasters > Project Raster tool in ArcToolbox and open it.

3. Set the input raster to be projected.

4. Enter the name and location for the output raster.

5. Enter the output coordinate system by Selecting it or Importing it from another feature class or rasters.

6. If a green button appears by the Geographic Transformation box, a datum transformation is required. Choose one from the drop down list.

7. Set the resampling technique to NEAREST if the raster contains discrete or categorical data. Set it to BILINEAR if the raster contains continuous data.

8. Set the output cell size.

9. Click OK.

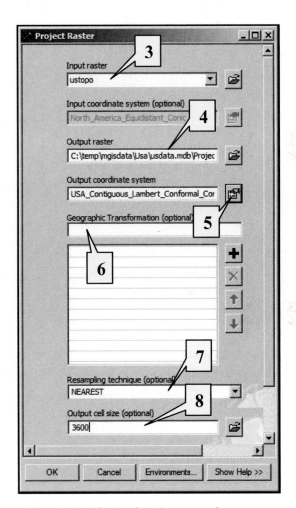

Fig. 11.29. The Project Raster tool

Chapter 12. Basic Editing

Objectives

➢ Understanding topological errors

➢ Using snapping to ensure topological integrity of features

➢ Adding features to map layers using basic editing functions

➢ Using the sketching tools and context menus to precisely position features

➢ Entering and editing attribute data

Mastering the Concepts

GIS Concepts

Editing can modify and update existing feature classes or to create entirely new ones. If a housing subdivision is added to a city, the new roads must be added to the city's roads feature class. Likewise, new parcels, sewer lines, and other infrastructure need to be added to the city database to ensure that it is up to date. Entirely new feature classes can be created, for example, if the city planning department decides to create a map of garbage collection zones where none had existed before.

When editing, care must be taken to create and maintain topological integrity between features. If two parcels share a common boundary, then the boundaries should match exactly. Line features that connect in the real world, such as streets or water lines, should connect in the feature class. Lines that cross each other should intersect at a node. Lines and polygon boundaries should not cross over themselves. These basic rules must be kept to ensure the logical consistency of features so that they properly represent the relationships of their real-world counterparts.

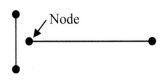

A dangle—the two lines fail to connect.

Topological data models permit the user to test, locate, and fix topological errors. However, topology is even more important when editing spaghetti models because the user must manage it without help. Basic editing includes two capabilities that aid in maintaining topological integrity, snapping and coincident boundary editing.

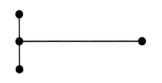

Correct topology—the horizontal line intersects the vertical one, creating three lines.

Snapping features

When creating line features that connect to each other, such as roads, one must take care that the features connect properly (Fig. 12.1). Lines that fail to connect, either by undershooting or overshooting, are called **dangles**. Although you may not be able to see the gap between the lines, the gap will exist unless the endpoints of each line (nodes) have exactly the same

Fig. 12.1. Topological relationships between lines

coordinate values. Even though the map may look correct, certain functions, such as tracing networks or locating intersections, will not work properly. It is impossible to intersect lines properly by simple digitizing.

Snapping ensures that the nodes of lines and the vertices of polygons match. When snapping is turned on, it affects features that are being added or modified. If you place the cursor within a specified distance of an existing node or vertex, then the new feature is automatically snapped to the existing one—that is, the coordinates are matched at that one point (Fig. 12.2).

Care must be taken in specifying the snap distance. If it is too small, then you will have difficulty making features snap. If it is too large, then you may find yourself constantly snapping to objects when it isn't needed. Three types of snapping can be set independently for each layer (Fig. 12.3).

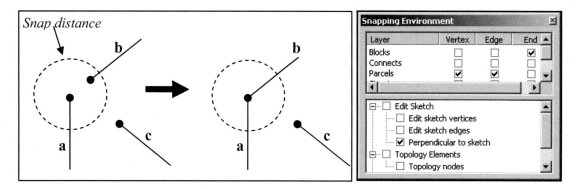

Fig. 12.2. Snapping. Line "b" snaps to "a" because it falls inside the snap distance. Line "c" remains unconnected.

Fig. 12.3. Setting the snapping environment

End snapping is the most restrictive, as it only allows a new line to be snapped to the endpoints of an existing line. End snapping can ensure that new streams connect to the ends of existing streams.

Vertex snapping allows the endpoints of the new line to be snapped to any vertex in an existing line. It can ensure that adjacent parcels connect only at existing corners.

Edge snapping constrains the feature being added to meet the edge of an existing feature. In this case, the vertex being added could be placed anywhere along an existing feature. Edge snapping can ensure that a street ends exactly on another street.

Figure 12.4 illustrates how snapping changes depending on which type of snap is set. The vertical line already exists in the layer, and the horizontal line is being added. The dashed circle shows the snap distance. If end snapping is turned on, then the new line will snap to the endpoint of the existing line (Fig. 12.4a). If

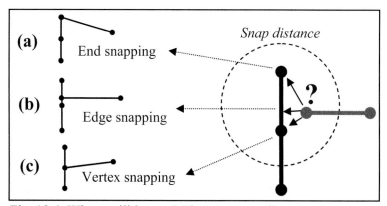

Fig. 12.4. Where will it snap? The snap type dictates where the new horizontal line will end.

426

edge snapping is turned on, then the line can be snapped anywhere between the end and the vertex (Fig. 12.4b). If vertex snapping is set, then the new line will connect to the closest vertex or end (Fig. 12.4c).

If more than one kind of snapping is turned on, then the most inclusive one takes precedence. To ensure snapping only to ends, therefore, both edge and vertex snapping must be turned off.

Creating adjacent polygons

Two adjacent polygons should share the same boundary. In shapefiles and geodatabases the boundary gets stored twice, once for each polygon. However, the vertices should exactly match in both features. If this condition holds, the polygons are said to share a coincident boundary (Fig. 12.5). If they fail to match, gaps or overlaps are generated, which constitute topological errors and may cause problems during display or analysis.

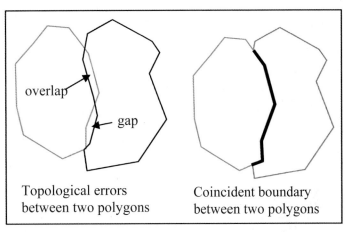

Fig. 12.5. Topological relationships between polygons

About ArcGIS

The editing process begins by opening the Editor toolbar. Next, you must turn on the editing functions and identify which folder or geodatabase will be available for editing. Only one folder or geodatabase can be edited at a time, but all the feature classes in the folder or geodatabase can be edited together. Setting several options, such as snapping, makes editing easier. Then you are ready to start editing features.

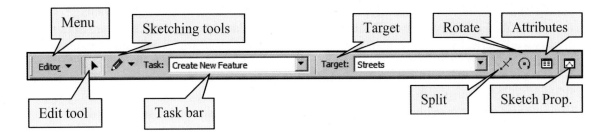

Fig. 12.6. The Editor toolbar contains many editing options and functions in one neat package.

The Editor toolbar

The Editor toolbar controls the process of editing (Fig. 12.6). The first button on the left shows the editing menu, used for turning editing on and off, setting options, and performing certain types of edits. The Edit tool selects one or more features for editing. The Sketch tool enters vertices when creating or modifying features. The Task bar controls how the Sketch tool works and performs a variety of editing functions, such as creating new features or modifying existing ones. The target window specifies which layer the edits will affect. The Split and Rotate buttons provide access to common editing commands. The Sketch Properties button is used to examine vertices of sketches. Finally, the Attributes button launches an editing menu for the feature attributes. In the next few sections, we will examine how these editing functions work.

About editing

What can you edit?

ArcGIS offers different levels of editing capability, depending on whether you've purchased ArcView, ArcEditor, or ArcInfo. ArcView can edit shapefiles and personal or file geodatabases only. ArcEditor and ArcInfo can also edit SDE databases, coverages, geometric networks, and feature topology. All of these levels are accessed through the same ArcMap interface.

ArcMap can edit several layers at once, as long as they are all in the same folder or geodatabase. This capability makes it easier to view and edit related layers simultaneously or to copy features from one layer to another.

> **TIP:** Coverages cannot be edited in ArcMap. Use the ArcEdit program in Workstation ArcInfo instead, or convert the coverages to shapefiles or geodatabases for editing.

Editing and coordinate systems

ArcMap can edit layers with different coordinate systems from the data frame. The edits will automatically be converted to the coordinate system of the layer being edited. Consider editing a shapefile of roads stored in a GCS while displaying the roads with a digital orthophotoquad (DOQ) in a UTM projection. The roads coordinates will be converted into decimal degrees before they are put into the shapefile.

Performance and reliability

Although ArcMap has many useful capabilities, such as editing across coordinate systems and editing multiple files simultaneously, one must be somewhat cautious in taking advantage of these benefits. Editing coordinate features can be a complicated process, and software does not always work perfectly under arduous conditions. If an editing scenario is demanding extensive system resources by having many files open or by doing many projection transformations during an edit session, system performance and reliability may suffer. Moreover, because spatial data files are complex constructions, the price of an editing glitch may be the loss of data integrity; an ill-timed system glitch might not only lose recent changes but also corrupt the entire data set. Thus, you are encouraged to use the following procedures while editing:

> ➢ **Always have a backup copy of the file being edited,** stored elsewhere on the disk or on a different medium. This precaution is especially important when working on data sets that would be expensive or impossible to replace should something go wrong.

> ➢ Shapefiles are the simplest and most robust type of data file to edit and the hardest to corrupt. When feasible, create and enter data as shapefiles first and then, if necessary, convert them to geodatabases or coverages later. However, geodatabases offer special capabilities that the user cannot take advantage of when using shapefiles.

> ➤ Limit the number of open files. A prolonged editing session should be conducted in its own map document and include only the layers that are needed.

> ➤ Save edits frequently in case of a system glitch. ArcMap does NOT automatically save edits at regular time intervals.

How editing works

An editing session must be initiated before any changes to a file can be made. This requirement protects the user from accidentally making changes to a file without realizing it. Also, because ArcMap can only edit within one directory or one geodatabase at a time, opening the session establishes the folder or geodatabase being edited.

On the Editor toolbar, a drop-down box specifies the **target layer**. Any feature created will be placed in this layer. The type of target layer also dictates the types of edits that can be made. You can create a new feature in any kind of layer (point, line, or polygon), but the AutoComplete Polygon task would be dimmed if the target layer contains points or lines.

Basic editing uses three components of the Editor toolbar, the Edit tool, the Sketch tool, and the Task bar (Fig. 12.7). The Task bar specifies which editing action is taking place, such as creating a new feature or reshaping an existing feature. The Sketch tool adds the vertices to a **sketch**, a provisional feature being added. The Edit tool selects features for the next editing task.

During editing you can access two context menus to aid in certain tasks or change edit settings (Fig. 12.8). The **Vertex menu** appears when you right-click on the sketch (Fig. 12.8a). This menu provides functions for adding, deleting, or moving vertices. The **Sketch menu** appears when you right-click off the sketch (Fig. 12.8b). This menu provides functions for specifying exact angles, lengths, distances, and more.

About the sketching tools

The sketching tools create a preliminary shape of a feature, analogous to an artist sketching a figure lightly in pencil prior to inking in the final lines of the picture. This

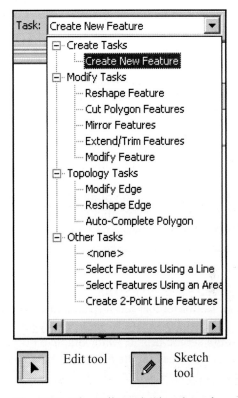

Edit tool   Sketch tool

Fig. 12.7. The Edit tool, Sketch tool, and Task bar work together.

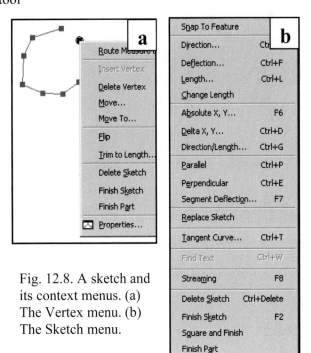

Fig. 12.8. A sketch and its context menus. (a) The Vertex menu. (b) The Sketch menu.

sketch is not actually added to the target layer until it is "finished." In Figure 12.8a, the Sketch tool (looks like a pencil) has been used with the Create New Feature task to enter the vertices of the polygon. When it is complete, the user right-clicks the sketch to show the Sketch menu and chooses Finish Sketch. The sketch then becomes a feature.

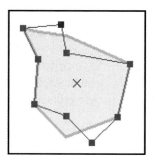

Sketches also appear when reshaping or modifying an existing feature. Figure 12.9 shows an original feature as a shaded polygon, and the lines and square vertices show a sketch that will become the new shape. Reshaping permits the user to change vertex locations, insert new vertices, or delete old ones. The sketch converts to a feature when the user finishes making changes.

Fig. 12.9. Modifying a feature

The Sketch tool is one of nine different sketching tools that are used to enter vertices for sketches and can be chosen by clicking the black drop-down arrow next to the current sketching tool (Fig. 12.10). Each sketching tool has its own way to enter vertices for a sketch. The functions of the other tools are explained in Chapter 13.

Adding adjacent polygons

The Editor toolbar has a special task for adding adjacent polygons. The Create New Feature task is first used to enter a complete polygon with no neighbors. To add an adjacent polygon, you would set the task to AutoComplete Polygon and digitize only the new part of the polygon. The editor ensures that the polygons share a coincident boundary and are free from topological errors including gaps and overlaps (Fig. 12.11).

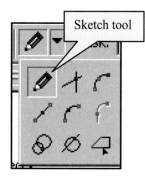

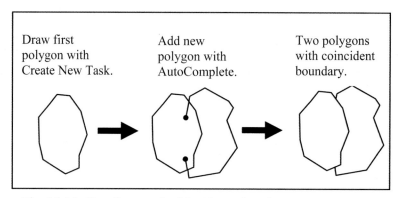

Draw first polygon with Create New Task.

Add new polygon with AutoComplete.

Two polygons with coincident boundary.

Fig. 12.10. Nine sketching tools

Fig. 12.11. Creating a coincident boundary between two adjacent polygons with the AutoComplete task

Editing attributes

Editing features often includes editing their attributes. The user can modify attributes in two ways. In the first way, edits are typed directly into an attribute table, as discussed in Chapter 4. The second method uses the Attributes window to edit the values of many features at once.

The Attributes window is accessed through a button on the Editor toolbar (Fig. 12.12). The left side of the editor shows the selected records; if no features or records are selected, then both areas will be blank. The right side of the editor shows the attributes of the feature highlighted on the left. You can edit fields for a single feature or for multiple features with this tool.

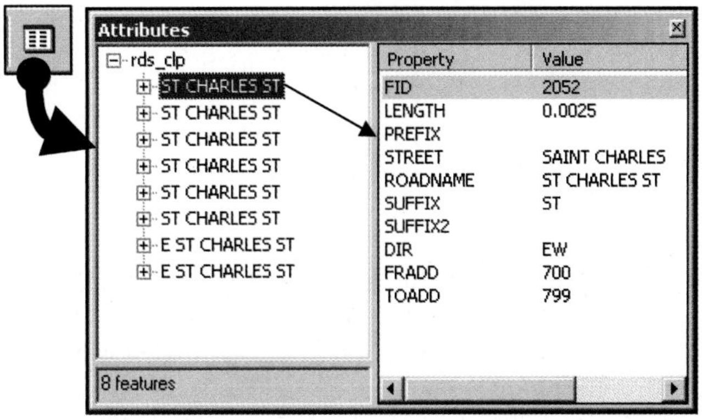

Fig. 12.12. The Attributes window can display and edit many records at once.

Summary

> Editing creates and maintains features in a feature class. Care must be taken to ensure topological integrity when editing.

> Snapping ensures that features connect properly with each other by automatically matching the coordinates of features within a certain distance of each other (the snap distance). Features may be snapped to endpoints, vertices, or edges.

> When adding adjacent polygons, you must ensure that they have coincident boundaries by using snapping or the AutoComplete task.

> Editing can create or modify map features and their associated attributes, from simple changes to a few features to creating an entirely new layer.

> Editing sessions are initiated and controlled using the Editor toolbar.

> You may edit multiple layers of data, but only one geodatabase or folder may be edited at a time.

> Edits are stored in the same coordinate system as the layer, regardless of the coordinate system of the data frame. However, editing is streamlined when the layer and frame coordinate systems match so that no reprojection need take place.

> Basic editing uses the Edit tool, one or more of the sketching tools, and the Task bar. The Task bar indicates the type of edit being performed, the Edit tool selects features to edit, and the Sketch tool provides the input points and segments.

> The Sketch tool creates preliminary figures, allowing them to be shaped and modified until the desired shape is reached. Finishing the sketch creates the new feature.

> Different sketching tools provide a variation of functions for placing vertices. Context menus associated with the sketching tools add even greater functionality.

TIP: Coverages cannot be edited in ArcMap. They have an older topological data model and cannot utilize the topology editing tools. Use the ArcEdit program in Workstation ArcInfo instead, or convert the coverages to shapefiles or geodatabases for editing.

Chapter Review Questions

1. List the main types of topological errors.

2. Which type of snapping (end, vertex, or edge) would work best in the following situations? (a) digitizing streams, (b) digitizing parcels, (c) digitizing streets, (d) digitizing traffic lights at intersections, (e) digitizing stream gages on streams.

3. How are editing functions and options accessed in ArcMap?

4. If a data layer is in UTM coordinates and the data frame is set to State Plane coordinates, in which coordinate system will the edits be stored?

5. True or False: Because you can Undo edits, it is not necessary to make a backup copy of any data set before editing. Explain your answer.

6. When you draw a new feature, how do you specify which layer it belongs to?

7. What are dangles and how are they prevented?

8. What is the purpose of the AutoComplete task? What errors does it prevent?

9. How many context menus can you access while sketching? How do you control which one you get?

10. How do you change the attributes of many records at once in the Attribute Editor?

Mastering the Skills

Teaching Tutorial

The following examples provide step-by-step instructions for doing basic tasks and solving basic problems in ArcGIS. The steps you need to do are highlighted with an arrow ➔; follow them carefully. Click on the video number in the VideoIndex to view a demonstration of the steps.

➔ Start ArcMap and open the map document ex_12a.mxd in the MapDocuments folder.

➔ Use Save As to rename the document and remember to save frequently as you work.

This map document contains layers for a neighborhood in Rapid City, including street edges, parcels, buildings, and water connectors. You will also create a feature class of water mains and laterals. These feature classes are stored in a geodatabase called eastpat. You will create some empty feature classes in which to practice before editing the "real" layers. When creating the new files, you'll assign them the same coordinate system as the streets feature class in eastpat.

1➔ Click the icon on the main menu to open ArcToolbox.

1➔ In ArcToolbox, navigate to the Data Management Tools > Feature Class > Create Feature Class tool.

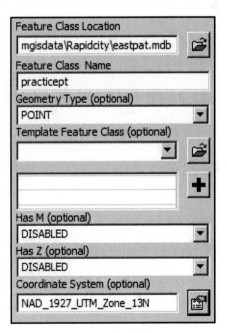

2➔ Click the Browse button for the Feature Class Location. Navigate to the mgisdata\Rapidcity folder and select the eastpat geodatabase (Fig. 12.13). Click Add.

2➔ Type the Feature Class Name practicept.

2➔ Set the Geometry Type to POINT.

2➔ Leave the Template Feature Class blank.

2➔ Scroll down, if necessary, to locate the Coordinate System box. Click the Browse button and then click the Import button.

2➔ Navigate to mgisdata\Rapidcity\eastpat and select the streets feature class. Click Add. Click OK to close the window.

2➔ Click OK to finish creating practicept.

Fig. 12.13. Creating practicept

The new feature class is automatically added to the map document. Now create two additional feature classes, one for lines named practiceln and one for polygons named practicepo. Be sure to set the correct geometry type (line or polygon) and the coordinate system for each.

1. In what coordinate system is the new feature class? _____

3➔ Repeat the preceding steps to create feature class practiceln. Make sure that the feature geometry is set to POLYLINE and that you set the coordinate system.

4➡ Create the feature class **practicepo**. Make sure that the feature geometry is set to POLYGON and that you set the coordinate system.

Some simple editing functions

First we need to set up our editing session.

5➡ Close ArcToolbox.

5➡ Turn off the Orthophoto layer for now.

5➡ If the Editor toolbar is not visible, click the Editor button.

5➡ From the Editor toolbar, choose Editor > Start Editing.

Points are the least complex features to edit, so we will begin with those.

6➡ Check the target layer and make sure it is set to **practicept**.

6➡ Check the Task bar and make sure it says Create New Feature.

6➡ Click the Sketch tool.

6➡ Click near the middle of one of the blocks defined by the Streets layer to add a point.

The point appears as soon as you click the mouse. Since points have only one "vertex," you do not need to finish the sketch or anything else. Notice that the point is selected.

7➡ Add a point to the rest of the blocks by clicking near the center of each in turn.

The last point added is selected. As you add each new point, the previous selection gets unselected. Now you should have seven unselected points and one selected point (Fig. 12.14).

7➡ Open the attribute table for the **practicept** layer.

Notice that the final point is selected both in the map and in the table. Our next step is to add a field to this table and edit its attributes. However, we cannot add a field while a layer is being edited, so we must first stop editing and save our edits so far.

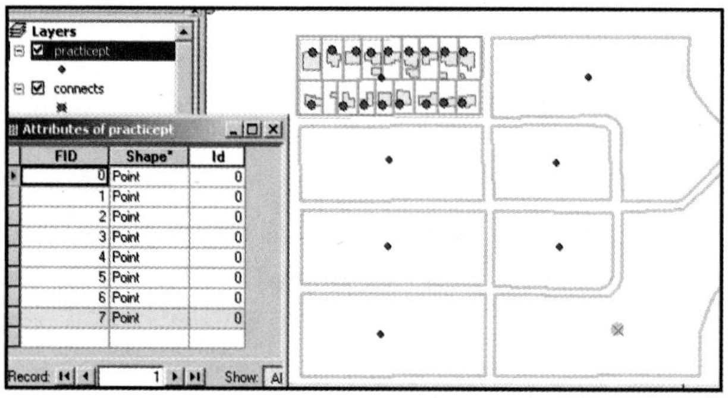

Fig. 12.14. Adding points to each block. The final point added is selected in both the map and the table.

8➡ Choose Editor > Stop Editing from the Editor toolbar.

8➡ Say Yes when prompted to save the edits.

Notice that the selection disappears when you stop editing. Now add the new field to the table.

9➡ In the table, click Options > Add Field.

9➡ Name it **Name**, set the field type to Text, and set its Length to **10**. Click OK.

9➜ Resize the table so that only the fields are visible. It will be easier to see the map with the smaller table.

9➜ Move the table so that you can better see the map.

Now turn editing back on and enter the block names.

10➜ Choose Editor > Start Editing from the Editor toolbar.

10➜ From the Selection menu, set the Selectable Layers to just the practicept layer.

10➜ Click the Edit tool on the Editor toolbar.

10➜ Click on the point in the upper-left block to select it.

10➜ Click in the highlighted record of the Name field and enter **BLOCK 1**.

It would be nice to see our handiwork, so let's turn the labels on for practicept.

11➜ Open the properties for practicept and click the Labels tab.

11➜ Check the box to turn on the labels and then set the Label Field to Name.

11➜ Change the symbol to Arial 12-point bold.

11➜ Click Placement Properties and choose to place the label on top of the point. Click OK and OK.

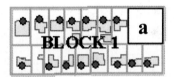

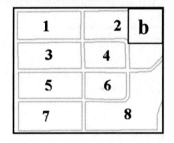

The block name should appear in the middle of the first block (Fig. 12.15a). Now let's add the others.

12➜ Click to the left of the second row in the table to highlight it. The corresponding point will be highlighted in the map.

12➜ Enter the name of the block again using the numbers in Figure 12.15b as a guide.

12➜ Repeat until every point is labeled. You may need to click the Refresh button to make the labels appear.

Fig. 12.15. Labeling the blocks by number

13➜ When finished adding the block names, close the table.

13➜ Click the practicept layer twice and rename it **Blocks**.

Now save the edits made so far but don't close the editing session.

13➜ Choose Editor > Save Edits from the Editor toolbar.

All of the edits added to practicept are now permanently stored in the file. Just for good measure, save the map document also. Note that saving the map document does NOT save the edits.

13➜ Choose File > Save from the main menu bar.

Next we switch the target layer and start creating lines.

14➜ Change the target layer on the Editor toolbar to practiceln.

14➜ Zoom in to Block 3.

14➜ Turn off the Blocks layer for now.

Whenever you start a new editing task, it is good practice to check the target layer, task, and Sketch tool to make sure they are correct for the next edits.

15➜ Make sure that the target layer is practiceln, the Task bar reads Create New Feature, and the Sketch tool is clicked.

15➜ Click somewhere on the screen to add the first vertex of a line.

15➜ Continue clicking until you have added five or six vertices.

15➜ Double-click to end the sketch and create the first line.

15➜ Add several more lines until you are comfortable with this task. Don't let them cross the streets for now.

Well, that's not so hard, is it? Now use the simple techniques to delete, move, and copy features.

16➜ Set the Selectable Layers back to All layers.

16➜ Click on the Edit tool on the Editor toolbar.

16➜ Click one of the new lines to select it.

16➜ Press the Delete key.

TIP: Sometimes features do not disappear when you press Delete. If this happens, click the Edit tool on the Editor toolbar and press Delete again. The Delete key does not work with some of the cursor tools, but it always works if the Edit tool is active.

17➜ Select another of the new lines.

17➜ Place the cursor on top of the selected line in order to make small crosshairs appear.

17➜ Click and drag the line to a new location.

17➜ Move some of the other lines to get the hang of this task.

It is also easy to make copies of features.

18➜ Select one of the lines.

18➜ Choose Edit > Copy from the main menu bar, or press Ctrl-C.

18➜ Choose Edit > Paste from the main menu bar, or press Ctrl-V. The copy will be placed exactly on top of the existing feature.

18➜ Move the selected feature to verify that there are two identical lines now.

You can copy features from one layer to another. Both layers must have the same feature type. Always set the target layer to the target, not the source, before beginning a copy.

19➜ Change the target layer to Streets.

19➜ Select one of the lines from practiceln.

19➜ Press Ctrl-C and then Ctrl-V.

19➔ Click and drag the new feature to a different location.

19➔ Click anywhere off all the lines to clear the selected features.

Notice that the original line has the practiceln symbol, but the new line has the thick pink symbol of Streets. However, we don't want this line in Streets, so get rid of it.

20➔ Select the pink line in Streets and press the Delete key.

20➔ Change the target layer back to practiceln.

20➔ Draw a box around the remaining lines with the Edit tool and delete them also.

Working with snapping

Snapping ensures that all the features that are supposed to touch actually do. It helps prevent topological errors between features. Snapping is flexible and can be administered precisely during sketches. Let's see how it works. First, we check the snap tolerance and turn snapping on for the desired layers.

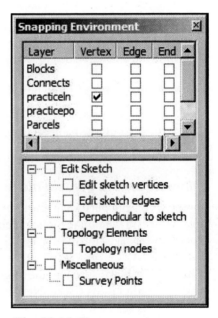

Fig. 12.16. Turn on vertex snapping for the practiceln layer.

21➔ Choose Editor > Options and click the General tab.

21➔ Make sure that the snap tolerance is set to 15 pixels. This distance is larger than normal, but it will better demonstrate the technique. Click OK.

21➔ Choose Editor > Snapping from the Editor toolbar.

21➔ Check the Vertex box for the practiceln layer (Fig. 12.16).

21➔ Dock the Snapping Environment window underneath the Table of Contents for easy access. Click and drag the section edges to adjust the sizes of the sections both within and between the windows.

22➔ Click on the Sketch tool.

22➔ Use the Sketch tool to draw a line with distinct angles and obvious vertices. Don't connect the ends. Double-click to finish the line.

22➔ Move the cursor toward one of the vertices and watch it jump to the nearest vertex when it gets close.

The cursor stays at the vertex until you move at least 20 pixels away. If you click to add a vertex while the cursor is "frozen" to the vertex, then the resulting new vertex will exactly match the existing one. Try it.

22➔ Click on one of the vertices to begin a new line.

22➔ Create another vertex that is off the existing line.

22➔ Come back and snap to another vertex of the first line.

22➔ Double-click to finish the sketch.

Now let's see how the different types of snapping operate. You have used vertex snapping so far.

23➔ In the Snapping Environment window, uncheck the Vertex box and turn on End snapping for practiceln.

23➔ Move the cursor over the different vertices. Now snapping only occurs at the endpoints of the lines, not at the vertices.

23➔ Turn off End snapping and turn on Edge snapping for practiceln.

23➔ Move the cursor over the lines and notice that snapping occurs anywhere along the line. Make the cursor follow along the edge.

23➔ Make a sketch that is snapped to one of the previous lines at several locations. Finish the sketch.

We will make extensive use of snapping during the rest of the chapter.

24➔ Uncheck all the boxes in the Snapping window but leave it available.

24➔ Choose Editor > Options from the Editor toolbar.

24➔ Click the General tab and set the snap tolerance to 10 pixels. This value works well for most editing.

Using the sketch context menus

Now that you can perform some simple editing functions, let's look at ways in which the Sketch menu enhances editing, such as correcting mistakes during a sketch.

25➔ Make sure that the target layer is practiceln, the edit task is Create New Feature, and the Sketch tool is selected.

25➔ Enter some vertices of a new line. Notice that the last vertex is red and the other vertices are green.

25➔ Right-click on top of the last red vertex to open the vertex context menu (Fig. 12.17).

25➔ Examine the menu entries for a moment. This context menu is called the Vertex menu.

25➔ Choose Finish Sketch.

Fig. 12.17. The Vertex menu

Choosing Finish Sketch from the context menu does the same thing as double-clicking at the end of the sketch. Double-clicking is faster, but the Vertex menu offers other useful functions.

26➔ Click on the map to start a new sketch and then add some vertices.

26➔ Right-click the last red vertex to open the Vertex menu.

26➔ Choose Delete Vertex from the Vertex menu.

Notice that the last vertex is deleted and the second-to-last one is now red. This menu can delete any of the previous vertices.

26➜ Right-click on any of the green vertices and choose Delete Vertex.

26➜ Right-click another vertex and examine the menu again.

Now you might want to get rid of the menu without actually selecting one of its entries.

26➜ Click on the blue window bar at the top to get rid of the context menu.

Notice that all this time we are still in the middle of creating this sketch. While sketching, you can carry out quite a few actions with no negative effects. Let's keep going.

27➜ Right-click on a segment of the sketch *between* two vertices.

27➜ Choose Insert Vertex from the context menu.

27➜ Watch the new vertex appear where you had right-clicked.

27➜ Right-click another segment of the sketch between two vertices.

The Vertex menu appears again. This time, notice how Insert Vertex is available and Delete Vertex is dimmed. Before, when you clicked ON a vertex, Delete Vertex was available and Insert Vertex was dimmed.

Remember, clicking ON the sketch gives the Vertex menu. A different menu appears if you click OFF the sketch.

27➜ Click the blue bar to make the menu go away.

27➜ This time, right-click OFF the sketch and somewhere on the map. The Sketch menu appears (Fig. 12.18).

27➜ Examine the entries in this longer menu.

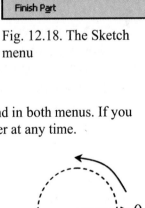

Snap To Feature	▶
Direction...	Ctrl+A
Deflection...	Ctrl+F
Length...	Ctrl+L
Change Length	
Absolute X, Y...	F6
Delta X, Y...	Ctrl+D
Direction/Length...	Ctrl+G
Parallel	Ctrl+P
Perpendicular	Ctrl+E
Segment Deflection...	F7
Replace Sketch	
Tangent Curve...	Ctrl+T
Find Text	Ctrl+W
Streaming	F8
Delete Sketch	Ctrl+Delete
Finish Sketch	F2
Square and Finish	
Finish Part	

Fig. 12.18. The Sketch menu

Some of the entries, such as Finish Sketch and Delete Sketch, are found in both menus. If you make a mistake as we continue, you can delete the sketch and start over at any time.

27➜ Choose Delete Sketch from the menu.

28➜ Start a new sketch with several vertices.

28➜ Right-click off the sketch to open the Sketch menu.

28➜ Choose the second entry, Direction.

28➜ Type **90** in the Direction box and press Enter.

As the mouse moves, notice that the cursor is constrained along a vertical line. You have forced your next segment to follow the angle you have specified. Angles are measured counterclockwise from the east (Fig. 12.19).

Fig. 12.19. Angles are measured in degrees from horizontal.

28➔ Click along the line to enter a vertex. The new segment is exactly vertical.

28➔ Right-click off the sketch and choose Direction again from the menu.

28➔ Type **-45** in the box and press Enter.

28➔ Click along the line to enter a vertex.

28➔ Double-click to finish the sketch.

2. What angle would you enter to create a line going S25°W? _____

Next, let's create a segment of a specific length.

29➔ Start a sketch and add several vertices.

29➔ Right-click off the sketch and choose Length from the Sketch menu.

29➔ Type **30** into the box and press Enter.

3. In what units are these distance values? _____

29➔ Move the cursor around, noticing how the last segment always remains the same length.

29➔ Click somewhere to enter the vertex.

> **TIP:** If the last segment goes off the page or is too small to see, try entering a different length. Simply right-click off the sketch again and type in a different length value.

The Sketch menu also combines the length and direction functions to enter a line with a particular bearing and direction. We will use this feature to enter the next segment of this same sketch.

30➔ Right-click off the sketch and choose Direction/Length.

30➔ Type **45** for the direction and **30** for the length (you can use different values, if necessary, to keep the sketch on the screen). Use the Tab key or mouse to switch between the boxes. Click Enter when both boxes are filled.

30➔ Notice you didn't even have to click to enter the vertex. Double-click now to finish the sketch.

> **TIP:** The angle box contains the current angle when you right-clicked. You can use this feature to measure an existing angle.

To learn about the other options on the Sketch menu, consult the Skills Reference.

Creating polygons

As our final task before we start actually editing some layers, we need to practice editing polygons, first as separate entities and then as adjacent features. First we do a separate polygon.

31➔ Select and delete the lines just created.

31➜ Set the target layer to practicepo.

31➜ Make sure the task is set to Create New Feature, and click the Sketch tool.

31➜ Start a polygon by adding several vertices. Notice how the shape remains closed, with the last vertex always connected to the first.

31➜ Double-click the last vertex to finish the sketch.

31➜ Try adding several more polygons that don't touch each other, for practice.

Next we will create some adjacent polygons. It is important to use the AutoComplete Polygon task to do this so that the polygons have a coincident boundary. Vertex snapping helps match the polygon vertices.

32➜ Turn on vertex snapping for practicepo in the Snapping Environment window.

TIP: When using AutoComplete, always start and end well inside an existing polygon. The dangles will automatically be deleted.

32➜ Change the Task to AutoComplete Polygon.

32➜ Make sure the Sketch tool is clicked.

32➜ Click INSIDE an existing polygon to start the sketch (Fig. 12.20).

32➜ Add more vertices to define the polygon.

32➜ Add the last vertex INSIDE the existing polygon and double-click to finish the sketch.

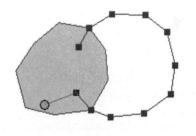

Fig. 12.20. Adding an adjacent polygon with AutoComplete

You can add polygons that start and end inside two different features, as long as the space is entirely enclosed (Fig. 12.21).

33➜ Begin a new sketch inside one of the two adjacent polygons and end it inside the other polygon. Finish the sketch (Fig. 12.21a).

Splitting provides another way to create adjacent polygons with coincident boundaries. In this task the Sketch tool cuts a polygon in pieces.

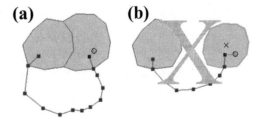

Fig. 12.21. Starting and ending a sketch inside two polygons: (a) works; (b) does not work because sketch is not closed

34➜ Select one of the polygons with the Edit tool (Fig. 12.22).

34➜ Change the task to Cut Polygon Features.

34➜ Click the Sketch tool.

34➜ Start a sketch OUTSIDE the selected polygon and add vertices of a line that goes through the polygon.

34➜ Add the last vertex OUTSIDE the polygon and double-click to finish the sketch.

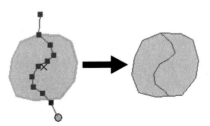

Fig. 12.22. Cutting a polygon

Cutting works on multiple polygons, too.

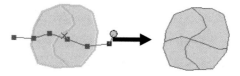

35➔ Check that the two polygons you just created are still selected. If not, select them (Fig. 12.23).

35➔ Sketch another line through both.

35➔ Finish the sketch.

Fig. 12.23. Cutting multiple polygons

Now you have all the basic skills to do some real editing. First, let's do a little cleanup work.

36➔ Choose Editor > Stop Editing from the Editor toolbar. Choose No when prompted to save the changes.

36➔ Remove practiceln and practicepo from the map document.

36➔ Zoom back out to the full extent of the Streets layer.

36➔ Save the map document.

Adding buildings

Your task is to continue developing the database started for this small area of Rapid City. You will digitize the buildings, parcels, and water connections. We will begin with the buildings.

37➔ Zoom in to Block 1 (upper left) and examine the different features present in this neighborhood survey. Pay particular attention to the buildings.

4. Do the buildings need to be snapped to any other features? _____

One feature of these buildings is readily visible—most of the corners are square. We can use one of the snapping options to make this requirement easy and exact.

37➔ Pan to Block 3 below Block 1 so that the entire block is visible.

37➔ Turn on the Orthophoto layer to show the houses. (They are fuzzy but clear enough to digitize approximately.)

37➔ Choose Editor > Start Editing from the Editor toolbar.

37➔ Make sure that the target layer is Buildings. Set the task to Create New Feature.

37➔ Open the Snapping window. Check the box near the bottom labeled Perpendicular to Sketch.

This last step ensures that all the building corners will be square. Now we are ready to start.

38➔ Start with the clearest square building, second from the left on the top row.

38➔ Click the Sketch tool and click on the upper-left corner to start the sketch (Fig. 12.24).

Fig. 12.24. Adding the house

For these buildings, we want all the lines to go north-south or east-west. So we will constrain the angle of the first segment, the west wall of this building, to be 90 degrees.

38➔ Right-click off the sketch and choose Direction.

38➔ Type **90** in the box and press Enter.

38➔ Add the second vertex at the lower-left corner.

Now that we have the first segment established, the remaining segments can easily be made perpendicular by using snapping.

38➔ Add the third vertex, making sure the segment is perpendicular.

38➔ For the fourth vertex, right-click where you want to put it and choose Square and Finish from the Sketch context menu. Finish ensures that the last corner is also a right angle (Fig. 12.24).

Now that we have one vertical line, we can use it to make all the others vertical or horizontal. Using Perpendicular or Parallel is easier and faster than typing an angle each time.

39➔ Start the next sketch by clicking on the upper-left corner of the next house to the left.

39➔ Right-click on the west wall of the house you previously digitized and choose Parallel from the context menu.

39➔ Enter the second vertex of the current sketch along the vertically constrained line.

39➔ Add the rest of the vertices except for the last one. Then right-click off the sketch and choose Square and Finish.

39➔ Finish creating the 8 houses on the western end of this block. Follow the house shapes as best you can (Fig. 12.25).

39➔ Choose Editor > Save Edits from the Editor menu.

Fig. 12.25. Follow the house shape as best you can.

Next we will add the sewer connections to the fronts of the houses. We don't know exactly where they go, so we will just enter a point on the front edge of the house but not in the driveway. We'll use snapping to make sure that the connections fall exactly on the house edge.

40➔ Turn off perpendicular snapping and turn on edge snapping for Buildings.

40➔ Change the target layer to Connects.

40➔ Make sure the task is Create New Feature and click the Sketch tool.

40➔ Add a water connection point to each house (Fig. 12.25).

40➔ Save the edits.

Creating parcels

Next we will create the parcel polygons. Of course, we can't tell exactly from the picture where each property line falls, but for practice an approximation will suffice. Parcels are adjacent polygons, so we must ensure that all of them have coincident boundaries. First, though, we need to add a field to contain the parcel number. We must stop editing to add a field to the table.

41➔ Choose Editor > Stop Editing. Say Yes if prompted to save the edits.

41➔ Open the attribute table for the Parcels layer.

41➔ Use the Options button to add a text field called Parcel_ID with a length of 8 characters.

41➔ Close the table and choose Editor > Start Editing.

Now let's begin adding parcels. We will outline the block boundary and then cut up the large block into smaller parcels. Snapping ensures that the block boundary matches the pink street boundary, avoiding gaps between different layers in the data set.

42➔ Set the target layer to Parcels.

42➔ Make sure that the task is set to Create New Feature.

42➔ Make sure the boundary of the entire block is visible in the map.

42➔ Open the snapping window and turn on both edge and vertex snapping for Streets and for Parcels.

42➔ Choose the Sketch tool and sketch the block boundary, snapping to each corner of the pink Streets layer. Finish the sketch.

42➔ Turn off the Streets layer.

Next we cut this large polygon up into smaller parcels. The polygon must be selected first.

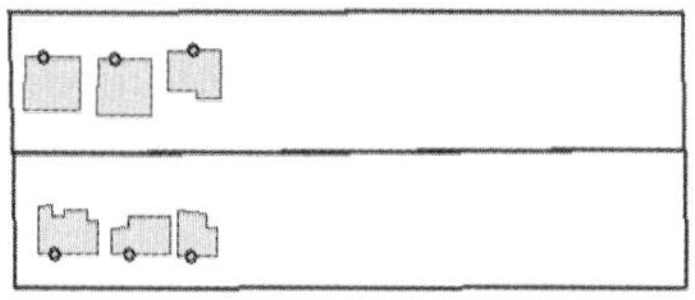

43➔ Make sure that the block boundary is still selected. If not, select it.

Fig. 12.26. Cut the block in two

43➔ Change the task to Cut Polygon Features.

43➔ Choose the Sketch tool.

43➔ Starting on or outside the boundary, digitize a horizontal line that follows the white fence line between the upper and lower houses (Fig. 12.26).

43➔ Finish the sketch outside the boundary on the other side. The polygon will split into two sections.

Now create a roughly vertical line to delineate the first two parcels on the left from the two selected polygons. Change direction by doing a "dogleg" along the centerline. Edge snapping will make sure the dogleg matches the existing line.

44➔ Turn off the Orthophoto to see better.

44➔ Make sure both upper and lower sections are still selected.

44➔ Add a vertex on the upper polygon between the first two houses (number 1 in Figure 12.27).

44➔ Add a second vertex (2) on the horizontal line.

44➔ Add a third vertex (3), still on the horizontal line but a little to the east, to align the final segment between the houses.

44➔ Add a final vertex (4) on the bottom line.

44➔ Double-click to finish the sketch.

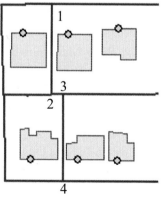

Fig. 12.27. Enter the vertices in the order shown.

45➔ Finish creating the parcels for the rest of the houses by using the same technique.

45➔ Save the edits.

Using the Attribute Editor

We created all the parcels at once because the editing was easier that way. Now we need to go back and add a parcel number for each one. The Attributes window facilitates this task.

46➔ Select the parcel in the upper-left corner of the block.

46➔ Click the Attributes button.

46➔ Position the Attributes window so that you can see all the parcels.

46➔ Click the Parcel_ID field on the right side of the window, and type in the parcel number **3-129876** (Fig. 12.28).
The 3- represents the block number.

47➔ Select the next parcel on the right.

47➔ Type in the parcel-id **3-129877**.

47➔ Continue adding the parcel-ids by going clockwise around the block and increasing by 1 each time.

Attributes			
⊟ parcels		Property	Value
⊞ 0		FID	18
		Id	0
		Parcel_ID	3-129876

Fig. 12.28. Adding parcel IDs with the Attributes window

TIP: You can highlight the first part of the number, **3-1298**, copy it using Ctrl-C, and paste it into each field using Ctrl-V. Press Enter for each record before going to the next.

Now all of these parcels belong to Block 3, so we will set the ID field to the block number. We can do this for all the parcels at once.

48➔ Choose Selection > Set Selectable Layers from the main menu bar and make Parcels the only selectable layer.

48➔ Use the Edit tool on the Editor toolbar to select all the parcels in Block 3.

48➔ Notice all the parcels are now shown in the Attributes window with a value of 0.

48➔ Click the entry "parcels" at the top of the list (Fig. 12.29). The attributes on the right go blank.

48➔ Click on the Id field and enter **3**. Press Enter. Notice that all the values on the left also change to 3.

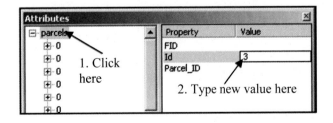

Fig. 12.29. Changing all of the parcel IDs at once

The block ID may not be the most useful value to display in the attribute editor. We can easily change what is shown in the list.

49➔ Open the Properties window for the Parcels layer.

49➜ Click the Fields tab and change the primary display field to Parcel_ID. Click OK.

49➜ Select all the parcels again to show the parcel IDs.

Now the parcel IDs are shown in the Attribute Editor. It is easy to find out which parcels match which records.

50➜ Click on a parcel record in the list on the left. The corresponding parcel on the map will briefly flash.

50➜ Right-click a record in the list to get a context menu (Fig. 12.30).

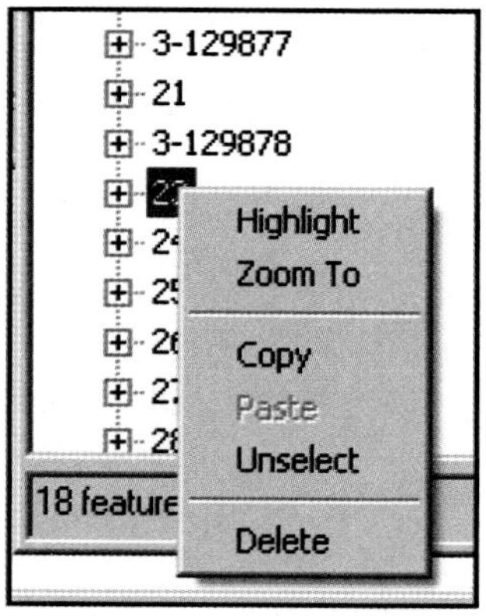

The context menu can highlight a feature (flash it), zoom to it, copy it, unselect it (it will disappear from the Attribute Editor), or delete it from the file.

50➜ When finished with the Attributes window, close it.

50➜ Stop Editing and save the edits.

50➜ Save the map document.

➜ Close ArcMap.

Fig. 12.30. Examining a record

This is the end of the tutorial.

More skills

Consult the Skills Reference section of this chapter to learn to do the following:

➢ Learn about more Sketching tools for adding new vertices

➢ Rotating features and splitting lines

➢ Creating multipart features

➢ Switching to another folder or geodatabase for editing

Exercises

1. Add the buildings for Block 8 to the buildings feature class in the mgisdata\Rapidcity\eastpat geodatabase.

2. Add water connectors to each building.

3. Digitize the parcels for Block 8, making sure that they all have coincident boundaries.

4. Enter the ID and Parcel_IDs for the parcels in Block 8. They will all begin with 8-, but you can make up the rest of the number. Have them increase by 1 as you go around the block clockwise.

5. Create a water mains feature class and include a text attribute field for the type (Main or Lateral). Add water mains along the streets of Block 8 and set their attribute type to Main.

6. Add water laterals to the water mains layer in Block 8, connecting each house to the water mains. Set their attribute type to Lateral.

7. Create a layout showing all the elements you digitized in Block 8. Label each parcel with the parcel number. Show the water mains with a different symbol than the laterals. For clarity, do not include the orthophoto in your layout (or include it but make it about 70% transparent). Print the layout.

Challenge Problem

Open ex_12b.mxd in the mgisdata\MapDocuments folder. It shows a portion of an image of agricultural land in eastern South Dakota. You will create and digitize some layers based on this photo. Put the layers in the folder mgisdata\Sdakota\farm (the orthophoto is already there). Be sure to use appropriate snapping to ensure topological integrity as well as the Autocomplete or Split tools for coincident polygon boundaries. Zoom in to the areas that you're digitizing to achieve accurate work. A working scale of 1:1,000 to 1:2,000 should be adequate.

1. Create three new shapefiles, roads (line), rivers (line), and crops (polygon). Import the coordinate system from the orthophoto.
2. Digitize the roads as single lines. (Hint 1: Try the Arc tool where the roads curve.)
3. Digitize the rivers, showing both shores.
4. Create crop polygons. Use your best judgment here to find homogeneous areas to classify as separate fields. The field boundaries next to rivers should be coincident with the river shores. You don't have to do EVERY field but make at least 25 different fields.
5. Create an attribute field for the crops called Crop and assign a crop to each field (e.g., corn, soybeans, wheat, and fallow). Just make them up.
6. Create a layout showing your work. Include labels for the crops. Print the layout.

Skills Reference

Beginning an editing session 448

Setting snapping options 449

Selecting features for editing..449

Moving features...450

Rotating features...450

Deleting features 451

Creating new features 451

Using sketch context menus 452

Creating multipart features...455

Creating features with square corners 455

Creating adjacent polygons 456

Using the Attributes window..457

Saving changes during an edit session 458

Stopping an edit session 458

Switching to another folder or geodatabase 458

Beginning an editing session

1. Open ArcMap and add the data layers you wish to edit. For best results, add only as many layers as needed.

2. If the Editor toolbar is not already displayed, click the Edit button on the main menu bar to launch the Editor menu.

3. Choose Editor > Start Editing from the Editor toolbar.

4. If data from more than one folder or geodatabase appears in the map document, you will be prompted to choose which one to open for editing.

5. If more than one coordinate system is represented in the edit folder or geodatabase, a warning will be given. Generally it is best to stick to one coordinate system; however, there are some situations where it is appropriate to have more than one. Just be careful.

You are now ready to begin editing.

Setting snapping options

1. Choose Editor > Options from the Editor toolbar.

2. Click the General tab (Fig. 12.31).

3. Choose units for the snapping tolerance.

4. Enter the snapping tolerance value.

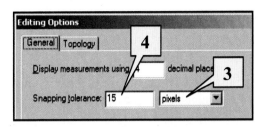

Fig. 12.31. Setting the snapping tolerance

TIP: Pixel units usually work best because then the snap distance on the screen remains constant regardless of scale.

5. Choose Editor > Snapping from the Editor toolbar.

6. Check the boxes for any layers to which you want to snap, according to the type of snap you want, vertex, edge, or end (Fig. 12.32).

7. Click and drag a layer up or down the list to change the snapping priority. Layers in the top of the list will be snapped to first.

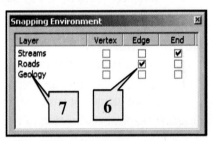

Fig. 12.32. Turning on snapping for layers

TIP: Keep the Snapping Environment window open to facilitate making changes to the snapping settings as you edit.

Selecting features for editing

All of these techniques respond to the options used for regular interactive selection, including setting the selectable layers, changing the selection method, or changing the selection options, as described in Chapter 6. Features may also be selected for editing using the Select by Attributes and Select by Location.

Selecting features with the Edit tool

1. Click the Edit tool on the Editor toolbar.

2. Click the feature to be selected or click and hold down the mouse button to drag a box around the features to be selected.

3. To select multiple features, hold down the Shift key and click on each feature in turn.

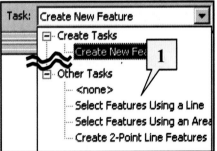

Fig. 12.33. Using the Task bar for selecting features

Selecting features with a line

1. Choose Select Features Using a Line from the Task bar (Fig. 12.33).

2. Choose a sketching tool.

3. Click on the map to begin drawing a line; double-click to end the line. All features that intersect the line will be selected.

Selecting features with a polygon

1. Choose Select Features Using an Area from the Task bar.

2. Click the Sketch tool.

3. Click on the map to begin drawing a figure; double-click to end it. All features that intersect the area will be selected.

Moving features

1. Select the feature(s) to be moved by using the Edit tool on the Editor toolbar.

2. Place the cursor over the objects until crosshairs appear.

3. Click and drag the feature to a new location.

Rotating features

1. Select the feature(s) to be rotated by using the Edit tool on the Editor toolbar or by using another selection method (Fig. 12.34).

2. Click the Rotate tool on the Editor toolbar.

3. Click and draw anywhere on the screen to rotate the features about their center of rotation (marked with an X).

4. To change the center of rotation, place the cursor over the X. It will change to a crosshairs symbol. Click and drag the center to a new location.

5. Click and drag anywhere outside the features to rotate them about the new center.

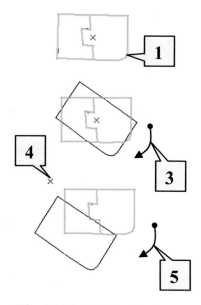

Fig. 12.34. Rotating features

Deleting features

1. Select the feature(s) to be deleted by using the Edit tool on the Editor toolbar or by using another selection method.

 2. Make sure the Edit tool on the Editor toolbar is clicked.

3. Press the Delete key to delete the features.

Creating new features

1. In the Editor toolbar, set the target layer to which the new feature will belong (Fig. 12.35).

2. In the Editor toolbar, choose Create New Feature from the Task bar.

3. Choose a sketching tool.

4. Sketch the feature by clicking on its vertices or by using one of the special techniques described in the steps that follow.

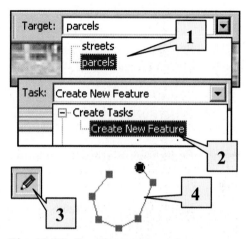

Fig. 12.35. Creating a new polygon

You may use the context menus or switch between different sketching tools while creating the vertices of a sketch, as described in the next two sections.

5. To delete or edit the vertex of a sketch, right-click on a vertex to bring up the vertex context menu (Fig. 12.36).

6. To add a vertex to a sketch, right-click on a segment of the sketch between two vertices.

7. To bring up the Sketch menu, right-click the screen at a location off the sketch.

8. To start again, right-click to raise one of the menus and choose Delete Sketch.

9. When the sketch is complete, right-click it and choose Finish Sketch or simply double-click on the last vertex.

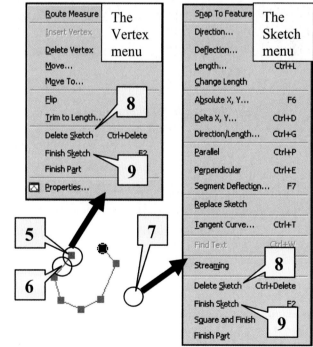

Fig. 12.36. Accessing the Vertex and Sketch menus while creating a sketch

Using sketch context menus

The following functions can be accessed during a sketch by right-clicking anywhere off the sketch.

Entering absolute *x-y* locations

1. Right-click off the sketch to open the Sketch menu.

2. Choose Absolute X,Y from the menu.

3. Enter the *x-y* coordinates and press Enter (Fig. 12.37). The coordinates must be in the same coordinate system as the data frame.

Fig. 12.37. Absolute X,Y

4. Finish the sketch, or switch to another sketching tool and keep sketching.

Using offsets from a previous location

1. Enter at least one vertex of the sketch.

2. Right-click off the sketch to open the Sketch menu.

3. Choose Delta X,Y from the menu.

4. Enter the distance change in the *x* and *y* directions in map units (Fig. 12.38). Press Enter.

Fig. 12.38. Delta X,Y

5. Finish the sketch, or switch to another sketching tool and keep sketching.

Creating a segment at a specified angle

1. Enter at least one vertex of the sketch.

2. Open the Sketch menu and choose Direction.

3. Type in the desired angle in degrees and press Enter (Fig. 12.39).

4. The segment is now constrained at the desired angle. Click at the desired distance to enter the point.

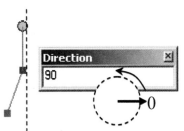

Fig. 12.39. Constraining the segment angle

5. Finish the sketch, or switch to another sketching tool and keep sketching.

TIP: The angle box contains the angle that was established when you right-clicked. You can use this feature to measure an existing angle.

Creating a segment of specific length

1. Enter at least one vertex of the sketch.

2. Open the Sketch menu and choose Length.

3. Type the length in map units in the box and press Enter (Fig. 12.40).

4. The segment is constrained to a set length. Click at the desired location to create the next vertex.

5. Finish the sketch, or switch to another sketching tool and keep sketching.

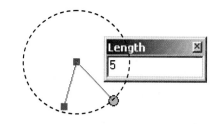

Fig. 12.40. Constraining the sketch segment to a set length

Creating a segment of set angle and length

1. Enter at least one vertex of the sketch.

2. Open the Sketch menu and choose Direction/Length.

3. Type in the desired angle and length and press Enter (Fig. 12.41).

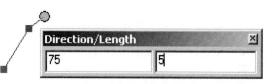

Fig. 12.41. Constraining the angle and the length of a segment

4. Finish the sketch, or switch to another sketching tool and keep sketching.

Creating a deflected segment

Use this option to create a new segment at a specified angle from the last segment in the sketch.

1. Enter at least one segment of the sketch.

2. Open the Sketch menu and choose Deflection.

3. Enter the angle in degrees that the new segment will be deflected from the current segment (Fig. 12.42). Press Enter.

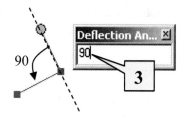

Fig. 12.42. Deflecting a vertex from the previous segment

4. The new segment is now constrained along a line. Click at the desired distance to enter the new vertex.

5. Finish the sketch, or switch to another sketching tool and keep sketching.

Deflecting a segment from a feature

Use this option to create a new segment at a specified angle to an existing segment.

1. Enter at least one segment (two vertices) of the sketch.

2. Right-click *on an existing feature segment* (the side of this building) to open the context menu and then choose Deflect Segment.

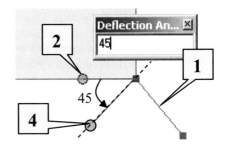

3. Type in the angle of deflection from the existing segment and press Enter (Fig. 12.43).

4. The new segment is now constrained at the specified angle (45 degrees) to the existing segment (the south side of the building). Click at the desired length to create the new vertex.

Fig. 12.43. Deflecting at an angle to an existing segment

5. Finish the sketch, or switch to another sketching tool and keep sketching.

Creating parallel or perpendicular segments

1. Enter at least one vertex of the sketch (Fig. 12.44).

2. Right-click *on an existing feature segment* (the side of this building) to open the context menu and then choose Parallel or Perpendicular.

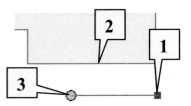

3. The new segment will be constrained in the chosen direction. Click to enter the new vertex at the desired location.

Fig. 12.44. Creating a segment parallel to an existing feature

4. Finish the sketch, or switch to another sketching tool and keep sketching.

Creating multipart features

Multipart features include more than one contiguous feature. This technique is used to create a single feature from multiple polygon areas such as Hawaii, from multiple segments forming a single street, or from multiple points forming a single feature.

1. Sketch the first part of the feature using any of the sketching tools and context menus (Fig. 12.45).

2. Right-click on or off the sketch and choose Finish Part.

3. Add as many parts as desired, ending each with a Finish Part.

4. On the last part, right-click on or off the final sketch and choose Finish Sketch.

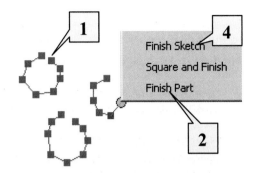

Fig. 12.45. Creating multipart features

Creating features with square corners

1. Set the target to a line or polygon layer.

2. Choose Editor > Snapping from the Editor menu.

3. Check the box labeled Perpendicular to Sketch (Fig. 12.46).

4. Click to add vertices. Each segment will be constrained to be perpendicular to the last segment.

5. To add the last vertex, right-click *off* the sketch and choose Square and Finish from the Sketch menu.

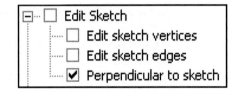

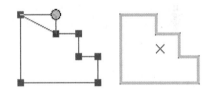

Fig. 12.46. Creating figures with square corners

Creating adjacent polygons

You can create adjacent polygons with coincident boundaries in two different ways by using the AutoComplete task or by creating a large polygon boundary and cutting it up into sections.

Using the AutoComplete task

1. Set the target layer.

2. Set the task to Create New Feature (Fig. 12.47).

3. Use one or more sketching tools to create the first polygon.

4. Change the task to AutoComplete Polygon.

5. Sketch the next polygon. Be sure to start and end inside the adjacent polygon or inside two adjacent polygons.

6. Finish the sketch to create the new polygon.

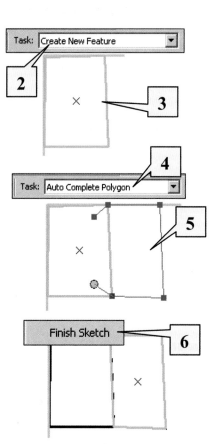

Fig. 12.47. Creating adjacent polygons with the AutoComplete task

Using the Cut Polygon Features task

1. Set the target layer.

2. Use the Edit tool on the Editor toolbar to select the polygon(s) to be cut (Fig. 12.48).

3. Set the task to Cut Polygon Features.

4. Sketch a line that will form the boundary between the new polygons. Important: The line must start and end **outside** of the existing polygons.

5. Finish the sketch. The new polygons will be created.

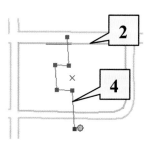

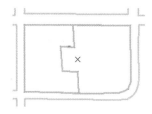

Fig. 12.48. Cutting a polygon

Using the Attributes window

1. Select the feature(s) to edit.

2. Click the Attributes button on the Editor toolbar.

To edit a single feature

3. Click the record to edit in the list on the left side (Fig. 12.49). Its attributes will be shown on the right.

4. Click on the attribute field to edit and then type in the new value. Press Enter.

5. Keep editing different fields until you are satisfied.

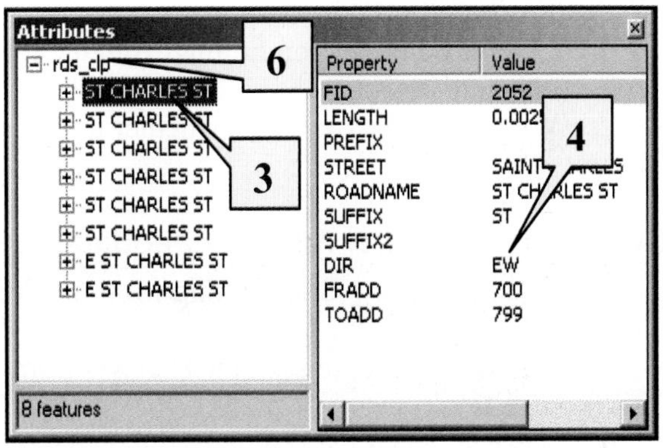

Fig. 12.49. The Attributes window for editing

To edit a field for all the features

6. Click the layer name at the top of the list. The fields on the right will go blank.

7. Click the field to edit and then type in the new value. The new value will replace the existing values for each record in the selected set.

Managing records

1. Clicking an entry in the list causes the feature to flash briefly on the map.

2. Right-click one entry in the list to bring up a context menu (Fig. 12.50).

3. Highlight causes the feature to flash. Unselect removes the feature from the selected set (and from the Attribute Editor). Delete deletes the feature from the layer.

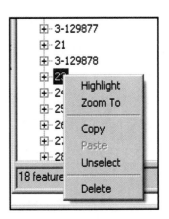

Fig. 12.50. Examining a record in the Attributes window

Changing the display field in the Attribute Editor

1. Double-click the layer in the Table of Contents to open its Properties window.

2. Click the Fields tab.

3. Change the primary display field to the desired attribute.

4. The change will be applied to the Editor when you make the next selection.

Saving changes during an edit session

1. Choose Editor > Save Edits from the Editor toolbar.

2. Continue editing.

Stopping an edit session

1. Choose Editor > Stop Editing from the Editor toolbar.

2. When prompted, indicate whether to save any current changes or discard them.

Switching to another folder or geodatabase

1. Choose Editor > Stop Editing from the Editor toolbar.

2. Choose whether to save the changes from the current session.

3. Make sure that the layers from the new folder to be edited appear in the Table of Contents.

4. Choose Editor > Start Editing from the Editor toolbar.

5. Select the new folder or geodatabase to edit.

Chapter 13. More Editing Techniques

Objectives

➢ Understanding topology and learning to preserve topological rules during editing

➢ Using geoprocessing functions during editing: merge, union, intersect, clip, and buffer

➢ Modifying shapes of existing features

➢ Editing shared lines and polygon boundaries

➢ Editing annotation

Mastering the Concepts

GIS Concepts

Chapter 12 presented some basic techniques for creating new features in a data layer. In this chapter, we examine additional ways to form and modify features with an emphasis on maintaining correct topology during editing.

Topology

Topology concerns how the features are spatially related to each other. The four types of spatial relationships introduced in Chapter 1 included adjacency, connectivity, overlap, and intersection. One major goal during editing, other than simply getting features into a feature class, involves ensuring the **logical consistency** of features, in other words, making sure that features are free of geometric errors and that their topological relationships are adequate for the purpose intended. Some of the topological rules commonly enforced through editing are shown in Figure 13.1.

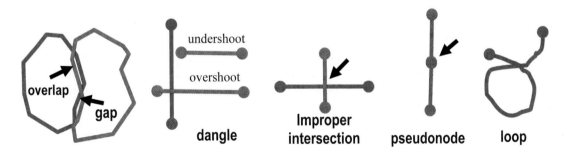

Fig. 13.1. Topological errors can occur when editing is not carefully performed.

(1) Adjacent polygons must share a coincident boundary that is exactly the same for both. There are no gaps or overlaps between adjacent polygons.

(2) Lines should always end on other lines. Failure of two lines to meet is called a dangle. Dangles can occur when one line is not quite long enough to meet the other (undershoot) or when one line crosses too far over the other (overshoot).

(3) Lines that intersect should always each have a node (endpoint) at the intersection. Lines crossing without nodes are termed improper intersections.

(4) Nodes should exist only where three or more lines intersect. A node where only two lines meet is called a pseudonode.

(5) Lines or polygon boundaries should not cross over themselves and form loops.

(6) There should not be duplicate copies of any points, lines, or polygons.

Perfectly legitimate exceptions can occur to topological rules. For example, a dead-end street by definition does not meet another road and must be a dangle. A highway overpass can cross over a street without intersecting it. Polygons showing areas sprayed with pesticide on different dates might have gaps or overlaps between them.

The effects of topologic errors can vary. In some cases they are merely a visual nuisance. At times they can be a legal liability, such as an overlap between two parcels. In other cases they contribute to undesirable outcomes, like the slivers than can be formed during geoprocessing operations. In the worst case, they can cause a failure of the data set for its purpose—a street network containing improper intersections will not properly route traffic. Each user must evaluate the purpose of a data set and establish the topological rules needed. In some cases these rules will include more than the six listed above. However, these six should be considered a basic level of integrity to which every data set should conform (not including unavoidable exceptions). Topological integrity is established and maintained in several ways.

First, careful editing of features minimizes the occurrence of errors at the outset. Most GIS software includes tools for ensuring basic topological integrity during the editing process, such as snapping and the Auto-Complete Polygon discussed in the previous chapter.

Second, some tools can be used to find errors and fix them. Many topological errors are difficult to find by mere inspection but require software algorithms to test and evaluate features at the x-y coordinate level. These tools are generally run on feature classes after editing is complete. The errors that are found are fixed by additional editing, which can create new errors. So the process is repeated until all errors have been found and fixed. These routines are extremely important in developing a **logical consistency** report for the feature class metadata. This report describes the procedures used and tests applied to a data set to ensure that it meets the topological rules above.

Third, some GIS systems contain advanced capabilities for defining topological rules and ensuring that they are met. The ArcGIS geodatabase model is one of these. In Chapter 9 you were introduced to network topology and some of the solvers that evaluated connectivity. Feature data sets within geodatabases can also contain **planar topology**, which expands the use of rules to model a variety of topological relationships, not just within a single layer but between layers as well. For example, the Must Not Overlap rule applies to adjacent polygons within a single feature class. Other rules specify relationships between feature classes, such as the Must Cover Each Other rule that describes the relationship between counties and states in which every bit of the state area must belong to a county and no part of a county should lie outside the state.

Figure 13.2 shows typical topological errors occurring between layers. The boundary of the Pine Ridge Indian reservation (stippled area) should match the boundary between Shannon and Bennett counties, but it does not. Neither does it match the South Dakota–Nebraska boundary.

Creating planar topology makes it easier to locate and eliminate boundary errors by using special Topology tools while editing. This approach is helpful because many errors are too small to see at normal viewing scales—in Figure 13.2 the county boundaries appear to match the state boundaries but might not actually do so. Such errors typically occur because data come from different sources or because some people do not understand or use snapping well.

Using planar topology requires an ArcEditor or an ArcInfo license, so it is not further considered in this text. Users interested in topology should read the appropriate sections in *Editing with ArcMap*, one of the digital books that come with ArcGIS. Users with an ArcView license have much more limited tools for working with topology and must take responsibility during editing to keep features as correct as possible.

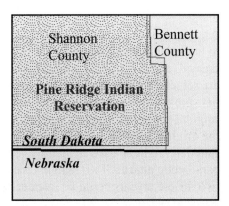

Fig. 13.2. Topological errors between layers

Combining features

Chapter 7 introduced overlay techniques, such as intersection, union, and merge. In the geoprocessing tools, these functions are applied to entire feature classes. During editing these functions can be applied to individual features on an interactive basis and often can aid in maintaining correct topology. The editing functions perform the same operations on features as the geoprocessing ones do, but the attributes are treated differently. Instead of combining all the attribute fields from both inputs, the output feature simply has the same attribute fields as the target layer.

When using these combinations, the user should be sensitive to how attributes of the features are handled. In cases of a merge, union, or intersect, the resulting features will be given the attributes of one of the original features. In a geodatabase, the attributes of the feature selected FIRST will be copied to the output feature. For shapefiles and coverages, the feature with the lowest feature-id will be copied. If the feature attributes matter, then the user must pay attention to which attributes are being copied and must correct any attributes that were not copied as desired.

Merge

A merge takes one or more features and combines them into a single feature (Fig. 13.3). If the two features are adjacent, then the boundaries between them are removed. If the two features are separate, then a multipart feature will be created. This function might be used frequently to update a parcels map when an owner has purchased two adjacent parcels and combined them into one.

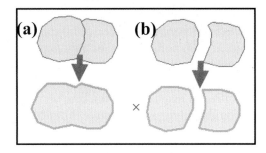

Fig. 13.3. (a) Merging adjacent polygons to create a new feature. (b) Merging separate polygons to form a multipart feature.

Union

A union performs the same operation as a merge, except that the original two features remain unchanged and a new feature is created in addition to the originals.

Intersection

An intersection creates a new feature from an area common to both original features (Fig. 13.4). A new feature is created, and the original ones are maintained. If two features are selected and an intersection is performed, the resulting new polygon will consist only of the areas shared by the original polygons. The new feature is created in addition to the original features, which are not changed or deleted. This function might be used to identify repeat infestations of pine bark beetle attacks. Two features representing infestations in two different years could be intersected to reveal the area attacked twice.

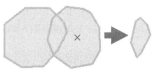

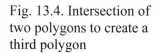

Fig. 13.4. Intersection of two polygons to create a third polygon

Clip

A clip behaves in a cookie-cutter fashion. If one feature lies over another, the underlying feature will be cut along the boundaries of the overlying feature (Fig. 13.5a). Two options may be specified, preserving the area common to both features (Fig. 13.5b) or discarding the area common to both features and retaining what is outside the clip polygon (Fig. 13.5c). In either case, the feature used for clipping is retained unchanged, and the feature underneath is modified. As shown in Figure 13.5c, clipping is one way to create a "donut" polygon. Clipping is useful anytime a polygon must have internal boundaries, such as a residential area with a park in the middle. It is also useful for creating an island polygon inside another. Clipping is crucial to enforcing the "must not overlap" rule by removing sections of a polygon that lie underneath another polygon.

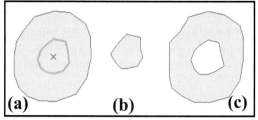

Fig. 13.5. Clipping. (a) Ready to clip the outer polygon with the selected polygon. (b) Option to preserve the common area. (c) Option to discard the common area.

Buffering features

A buffer delineates the area within a specified distance of a feature and can be created from points, lines, or polygons (Fig. 13.6a–c). The output may be lines or polygons. The buffer distance must be specified in map units. In the case of buffering polygons, a negative distance may be used to reduce the size of the feature (Fig. 13.6d). Buffers are useful for such tasks as identifying setbacks from parcels, finding drug-free zones around schools, or creating road widths from a set of centerlines.

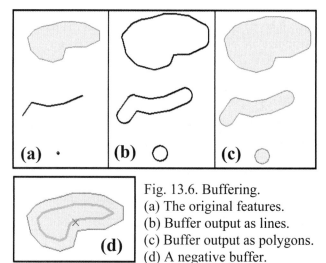

Fig. 13.6. Buffering.
(a) The original features.
(b) Buffer output as lines.
(c) Buffer output as polygons.
(d) A negative buffer.

Complex polygon topology

Some map types, such as geology and soils, have a complex interweaving of different units. It takes some forethought and planning to effectively create the polygons with correct topology with a minimum of effort. Many of the techniques learned in Chapters 12 and 13 can help.

Consider the geology map shown in Figure 13.7. It would be easy to get "painted into a corner" and not be able to finish the sequence of polygons. For example, if one digitized polygons 2 and 5 first, polygon 3 would be a hole that would be a nuisance to fill. The Auto-Complete Polygon tool can only enter another polygon at the start and end of the boundary being created. An island polygon, such as 13, cannot use Auto-Complete at all because it lies completely inside another polygon. It must be added after the polygons around it, and clip must be used to eliminate the overlapping area underneath. Here are two ways to approach the editing. Both methods should avoid gaps and overlaps and result in topologically correct polygons.

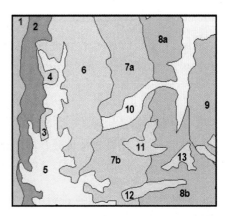

Fig. 13.7. Complex editing

Divide and conquer

Start by creating the outline of the entire map. Then use Cut Polygon Features to carve out polygons 1, 2, and 6 (creating a fake boundary between 6 and 7b that goes through 5). Then create 7 as a single unit, 8 as a single unit, and 9. Finally, use Create New Feature to create 5, 10, 11 to 13, and the remaining islands on top of the larger units. After creating each island, perform a clip to remove the polygon layer underneath.

Adding territory

Start with Create New Feature to create polygon 1. Switch to Auto-Complete polygon and add 2 and 3. Add 4, temporarily closing the narrow neck between it and 6. Add 5 and 6. Go back and merge 4 and 6. Add 7a and 7b, leaving a hole where 10 is. Add 11 and 12 using Auto-Complete. Add 8a, 8b, and 9. Finally close off 10. Go back to Create New Feature and create the islands, making sure to clip out the underlying polygon areas beneath each one.

Stream digitizing

Two methods can be used to place vertices for lines and polygons. So far you have used the point-by-point method in which each vertex is placed with a click (Fig. 13.8a). **Stream digitizing** offers another approach. The user chooses a tolerance that controls how far apart each vertex should fall. The first vertex is added with a click. Then the user smoothly moves the mouse pointer over the line to be digitized. Each time the tolerance distance is covered, a new vertex is added automatically (Fig. 13.8b). Stream digitizing can be faster than point-by-point digitizing, but it does have some drawbacks.

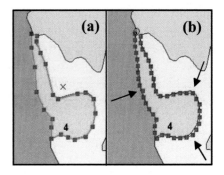

Fig. 13.8. A polygon captured with (a) point-by-point and (b) stream digitizing

First, in point-by-point digitizing the user only adds as many vertices as needed. Straight edges require fewer vertices and curved edges need more closely spaced vertices. In stream digitizing, one tolerance is used for both. Thus, straight lines often have more vertices than needed and may not be as straight, and curved sections may not have enough vertices. As a result, the features may be less accurately captured, and the file size is larger due to the extra vertices. The choice of stream tolerance is critical for good results. The user can turn stream digitizing off and on during the editing of a single feature, but that tends to slow down the digitizing.

Second, when creating a line point by point, the user can place a vertex exactly on the inflection points of the feature to capture the shape with greatest accuracy (Fig. 13.8a). In stream digitizing, the vertex falls when it reaches the set tolerance and may be offset from the actual inflection.

The third problem is that a slight movement off the line will create unwanted vertices that the user must go back and clean up later. A steady hand is definitely required for stream digitizing. In Figure 13.8b, the sketch shows three places where the curve deviates from the feature being digitized (arrows).

Most users find that they develop a preference for one kind over the other. The style of the features being digitized affects the results as well. Feature classes with many long straight lines will be much more efficiently digitized using the point-by-point method. Feature classes with many smooth curves can benefit from stream digitizing.

About ArcGIS

We now turn to a discussion of some additional tools and techniques for editing in ArcMap.

Using different sketching tools

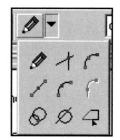

So far, we've used the Sketch tool to enter vertices of features. However, several different sketching tools expand the editing capabilities. If the black arrow on the current sketching tool is clicked, the menu expands to show more tools (Fig. 13.9).

The Distance-Distance tool finds locations that fall a specified distance from two points. Imagine that the owner of a house left a note that the septic tank lies 50 feet from the NE corner of the house and 32 feet from the NE corner of the tool shed. The tool could use this information to pinpoint the two possible locations of the tank when it becomes necessary to dig it up for repairs (Fig. 13.10a).

Fig. 13.9. The nine sketching tools

The Arc tool creates true a **parametric arc** of set radius. The user enters the start point, the point through which the curve passes, and the endpoint, as shown by the patio in Figure 13.10b.

The Intersection tool finds the intersection point of two lines, such as the location where the two sides of a parcel would intersect if the cutout did not exist (Fig. 13.10c).

The Trace tool creates new features, which follow along existing features, either directly on top of them or with a specified offset. For example, the second side of a street could be created from the first side, with the exact width in between (Fig. 13.10d).

Additional sketching tools perform other specialized functions. The Midpoint tool will enter a new point midway between two points entered by the user. The Endpoint Arc tool enters a different kind of curve than the Arc tool. The user enters the two endpoints of the arc and then types in the radius of the arc. The Tangent tool adds a new sketch segment that is tangent to the previously sketched segment. The Direction-Distance tool allows the user to create a new vertex using a bearing direction from a known point plus a distance from another point.

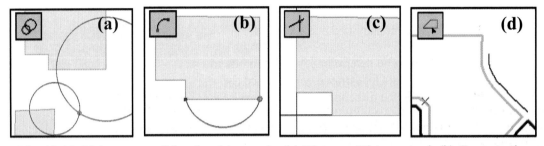

Fig. 13.10. Using some of the sketching tools. (a) Distance-Distance tool. (b) Curve tool. (c) Intersection tool. (d) Trace tool.

Changing existing features

Once a polygon or line has been created, it can be changed by eliminating unwanted vertices, adding vertices, or moving existing vertices. Other modifications include flipping lines, trimming them, extending them, and performing a variety of other editing functions. These functions are useful for cleaning up errors from stream digitizing.

Modifying features

The Task bar contains an option for modifying existing features by adding, deleting, or moving vertices. The Modify Feature task causes the sketch of the feature to appear. When the edits are complete, the feature is updated to the current form of the sketch. Figure 13.11 shows how an original circular polygon is modified by moving vertices in the sketch.

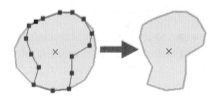

Fig. 13.11. Modifying a feature

Reshaping features

The reshaping task reenters a feature or a portion of a feature by using a new sketch to define the revised shape. The sketch must start and finish on the original feature, so it is helpful to have vertex or edge snapping on. When the sketch is finished, the original feature is modified to follow the sketch. Figure 13.12 shows the process of reshaping parts of lines and polygons.

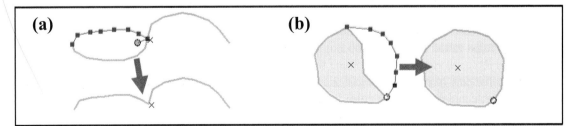

Fig. 13.12. Reshaping features. (a) Reshaping part of a line. (b) Reshaping a polygon.

465

Flipping lines

As we found in the chapter on geocoding, lines can have a direction. Lines begin at the "from" node and end at the "to" node. Normally the direction of a line generates little concern, and either direction works just as well. In some instances, however, the direction matters. In geocoding, it matters because it defines which way the addresses increase along the street. A streams feature class used for network analysis uses the line direction to encode the flow of water. A similar situation would hold for constructing a network of water pipes or sewers in which the direction of flow must be constrained and recorded.

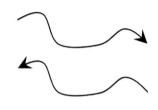

The direction of lines may be determined by drawing them with arrow-ended symbols. Once drawn, you can see the direction of flow and flip the direction, if necessary (Fig. 13.13).

Fig. 13.13. A flipped line

Topological editing

Building planar topology with rules for a feature dataset and using it during editing requires an ArcEditor or an ArcInfo license. However, users with ArcView licenses or users who are editing shapefiles can use a function called **map topology** to edit features with shared edges or vertices. Map topology creates temporary relationships between features so that they can be edited together. Its purpose is to preserve existing coincident boundaries and nodes.

Topological editing uses a **cluster tolerance** to help enforce snapping and coincident boundaries. When two vertices of features being edited fall closer together than the cluster tolerance, the vertices are made coincident. Setting the cluster tolerance requires great care. If too large, it will change coordinates unnecessarily. If too small, it does not help prevent topological errors. However, it is better to err on the side of too small than too large. The default tolerance is designed to preserve the accuracy of features rather than do extensive corrections and should be used unless you have reason to do otherwise.

The user creates map topology by selecting which feature classes will participate in it. The topology is created on the fly for the set of features currently in the data view. The endpoints of lines are called nodes, and lines or polygon boundaries are called edges. The user selects the node or edge to be edited using the Topology Edit tool and then selects one of the topology tasks from the Task bar.

The Topology Edit tool is similar to the normal Edit tool, except that it selects shared boundaries for editing. When using the Topology Edit tool to select a feature or part of a feature, all other features sharing that boundary are also affected by the edits.

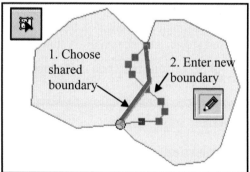

1. Choose shared boundary

2. Enter new boundary

Figure 13.14 shows the use of the Topology Edit tool and the Reshape Edge task to change the boundary between two polygons. First, the shared boundary is selected using the Topology Edit tool. The selection color of the Topology Edit tool is

Fig. 13.14. Reshaping a polygon

purple to distinguish the selection from those made with the Edit tool. The Sketch tool is then used to draw the new boundary between the polygons. When the sketch is finished, the new boundary replaces the old one, and the change is applied to both polygons.

The Topology Edit tool is also convenient when working with connected line features, such as roads. If one road node is moved, the roads attached to the node move also. In Figure 13.15, the road node was selected with the Topology Edit tool and then moved downward. The attached vertices of the other lines are also moved when the sketch is finished.

The Topology Edit tool can in many cases be used in place of the Edit tool when editing. It can be applied to moving features, reshaping them, and modifying them. However, the map topology must be created prior to using the Topology Edit tool.

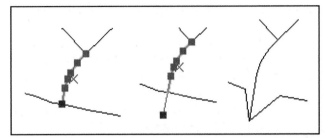

Fig. 13.15. When editing shared nodes, the node is moved, and the attached roads follow.

Editing annotation

In Chapters 1 and 3 you learned about dynamic labels and creating annotation that exists as simple text within a map document. Annotation can also be stored inside a geodatabase as a feature class, in which case it can be used in multiple map documents. Annotation feature classes can only be modified using the Editor, which is why a discussion of feature annotation has waited until this chapter. The tutorial covers the Annotation Editor toolbar and the special functions and tools available to create and modify annotation.

Summary

> Careful editing helps create and maintain topological integrity between features. Topological rules establish requirements for feature adjacency, connectivity, overlap, and intersections.

> Planar topology establishes rules about the spatial relationships within and between layers, and can be used to help locate and eliminate errors. Building and editing with planar topology requires an ArcEditor or an ArcInfo license.

> Features may be combined to create new ones using merge, union, intersect, and clip. Care must be taken to ensure that attributes are correctly copied during these transactions.

> New features can be created by buffering points, lines, or polygons.

> Strategic planning helps when digitizing complexly arranged polygon feature classes.

> Stream digitizing can help create smooth curved boundaries, but it has some drawbacks.

> The sketching tools provide nine different ways to create vertices.

> Modifying and reshaping are two ways to change existing features.

> ➤ Temporary topologic relationships, called map topology, may be used when editing with an ArcView license, allowing features that share vertices or boundaries to be edited simultaneously with the Topology Edit tool.

Chapter Review Questions

You may need to consult the Skills Reference section to answer some of these questions.

1. Define each of the following: dangle, overlap, gap, improper intersection.

2. What is meant by the term logical consistency?

3. What is the difference between modifying a feature and reshaping a feature?

4. What is the difference between a union and a merge?

5. What determines the direction of a newly created line?

6. Examine the drawing. In (a), the polygons were created one after the other using Create New Feature. In (b), the first polygon was created as a donut and then filled using the Trace tool. They look identical, but they are not. What is the difference? (**Hint:** What is the area of the outer polygon in each case?)

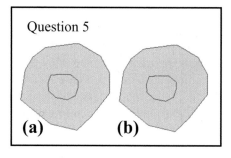

Question 5

(a) (b)

7. How do you establish whether the output feature of a buffer command is a line or a polygon?

8. What are the two types of output features possible when you divide a line?

9. Which sketching tool can be used to fill a hole in a polygon?

10. Explain the differences between map topology and planar topology in a geodatabase.

Mastering the Skills

Teaching Tutorial

The following examples provide step-by-step instructions for doing basic tasks and solving basic problems in ArcGIS. The steps you need to do are highlighted with an arrow ➔; follow them carefully. Click on the video number in the VideoIndex to view a demonstration of the steps.

➔ Start ArcMap and open ex_13.mxd in the mgisdata\MapDocuments folder.

➔ Add the practice layers from last chapter, practicept, practiceln, and practicepo.

➔ Use Save As to rename the document and remember to save frequently as you work.

Using other sketching tools

First, we will practice using some of the different sketching tools. Figure 13.16 shows the block numbers from the last lesson as a reminder to help navigate to the right places during this tutorial.

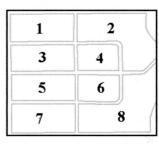

Fig. 13.16. Block numbers

1➔ Choose Start Editing from the Editor toolbar.

1➔ Zoom in to Block 1, showing the buildings and parcels.

1➔ Zoom in to the third house from the left in the lower row (Fig. 13.17).

Imagine drawing a patio nestled in the cutout on the NE side of the house. The patio is flush with the edges of the house. Use the Intersection tool to lay out this patio precisely.

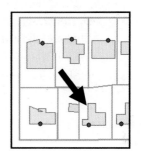

Fig. 13.17. Zoom to this parcel

1➔ Choose Editor > Snapping from the Editor toolbar. Dock and adjust the Snapping Environment window under the Table of Contents, if necessary.

1➔ Turn on Vertex snapping for the buildings layer.

2➔ Make sure that the target layer is practiceln, the task is Create New Feature, and the Sketch tool is clicked.

2➔ Place the first vertex at point 1 (Fig. 13.18), making sure it snaps to the corner.

We want the next vertex to lie at the intersection of the two building walls enclosing the patio. We will use the Intersection tool for this task.

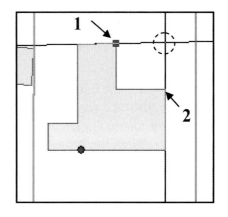

Fig. 13.18. Finding the intersection of the east and north walls

 2➔ Switch to the Intersection tool using the drop-down box next to the Sketch tool.

2➔ Place the cursor over the east wall of the house. A black line will appear, indicating the direction of the wall. When the correct line appears, click. The line remains.

2➔ Place the cursor over the north wall so that its line intersects the first. Click again.

The point is automatically placed at the intersection when you click the second line. Finally, we need to add the third point at 2.

2➔ Switch back to the Sketch tool.

2➔ Move the cursor toward point 2 until it snaps to the corner of the building.

2➔ Double-click to enter the third vertex and complete the sketch (Fig. 13.19).

Fig. 13.19. The patio

Next we'll use the Arc tool to make a curved patio.

3➔ Select and delete the first patio.

3➔ Switch to the Arc tool.

3➔ Start the patio again at point 1 of the house (Fig. 13.20).

3➔ Click another point at the approximate intersection of the north and east walls. The arc will pass through this point.

3➔ Double-click on point 3 to add the third vertex and finish the sketch (Fig. 13.20).

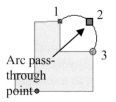

Fig. 13.20. Sketching an arc

Consult the Skills Reference to learn about the other Sketching tools.

Stream digitizing

Let's experiment with stream digitizing on the roads. We will be careful to create proper topology by making sure that we have a node at every road intersection.

First we must select a stream tolerance, starting with 30 meters and adjusting it later, if necessary.

4➔ Zoom to the extent of the orthophoto and locate the neighborhood of Star Village north and west of the eastpat neighborhood (Fig. 13.21). Zoom in to this area.

5➔ Choose Editor > Options on the Editor toolbar.

5➔ On the General tab in the Stream Mode box, change the stream tolerance to 30. Click OK.

5➔ Change the target layer to practiceln.

5➔ Turn on end snapping for practiceln.

Fig. 13.21. Locating Star Village

6➔ Click the Sketch tool.

6➔ To place the node, click once on the north end of the road connecting Star Village to the other streets.

6➜ Right-click on the display area and choose Streaming from the Sketch menu. Notice that the F8 key is a shortcut to toggle Streaming off and on.

6➜ Click once more to begin streaming. Add a vertex on the road near the first.

6➜ Carefully move the cursor along the street centerline. Points will be added as you move. No clicking to add vertices is needed.

6➜ Stop moving when you get to the next intersection and click once to add the final vertex. Then right-click and choose Finish Sketch.

Chances are that your first attempt at streaming did not turn out well. You can delete the feature and try again. The second time you will not need to toggle Streaming on. Just click once to begin the feature, stream along it, and then right-click and finish the sketch (or double-click). When you are satisfied with the first street, go on.

7➜ Place the cursor on top of the southern end of the road you just digitized, making sure you snap to it. Click to begin streaming.

7➜ Stream digitize eastwards. Stop when you get to the end of the road.

Hopefully it went better this time. However, you forgot something. Topology rules state that lines should always intersect at a node. When you add the curved road north of this street, even if you snap to the existing road, the nodes will be absent. You could have stopped and restarted the line at each intersection. Instead, you can fix it now.

8➜ Click the Split tool and place a node where the undigitized street will meet the digitized one. Both sides become unselected after the split.

8➜ Click the Edit tool, if necessary, and select the eastern half of the street.

8➜ Click the Split tool again and add another node at the other intersection.

9➜ Click the Sketch tool. Begin at one end of the curved street, making sure you snap to the node you added.

9➜ Stream digitize around the curve. Make sure you snap to the other end and finish the sketch.

10➜ Click on the upper end of the next road east, the straight one that runs north-south.

10➜ Click F8 to toggle Streaming off. Point by point works better for straight lines.

10➜ Add the next vertex where the small street heads east.

10➜ Add another vertex where the east-west road meets the north-south street, making sure you snap to its end.

10➜ Add a final vertex at the end of the road and double-click to finish the sketch.

11➜ Click the Split tool and split the new line where it meets the east-west street.

11➜ Select the upper part. Turn on vertex snapping.

11➜ Snap to the vertex in the upper part and split the road.

11➜ Add the final small road heading eastwards.

> **TIP:** An alternate way to ensure proper road topology is to stop the feature each time you encounter an intersection. Otherwise, you can do splits after a group of roads are added.

→ Add the remaining roads in Star Village (Fig. 13.22). Use vertex, end, or edge snapping to make sure features don't have dangles but don't worry about nodes at intersections for the moment. Use either stream or point digitizing.

→ Click the Edit tool. Select each road in turn. If it requires a split at an intersection, use the Split tool to add it. Continue until all the roads have proper topology.

Fig. 13.22. Star Village roads

Maintaining good topology requires effort. If these roads were in a feature dataset with topological rules, then the intersections could be corrected later. However, if the feature class is a shapefile or the user only has an ArcView license, then the topology must be a concern through the entire editing process.

Changing polygons

Next, we will go back and modify some of the parcels we created in the last chapter. Begin by entering the parcel boundaries of an elementary school.

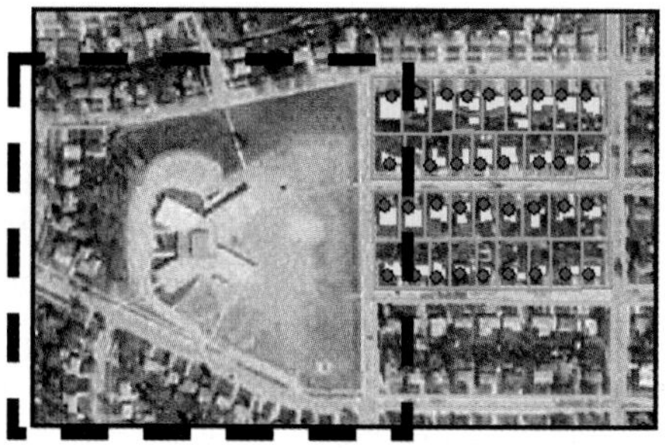

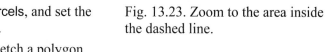

12→ Zoom to the extent of the streets layer and then pan into the open block to the west, which has a star-like school building in the center (Fig. 13.23).

12→ Set the target layer to parcels, and set the task to Create New Feature.

Fig. 13.23. Zoom to the area inside the dashed line.

12→ Using the Sketch tool, sketch a polygon around this entire block, bounded by the street edges. Finish the polygon.

The side of the street is rather difficult to see on the west edge of the parcel. Let's zoom in and try to do a better job.

13→ Zoom in as close as possible to the west edge of the new parcel but still showing the entire street.

13→ Turn on vertex snapping for parcels.

13→ Change the task to Reshape Feature.

13→ Select the parcel, if necessary.

13→ Switch to the Sketch tool.

14→ Click on the parcel polygon at the north end of the bounding street.

14→ Click new vertices along the street, defining a sketch.

14→ At the south end of the street, double-click one last vertex on the corner of the parcel to finish the sketch and create the new edge.

➔ Experiment until the line is satisfactory. Try starting and stopping the Reshape task in different locations along the polygon.

TIP: It is usually helpful to use vertex or edge snapping when reshaping features.

That is how to reshape a standalone feature, but shared editing works best when reshaping a boundary between adjacent polygons.

15➔ Click the blue Back to Previous Extent arrow to return to the previous extent.

15➔ Turn on edge snapping for the parcels layer.

15➔ Change the task to Cut Polygon Features, and cut the block in two with a straight north-south line to the west of the school.

Now let's redraw this straight line to more closely match the real parcel boundary. In order to edit it as a shared boundary, however, you must first create a map topology.

16➔ Choose Editor > More Editing Tools > Topology from the Editor toolbar. The Topology toolbar will appear.

16➔ Click on the Map Topology icon on the Topology toolbar.

16➔ Check the box to have the parcels layer participate in the map topology. Leave the cluster tolerance at the default value. Click OK.

17➔ Change the task to Reshape Edge under the Topology Tasks group in the Task bar.

17➔ Turn on vertex snapping for parcels.

17➔ Click the Topology Edit tool on the Topology toolbar.

17➔ Click on the north-south line in the middle of the block (Fig. 13.24). It will turn purple to indicate it is a shared boundary.

17➔ Click the Sketch tool.

17➔ Click the north end of the boundary to start reshaping. Be sure to click ON the end.

17➔ Enter new vertices that go behind the houses, as shown in Figure 13.24.

17➔ Click again on the south end of the shared boundary to end the sketch. Be sure to click ON the end.

17➔ Right-click and choose Finish Sketch.

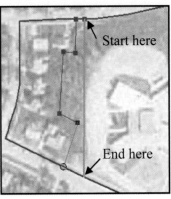

Fig. 13.24. Reshaping the shared boundary

Next we will learn how to modify a sketch, instead of reshaping it.

18➔ Change the target layer to buildings.

18➔ Change the task to Create New Feature and click the Sketch tool.

18➔ Create a building polygon for the large building in the center of the block. Be sloppy about it.

18➜ Change the buildings symbol to a hollow red line to show the photo underneath.

18➜ Zoom in to the building in order to see it as clearly as possible.

As intended, the building does not exactly match the photo. Let's improve the fit.

19➜ The building should still be selected; if not, select it.

19➜ Change the task to Modify Feature. The building sketch appears.

19➜ Make sure the Edit tool is clicked on the Editor toolbar.

19➜ Move the cursor over a vertex of the building. When it changes to a crosshair, click and drag the vertex to a new location.

19➜ Right-click another vertex and choose Delete Vertex from the Sketch menu.

TIP: For best results, the Edit tool should be active while modifying the feature.

20➜ Right-click in the middle of the segment and choose Insert Vertex to replace the vertex just deleted.

20➜ Move the vertex back to its original location.

20➜ Continue moving, adding, or deleting vertices until you are satisfied with the shape of the building.

20➜ Right-click on the sketch and choose Finish Sketch.

Notice that the task changes back to Create New Feature (or possibly something else). Now clean up and prepare for the next task, changing line features.

21➜ Choose Save Edits.

21➜ Change the building symbol back to the solid beige color.

21➜ Zoom to the extent of the roadlines layer and turn it on.

Changing lines

The roadlines layer represents the centerlines of the roads, but they do not match the orthophoto roads. Editing these roads will provide a better match with the photo and the rest of the data.

First, we will move all of the roads as a group to get the best average fit. Examine the roads and note the direction and distance of offset in different locations of the map. Note that the offset is generally to the west and slightly south of the true location. We will move all the streets to a better position, being careful not to move any other features.

22➜ Choose Selection > Set Selectable Layers from the main menu bar and make roadlines the only selectable layer.

22➜ Change the target layer to roadlines.

22➜ Click the Edit tool and select ALL of the road lines. Don't miss any!

23➜ Zoom in to the lower left corner of the roadlines layer.

23➜ Click on the Edit tool.

23➜ Click and drag the lower left road features to the right and slightly up so that the corner falls in the exact middle of the road in the photo (Fig. 13.25). Move it several times, if necessary, to get as close as possible.

23➜ Zoom back to the extent of the roadlines layer again.

Already the road locations look much better, but some areas still need adjustment, such as the lower right near the corner of Block 8.

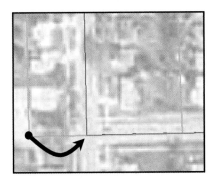

Fig. 13.25. Move the road corner to the center of the road in the photo.

24➜ Clear the selected features so that nothing is selected.

24➜ Zoom in to the roads extending down from the lower right corner of Block 8.

24➜ Turn off snapping for parcels and turn on end snapping for roadlines.

The view should look similar to the picture in Figure 13.26. The following directions refer to the street segments according to the lettering scheme indicated. Since these features must connect at their ends, always use the Topology Edit tool during these edits. First, however, we must add the road lines to the map topology.

25➜ Click the Map Topology button on the Topology toolbar. Uncheck parcels and check roadlines. Click OK.

25➜ Choose the Topology Edit tool.

25➜ Select the node at the intersection of streets A, B, and C. (If you get an edge, hold down the N key to ensure that a nodes must be selected.)

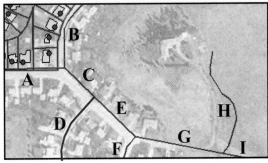

Fig. 13.26. The vicinity of East Oakland St.

25➜ Click and drag the node to the intersection point of the streets in the Orthophoto.

25➜ Select the node at the other end of segment C and move it to the intersection on the Orthophoto of segments D and C/E.

Street C moves to its new location. Because you used the Topology Edit tool, the streets connected to C moved also, so already this section looks much better.

TIP: To have attached features proportionally stretched when moving nodes, check the box on the General tab of the Editing > Options menu. This option provides easier adjustments but causes greater changes to coordinates.

25➜ Repeat the previous steps with the lower end of Street E to match it to the intersection in the photo.

Examining Street B up close, notice that its last segment flattens out and does not follow the photo street as well as it could. Use Reshape to fix it.

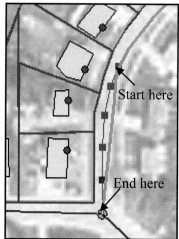

26➜ Turn on edge snapping for roadlines.

26➜ Zoom in to the lower end of Street B where the flat section is (Fig. 13.27).

27➜ Click Street B with the Topology Edit tool.

27➜ Choose Reshape Edge from the Topology Tasks in the Task bar.

27➜ Click the Sketch tool.

27➜ Click above the flat section to start the reshaping sketch.

27➜ Enter new vertices along the street in the photo.

27➜ Double-click at the intersection of Streets A, B, and C to end the sketch.

Fig. 13.27. Reshaping Street B

TIP: If the sketch does not result in a changed shape, try starting the reshape sketch from the other end and be sure to cross over the original feature at the other end.

Next, Street F has a bit of a kink where it runs into Streets E and G.

28➜ Zoom to the previous extent and zoom in to the intersection of F, E, and G.

28➜ Change the task to Modify Edge in the Topology Tasks.

28➜ Select Street F with the Topology Edit tool.

28➜ Add, delete, and move vertices until the kink is gone. (You cannot move the end vertex because it is shared by other features.)

28➜ Right-click the sketch and choose Finish Sketch.

28➜ Save your edits.

More techniques for polygons

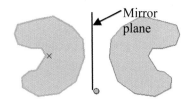

We could continue adjusting roads, but instead we will now explore some of the geoprocessing-style functions.

29➜ Turn off the practiceln layer.

29➜ Turn off all snapping.

Fig. 13.28. Mirroring a feature

29➜ Change the target layer to practicepo and turn it on. Turn the photo off.

29➜ Zoom to the extent of the roadlines layer and then pan left to the open area.

First we'll learn how to mirror a polygon.

30➜ Make sure the task is Create New Feature, and the Sketch tool is selected.

30➜ Create an irregular polygon with an asymmetric shape (Fig. 13.28).

30➜ Make sure the polygon is selected.

31➜ Choose the Mirror Features task from the Editor toolbar.

31➜ Make sure the Sketch tool is selected.

31➜ Enter the start and end vertices of a line that defines the mirror plane (i.e., the line across which the figure is mirrored). The feature is mirrored as soon as you enter the second point of the line.

Now we will experiment with combining features.

32➜ Set practicepo to be the only selectable layer.

32➜ Select the two polygons with the Edit tool.

32➜ Choose Editor > Merge from the Editor toolbar.

32➜ Select the first polygon in the list and watch it flash. The other polygon will flash if it is selected. Select the first and click OK (Fig. 13.29).

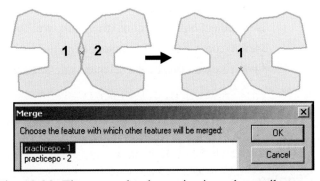

Fig. 13.29. The merged polygon is given the attribute values of the polygon selected in the Merge window.

TIP: The attribute values of the merged polygon will become the same as the attributes of the selected polygon in the window. (The polygon list shows the primary display field for the layer.)

No change may be visible, but the multiple polygons are now in fact a single feature.

32➜ Click off the polygons with the Edit tool to clear the selection.

32➜ Now click on ONE of the polygons. They are now a single multipart feature.

32➜ Open the practicepo attribute table and notice that it contains only the single feature (unless you have other polygons from a previous session, in which case clicking Show Selected will demonstrate that these two shapes form one multipart polygon).

Both polygons are highlighted when you click one, showing that they are a single multipart feature. Moreover, the table contains only a single record for both.

33➜ Resize the table and place it to one side so that you can see both it and the polygons.

33➜ Set the task back to Create New Feature and click the Sketch tool.

33➜ Create a new polygon such that it overlaps both parts of the merged polygon (Fig. 13.30).

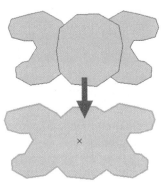

Fig. 13.30. Merging a new polygon with the merged mirror polygons

33➔ Select all of the features with the Edit tool.

33➔ Choose Editor > Merge from the Editor toolbar.

The boundaries disappear and now all the features form a single merged feature.

1. How many polygons are selected now? _____

Next we will see how Intersect works.

34➔ Choose Edit > Undo Merge from the main menu bar.

34➔ Select both polygons (there are two, the merged mirror features and the one on top).

34➔ Choose Editor > Intersect from the Editor toolbar.

34➔ Click on one of the selected intersection areas and drag them below the original features.

2. Are the two areas separate polygons, or are they part of a multipart polygon? _____ How can you tell? _____

Clipping

Finally, let's examine the clip function.

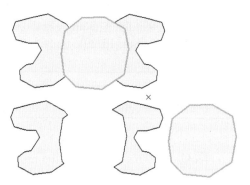

35➔ Delete the intersection result.

35➔ Select the single polygon area overlapping the merged mirror polygon (Fig. 13.31).

35➔ Choose Editor > Clip from the Editor toolbar.

35➔ Keep the buffer at zero. Choose to discard the area that intersects and click OK.

Fig. 13.31. Clipping with the circle polygon

35➔ Click and drag the clip polygon out of the way to see the result.

36➔ Select and delete all of the polygons in the view.

36➔ Close the attribute table.

Clip is highly useful for creating island polygons that are surrounded by other polygons. Imagine that you have a geological unit made of carbonate rock surrounding a small stock of granite.

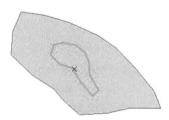

37➔ Use Create New Feature to create the carbonate polygon as shown in Figure 13.32. Finish the sketch.

37➔ Create another new polygon inside the carbonate, representing the stock. Finish the sketch.

Fig. 13.32. A granite stock inside a carbonate unit

Now the stock is covering part of the carbonate, like a wedding cake. This is an overlap, a topological error. We use clip to remove the carbonate under the stock.

37➜ Choose Editor > Clip.

37➜ Choose to Discard the area that intersects. Click OK.

37➜ Just to check, move the still selected stock polygon away from the carbonate. The carbonate does indeed have a hole in it.

37➜ Choose Edit > Undo Move to return the stock to its location.

You can delete the stock if what you really want is a hole in a polygon (although in many cases this is a topological error). Finish Part is a faster way to create a polygon with a hole inside (a donut).

38➜ Make sure the Create New Feature and the Sketch tool are selected.

38➜ Begin a sketch and create a large circular polygon area. Don't finish it yet (Fig. 13.33).

38➜ After adding the last vertex, right-click and choose Finish Part.

38➜ Click inside the first sketch and create another ring sketch outlining the inside of the donut.

38➜ After completing the second ring, right-click and choose Finish Sketch.

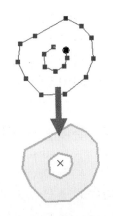

Fig. 13.33. Creating a multipart donut polygon

Buffering

Buffers create new features around existing ones. Suppose we want to create street edges from the road centerlines. Assuming that the average road width in this part of town is 15 meters, we will buffer half that distance, 7.5 meters.

39➜ Zoom to the extent of the roadlines layer.

39➜ Make roadlines a selectable layer, keeping practicepo selectable also.

39➜ Make sure that practicepo is the target layer.

39➜ Delete any polygons from practicepo that might still be in the view.

40➜ Use the Edit tool to click and drag a box around all of the roads to select them.

40➜ Choose Editor > Buffer from the Editor toolbar.

40➜ Type the value 7.5 in the distance box and press Enter.

Fig. 13.34. Streets created by buffering the road lines

TIP: When buffering, you may get an error message stating that the units or measures are out of bounds. If this happens, save your edits, stop editing, save the map document, and close ArcMap. Start ArcMap again and reopen the map document. This routine usually fixes the problem.

Since each segment is buffered individually, the buffers intersect at the corners. Open corners are more attractive and realistic and can be created by merging the buffers (Fig. 13.34).

41➔ Make sure the buffers are still selected.

41➔ Choose Editor > Merge from the Editor toolbar. Select any of the polygons in the list and click OK.

41➔ Save all edits.

TIP: You can also buffer points and polygons. To create a *negative* buffer that *shrinks* a polygon, enter a negative value for the buffer distance.

Editing annotation

Annotation can be stored in a feature class, which makes it available for use in many map documents. We will create road name annotation for the roadlines layer and edit it.

42➔ Zoom to the roadlines layer and turn off all other layers. Close the Topology toolbar if it is still open.

42➔ Open the Labels tab of the roadlines layer properties.

42➔ Create labels using the ROADNAME field with 8-point Arial font.

43➔ Right-click the Layers data frame name and choose Convert Labels to Annotation.

43➔ Choose to store the annotation in a database. Notice the Destination box; the annotation will be saved in the eastpat geodatabase with the name roadlinesAnno.

43➔ Click Convert. The roadlinesAnno feature class is added to the data frame.

TIP: If there are any unplaced labels, they will be placed in an Overflow window. You can place them individually as you learned in Chapter 3.

44➔ Set the target layer to roadlinesAnno.

44➔ Right-click the gray area on the main menu bar and open the Annotation toolbar.

44➔ Click the Edit Annotation tool on the Annotation toolbar.

44➔ Select E OAKLAND ST.

When annotation is selected, there are four handles within it (Fig. 13.35). The shape of the cursor changes when it is placed on top of a handle, indicating the action that will occur on a click and drag. The two blue handles on the ends rotate the annotation. The red triangle enlarges or shrinks the text. The black cross allows the annotation to be moved.

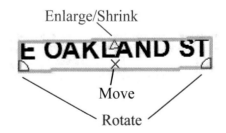

Enlarge/Shrink

Move

Rotate

Fig. 13.35. Action points on selected annotation

45➔ Place the cursor near the top middle until it changes to a two-ended arrow. Click and drag to enlarge or shrink the text. Leave it large.

45➜ Place the cursor near the bottom middle until it changes to a four-ended arrow. Click and drag to move the text.

45➜ Place the cursor near the end until it changes to a rotate tool. Click and drag to rotate the text.

45➜ Shrink the text back to its approximate original size and move it back to its starting location.

46➜ Right-click E OAKLAND ST and choose Attributes (Properties in version 9.2).

46➜ Examine the options on the Annotation tab (enlarge the window, if necessary, to see the font buttons). You can set the font properties and retype the text, if necessary.

46➜ Examine the attributes tab. Each piece of annotation can have all these properties set individually.

46➜ Return to the Annotation tab. Click the drop-down box, choose Default for the symbol, and click Apply. The text will go back to original size.

The Attributes window is the same one you encountered in Chapter 12. It has different tabs when you are working with annotation.

47➜ The Edit Annotation tool should still be active. Hold down the Shift key and click on E ST PATRICK ST to select it also. Now both annotation features appear in the Attributes window.

47➜ Click on the roadlinesAnno entry at the top of the window on the left side of the Attributes window.

47➜ Click the B button for boldface type and click Apply. Both annotation items are updated to Bold font.

47➜ Close the Attributes window.

Now we will create new annotation for the streets in the southeast corner.

48➜ Click the Straight Annotation button.

48➜ Type **SIDNEY DR** in the Text box on the Annotation toolbar.

48➜ Click on the center of **SIDNEY DR** shown in Figure 13.36.

48➜ Rotate the text to the desired position and click again to place it.

48➜ If you don't like where it was put, press the Delete key to get rid of it and then place it again. (You don't need to type the text again.)

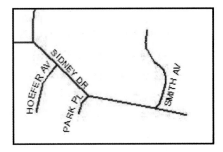

Fig. 13.36. Street names in the southeast corner of the map

49➜ Click in the Text box and type **ASPEN ST**.

49➜ Place the annotation on the unlabeled road that goes north from Maywood Dr.

50➜ Click the Edit Annotation tool on the Annotation toolbar.

50➜ Click on E OHIO ST to select it. Move it to the center of the block.

50➜ Right-click E OHIO ST and choose Copy.

50➜ Right-click anywhere and choose Paste. A copy of E OHIO ST will be placed on top of the original.

50➜ Click and drag the copy to the eastern part of the street between Walnut Dr and Smith Av. Rotate and adjust its position until you are satisfied.

Flipping annotation, like flipping a line, makes it go in the opposite direction.

51➜ Select BALSAM AV with the Edit Annotation tool.

51➜ Right-click it and choose Flip Annotation. Now it matches the other nearby annotation for easier reading.

Placing curved annotation is slightly more complicated, but not much.

52➜ Change the Construction drop-down box on the Annotation toolbar to Curved.

52➜ Type HOEFER AV in the Text box on the Annotation toolbar.

52➜ Click at the end of Hoefer Av as shown in Figure 13.36. Place another vertex at the bend and a final one at the end. Double-click to end the sketch.

52➜ If you don't like the result, delete it and try again.

Annotation can be made to follow a particular feature, either curved or straight.

53➜ Select E MEADE ST with the Edit Annotation tool.

53➜ Right-click the segment of E Meade St underneath the E ST FRANCIS ST annotation and choose Follow This Feature. The text is moved to the point you clicked on and rotated to follow the street angle.

53➜ Place the cursor on the text until you see the four-ended arrow. Click and drag the text above and below the street, and then leave it above.

54➜ To make new annotation that follows a feature, click the Straight Annotation tool on the Annotation toolbar.

54➜ Change the Construction Method to Follow Feature.

54➜ Type City Bus Route in the Text box.

54➜ Click along the inclined segment of E Meade St to select the feature on which to place the annotation.

54➜ Move the mouse until you are satisfied with the position and then click again to place the annotation.

54➜ Click the Attributes button on the Editor toolbar. Change the font to italic and click Apply. Close the Attributes window.

55➜ Click the Edit Annotation tool and select MAYWOOD DR.

55➜ Press the O key on the keyboard to open the Follow Feature Options.

55➜ Fill the button for Curved and click OK.

55➜ Right-click on the center of Maywood Dr and choose Follow This Feature.

55➔ While the annotation is still selected, place the cursor on it and click and drag it above, below, or along the road to get the best placement and curvature. Release the mouse button when you are satisfied with the placement.

You can also add annotation with a leading line.

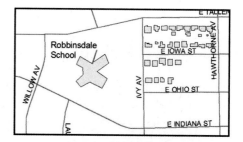

Figure 13.37. Stacked annotation

56➔ Turn on the buildings layer.

56➔ Select the Leader Line tool.

56➔ Type **Robbinsdale School** in the Text box.

56➔ Click on the large school building. Click again a little ways away to place the annotation (Fig. 13.37).

56➔ Click the Edit Annotation tool. Move the cursor over the text to the Enlarge/Shrink handle and enlarge the text slightly.

You can stack annotation for better placement.

57➔ Click on the Robbinsdale School annotation to select it.

57➔ Right-click and choose Stack.

57➔ Open the Attributes window and change the annotation to left-justified text. Click Apply and close the Attributes window.

57➔ Click and drag the annotation to a good position as shown in Figure 13.37.

Because annotation is a feature class, you must save the edits to make them permanent.

58➔ Choose Editor > Save Edits.

58➔ Choose Editor > Stop Editing.

➔ Exit ArcMap. You don't need to save your changes.

This is the end of the tutorial.

More skills

Consult the Skills Reference section of this chapter to learn to do the following:

➢ Use more of the sketching tools, such as the Trace tool

➢ Copy lines parallel to each other with a specified offset distance

➢ Divide lines into pieces or place points a specified distance along a line

Exercises

For the first three questions, use your practicepo or practiceln shapefile to create the shape shown. Take care to make the feature parts the same size and/or equidistant if they look that way in the example. Also list the commands and tools you used to create it. **Capture** your figures.

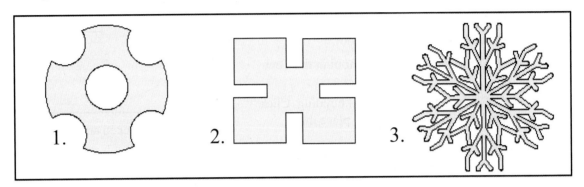

4. Bring the mgisdata\Rapidcity\geologyclip1.img image into ArcMap and digitize the geology into a shapefile with the same coordinate system as the image. Include a field for the geology unit code. Use the techniques learned in the last two chapters to ensure that there are no gaps or overlaps between polygons. After you are done, calculate the areas of the features (what are the units?) and find the total. Convert it to square miles and include it with your answer to check whether you have overlaps. Create a layout showing the map with pastel colors, labels, and the Rapid City roads and onsite systems on top.

5. Finish adjusting the roads in the roadlines feature class to match the Orthophoto.

6. Finish editing the roadlinesAnno annotation for the best possible appearance and legibility.

7. Create a layout showing your adjusted roads and annotation.

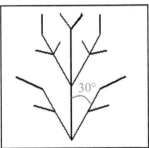

Hint on Exercise 3:
Create half of this first.

Challenge Problem

Design and construct a system of water mains for the neighborhood included by the road lines. Establish a realistic flow of water through the network by having the lines in the correct direction. Every building must have reasonably close access to water. Create a layout showing your work, complete with arrows on the lines. Put the orthophoto (60% transparent) in the background.

Skills Reference

Using the sketching tools 485

Modifying a feature 489

Reshaping a feature 490

Flipping a line..490

Mirroring features..491

Dividing a line..491

Splitting lines exactly 492

Moving a feature an exact distance 492

Using Copy Parallel..492

Trimming lines 493

Extending lines 493

Merging features..493

Union of features 494

Intersection of features 494

Clipping features 494

Creating donut polygons 495

Buffering features..496

Editing shared features 496

Editing annotation 498

Using the sketching tools

The following techniques may be used to create a point or to place the next vertex of a line or polygon. **Note that in each case below, you have already selected the target layer and the task.** Each sketching tool can be used at the beginning, middle, or end of a sketch. You can also use multiple tools in succession during a single sketch.

Finding intersections

1. Click the small black arrow on the current sketching tool and choose the Intersection tool.

2. As you move the tool over a line on the screen, its extension will be indicated. Click the first line on the south side (1 in Fig. 13.38).

3. Click the second line on the west side. Now the point of intersection will be visible where the extensions of the two lines cross.

4. Click near the intersection to enter the point or vertex.

5. Finish the sketch, or switch to another sketching tool and keep sketching.

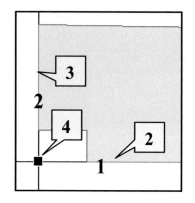

Fig. 13.38. Placing a vertex at the intersection of two lines

Using distance-distance intersection

1. Click the small black arrow on the current sketching tool and choose the Distance-Distance tool.

2. Click at the center of the first distance circle and move the mouse out to expand the circle (Fig. 13.39).

3. Press the D key to bring up the Distance window, and type in the exact distance in map units. Press Enter to close the window.

4. Click at the center of the second circle, and type D to enter a distance for it. Press Enter to close the window.

5. Move the cursor over one of the two intersection points (a blue circle will appear). Click to create the point or vertex.

6. Finish the sketch, or switch to another sketching tool and keep sketching.

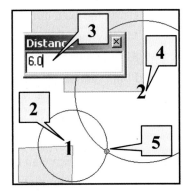

Fig. 13.39. Finding the intersection point of two distance circles

Creating a parametric arc

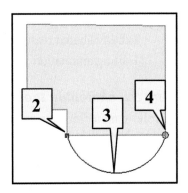

1. Click the small black arrow on the current sketching tool and choose the Arc tool.

2. Click the starting point of the curve (Fig. 13.40).

3. Click the point through which the curve will pass.

4. Click the endpoint of the curve.

5. Continue adding pass-through points and endpoints to create multiple curves, if desired.

6. Finish the sketch, or switch to another sketching tool and keep sketching.

Fig. 13.40. Sketching a parametric curve

Tracing along features

1. Click the Edit tool and select the feature to be traced (Fig. 13.41).

2. Click the small black arrow on the current sketching tool and choose the Trace tool.

3. To offset from the original feature, press the letter O key on the keyboard to specify an offset distance in map units. Type the distance into the box and press Enter. (If the trace appears on the wrong side, press O again and make the distance negative.) If desired, also change the style of the corners.

4. Click on the feature to begin tracing and move the mouse along the feature to create the trace.

5. Click again to finish the trace and create the sketch.

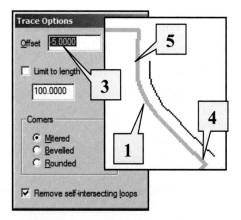

Fig. 13.41. Creating a sketch by tracing

6. Finish the sketch, or switch to another sketching tool and keep sketching.

> **TIP:** Features have ending locations according to how they were digitized. A trace stops when it reaches the end of the feature. To continue tracing, click at the stopping point to end the current trace and then click again at the same point to begin a new trace.

Using the Endpoint Arc tool

With this tool the user specifies the beginning, end, and radius of an arc.

1. Click on the Endpoint Arc tool.

2. Enter the first point of the arc (Fig. 13.42).

3. Enter the second point of the arc.

4. Move the mouse to the desired size.

5. To specify an exact radius, press the R key, type in the radius, and press Enter.

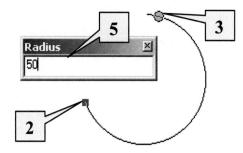

Fig. 13.42. Using the Endpoint Arc tool

Using the Midpoint tool

The Midpoint tool is useful for entering a vertex at the halfway point of a specified distance. Imagine that you wish to split an existing parcel exactly in half.

1. Turn on vertex snapping for the parcel layer to ensure that the exact midpoint is found.

2. Select the parcel polygon to split and set the edit task to Cut Polygon Features (Fig. 13.43).

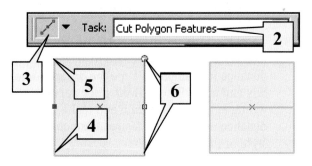

 3. Choose the Midpoint tool.

4. Click on the location representing the start of the measuring line to be halved.

Fig. 13.43. Using the Midpoint tool to split a parcel in half

5. Click on the other end of the measuring line. A point is entered at the midpoint of the measuring line.

6. Click to start and end another measuring line on the other side of the parcel, thereby entering a second vertex.

7. Right-click and choose Finish Sketch.

Using the Tangent Curve tool

This tool creates an arc that is tangential to the previous sketch segment and is useful for creating smooth curves for roads or other features. At least one sketch segment must be present for the tool to be active.

1. Use Sketch or another sketching tool to create one or more sketch segments (Fig. 13.44).

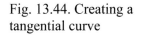 2. Click the Tangent Curve tool.

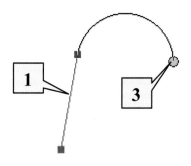

3. Move the mouse to the desired endpoint of the tangential curve and click to create the curve.

Fig. 13.44. Creating a tangential curve

Using the Direction-Distance tool

This tool creates a new vertex using a bearing direction from a known point plus a distance from another point. Imagine digitizing a service road that stops 20 feet from a well.

1. Use another tool to enter the first part of the line (Fig. 13.45).

2. Choose the Direction-Distance tool.

3. Click on the well to establish the rotation point of the direction segment. A temporary line appears.

4. Rotate the line to the desired direction and click, or press the D key to type a specific direction angle and press Enter.

5. Click on the well to define the center of the circle.

6. Move the mouse to the desired radius and click, or press R to type a specific radius and press Enter.

7. Click on one of the two possible intersections to create the vertex.

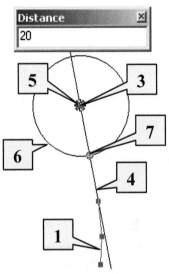

Fig. 13.45. Using the Direction-Distance tool

Modifying a feature

1. Choose Modify Feature from the Task bar in the Editor menu.

2. Select a feature to be modified with the Edit tool. The sketch of the feature will appear (Fig. 13.46).

3. To delete a vertex, place the cursor on the vertex, right-click, and choose Delete Vertex from the Sketch menu.

4. To add a vertex, place the cursor over the spot to add the vertex, right-click, and choose Add Vertex.

5. To move a vertex, place the cursor on top of the vertex to be moved until a crosshair symbol appears. Click and drag the vertex to a new location.

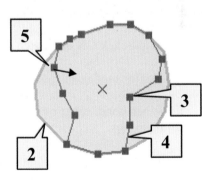

Fig. 13.46. Modifying a feature

6. When done modifying the vertices, right-click ON the sketch and choose Finish Sketch from the context menu.

TIP: For best results, the Edit tool should be active while modifying the feature.

Reshaping a feature

1. Choose Reshape Feature from the Task bar in the Editor menu.

2. Select a feature to reshape.

3. Click on the Sketch tool.

4. Click ON the feature at the start of the section to be reshaped (Fig. 13.47).

5. Enter vertices defining the new shape. Use the other sketching tools and context menus as needed.

6. Double-click ON the feature at the end of the section being reshaped, or right-click and choose Finish Sketch.

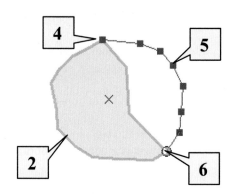

Fig. 13.47. Reshaping part of a polygon boundary

TIP: It is very helpful to use snapping when reshaping features. If the start points and endpoints of the new sketch do not fall exactly on the feature, the reshaping will not be performed.

Flipping a line

1. Choose Modify Feature from the Task bar on the Editor toolbar.

2. Click the Edit tool and select the line to be flipped. The line's sketch will appear. The red vertex shows the end of the line.

3. Right-click ON the sketch and choose Flip from the Sketch menu.

4. To flip another line, select it and repeat step 3.

5. When done, either right-click the sketch to finish it or clear the selection by clicking on the screen.

Displaying line directions

1. Click the line layer's symbol in the Table of Contents.

2. In the Symbol Selector, scroll to the bottom of the list to find the arrow symbols.

Arrow at End   Arrow at Start   Arrows at Start and End

3. Choose the symbol with an arrow at the start of the line or at the end of the line (Fig. 13.48).

Fig. 13.48. Arrows in the Symbol Selector for showing line directions

4. Change the color or width as desired. Click OK.

Mirroring features

1. Choose Mirror Feature from the Task bar on the Editor toolbar.

2. Select the feature to be mirrored (Fig. 13.49).

3. Click the Sketch tool.

4. Enter two endpoints of a line that defines the mirror plane (i.e., the line across which the feature is mirrored).

5. After the second click, the mirror feature will be created.

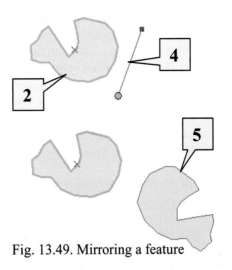

Fig. 13.49. Mirroring a feature

Dividing a line

This task makes it possible to place points along a line, either at equal intervals defined by the user or evenly spaced along the available distance. If the target layer contains points, then point features along the line will be placed in the target. If the target layer is a line layer, then equal line segments will be created and placed in the target as lines.

1. Check the target layer type. If it is lines, then line segments will be created. If it is points, then points will be created.

2. Select the line to divide (Fig. 13.50).

3. Choose Editor > Divide from the Editor toolbar.

4. Choose to place a set number of points evenly along the line or to separate points by a set distance. Click OK.

5. Usually you will want to check the box to delete the original selected feature.

6. The new features will be placed in the target layer.

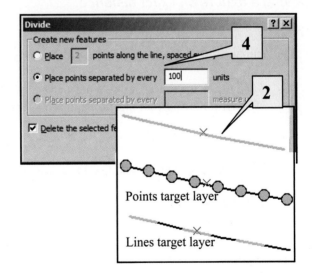

Fig. 13.50. Placing points along a line

Splitting lines exactly

You can split lines by clicking the Split button and pointing at the location to split, as in Chapter 11. This method can split a line at a specified distance from the start point or endpoint.

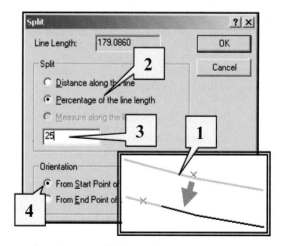

1. Select a line to split (Fig. 13.51) and choose Editor > Split.

2. Fill the button to choose distance or percentage measurement.

3. Enter the distance or percentage.

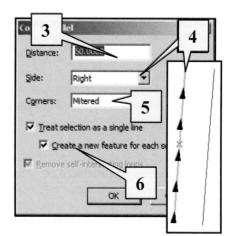

4. Choose the direction of the split (from start or end).

Fig. 13.51. Splitting a line at an exact location along it

5. Click OK. The line will be split in two with both pieces remaining selected.

Moving a feature an exact distance

1. Select the feature to be moved (Fig. 13.52).

2. Choose Editor > Move from the Editor toolbar.

3. Type the change in *x*- and *y*-coordinates as an offset from the current location and press Enter.

Fig. 13.52. Moving a line an exact offset

Using Copy Parallel

This function copies a feature and places it parallel to the original at a specified distance.

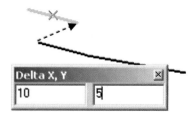

1. Select the feature to copy (Fig. 13.53).

2. Choose Editor > Copy Parallel from the Editor toolbar.

3. Type the distance offset.

4. Examine the arrows on the original line and determine if the new one should be placed to the left or right.

5. Change corner style if desired.

Fig. 13.53. Making a parallel copy

6. Check the boxes to treat the selection as a single line and to create a new feature for each selected feature. (Usually you'll want both of these.) Click OK.

Trimming lines

1. Choose Extend/Trim Features from the Task bar in the Editor menu (Fig. 13.54a).

2. Select the line to be trimmed.

3. Click the Sketch tool.

4. Enter two endpoints of a line crossing the feature and finish the sketch.

5. The feature will be trimmed where it crosses the line.

TIP: If the wrong side of the line disappears, choose Edit > Undo from the main menu bar (or press Ctrl-Z) to restore the original line. Then repeat step 4 but enter the line in the opposite direction.

Extending lines

1. Choose Extend/Trim Features from the Task bar of the Editor toolbar.

2. Select the line to extend (Fig. 13.54b).

3. Click the Sketch tool.

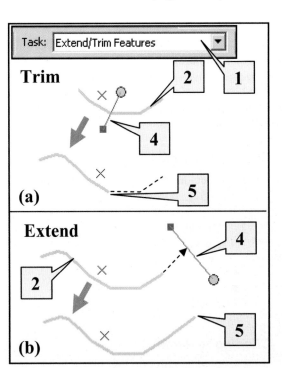

Fig. 13.54. Trimming or extending lines

4. Enter two endpoints of a line defining the stopping point of the extension.

5. The feature will be extended to meet the line.

Merging features

1. Select at least two features to be merged (Fig. 13.55).

2. Choose Editor > Merge from the Editor toolbar.

3. Select the feature into which the others will be merged. The merged feature will be given the attributes of the selected feature. The value shown in the window is the primary display field for the layer.

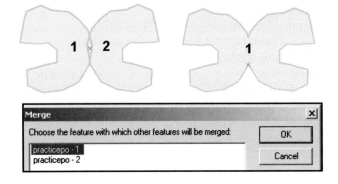

Fig. 13.55. Merging features

4. Click OK.

Union of features

1. Select at least two features to union (Fig. 13.56).

2. Choose Editor > Union from the Editor toolbar.

3. The union feature remains selected and on top of the original figures. Move or delete one set.

Fig. 13.56. Union of features

Intersection of features

1. Select at least two features to intersect (Fig. 13.57).

2. Choose Editor > Intersect from the Editor toolbar.

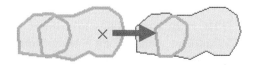

Fig. 13.57. Intersecting two polygons

3. The new feature remains selected on top of the original polygons. Move or delete one or both sets.

Clipping features

1. Select a polygon that lies at least partially on top of another polygon (Fig. 13.58).

2. Choose Editor > Clip from the Editor toolbar.

3. If desired, apply a buffer to the clipping polygon by entering the value in map units.

4. Choose whether to preserve the overlapping area or to discard the overlapping area.

5. Click OK.

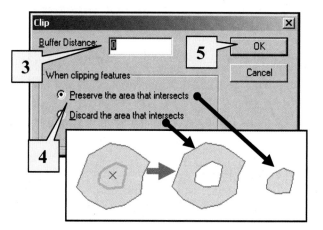

Fig. 13.58. Clipping with a polygon

6. The clipping polygon will remain on top of the original feature. You may move or delete it.

Creating donut polygons

Donut polygons have an empty hole in the middle and are created in two ways.

Creating multipart donuts

1. Set the task to Create New Feature, and choose one of the sketching tools.

2. Enter one ring of the donut polygon feature. On the last vertex, right-click and choose Finish Part from the context menu (Fig. 13.59).

3. Enter the next ring of the donut polygon feature and choose Finish Part.

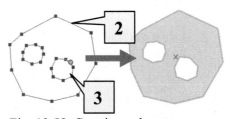

Fig. 13.59. Creating a donut or "Swiss cheese" polygon

4. Keep adding parts as long as needed. When the last part is done, choose Finish Sketch instead of Finish Part.

Using Clip to create donuts

1. Set the task to Create New Feature and choose the Sketch tool.

2. Sketch the exterior ring of the donut polygon and finish the sketch (Fig. 13.60).

3. Sketch the interior ring of the donut polygon and finish the sketch.

4. Select the interior ring, if necessary, and choose Editor > Clip from the Editor menu.

5. Choose the option to discard the area that intersects. Click OK.

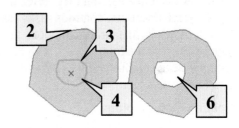

Fig. 13.60. Creating a donut by clipping

6. The clip polygon remains selected. You may keep it or delete it.

To fill a hole with another polygon

1. Set the task to Create New Feature and select the polygon around the hole.

2. Choose the Trace tool.

3. Click on the boundary of the hole to begin tracing.

4. Trace all the way around the hole. Click again to end the trace.

5. Right-click the sketch and choose Finish Sketch.

Buffering features

1. Make sure the target layer is a polygon or line layer. The buffer output type (line or polygon) will match the type of the target layer.

2. Select the feature(s) to be buffered. It may be a point, a line, or a polygon (Fig. 13.61).

3. Choose Editor > Buffer from the Editor toolbar.

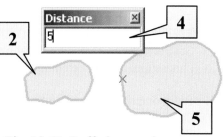

Fig. 13.61. Buffering a polygon

4. Type the buffer distance in map units and press Enter.

5. The new buffer will remain selected on top of the original feature. Move the buffer or delete the original, if necessary.

Editing shared features

To edit **shared features** that have some of the same nodes or boundaries, you must first build a map topology. Then shared features are selected using the Topology Edit tool. The selected feature will appear purple. Once selected, the shared feature may be modified, reshaped, or moved. All features sharing that vertex or boundary will be affected by the edits. The following examples represent two common tasks associated with shared editing, but there are many more. Consult the Help files or digital books that come with the software for more information about map topology and editing. Users with ArcEditor or ArcInfo licenses have even more topological editing tools available.

Creating map topology

1. Choose Editor > More Editing Tools > Topology from the Editor toolbar. The Topology toolbar will appear.

2. Click on the Map Topology icon on the Topology toolbar.

3. Check the boxes for the layers to participate in the map topology (Fig. 13.62).

4. You should normally leave the cluster tolerance at the default value. Click OK.

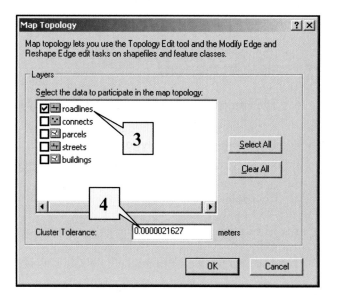

Fig. 13.62. Creating map topology

Reshaping a common boundary

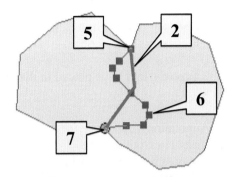

1. Turn on vertex or edge snapping, and click on the Topology Edit tool.

2. Click the shared boundary to edit. It will be selected in a purple color to show that it is shared (Fig. 13.63).

3. Choose Reshape Edge from the Topology Tasks group in the Task bar on the Editor toolbar.

4. Click the Sketch tool.

5. Click the beginning of the shared boundary to start entering the sketch.

Fig. 13.63. Reshaping a shared boundary

6. Enter vertices defining the new shape of the boundary.

7. End the sketch on the other end of the shared boundary and double-click to finish it.

8. The boundary will change to the shape of the sketch.

Moving shared nodes

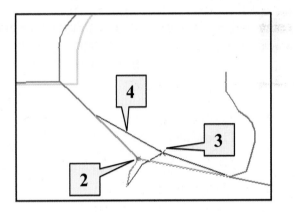

1. Click the Topology Edit tool.

2. Hold down the N key and click on a node at the intersection of the lines to select it. The N key ensures that only a node is selected (Fig. 13.64).

3. Click and drag on the node to move it to its new location.

4. As you move the node, the movement of the shared features will be shown.

5. When you release the mouse button, the node and its attached lines will go to their new places.

Fig. 13.64. Moving shared line features

TIP: To have attached features proportionally stretched when moving nodes, check the box on the General tab of the Editing > Options menu. This option often provides better adjustment of features.

Editing annotation

1. Create annotation from dynamic labels as described in the Skills Reference of Chapter 3.

2. Choose Editor > Start Editing from the Editor toolbar. Open the Annotation toolbar, if necessary (Fig. 13.65).

3. Type the text to be placed in the text box.

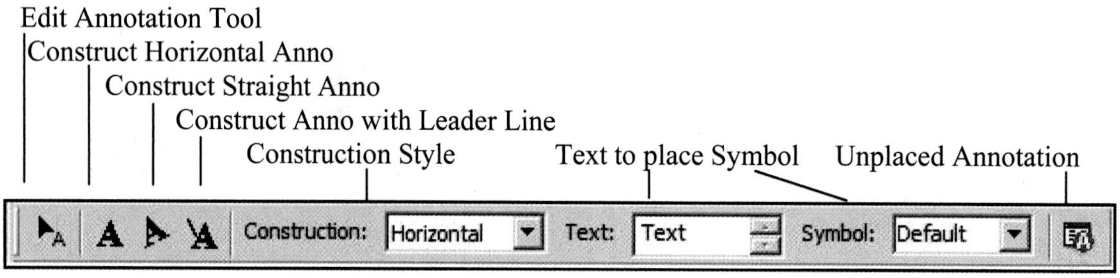

Edit Annotation Tool
Construct Horizontal Anno
 Construct Straight Anno
  Construct Anno with Leader Line
   Construction Style   Text to place Symbol Unplaced Annotation

Figure 13.65. The Annotation toolbar editing functions

Adding new annotation labels

4. Click the desired construction button on the Annotation toolbar (Horizontal, Straight, or Leader Line).

5. For horizontal annotation, click on the map once to place it. For Straight annotation, click once to enter the anchor point and again to establish the angle of rotation. For leader line annotation, click once to place the far end of the pointer and again to define the near end and place the label.

To place curved annotation

6. Choose the Straight Anno tool and change the Construction Method to Curved.

7. Click as many times as necessary to define the curve. Double-click to end and place the annotation.

Make existing labels follow features

8. Click the Edit Annotation tool and select the annotation.

9. Right-click the annotation and choose Follow > Follow Feature Options and choose Straight or Curved depending on the desired result.

10. Right-click the line to be followed and choose Follow This Feature. Place the cursor on the annotation and move it to the desired location along the line feature.

To edit existing annotation labels

11. Click the Edit Annotation tool and select the annotation. Colored handles appear on the annotation as shown in Figure 13.66.

12. Click and drag on the red handle to enlarge or shrink the annotation.

13. Click on one of the two blue handles to rotate the annotation.

14. Click on black handle to move the annotation to a different location.

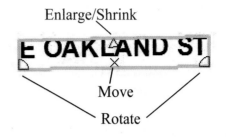

Fig. 13.66. Action handles on selected annotation

Modifying annotation properties

15. Select one or more pieces of annotation using the Edit Annotation tool.

16. Right-click the feature and choose Attributes from the context menu.

17. Change one or more of the attributes for the annotation.

For more information and techniques for editing annotation, consult the Annotation chapter of the ArcGIS Desktop pdf books.

Chapter 14. Geodatabases

Objectives

- ➢ Understanding the geodatabase model

- ➢ Creating a geodatabase

- ➢ Importing layers into a geodatabase

- ➢ Setting up a feature dataset and importing layers

- ➢ Setting up and using attribute domains

- ➢ Using split and merge policies

Mastering the Concepts

About ArcGIS

As we have learned, the ArcGIS software has a long history of development, and coverages were the first data model used. Later, shapefiles were developed for use in the ArcView package. The geodatabase model arrived with the release of ArcGIS 8. The new model offers several advantages over both coverages and shapefiles. Like coverages, it can store topological relationships between features for added functionality in modeling real-world phenomena, but the geodatabase is simpler in construction and more robust in general usage. Like shapefiles, the features of a geodatabase are simply constructed and difficult to corrupt but provide many more benefits than shapefiles. Another advantage of geodatabases over shapefiles is the automatic tracking of lengths and areas for features. Geodatabases also provide behavior and validation rules that control how features behave, how they can connect, and what attributes are allowed. A user can also set up planar topology for features and construct topological rules, such as specifying that a state boundary must always match county boundaries or that counties must never cross a state boundary. We've used feature classes extensively in this text, but this chapter describes the feature dataset and the geodatabase capabilities available to users with an ArcView license. More advanced and powerful features are available to those with an ArcEditor or an ArcInfo license; some of these are mentioned as well.

About geodatabases

The geodatabase model is constructed on the architecture of standard database systems and can be implemented with several commercially available database software packages. Three types of geodatabases may be created.

Personal geodatabases are designed for use by individuals or small workgroups and are stored in a single Microsoft Access file. This file is limited to 2 GB in size and works only in the Windows operating system. This book uses personal geodatabases.

File geodatabases are also designed for individuals and small groups, but each data set is stored as a separate file within a system folder, and each file can be up to one terabyte in size. File geodatabases can be accessed by different operating systems and are best for cross-platform operations, such as a company that supports Windows, Linux, and Unix computers.

SDE geodatabases store GIS data within one of several commercial relational database management systems (RDBMS), such as Oracle®, SQLServer®, and IBM Informix®. ArcSDE geodatabases are designed to meet security and management needs for large data sets accessed by many simultaneous users. Such complex systems are often called organizational databases, or Enterprise GIS. For example, at least three groups of users might need to edit parcel records in a large city database: the tax department, the city surveyors, and the deeds office. Instead of maintaining three sets of data, an ArcSDE geodatabase allows users to "check out" certain portions of data for editing and merges the changes back into the central database. Potential conflicts between simultaneous editing by different users are avoided. A software package called the Arc Spatial Database Engine (ArcSDE) is required to translate geographic data into a consistent and customized format in the large database.

Organizational databases have distinct advantages when the spatial data sets are large and/or are accessed by many different people. For example, such a system might be developed for the planning department of a large city, which may have dozens of people in several departments needing access to the information, both for reading and for editing. A DBMS/ArcSDE setup allows many people simultaneous access to data sets, provides security and permission handling, and manages cases of multiple simultaneous editing of the data. It also makes it possible to set up version tracking for added data security and protection. However, the additional work in setting up a DBMS only makes sense when these issues are present. In this chapter we will work exclusively with personal geodatabases, but most of the commands and functions are the same for the organizational version.

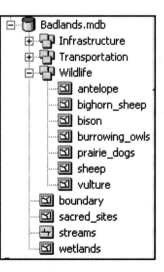

A geodatabase may contain a variety of objects (Fig. 14.1). Feature classes are used to store the spatial information. We've used these extensively in this book already. Geodatabases may also contain tables, layers, relationships, geometric networks, and feature datasets. A feature dataset is a collection of related feature classes with the same coordinate system. The Badlands database in Figure 14.1 contains three feature datasets: Infrastructure, Transportation, and Wildlife. The Wildlife feature dataset has been expanded to show the feature classes, including the home ranges of various species that live in the park. The geodatabase also contains four standalone feature classes: boundary, sacred_sites, streams, and wetlands. In this case the feature classes all share the same coordinate system, but they do not have to.

Fig. 14.1. A geodatabase with feature classes and feature data sets

---

TIP: Designing and creating good geodatabases involve many issues far beyond the scope of this book. Users planning to work with geodatabases should read more about them, particularly in the online documentation *Building a Geodatabase.*

Creating geodatabases

Geodatabases must be created as empty shells to which feature classes and other database objects are later added. Feature classes may be loaded from a variety of sources. Coverages and shapefiles may be imported into the geodatabase, either as stand-alone feature classes or as part of a feature dataset. INFO and dBase tables may be loaded. Users may also create empty feature classes and use editing to add features, as we did in Chapters 12 and 13. Rasters, tables, annotation, and relationships can also be stored inside a geodatabase. Often new feature classes are added to a geodatabase as a result of the operation of a tool or command.

The design and structure of a geodatabase, including the datasets and feature classes, the fields in the tables, the stand-alone tables, the relationship, and other objects, are together termed the database **schema**. The description of all the objects in a geodatabase can be stored and saved without any actual data and then reused to generate multiple geodatabases with the same structure. This approach offers many advantages for developing large complex databases with a well-designed and tested schema. A city planning department that is relatively new to GIS might choose to use a published schema and populate the database with the local information rather than trying to design the geodatabase from scratch.

Because they are based on a new data model, many databases are currently developed by importing shapefiles or coverages of existing data. This process is usually straightforward. The entire data set, including features, attribute tables, and coordinate systems, is transferred into the database. In the case of coverages, the attribute fields must be converted from the coverage field types to the geodatabase field types. This conversion is handled automatically although the user can customize the interpretation of certain fields if desired. The user can also request a coordinate system reprojection as part of the transfer process.

Unlike layer files, which store only a link to the source data, the geodatabase stores the feature data itself. Importing a shapefile creates a copy of the features inside the database, leaving the original features unchanged. The user then has two copies of the data.

Creating feature datasets

A feature dataset is a collection of feature classes that are related to each other in some way (Fig. 14.2). For example, a transportation feature dataset might contain feature classes for roads, railways, airways, and interstates. These features could be used to construct a geometric network so that the movement of people or goods along the network could be modeled. Similar models could be constructed for utilities by using feature classes representing water mains, water sources, sewer lines, and treatment plants. Feature datasets may also contain planar topological relationships between the feature classes, allowing for easier identification and correction of errors, such as gaps or overlaps between polygons. (See Chapter 13 for a description of planar topology.)

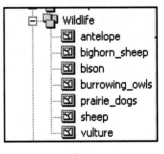

Fig. 14.2. A feature dataset contains related layers with a common coordinate system.

About the spatial reference

The feature classes in a feature dataset must all share the same spatial reference, which includes the coordinate system, the X/Y domain, and the resolution. Prior to version 9.2, the *x-y* coordinates were stored using four-byte integers, necessitating a trade-off between the precision of the data and the total range of coordinates stored. Users had to manually set the domain and precision to achieve the desired results. In version 9.2, the storage of coordinates was converted to high-precision integers, eliminating the need to trade precision with coordinate range. The software now chooses an appropriate set of values, which will be adequate in most circumstances, thus relieving the user of an arduous planning process. However, it remains true that the spatial reference of a data set cannot be modified once it has been specified.

The spatial reference includes two components, the domain and the resolution. The **X/Y Domain** is the range of allowable *x-y* values that can be stored in a feature class. Integers are used because the computer can process them several orders of magnitude faster than floating-point values; in the case of SDE data bases, they can also be compressed for more efficient storage space. The software converts the integers to real numbers on the fly when it needs them (e.g., when reporting the current *x-y* cursor location in ArcMap).

The **resolution** replaces the earlier concept of precision. It refers to an underlying coordinate grid to which values are snapped. The units are the same as those for the defined coordinate system. A resolution of 0.001 meter in a projected coordinate system stores values to the nearest thousandth of a meter. A resolution of 0.000001 in a GCS stores values to the nearest millionth of a degree. The default resolution and domain are determined based on the coordinate system chosen by the user and will be fine for most applications. Users requiring very high resolution, for example, in laying out a surveying grid, can increase the resolution at the expense of a smaller domain.

Using default values

One advantage of using geodatabases lies in setting up default values for attributes. For example, if a user is digitizing roads, residential streets (LOCAL) are usually the most common type and typically have two lanes and a speed limit of 25 mph. The user can set default values for each of these fields (Fig. 14.3). Then when each road is added, the road type, lanes, and speed limit are automatically set to LOCAL, 2, and 25, respectively. The information only needs

Field Properties		
Alias	TYPE	
Allow NULL values	Yes	
Default Value	LOCAL	
Domain	RoadType	
Length	3	

Fig. 14.3. Setting up default values for attributes can save much time during data entry.

editing if the new road differs from the default. A user could even set the defaults one way, digitize all the local roads, and then close the layer, change the defaults, open the layer again, and digitize the next group. This approach saves editing time and helps reduce attribute errors.

Setting up domains

One great advantage of geodatabases includes using **domains** for attributes. An attribute domain constrains the values that may be entered for a particular attribute. Such constraints help ensure correct data entry and keep out incorrect or extraneous values. For example, imagine an attribute field named PipeDiam that contains the diameter of water pipes. Water pipes don't come in an infinite variety of sizes; in fact, a town might only use 1-inch, 3-inch, 6-inch, and 12-inch pipes. Setting up a domain for the PipeDiam field prevents any user from mistakenly entering a 4-inch or 5-inch pipe. These constraints can enforce data entry rules in situations when more than one person is entering data and not everyone knows or can remember all the rules.

Domains come in two types. The first type is a **range domain**, which specifies the lowest and highest possible values but lets the values take any number in between. Range domains are only applicable to numeric data. For example, a range domain could be applied to a student Grade Point Average (GPA) field. A GPA ranges between 0 and 4.0. A range domain would allow any value between 0 and 4.0 but would exclude negative values or values greater than 4.0. The second type is a **coded domain**, which allows only certain values taken from a list. The PipeDiam domain mentioned previously is a coded domain because it allows only four specific values from a list: 1, 3, 6, or 12.

Domains are created and maintained in ArcCatalog as a property of the geodatabase rather than as a property of a single feature class or attribute field. Therefore, a domain can be used more than once, and it can be used in multiple feature classes. For example, a range domain called Percent, which allows values from 0 to 100, could be used in many feature classes whenever a percentage value is being stored in an attribute.

Figure 14.4 shows how the properties of a percentage domain would be set up in ArcCatalog. The field type is specified as short integer. The domain type is range, and the minimum value of 0 and the maximum value of 100 is entered. A domain can also have split and merge policies assigned to it, which we will cover in the following section.

Field Type	Short Integer
Domain Type	Range
Minimum value	0
Maximum value	100
Split policy	Default Value
Merge policy	Default Value

Fig. 14.4. Properties of a range domain

The field type of the domain must always match the field type of the attribute to which it will be assigned. Thus, a short integer domain can only be used with a short integer attribute field and not with a long integer or float field. Users must pay careful attention to the fields for which the domain is intended in order to select the appropriate field type for the domain.

An example of the PipeDiam domain is shown in Figure 14.5. The field type is short integer, and the domain type is set to coded values. The codes themselves are typed in below. The code is the value actually stored in the attribute field. The description appears in legends and tables during an ArcMap session, providing easily understood information to the people working with the attributes. However, the descriptions do need to fit well inside the menus and legends, so it is best to keep them reasonably short.

Domain Properties

Field Type	Short Integer
Domain Type	Coded Values
Split policy	Default Value
Merge policy	Default Value

Coded Values:

Code	Description
1	1-inch
3	3-inch
6	6-inch
12	12-inch

Fig. 14.5. Properties of a coded domain

Coded domains also provide a valuable way to combine the ease of numeric codes with the luxury of understandable text information for people. Land-use planning, for example, often uses numeric codes to indicate various zoning types, such as 32 = Residential and 45 = Commercial. Numeric codes are advantageous because they are less susceptible to typing errors and take less space to store. However, it is difficult for people to remember the meanings of many codes. With a coded domain, the zoning codes may be stored as integers, but the information presented to the user takes the form of the textual description (Fig. 14.6).

Split and merge policies

Domains offer a way to control the updating of attributes during a split or a merge operation. The default policy of making copies or entering blanks may be overridden by alternate and more useful actions. Split and merge policies are associated with each attribute domain. Setting up these policies correctly can save much time during editing.

Coded Values:		
	Code	Description
	12	Heavy Industrial
	14	Commercial
	16	Low Density Residential

Fig. 14.6. Using domains to link numeric land-use codes to easily interpreted descriptions

The merge policy

If we examine what happens during a merge, the usefulness of the merge policy immediately becomes apparent. Imagine two parcels that have been combined into one. Each parcel has its own attributes. After merging, there will only be one parcel. Which attributes will it have? By default in coverages or in shapefiles, the merged parcel will contain the attributes of the first feature found in the database, hardly a consistent and effective method. In geodatabases, it will contain the attributes of the first feature selected before merging. The merge policy overrides this haphazard method by specifying a different action for each field.

Three different policies may be assigned to fields: Default, Weighted Average, or Sum. If Default is chosen, the merged feature attribute will contain the default value established for that field. Weighted Average assigns the new value based on the relative areas of the two input features, and Sum assigns the new value as the sum of the values in the two input features. Let's examine the attribute fields in Figure 14.7 one by one.

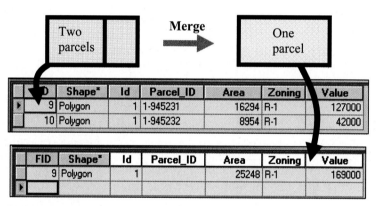

Fig. 14.7. Attributes from two parcels are updated according to the Merge policy when they are combined into one.

The Zoning field is a text field. By far the majority of parcels in most towns are residential, so it might make sense to have the default value be R-1 for zoning. This default is usually set at the time the attribute field is created. If the merge policy is set to default, then after the merge the new polygon will contain the default value for that field, R-1. The default value may be incorrect in some cases but can be edited, if necessary. A good default will be right more times than it is wrong and will save editing time. However, some managers prefer generating a blank field so that they can manually ensure that the parcel gets the right value rather than risk automatic assignment of a wrong value.

TIP: A user can cause a field to receive a blank during a split or merge by setting the policy to Default and leaving the default value blank.

The Area field is numeric and contains the area of the parcel in square feet. This field is an obvious candidate for the Sum policy. After the merge, the new feature will be assigned an area equal to the sum of the areas in the original polygons. The Value field would also logically be assigned Sum for the merge policy. A Weighted Average is similarly applied, except that the values are averaged instead of summed.

The Parcel_ID field is not a numeric quantity and so cannot qualify for Sum or Weighted Average. Furthermore, it is nominal data, so it is not practical to assign a domain code for every single unique ID. This attribute cannot effectively be assigned a domain. Instead, the normal rules for splitting and merging features without domains would apply. Notice that the Parcel_ID field is blank, signaling to the operator that a new code must be assigned.

It should be noted that merge policies in ArcGIS are not evaluated during editing. They exist so that programmers writing advanced applications can take advantage of them, but they will not help the average user. The split policies do work, however.

The split policy

The split policy is analogous to the merge policy, but it is applied when a single feature is split in two. The policy choices include Default, Duplicate, and Geometry Ratio. In the case of Default, the policy works the same as for merging. If Duplicate is used, then both new features retain the value of the original feature. If Geometry Ratio is used, then the new value is assigned based on the relative size of the original and split features.

Let's examine an example of splitting a parcel using the policies (Fig. 14.8). The Parcel_ID field has no domain and follows the standard rule of being duplicated in the resulting polygons. At least one of the new parcels will need a new number. The Area and Value fields use the Geometry Ratio policy. The proportion of areas in the two new polygons is 60-40; thus the smaller parcel is assigned 40% of the original area and value, and the larger parcel is assigned 60% of the original area and value. Finally, the Default policy is applied to the Zoning field, again assigning the most common value of R-1.

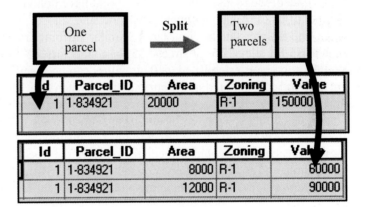

Fig. 14.8. The attributes of two parcels created by splitting one parcel according to the split policy

In deciding whether to use split and merge policies, one must consider the number of attributes involved and the amount of editing that a data layer receives. Setting up and applying domains to attributes takes some time and thought. Unless the layers are being frequently edited, then the effort may not be worth it. When appropriately used, however, domains can save a substantial amount of time and prevent many incorrect data entries.

About subtypes

Subtypes are another way to facilitate the entry and validation of attribute data. Subtypes may be set up for a particular feature class. Every feature in the layer then must belong to one of the subtypes. For example, the roads layer of a city might be constrained to contain only certain types of roads (Fig. 14.9). One field is chosen as the subtype attribute. The allowed road types might be alleys, streets, connectors, highways, and interstates. Then every road in the layer must belong to one of these types.

So far this setup sounds similar to a coded domain. However, subtypes provide added functionality. First, when a data layer is loaded into a map, the subtypes are automatically displayed with different symbols. Second, each subtype can have its own default values for attribute fields instead of having a single default for all features in the feature class.

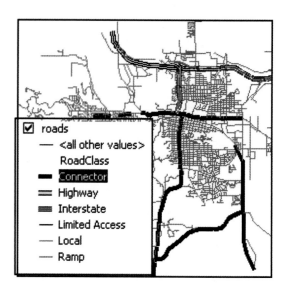

Fig. 14.9. Subtypes are formal classes of features within a layer.

In the case of the roads, each road type might have default values for certain attributes, such as number of lanes. An alley typically has one lane, a local street has two lanes, a connector has four lanes, and an interstate has four lanes. Similar attributes might be determined for speed limits, such as 15 for alleys, 25 for streets, 35 for connectors, 65 for highways, and 75 for interstates. When a subtype is established, default values can be specified for each of the types. These values are automatically entered in the attribute table when new features are created.

During an editing session, each subtype appears as a separate entity in the target list. To add local streets, the user would set the target to the Local subtype and begin adding streets. As they were digitized, each street would immediately receive the Local default values, two lanes, 25-mph speed limit, and so on. To enter a connector, the user would change the target layer to the Connector subtype, and subsequently added streets would receive the default values for connectors. Subtypes thus streamline both data entry and the assignment of attributes to features. Occasionally features may deviate from the default values, such as a connector having six lanes, but such exceptions may be corrected as needed.

> **TIP:** Subtypes can be created for geodatabase feature classes using ArcEditor or ArcInfo. ArcView can view subtypes, but it cannot create or edit them. Furthermore, *layers containing subtypes cannot be edited in ArcView*, even if the user doesn't want to change the subtypes. The layer will simply not be available for editing of any kind.

We will not study subtypes further in this chapter, but if you have an ArcEditor or an ArcInfo license, you may want to read more about them in the online documentation *Editing in ArcMap* and experiment with them on your own.

Summary

➢ A geodatabase is a new data model developed for ArcGIS. It uses commercially available standard database formats and technology.

➢ Geodatabases offer significant advantages over shapefiles and coverages, including simpler and more robust implementation, the storing of topology, geometric networks, and behavior and validation rules.

➢ Geodatabases are created as empty containers and then filled with feature classes, feature datasets, layers, tables, relationships, and other objects.

➢ Feature datasets contain related feature classes with the same spatial reference. The X/Y Domain specifies the range of *x-y* coordinates allowed in the dataset, and the resolution specifies the underlying grid to which features are snapped. Features outside the domain cannot be loaded into the dataset.

➢ Attribute domains define allowed values for attributes in geodatabases. Attribute domains may cover ranges of numerical values, or they may have specific coded text or numeric values.

➢ Domains also have split and merge policies, which define the assignment of attributes when features are merged or split.

➢ Subtypes allow features in a layer to be classified into specifically defined groups. Each group may have its own set of default attribute values.

➢ Subtypes may be set up with ArcEditor or ArcInfo, but they can only be viewed with ArcView.

> **TIP:** Designing and creating good geodatabases involves many issues far beyond the scope of this book. Users planning to work with geodatabases should read more about them, particularly in the online documentation *Building a Geodatabase*.

> **TIP:** Subtypes can be created for geodatabase feature classes using ArcEditor or ArcInfo. ArcView can view subtypes, but it cannot create or edit them. Furthermore, *layers containing subtypes cannot be edited in ArcView at all, for any reason*. Don't create subtypes if someone must be able to edit the layer with only an ArcView license.

> **TIP:** If the import layer has a different coordinate system than the feature dataset, then the features will automatically be reprojected as they are loaded.

> **TIP:** Domain policies and codes may be edited after the domain is created. However, the field type and domain type, once set, cannot be altered. To change them, you must delete the domain and create a new one with the desired properties.

Chapter Review Questions

You may need to consult the Skills Reference section to answer some of these questions.

1. List three advantages of the geodatabase model.

2. What is the function of a geodatabase schema?

3. How does a personal geodatabase differ from a file geodatabase and an SDE geodatabase?

4. What is the difference between a feature class and a feature dataset?

5. Why is a resolution of 0.001 appropriate for a parcels feature class stored in UTM but not for one stored in a GCS?

6. Can you have a geodatabase that contains one feature dataset in UTM Zone 13 and another feature dataset in South Dakota State Plane?

7. You have a forest feature class with polygons showing individual stands of trees. It contains the three attribute fields: TreeSpecies, CanopyCover, and Acres. TreeSpecies contains the dominant type of tree (e.g., ponderosa pine, aspen, etc.). Canopy cover gives the percentage of the stand covered by tree crowns. The Acres field contains the area of the stand. For each of these attributes, list the most appropriate split policy and merge policy.

8. For each of the following zoning attributes—Zoning, StreetAddress, Value, and PercentImperviousArea—state whether it would be suitable for a domain. If yes, then say whether a coded or range domain would be more appropriate. In each case, explain your reasoning.

9. Domains are set up as properties of a geodatabase rather than as properties of feature classes. Why is this arrangement an advantage?

10. How does one establish a default value for an attribute in a feature class? In a shapefile?

Mastering the Skills

Teaching Tutorial

The following examples provide step-by-step instructions for doing basic tasks and solving basic problems in ArcGIS. The steps you need to do are highlighted with an arrow ➜; follow them carefully. Click on the video number in the VideoIndex to view a demonstration of the steps.

In this examples section, we will build a geodatabase of Rapid City using the existing data layers in the mgisdata\Rapidcity folder.

1➜ Start ArcCatalog and navigate to the mgisdata\Rapidcity folder.

1➜ Right-click the Rapidcity folder and choose New > Personal Geodatabase.

1➜ Type in **rapidcity** as the name of the geodatabase and press Enter.

Adding coverages to a geodatabase

We will begin by bringing in two coverages, the city boundary and the land use data. Recall that coverages contain multiple feature classes, so we must specify polygons as the class to import.

2➜ Right-click the new rapidcity geodatabase and choose Import > Feature Class (single) tool (Fig. 14.10).

2➜ Click the Browse button for the Input Features, navigate to the mgisdata\Rapidcity folder, and double-click the landuse coverage to see its feature classes.

2➜ Select the polygon feature class and click Add to close the menu.

2➜ Enter **landuse** for the Output Feature Class Name. Leave the rest of the options as their defaults and click OK.

2➜ Click the + button next to the rapidcity geodatabase to expand its contents and click the Preview tab to see the polygons.

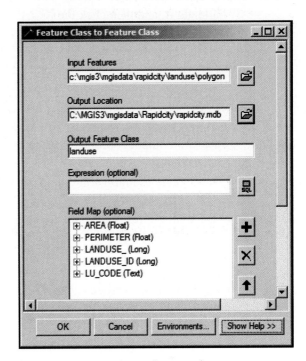

Fig. 14.10. Importing a feature class

Everything looks in order. Next we import the citybnd coverage.

3➜ Right-click the rapidcity geodatabase and choose Import > Feature Class (single).

3➜ Set the input feature class to the polygons of citybnd in the Rapidcity folder.

3➜ Set the output name to **citybnd**.

3➔ Scroll down to the lower half of the tool and examine the other inputs (Fig. 14.11).

4➔ Click the SQL button next to the box marked Expression. This Query Builder window, familiar from other chapters, allows the user to import only the records meeting certain criteria, such as only the residential parcels in a land-use feature class. Click Cancel to close the window.

4➔ The Field Map section lists the attributes in the feature class, showing the way the attributes will appear in the new feature class.

This section controls how the item names are translated from coverage data types to geodatabase types. This process happens automatically but can be customized if needed. Certain fields may be removed, and the order can be changed. For example, the AREA, PERIMETER, and CITYBND_ fields have counterparts in the geodatabase format and don't need to be carried over.

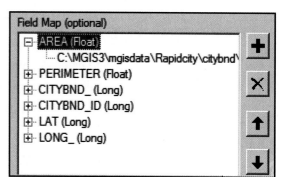

Fig. 14.11. Mapping item names from coverage format to geodatabase format

4➔ Click on the AREA field to highlight it, and then click the X button to delete it. (It will not be deleted from the original coverage.)

4➔ Also delete the PERIMETER and CITYBND_ fields from the map.

5➔ Click the Environments button. It opens the Environment Settings for the geoprocessing environment. You might use this menu under General Settings to specify an output coordinate system so that the new feature class is projected before it is placed in the geodatabase. Click Cancel to close the window.

5➔ When done examining the input boxes, click OK to import the citybnd feature class.

Often users have multiple feature classes to import into the same geodatabase. Another tool allows the imports to be set up and then run without supervision.

6➔ Right-click the rapidcity geodatabase and choose Import > Feature Class (multiple).

6➔ Click on the watersheds shapefile and drag it over to the Input Features box on the tool. This method is sometimes faster than using the Browse button. The feature class is placed in the list to process.

6➔ Click and drag the geologywest shapefile to the Input Features box.

6➔ Click OK to begin processing the feature classes.

6➔ Expand the geodatabase and examine the new feature classes.

Notice that when using the tool for importing multiple feature classes, you cannot set options individually for each one as you could with the single input tool.

Finally, let's create a new empty feature class to which we will later add water mains.

7➔ Right-click the rapidcity geodatabase and choose New > Feature Class.

7➔ Enter the feature class name **waterlines**.

7➔ Set the type to Line Features. You don't need M or Z values. Click Next.

7➔ Click the Import button, and assign the new feature class the same coordinate system as the landuse feature class already in rapidcity.mdb. Click Next.

7➔ Take the default value for the XY tolerance and resolution. They are estimated by the software based on the feature class spatial reference. Click Next.

Next we add the fields for the water lines: Linetype, Diameter, and Capacity (Fig. 14.12).

8➔ Click in the next empty box under SHAPE and enter the Field Name as **Linetype**.

8➔ Click in the Data Type box and select Text from the pop-up list. Then set the Length to 8 in the Field Properties.

8➔ Click the next empty box and create a **Diameter** field with the short integer type. Do not change the field properties.

Field Name	Data Type
OBJECTID	Object ID
SHAPE	Geometry
Linetype	Text
Diameter	Short Integer
Capacity	Short Integer

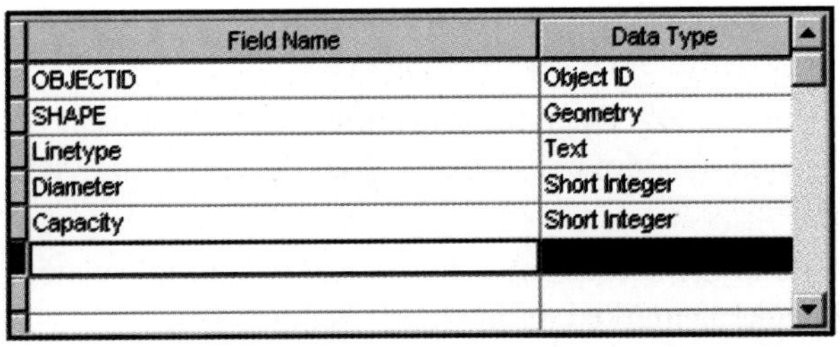

Fig. 14.12. Entering the fields for the water lines

8➔ Create a short integer **Capacity** field. Do not change the field properties.

8➔ Click Finish to create the feature class.

Using attribute defaults

We can streamline our data entry of the water lines by using default values. We will begin by entering water mains. Assume that the mains are made of 12-inch pipe and have a rating of 200 gallons per minute (gpm). Our first task involves setting these default values for the attributes.

9➔ In ArcCatalog, locate the waterlines feature class in the geodatabase.

9➔ Right-click the waterlines feature class and choose Properties.

9➔ Click the Fields tab.

9➔ Click on the Linetype field to select it.

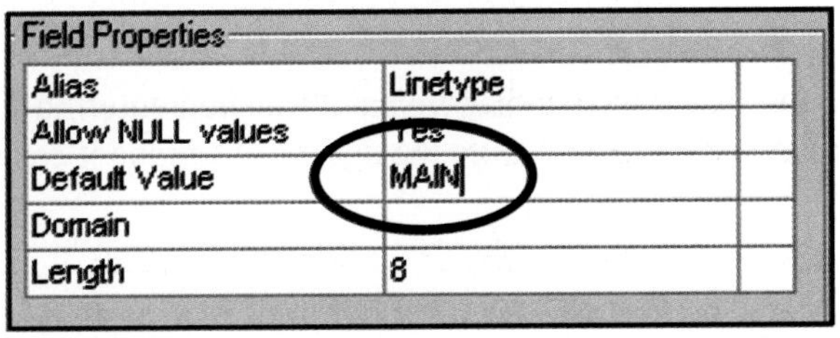

9➔ Type **MAIN** in the Default Value box (Fig. 14.13).

Field Properties		
Alias	Linetype	
Allow NULL values	Yes	
Default Value	MAIN	
Domain		
Length	8	

Fig. 14.13. Entering the default value MAIN for the attribute Linetype

9➔ Click the Diameter field and set the default value to 12.

9➔ Click the Capacity field and set the default value to 200.

9➔ Click Apply to establish the values and OK to close the window.

Now we will use these defaults as we digitize some water mains.

➜ Close ArcCatalog and start ArcMap with the document ex_14.mxd in the mgisdata\MapDocuments folder.

➜Use Save As to rename the document and remember to save frequently as you work.

10➜ Add the new waterlines feature class to the map.

10➜ Change the waterlines symbol to a thick line that contrasts well with the orthophoto.

11➜ Open the Editor toolbar, if necessary, and choose to Start Editing.

11➜ Make sure that the target layer is waterlines, the task is Create New Feature, and the Sketch tool is selected.

11➜ Open the Attributes window. It will be blank right now because nothing is selected. Move it so that it does not block the view of the map.

One of the sources of water in Rapid City is a water gallery and small treatment site next to Rapid Creek, as marked on the map. Begin digitizing water mains from this point.

12➜ Click on the water gallery marker to start a water main. Click on the road immediately to the right of the gallery and continue digitizing south down the road to the edge of the current display. Double-click to finish the sketch.

12➜ Examine the attributes of the new feature in the Attributes window.

Notice that the defaults of MAIN, 12, and 200 are automatically entered in the fields. Just imagine how much time this will save. If you happen to enter a different type of water line, simply change the default values in the Attributes window before going on.

13➜ Turn on edge snapping.

13➜ Add several east-west water mains going along those streets.

13➜ Stop editing and save the edits.

Now let's suppose that we have finished entering water mains for a while and want to create laterals. We can change the defaults to make it easier to enter these also.

TIP: Theoretically, you should be able to switch between ArcMap and ArcCatalog for tasks like these. However, the file locks do not always work properly, and removing the data set from ArcMap is not always sufficient. Experience has taught the author that making changes to feature classes is best performed with only one of the programs open at a time.

➜ Save the map document and close ArcMap.

14➜ In ArcCatalog, open the properties for the waterlines feature class.

14➜ Click the Fields tab.

14➜ Click the Linetype field and change the default value to LATERAL.

14➜ Click the Diameter field and change the default value to 3.

14➜ Click the Capacity field and change the default value to 10.

14➜ Click Apply and then click OK to close the window.

14➜ Close ArcCatalog so that it cannot interfere with editing.

15➜ Open ArcMap again with your most recent map document.

15➜ Start editing and turn on edge snapping.

15➜ Open the Attributes window.

15➜ Choose the Sketch tool and digitize several new water lines coming off the main ones (Fig. 14.14).

Notice that each of the new lines has the new default values of LATERAL, 3, and 10. You can give your water lines the same symbols as shown in Figure 14.14 if you wish.

Fig. 14.14. Creating different types of water lines with default attributes

15➜ Stop editing and save the edits.

➜ Exit ArcMap. You don't need to save the map document.

With ArcEditor or ArcInfo, you could have created subtypes for the waterlines layer. The laterals and mains would be two different subtypes. Using subtypes, you don't have to stop editing and change the defaults. Each subtype is a separate target layer and has its own defaults.

Creating a feature dataset

Our next step in building this geodatabase will be to create a feature dataset called Commerce that contains the gas stations and the restaurants we geocoded in Chapter 10.

The spatial reference is a critical property of a feature dataset because all feature classes inside it must share the same coordinate system. The user must always be careful to specify the coordinate system. ArcGIS will then calculate the appropriate XY Domain and Resolution based on that coordinate system. Once specified, the coordinate system cannot be changed.

16➜ Open ArcCatalog and examine the metadata for several of the layers in the mgisdata\Rapidcity folder.

1. What appears to be the coordinate system of these data layers? _____

17➜ Return to the Contents tab. Right-click the rapidcity geodatabase and choose New > Feature Dataset.

17➜ Type the name Commerce for the new feature dataset and click Next.

17➜ Expand the Projected Coordinate Systems entry, navigate to the NAD 1927 UTM Zone 13N coordinate system, and select it. Click Next.

17➜ This feature dataset will have no vertical coordinate system, so leave it set to <None>. Click Next.

17➜ Accept the default XY, Z, and M tolerances.

17➜ Keep the box checked to accept the default resolution and domain extent.

17➜ Click Finish.

Now we add some feature classes to the feature dataset.

18➔ Right-click the Commerce feature dataset and choose Import > Feature Class (multiple).

18➔ Click and drag the gas_stations shapefile in the mgisdata\Rapidcity folder to the Input Features box in the tool.

18➔ Also add the restaurants shapefile, rceats2.shp, created in Chapter 10. (If you didn't create one, load the rceats.shp file from the mgisdata\MapDocuments\Results folder instead.)

18➔ Click OK to start processing.

18➔ Examine the new feature classes in the feature dataset.

TIP: If the import layer has a different coordinate system than the feature dataset, then the features will automatically be reprojected as they are loaded.

Now create another feature dataset to hold transportation features. This time we will import the spatial reference from the Commerce feature dataset since we already have it established.

19➔ In ArcCatalog, right-click the geodatabase and choose New > Feature Dataset.

19➔ Name it **Transportation** and click Next.

19➔ Click the Import button, navigate to the mgisdata\Rapidcity\rapidcity geodatabase, and select the Commerce feature dataset to import from. Click Add. Click Next.

19➔ Leave the vertical coordinate system set to None. Click Next.

19➔ Accept the default tolerance, resolution, and domain settings. Click Finish.

20➔ Right-click the Transportation dataset and choose Import > Feature Class (single).

20➔ Import the rc_roads shapefile from the mgisdata\Rapidcity folder. Name the output feature class **roads**. Don't change the Field Map settings. Click OK.

Adding geodatabase layers from ArcMap

You can also use ArcMap to add layers to a geodatabase. This approach is beneficial when importing only a selection of the features or changing projections. We will use this capability to import the Rapid City schools from the feature class containing all schools in the state.

➔ Start ArcMap with a new, empty map.

21➔ Load the roads feature class from the rapidcity geodatabase and the orthophoto rceast_nw.sid from the mgisdata\Rapidcity folder.

2. What is the coordinate system of the roads? _____

22➔ Add the schools layer from mgisdata\Sdakota\southdakota.mdb.

22➔ Rename the layer **SD Schools**.

A coordinate system warning may appear when loading the schools. Let's check its coordinate system to try to identify the problem.

3. What is the coordinate system of the schools? _____

When ArcMap brought in the data, it automatically chose a transformation from NAD83 to NAD27, which could shift features by up to several meters. A point location on a large school lot will not be significantly affected, so we proceed with the default transformation.

23➔ Zoom to the extent of the roads layer.

23➔ Set the Selectable Layers to SD Schools only.

23➔ Use the Select Features tool to select the schools that fall within or near the area covered by the roads.

Now we can export these selected features directly to the geodatabase as a standalone feature class, taking advantage of the projection capability along the way.

24➔ Right-click the SD Schools layer and choose Data > Export Data.

24➔ Choose to export the Selected Features.

24➔ Choose to use the same coordinate system as the *data frame*.

24➔ Click the Browse button.

24➔ Use the bottom drop-down box and change Save As Type to File and Personal Geodatabase feature classes.

24➔ Navigate to the folder with your rapidcity geodatabase in it.

24➔ Type in the feature class name **schools** and click Save.

24➔ Click OK in the Export Data window.

24➔ Click No to add the feature class as a data layer.

25➔ Switch to ArcCatalog and examine the geodatabase to see the new layer.

25➔ Open its properties to check that the coordinate system has been changed to UTM NAD27. Close the Properties window when you are finished.

TIP: If the schools layer does not appear, it probably means that ArcCatalog did not update the display. Select the folder containing the geodatabase and choose View > Refresh from the main menu bar. If the schools still don't appear, try restarting ArcCatalog.

➔ Close ArcMap. You do not need to save your changes.

TIP: We close ArcMap because we are going to add domains to our geodatabase, and a geodatabase cannot be modified in ArcCatalog while it is open in ArcMap.

Setting up attribute domains

For the next task, we will create some attribute domains for the roads in Rapid City. Recall that using domains is one benefit of the geodatabase model over coverages and shapefiles. We will create four domains for the roads: RoadType, NumLanes, Direction, and StreetAbbrev. RoadType contains a text designation such as Highway or Local, NumLanes gives the number of lanes, Direction gives one of the eight directions (N, W, SE, etc.), and StreetAbbrev indicates the street type abbreviation (RD, ST, AVE).

4. Which type of domain, range or coded, would you suggest for each of these attributes?

___

The first domain, RoadType, will be a coded value domain based on a text field and will contain the values Interstate, Highway, Ramp, Connector, and Local. Note that in creating the domain, we must be careful that every road in Rapid City fits into one of these groups. Such a classification scheme is called complete. We will use a three-letter code for the stored value but enter a longer description, which will appear to the user. We should also consider the split and merge policies to use. If a street is split, its type will likely remain the same, so we should choose Duplicate for the split policy. In the case of a merge, the street types might not match, so we should choose the Default Value policy.

26➔ In ArcCatalog, right-click the rapidcity geodatabase and choose Properties. Click the Domains tab.

26➔ Type the domain name, **RoadType**, into the first box, and enter the description **Type of Road** (Fig. 14.15).

26➔ Set the Field Type to Text.

26➔ Set the Domain Type to Coded Values.

26➔ Set the split policy to Duplicate.

26➔ Set the merge policy to Default Value.

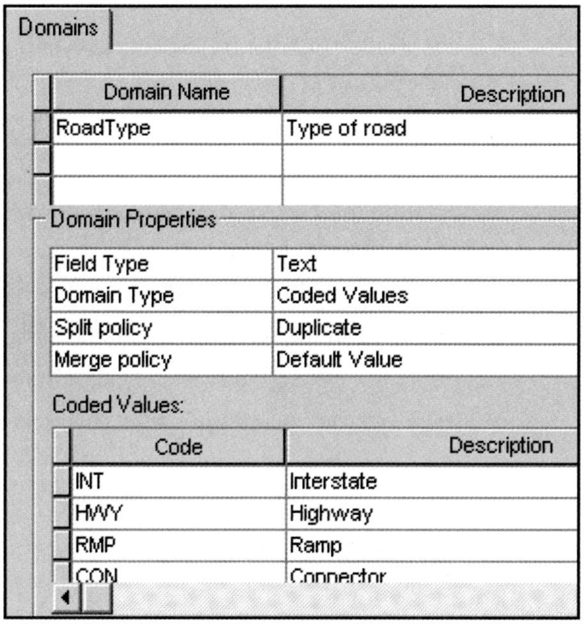

Fig. 14.15. Setting up the RoadType domain

27➔ Type the three-letter codes into the Code column and enter the description for each in the Description column: INT = Interstate; HWY = Highway; RMP = Ramp; CON = Connector; and LOC = Local.

27➔ Click Apply.

> **TIP:** ALWAYS, ALWAYS click Apply after EACH domain is added and before adding the next one. If you made an error in one domain but added more before clicking Apply, then it may not be clear which one has the error. You'll have to delete them all and start again.

The second domain, NumLanes, will be a short integer range domain and will allow values from 1 to 12. The split policy will be Duplicate, and the merge policy will be Default Value.

28➔ Enter the domain name, **NumLanes**, and the description, **Number of Lanes**.

28➔ Set the field type to Short Integer.

28➜ Set the Domain Type to Range.

28➜ Set the minimum value to 1 and the maximum value to 12.

28➜ Set the split policy to Duplicate and the merge policy to Default Value.

28➜ Click Apply. We do not need to enter any codes since this is a range domain.

Finally, we enter the third and fourth domains, Direction and StreetAbbrev. The Direction domain will be a coded domain with values of N, S, E, W, NW, NE, SW, and SE. The descriptions will be the same as the codes. The split policy will be Duplicate, and the merge policy will be Default Value. The StreetAbbrev domain will be a coded domain containing AVE, ST, RD, DR, BLVD, CT, LN, WAY, CIR, and PL, with the full word entered as the description (Avenue, Street, etc.). The split and merge policies will be Duplicate and Default Value.

TIP: You must *always* enter a description, even if it will be the same as the code. If the description is left blank, then the values will appear blank when you try to view them or edit them in ArcMap. ArcMap always displays the descriptions, not the codes.

29➜ Enter the Direction domain parameters and click Apply.

30➜ Enter the StreetAbbrev domain parameters and click Apply.

TIP: Domain policies and codes may be edited after the domain is created. However, the field type and domain type, once set, cannot be altered. To change them, you must delete the domain and create a new one. To delete a domain, select it and press the delete key.

Now that we have set up the domains, our next step is to create the attributes in the roads file. At the time we create them, we will assign the domains and also set the default values. The default road type will be Local, the default number of lanes will be 2, and the default speed limit will be 25 because most of the roads in town are local residential streets.

31➜ Click OK to close the Domains window.

31➜ Expand the Transportation feature dataset if necessary, right-click the roads feature class, and choose Properties.

31➜ Click the Fields tab.

31➜ Type the new field name, TYPE, in the first available box in the fields list (Fig. 14.16).

31➜ Click in the Data Type box and choose Text.

31➜ Type LOC for the default value.

31➜ Click the Domain box and choose RoadType.

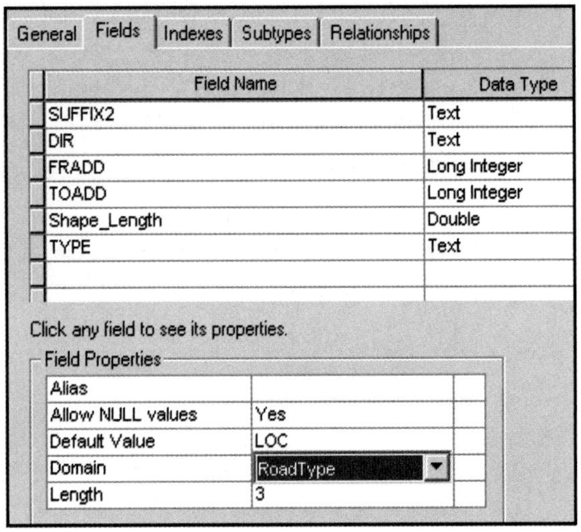

Fig. 14.16. Setting the domains and default values for a new field

31➔ Set the Length to 3.

31➔ Click Apply.

Next we create an attribute for the number of lanes.

32➔ Type the new field, LANES, in the next available box.

32➔ Set the field type to Short Integer.

32➔ Set the default value to 2.

32➔ Set the domain to NumLanes.

32➔ Click Apply.

TIP: Note that the list of domains available was different in the previous two examples. The domain field type must match the attribute field type; a short integer domain cannot be applied to a long integer attribute, for example.

Now we can assign domains and default values to some existing fields. The Direction domain actually applies to two fields, the PREFIX field and the SUFFIX2 field. One benefit of domains is that they may be applied multiple times.

33➔ Click the row for the field PREFIX to select it.

33➔ Leave the default blank.

33➔ Set the domain to Direction. Click Apply.

33➔ Select the SUFFIX2 field and set the domain to Direction. Do not set a default. Click Apply.

33➔ Select the SUFFIX field and set the domain to StreetAbbrev. Do not set a default. Click Apply.

Using domains when editing

Now we have set up our domains and are ready to see how they work.

➔ Close the Properties window and exit ArcCatalog so that it won't interfere with editing the roads.

34➔ Start ArcMap with a new, empty map, and load the roads layer from the rapidcity geodatabase.

34➔ Open the Editor toolbar, if necessary, and choose to Start Editing.

35➔ Use Select By Attributes to select all the roads where [STREET] = 'MOUNT RUSHMORE'.

35➔ Close the Select By Attributes window.

36➔ Click the Attributes button.

36➔ Click the top entry in the attribute list and examine its attributes.

First, notice that the SUFFIX field entries are listed as "Road" rather than RD, which is the code actually in the field.

36➔ Click in the space next to the SUFFIX field.

A coded value domain has a drop-down list showing the codes for that attribute (Fig. 14.17). This approach prevents the user from entering an inappropriate code, such as AV instead of AVE. The same holds for the SUFFIX2, PREFIX, and TYPE fields.

36➔ Click each of the other coded value fields and examine the choices.

Notice that most of the fields also include a <Null> entry. Whether this null choice appears depends on whether you answered "yes" or "no" to the Allow Null Values entry in the Fields properties in the previous section.

37➔ Click on the TYPE field and choose Connector.

37➔ Click on the LANES field.

Since LANES was assigned the NumLanes domain, a range domain, there is no list. Simply type in the value. No error checking is done at this point.

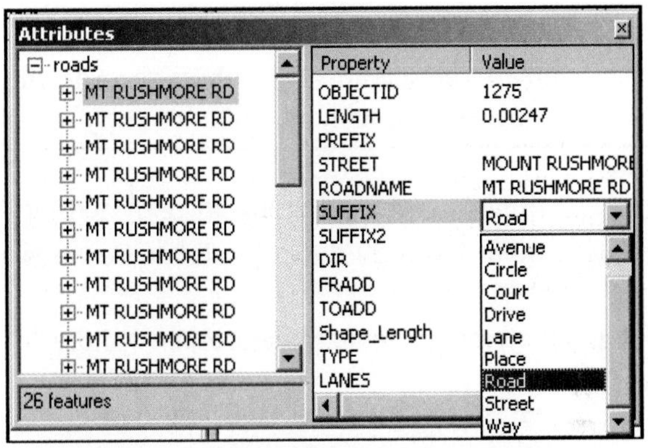

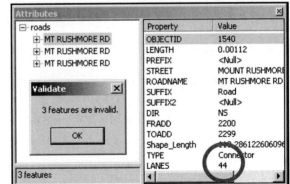

Fig. 14.17. Coded domains appear as drop-down lists during editing.

37➔ Enter the value 4 in the LANES field.

So far, we've only entered values for the first road. Let's assign the same values to all of the Mt. Rushmore roads.

38➔ Click the row labeled "roads" at the top of the Attributes window. The attributes go blank.

38➔ Click on the PREFIX field and choose <Null>.

38➔ Click on the SUFFIX2 field and choose <Null>.

38➔ Click on the TYPE field and choose Connector.

38➔ Click on the LANES field and enter 4.

Fig. 14.18. Validating attribute entries

Now let's see how to take advantage of the range domain error checking. It is done by means of a separate step called validation.

39➔ Click one of the segments of MOUNT RUSHMORE RD in the Attributes window.

39➔ Click the LANES field and enter the number of lanes, but pretend to make a typing mistake by entering 44 instead of 4 (Fig. 14.18).

Recall that the NumLanes domain has a range from 1 to 12. Now test for these attribute errors by using the validation function.

40➜ Choose Editor > Validate Features from the Editor toolbar.

If any errors are found, it reports the number in a dialog window and places the incorrect records in the selected set, ready for editing. Only attributes with domains will be tested.

TIP: Note that the PREFIX and SUFFIX2 fields were originally blank. If we had failed to apply the <Null> value specifically to these records, then all of the records would have shown up as errors. A blank field without a <Null> entry is considered an error.

40➜ Correct the wrong entry by setting the number of lanes back to 4.

40➜ Choose Editor > Validate Features again.

This time it reports that all features are valid.

➜ Stop editing and save your edits.

➜ Exit ArcMap. You don't need to save the map document.

Exploring the split and merge policies

In this section, we will experiment with using the split and merge policies to help manage parcel attributes during editing. First, however, we will import some parcels into our geodatabase and add some additional domains.

41➜ Start ArcCatalog and navigate to the directory containing the rapidcity geodatabase.

41➜ Import the feature class parcelval8 from mgisdata\Rapidcity\eastpat.mdb into the rapidcity geodatabase as a standalone feature class named parcelval.

41➜ Preview the attribute table of parcelval feature class.

5. Which one of these fields should not have a domain? Why not? _____

6. For each field that can have a domain, state whether a coded or range domain would be best.

    _____

7. For each field that can have a domain, state which split policy would work best.

    _____

8. For each field that can have a domain, state which merge policy would work best.

    _____

With these answers in mind, let's create the domains.

42➜ Right-click the rapidcity geodatabase and choose Properties.

42➜ Enter a new domain called ZoneCodes. Make it a text domain with coded values.

We will enter a few zoning codes and descriptions as examples: R-1 = Low Density Residential, R-2 = Med Density Residential, R-3 = High Density Residential, and C-1 = Commercial.

42➔ Set the split policy to Duplicate.

42➔ Set the merge policy to Default Value.

42➔ Enter the zoning codes as listed above.

42➔ Click Apply.

For the ID field, create a generic long integer ID field domain with a duplicate split rule, which could be used in many different feature classes.

43➔ Enter a new domain called LongIntDup with the description Generic Long Integer with Dup Splits.

43➔ Make it a range domain from 0 to 999999999.

43➔ Set the split policy to Duplicate and the merge policy to Default Value.

43➔ Click Apply.

Since ParcelValu and Area fields may range quite a bit in size, there is no particular benefit to setting a maximum—parcels can get large and expensive. Mainly we want the domains in order to use the split and merge policies. Both are also long integer fields, so let's simply create a single generic domain for both, called LongIntRatio, with a very wide range.

44➔ Enter a new domain called LongIntRatio with the description Generic Long Integer Values with Ratio Splits.

44➔ Make it a range domain from 0 to 999999999.

44➔ Set the split policy to Geometry Ratio and the merge policy to Sum Values.

44➔ Click Apply. Click OK to close the window.

Now we still need to assign the domains to the attributes in parcelval, so let's start with the ID field.

45➔ Right-click the parcelval feature class in the geodatabase and choose Properties.

45➔ Click the Fields tab and select the ID field.

45➔ Set the field alias to BlockID.

45➔ Change the default value to 0 (as a flag to identify blank records).

45➔ Set the domain to LongIntDup. Click Apply.

Now the Parcel_ID field has no domain, but we can still set a default value.

46➔ Click the Parcel_ID field to select it.

46➔ Set the default value to 0-000000. Click Apply.

46➔ Click the Area field and set the domain to LongIntRatio. Click Apply.

47➔ Click the ParcelValu field and set the domain to LongIntRatio. Click Apply.

47➜ Select the Zoning field. Set the domain to ZoneCodes and enter the default value as R-1. Click Apply.

47➜ Click OK to close the Properties window.

Now let's see how the split and merge rules work.

➜ Close ArcCatalog and start ArcMap with a new empty map.

48➜ Add the parcelval feature class from the geodatabase.

48➜ Label the parcel features with the Parcel_ID field in 10-point bold font.

> **TIP:** When setting up the labels, click the Placement Properties button, click the Placement tab, and choose the Try Horizontal First option. The labels will fit the polygons better.

48➜ Open the Editor menu, if necessary, and choose Start Editing.

Imagine that these parcels have been approved for a new subdivision being built. During the final site preparations, however, some site difficulties have surfaced, necessitating some changes in the layout of the parcels. The first parcel listed for change is the third from the right on the bottom row, currently labeled 8-453934. This parcel must be split in half.

49➜ Click the Edit tool on the Editor toolbar, and click on the parcel to be split.

49➜ Open the attribute table of parcelval, and click on the Show Selected button.

49➜ Resize and move the table to show the fields and the map at the same time.

49➜ Examine the values in the table.

Now let's split the parcel.

50➜ Turn on edge snapping for the parcelval layer.

50➜ Make sure that parcelval is the target layer, and change the editing task to Cut Polygon Features.

50➜ Use the Sketch tool to add a line cutting the parcel approximately in half. Double-click to finish the sketch.

Id	Parcel_ID	Area	Zoning	ParcelValu
8	8-453934	17438	Low Densit	174000

Id	Parcel_ID	Area	Zoning	ParcelValu
8	8-453934	8302	Low Densit	82843
8	8-453934	9136	Low Densit	91157

Fig. 14.19. The parcel attributes before and after a split

50➜ Examine the updated attributes in the table (Fig. 14.19). Yours may differ slightly.

Notice that two polygons are now selected. The Parcel_ID field had no domain and followed the ordinary split rule of copying the old value to both new features. We set a Duplicate rule for the BlockID, so the value 8 appears in both new polygons. The Area and ParcelValu fields, which have a Geometry Ratio rule, have been divided between the two polygons based on their respective areas. The Zoning value has defaulted to Low Density in both. All of these new fields are correct, except that a new parcel-ID must be added. Add the next highest value available in that block, 8-453949.

51➜ Click in the Parcel_ID field in the bottom record of the table and change the value to 8-453949.

It would make sense to try a merge next, but the merge policies don't actually work in the Editor. They exist so that programmers developing advanced applications can take advantage of them, but they will not help the average user. In the Editor, you simply pick which of the features will give its attributes to the new feature.

Id	Parcel_ID	Area	Zoning	ParcelValu
8	8-453914	6244	Low Densit	115000
8	8-453915	7486	Low Densit	121000

ID	Parcel_ID	Area	Zoning	ParcelValu
8	8-453914	6244	Low Density	115000

Fig. 14.20. Parcel attributes before and after a merge

Let's merge the third and fourth parcels from the left in the top row, 8-453914 and 8-453915.

52➜ Move the table so that you can see these two parcels.

52➜ Select the two parcels to be merged.

52➜ Briefly examine their attributes in the table.

52➜ Choose Editor > Merge from the Editor toolbar. If asked to choose which is to be merged, pick the 914 parcel. Its attributes will be used for the new polygon.

52➜ Examine the attributes again (Fig. 14.20).

➜ Split or merge a few more parcels for practice, if you wish.

➜ When finished, stop editing and save the edits.

This is the end of the tutorial.

➜ Exit ArcMap. You do not need to save the changes to the map document.

More skills

Consult the Skills Reference section of this chapter to learn to do the following:

➢ Deleting geodatabases and geodatabase feature classes

Exercises

In this set of exercises, you will redesign the southdakota geodatabase to contain several feature datasets and add domains to it. The original database is in SD State Plane South projection; the new one should use SD State Plane North.

1. Create a NEW geodatabase named **sdnorth** to contain the feature classes in the mgisdata\Sdakota\southdakota geodatabase.

2. Add a Water feature dataset to the geodatabase for water features. Use State Plane North NAD 1983 for the coordinate system. Put the hydrology layers from the southdakota geodatabase, including the rivers, lakes, and watersheds, in the Water dataset. Also include the prec_ann shapefile from the Sdakota folder.

3. Create another feature dataset called Administrative and include all of the boundary layers: cities, counties, federal lands, state, and urban areas.

4. Include the rest of the feature classes as standalone feature classes, making sure they are all also in the same State Plane North projection. (**Hint:** Use the environment settings on the importing tool to set the output coordinate system. Be sure to set it back to Same as Input when you are finished.)

5. **Capture** the layout of the geodatabase in the ArcCatalog window.

6. Using the restaurants layer in the rapidcity database, add two attributes for the type of restaurant (franchise or local) and the number of tables. Create attribute domains for the fields. **Capture** the windows showing the properties for each domain.

7. Enter the information for each restaurant. (A franchise includes a national chain, such as Burger King or Kentucky Fried Chicken. All others are considered local. Simply make up some values for the number of tables.)

8. Create a layout showing the type of restaurant, labeled with the number of tables.

Challenge Problem

Create a geodatabase with at least four feature classes and at least one feature dataset. The data you use may come from any source but should comprise a logical collection of related data. **Capture** the layout of your geodatabase in ArcCatalog and create a map layout showing your data set. On the layout, include the coordinate system chosen for your database and a graphic showing its contents (as in Fig. 14.1).

Skills Reference

Creating a geodatabase 527

Deleting a geodatabase 527

Deleting feature classes 527

Deleting feature datasets 527

Creating a feature dataset 528

Creating a feature class...529

Importing feature classes..530

Creating attribute domains 531

Setting default values or domains...532

Creating a geodatabase

1. In ArcCatalog, navigate to the folder in which the geodatabase will be created.

2. Right-click the folder and choose New > Personal Geodatabase from the menu.

3. The geodatabase appears in the display window to the right.

4. Type in the name of the new geodatabase and press Enter.

Deleting a geodatabase

1. In ArcCatalog, navigate to the folder containing the geodatabase.

2. Right-click the geodatabase and choose Delete.

3. Say Yes, when prompted, to confirm deleting the geodatabase.

Deleting feature classes

1. In ArcCatalog, navigate to the folder containing the geodatabase and expand it.

2. Right-click the feature class to be removed and choose Delete.

3. Click Yes, when prompted, to confirm deleting the feature class.

Deleting feature datasets

1. In ArcCatalog, navigate to the folder containing the geodatabase and expand it.

2. Right-click the feature dataset to be removed and choose Delete.

3. Click Yes, when prompted, to confirm deleting the feature dataset.

Creating a feature dataset

1. Determine the desired coordinate system and resolution for the feature dataset since it can't be changed once defined. ArcGIS will assign default values for the resolution and domain based on the coordinate system. If you have another feature class or feature dataset with the same properties, you can import it from that one.

2. In ArcCatalog, navigate to the geodatabase to contain the feature dataset.

3. Right-click the geodatabase and choose New > Feature Dataset.

4. Enter the dataset name (Fig. 14.21). Click Next.

5. Choose the coordinate system to use for the feature dataset. If you have another feature class to import the spatial reference from, choose Import. Locate the feature class, select it, and choose Add. Click Next.

6. If your dataset needs a vertical coordinate system (3D data only), select one. Click Next.

7. Examine the default tolerance. It is recommended to accept it.

8. It is recommended to accept the default resolution and domain extent. If you uncheck the box, you will be shown another panel in which to set them explicitly.

9. Click Finish to complete creating the feature dataset.

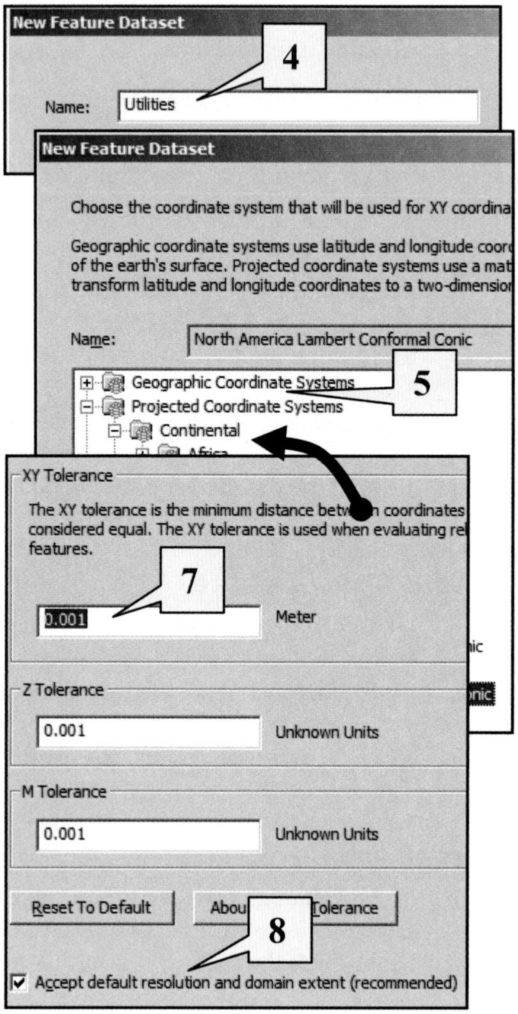

Fig. 14.21. Creating a new feature data set

TIP: Previous versions of ArcGIS required users to specify an appropriate trade-off between precision and the X/Y domain. This constraint has been removed in ArcGIS 9.2 by storing coordinates in high precision, so users may feel confident in accepting the defaults.

Creating a feature class

1. In ArcCatalog, right-click the geodatabase or feature dataset that will contain the feature class and choose New > Feature Class.

2. Enter the name of the feature class (Fig. 14.22).

3. Choose the geometry type. Click Next.

If you are adding to a feature dataset, the spatial reference will already be defined. If it is a stand-alone feature class, you must define the spatial reference.

4. Specify the coordinate system for the feature class, or click Import it to copy it from another data set. Click Next.

5. Examine the XY Tolerance. The default is fine for most applications unless very high precision is needed.

6. Accepting the default resolution is usually recommended. If you uncheck the box, you'll get another panel to enter the resolution. Click Next.

7. To add new attribute fields, click in the first available box in the Field Name list. Type in the name, set the Data Type, and change the field properties as necessary.

8. To add fields from another feature class or shapefile, click the Import button. Locate the feature class and select it. All fields in the layer will be added to the new one.

9. Click Finish to create the feature class.

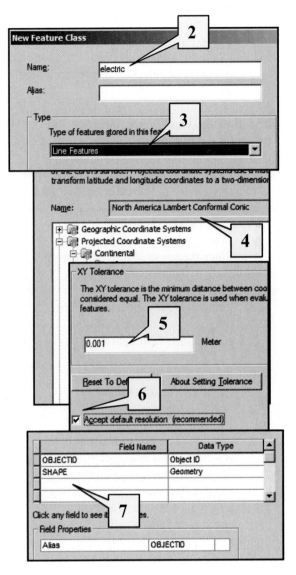

Fig. 14.22. Creating a new feature class

Importing feature classes

The following directions apply to importing feature classes as standalone classes or as parts of a feature dataset.

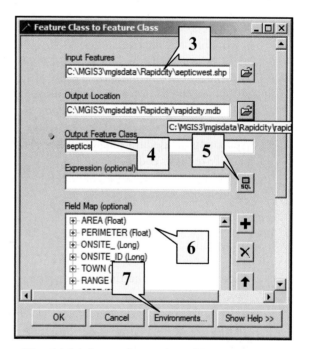

1. In ArcCatalog, navigate to the folder containing the geodatabase.

2. Right-click the geodatabase or feature dataset, choose Import, and select an import option based on the type of data to bring in (Fig. 14.23).

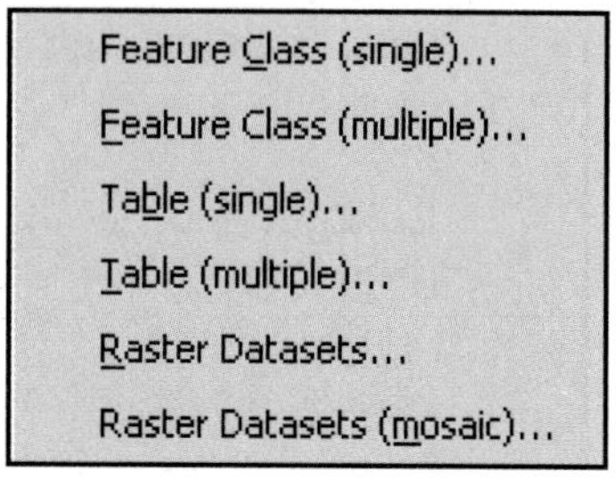

Fig. 14.23. Options for importing feature classes

The following directions assume that you have chosen the Feature Class (single) tool. The other choices will be similar.

3. Click the Browse button and navigate to the directory containing the data to import (Fig. 14.24). Select it and click Add.

4. Enter the name of the new feature class in the geodatabase.

5. Enter an SQL expression to bring only selected records into the geodatabase (optional).

6. Examine the fields list. Add or delete fields from the list, if necessary.

7. Click the Environments button to set the geoprocessing environment defaults, if desired. For example, use the coordinate system setting to project the data to a new coordinate system as it is imported.

Fig. 14.24. Importing a shapefile into a geodatabase

8. Click OK to start importing the feature class.

TIP: To import several feature classes at the same time, choose Import > Feature Class (multiple). The options to rename the feature class, perform selections, and edit field names are not accessible with the multiple feature class tool, however.

TIP: When importing pre-version 9.2 feature classes, you may receive a warning that the output feature resolution is larger than the input resolution. You need not be concerned in most cases.

Creating attribute domains

Attribute domains are properties of the geodatabase and are set at that level.

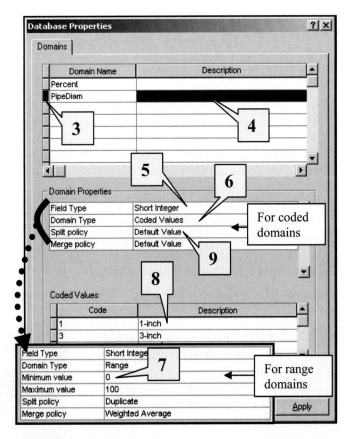

1. In ArcCatalog, navigate to the folder containing the geodatabase.

2. Right-click the geodatabase and choose Properties. The Domains tab is the only tab.

3. Click in the first available Domain Name box and enter the name of the domain (Fig. 14.25).

4. Enter a longer description of the domain for documentation purposes.

5. Set the attribute type (Short Integer, Long Integer, Text, etc.). The domain can only be applied to attributes of the same type.

Fig. 14.25. Setting up a domain

6. Click in the Domain Type box and choose a Range or a Coded Value domain.

7. For a Range domain, enter the minimum and maximum range values.

8. For a Coded Value domain, enter each value code and description in the lowest set of boxes. The descriptions should not be too long (<15 characters is probably best).

9. Click in the Split and Merge policy boxes and choose the type of policy desired.

10. Click Apply to create the domain. It will be checked for errors and added if OK.

11. To enter additional domains, repeat steps 3 through 10 or click OK if finished.

12. To delete a domain, click on the gray area to the left of its name to select it and press the Delete key.

TIP: Domain policies and codes may be edited after the domain is created. However, the field type and domain type, once set, cannot be altered. To change them, you must delete the domain and create a new one with the desired properties.

Setting default values or domains

1. In ArcCatalog, right-click the feature class containing the attribute and choose Properties.

2. Click the Fields tab (Fig. 14.26).

3. Select the field by clicking on the gray area to the left of its name. Its properties appear underneath.

4. Click the box for Null values and choose Yes to allow null values or No to disallow them.

5. Click in the Default Value box and type in the default value.

6. To assign a domain to an attribute, click in the Domain box and select a Domain from the list.

7. If desired, enter default values for additional items by repeating steps 3 through 5.

8. Click Apply to effect the changes without closing the window, or click OK to effect the changes and close the window.

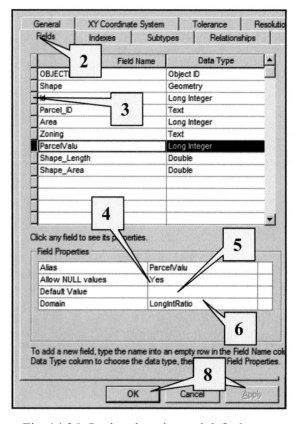

Fig. 14.26. Setting domains and default values for attributes

---

TIP: The Domain type must match the field type. A long integer domain cannot be applied to a short integer field, for example. Only valid domains will be listed for a field.

Chapter 15. Metadata

Objectives

➢ Understanding the layout and main sections of metadata

➢ Gaining a basic understanding of data quality issues

➢ Learning to use the metadata editor and templates to create metadata

Mastering the Concepts

GIS Concepts

Providing metadata is a fundamental part of creating GIS data sets. It is a professional obligation when data will be distributed to other users or the general public. Best management practices dictate that metadata should also be prepared for in-house data so that critical information about the source and processing of a data set will not be lost. Figure 15.1 shows the metadata developed for the radar image georeferenced in Chapter 11, which you will create in this chapter.

Preparing a 100% complete set of metadata is a daunting task. However, even partial metadata is better than none. This chapter provides a basic introduction to the most important sections of metadata and how they can be developed.

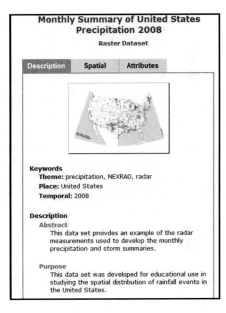

Fig. 15.1. Metadata example

The metadata standards

To facilitate sharing of metadata, which crosses over different GIS formats and programs, the Federal Geographic Data Committee (FGDC) has established a metadata standard that specifies the layout and information provided. The FGDC "is an interagency committee that promotes the coordinated development, use, sharing, and dissemination of geospatial data on a national basis" (www.fgdc.gov). Their Web site contains a wealth of useful information not only on metadata but on geospatial data sharing in general. The metadata standard is described in its entirety in publications offered by the FGDC, and these reports serve as the basis for all metadata questions and answers. The following publications are especially recommended for the beginner.

Geospatial Metadata Fact Sheet (2005). Brief explanation of what metadata is and how it is organized.

Geospatial Metadata Quick Guide (2005). The famous "Don't Duck Metadata" publication, which describes some of the basic types of information that should be included.

Top Ten Metadata Errors (2006). A short, fun read.

CSDGM Essential Metadata Elements (2008). The guide for those who have resigned themselves to producing metadata and want to achieve a minimum standard.

Content Standard for Digital Geospatial Metadata Workbook (2000). Exhaustive and too much for a beginner but an excellent reference for the serious metadata enthusiast. It does contain some good examples that beginners might find helpful.

The metadata standard lays out the specific information fields that should be included and defines which ones are optional and which are mandatory. It also specifies the exact format and organization of the information so that programs and metadata interpreters can always find the information in the same place and so that the metadata is easily transportable from system to system.

Some of the information in metadata is easy to understand and document, such as who produced the data. Many of the terms used in the standards and literature presuppose an understanding of geospatial data topics, such as logical consistency or completeness. Expertise in these areas will be developed by users as they travel the path to becoming geospatial professionals. Merely trying to understand these terms as you try to produce metadata will make you a better creator of data as well. Some sections of metadata are not applicable to a particular data set and can be ignored. As you gain experience reading and creating metadata, you will understand more and more about geospatial data, where it comes from, what it is good for, and what happens to it during its lifetime. You will also gain experience in evaluating the quality of data and deciding whether it is good enough for a particular project or application.

Data quality issues

Assessing the quality of a data set is not a simple judgment of good or bad but is a complex analysis of whether a particular data set is fit for an intended purpose. A data set that is inadequate for one application might be completely suitable for another. The data quality information in metadata is provided so that a potential user can evaluate the data set in light of what he or she wants to do with it. During this process, six key issues must be considered: 1) lineage, 2) positional accuracy, 3) attribute accuracy, 4) logical consistency, 5) completeness, and 6) temporal accuracy. The sections below discuss these issues and provide examples of how a user might document the information in the metadata.

Lineage

Lineage is concerned with documenting the original **source** of the data set and recording all of the operations and transformations that have occurred between the source and the final product. Questions to be answered about the data include who collected the data, how and why it was collected, what was the original scale or accuracy, how it was converted into digital form, and what operations have been performed on it since.

The metadata includes sections on the **process steps**, which list the sequence of operations that have been performed in producing the data. In particular, the impacts of any of the processing steps on the attribute or positional accuracy need to be recorded. For example, if the lines are generalized to the nearest 500 meters to prepare a detailed states map for a faster-drawing national map, that information should be recorded. If the feature class is the result of intersecting two layers with an XY Tolerance of 25 meters, then that information, with the potential impact on the feature accuracy, must be noted.

> ➤ You digitize published maps of aquifer outcrop areas from two adjacent quadrangles generated by the U.S. Geological Survey. You merge the two quadrangles and match the edges between them. In the metadata, you include the publication information and a complete citation for both maps. You record a process step for digitizing the maps, including the RMS errors reported for registering the maps on the digitizer. You report a second process for edge-matching and merging the two maps, including the RMS error associated with that step.

Positional Accuracy

Positional accuracy is an assessment of how closely features in the data set correspond to their actual locations in the real world. Impacts on positional accuracy occur at several stages in the development of a data set: 1) errors present in the source data or in the original data collection, 2) errors associated with transformations or georeferencing, and 3) errors associated with subsequent processing.

(1) Source data errors are generated many ways, but the most common are: by digitizing a paper map or scan of a paper map, by surveying, by photogrammetry or remote sensing, or by GPS measurement. Each of these methods has different sources of error that impact the positional accuracy.

Standard practice assumes that maps are accurate to within one line width, or approximately 0.5 mm. This value can be converted to ground units using the scale, as demonstrated in Chapter 2. A 1:24,000 scale map has an effective resolution of 12 meters, a 1:100,000 scale map has an effective resolution of 50 meters, and so on. If a map is digitized, its accuracy can be estimated from the scale. This estimate serves as a minimum, however, for it neglects potential inaccuracies in assembling the map data. Some phenomena or objects can be less precisely located than others. A road is easy to locate, but a soil map has an inherent uncertainty in placing the boundary between two soils types that grade into one another.

Surveying and photogrammetric methods follow rigorous standards during development and are usually subjected to quality assurance and control steps along the way. In general, base data sources, such as topographic maps and derived digital products, can be assumed to conform to the National Map Accuracy Standards. These standards require testing by comparing at least 20 checkpoints composed of well-defined and locatable points to corresponding locations on a reference map of higher accuracy. No more than 10% of the points can be off by more than 1 in 10,000 units. In practice, few maps are actually tested due to the time and cost constraints associated with mapping projects.

The accuracy of remotely sensed information varies greatly due to the differences in resolution between satellites and the differences in geometric correction that have been applied. GPS units also vary widely in their accuracy.

(2) A paper map or a scanned digital image must be georeferenced, transforming the digitizer, scanner, or screen coordinates into a real world coordinate system (see Chapter 11). The RMS error associated with this transformation should be recorded and included with the metadata.

(3) Various geoprocessing operations, such as generalizing lines or intersecting feature classes, can degrade positional accuracy. Processing tolerances, such as the XY Tolerance used when intersecting to remove slivers or a cluster tolerance applied when correcting topology errors,

provide an estimate of the impact. These tolerances should be recorded when listing the process steps in the metadata.

> You digitize the power lines on a 1:24,000 scale topographic map. You infer the source map accuracy to be 12 meters. The RMS error reported when you registered the paper map on the digitizer was 4 meters. Report both of these in the metadata.

> You go out in an ATV and collect GPS data on the location and extent of prairie dog towns. Previous testing indicated that your GPS unit has an accuracy of 10 meters, and you estimate that you can drive within 5 meters of the town boundary without damaging the burrows. Report the accuracy as 15 meters.

Attribute accuracy

Attribute accuracy assesses how well the values in the attribute fields correspond to the true values in the real world. Does the label on Rapid Creek represent its real name? Does the 40% crown cover value of a tree stand really reflect the actual density of the crown foliage? Because there are so many different kinds of attributes, assessing accuracy is complex, variable, and unguided by any sort of standard methodology. However, some of the issues that should be considered are as follows.

For categorical data, attention should be given to the classification scheme. How detailed is it? Are the crown-cover categories divided into 10 groups of 10% each or into 3 groups? Do the land-use categories represent all possible land-use types? Do the geologic units provide sufficient detail to locate potential gold horizons or are they simply grouped by age or gross rock type?

Issues related to all data types include the following: What is the expected rate of error in assigning values to the field? How often is something misclassified or wrongly recorded? How much heterogeneity exists within the measurement unit? The sand percentage in a soil polygon, for example, cannot be expected to be uniform over the entire area. Have any attempts been made to assess the variability within a unit?

If any information on attribute accuracy is known or if any tests have been conducted to assess the accuracy, these should be reported in the metadata. If nothing is known, then it is acceptable to put Unknown. Here are some examples.

> A land-use map has been generated from a Landsat satellite image. You have followed industry practice and tested the classification against a set of ground data points and developed the overall accuracy, producer's accuracy, and user's accuracy for the classification. Include these three values in the attribute accuracy report.

> Your survey agency has surveyed all the counties in the state on their approval of the governor's performance and is reporting an approval rating for each county. The survey method has an accuracy of +/− 3 percentage points. Include this accuracy in the metadata.

> You have digitized a geologic map from a plate in a master's thesis. The author did not include any accuracy assessment, and you have no funds to evaluate the accuracy of the map. Put Unknown in the attribute accuracy report in the metadata but do include the scale of the original map.

Logical consistency

The logical consistency is a measure of how well the features in the data model correspond to their counterparts in the real world and of how well the data model is able to model the relationships in the real world. This is another complex and difficult area that is rarely tested in full. However, it is usually possible and desirable to test the internal logical consistency of a data set and how well it fulfills the topological rules that have been defined for it. These topological rules were described in Chapter 13. Polygons should not have gaps or overlaps, lines should intersect at nodes and not cross each other, and so on. The logical consistency report can include the topological rules that were tested and the results of the test.

> You develop a planar topology for a Transportation feature data set and check for dangles, self-intersection, and improper intersections. All identified errors were resolved. Include a statement of the topological rules that were tested, and report that the data set does not contain violations of those rules in the logical consistency report.

> You digitized a geological map as a shapefile using snapping and AutoComplete polygon, and you are reasonably sure that most of the polygons have no gaps or overlaps. However, you don't have an ArcEditor license and cannot test. Report that snapping and coincident boundary protocols were followed during digitizing and that the data layer is free or nearly free of gaps and overlaps.

Completeness

The completeness of a data set refers to how well it has captured every possible instance of the objects in the data set. Completeness for polygons is usually easy to assess; do the polygons completely cover the area of interest or not? Lines and points can be less certain. Did the department of transportation digitize every known road in its GPS survey, or could it have missed some? Did it digitize only public roads, or are private roads also included? Did it include gravel and dirt tracks or only paved roads? Does the table of oil and gas wells in the state include every known well, or have some historical records been lost or misplaced prior to the assembly of the data?

> You assembled a database of septic systems by identifying houses on digital orthophotos taken five years ago. You realize that you may have missed systems for houses that were built after the photos were taken or for houses under heavy tree cover that were not visible in the photo. State these limitations in the completeness report.

Temporal accuracy

Some geospatial data, such as a map of greenness condition or a NEXRAD radar rain map, represent ephemeral conditions valid for a few hours or days. Other data sets represent phenomena that change more slowly, and data sets might be considered valid for several years. Population, for example, is usually referenced to a Census year or census update. Land-use data changes slowly as cities grow and new developments spring up. Some data can be considered virtually invariant. Geology, major road networks, rivers, and most international boundaries change infrequently on the scale of human affairs.

For data that will change significantly over the life time of the data set, it is important to designate how frequently the data provider intends to update the information. Some data is undergoing continual revision and is considered to be in progress, with updates planned at irregular or regular intervals. Census data, for example, is gathered every 10 years, and estimated updates are available every 4 to 5 years for some portions of the data set. However, a lot of GIS data is simply produced once, and no updates are ever planned.

Two sections in the metadata refer to the time period, that of the source and that of the data product being documented. These two are often the same but could be different. For example, the city planning office might have digitized the original platting records that were updated manually until 1995 to produce their parcels feature class (the source), but continued updates bring the feature class time period to within the past year (the product).

➢ Your office performed a GPS survey during 2004 and 2005 to locate and classify the existing off-road trails in the national forest, both official and unofficial. No funds are provided to locate and map new trails created (illegally) in the future. The temporal period of both the source and the product is 2004–2005, and no updates are planned.

➢ Your office maintains the county parcel records that were digitized from plat maps by a consultant in 2001. Every sale or modification is reported to the county as required by law through the office of deeds and land titles. You process the updates and make changes to the internal database as they come in. You push the updates to the public version of the database on your Web site every 3 months. In this case, the time period of the source is 2001, the time period of the data set is the date it was last pushed to the public site, and the update frequency is quarterly.

The metadata format

FGDC metadata is usually stored either as a simple text file or as a text file in eXtensible Markup Language (XML). XML is similar to the more familiar HTML used to develop Web pages. In addition to the basic information fields, it also contains formatting tags that tell a program how to read and display the information.

The metadata content standard divides the information into seven main sections.

1. Identification

This section contains basic information about the original source of the data, who created it, what it contains, and any restrictions that have been placed on its use. The *Abstract, Purpose,* and *Supplemental Information* communicate what the data set contains and notifies the user of any important information at the top. The *Currentness Reference* indicates the time period that the data represent, and the *Status* says how frequently it is updated. The *Data Set Credit* and *Citation Details* provide information on who created the data product and how it should be cited in the literature. The *Spatial Domain* and *Keywords* help anyone searching for data to determine the location and the content of the data. The *Access Constraints* and *Use Constraints* specify who can use the data and whether it can be given to others and usually contain standard liability and release statements to protect the data provider from lawsuits.

2. Data Quality

This section provides information on the important data quality issues discussed in the previous section. Sections are included on the *Logical Consistency* and the *Completeness* of the data set as a whole. There is a section to report on the *Attribute Accuracy* and the *Positional Accuracy.* The original scale and accuracy of the *Source* data must be given, including the temporal accuracy; the accuracy of the source may be different (better or worse) than the final data set. A complete *Source Citation* should also be included. Finally, there is a *Process Step* section that records the processing steps between the source and final product and the impact of these steps on the accuracy of the data.

3. Data Organization

This section contains information on the format and organization of the data, such as whether it is vector or raster data. ArcCatalog fills out this section automatically.

4. Spatial Reference

This section contains the complete spatial reference information on the coordinate system, including the GCS used, the map projection and its parameters, and the map units. ArcCatalog fills out this section automatically.

5. Entity Attribute

This section describes the attribute fields present in a vector data set or the Value field in a raster data set. This section is critical for informing data users of information, such as what the land-use codes represent or whether the AREA field is in square kilometers or square miles. The user can choose one of two methods to describe the attributes. The *Overview Description* can be used when all of the fields are similar or self-explanatory and can be described together in a paragraph or two. For example, a city feature class containing decadal population values since 1800 is easy to explain. A land-use map with only a couple fields might also fall into this category. The *Detailed Description* is used when each field requires a different explanation. The metadata creator enters separate information for each field.

6. Distribution

This section describes how the data are made available, by whom, and under what conditions. It states who the *Distributor* is and provides information for contacting that person or organization. It describes the format of the data (DVD, CD, FTP download) and the *Standard Process* by which a user should request the data, such as the ordering information or download instructions. It also reports the *Available Time Period* when the distributor intends to make the data available.

7. Metadata Reference

This final section is metadata about the metadata, including who developed it, in what format it is stored, and if any restrictions are placed on the distribution of the metadata, as opposed to the distribution of the data itself (usually the metadata are freely distributable even if the data set is not). Some of this section is filled out by ArcCatalog.

Acquiring metadata

Many projects utilize data sets that come from another organization, which, if it follows industry practice, has already supplied metadata for the data set. In this case, implementing the metadata becomes easy. The user need only ensure that the metadata is properly stored with the data set in ArcCatalog and that any subsequent processing or modification after acquisition has been recorded in the Process Step section.

For example, if the user has downloaded wetlands from the National Wetland Inventory and extracted the features in a particular county for a project, he or she only needs to record the extraction. This information would go into the Completeness Report and the Process Step sections of the metadata. The rest of the metadata need not change. In fact, it should not change;

it is important to retain the originators of the data set so that they receive proper credit and so that the prior information about the data is not lost.

However, this process may be complicated by the fact that not all organizations store the metadata as part of the data set, as ArcGIS does. The metadata may exist in a completely different document close to, but not with, the data that it represents. It might be somewhere on the Web site where the data are downloaded, in another folder on its source DVD, or occasionally even further afield. In many cases the separate file will be in a standard FGDC format (text or XML) and can be easily imported to ArcGIS. Sometimes, however, the format is different or not quite correct, and the user will have a more difficult time implementing the metadata. After working with metadata for a while, you will learn to recognize the formats that are suitable and that can be imported to ArcGIS with a minimum of fuss.

Occasionally you will encounter a standard data set for which metadata are not provided. For example, if one downloads the Digital Line Graph (DLG) 1:24,000 product from the USGS EROS Data Center, none of the metadata is actually available on site as a text or XML file (at least not that the author has been able to locate after much searching). However, generic DLG metadata files are floating around on the Web, and it is possible to find a suitable one with basic information that can be adapted for your purpose.

Metadata development

For data sets developed by you or your organization, the metadata process must start at the beginning. Although entering the metadata may be one of the final steps in producing the data set, the process begins far earlier when the data are first being assembled. Critical information available at the start of the process and throughout its duration must be recorded so that it can be added to the metadata when the time comes.

Imagine that you have scanned and are digitizing an historical map of the Lewis and Clark journey. You will need to record as much information as you can about the map itself: its author, year of publication, citation information, scale, medium on which it was drafted, and so on. Is there any information on the positional accuracy of the features on the map? For what time period was the source data relevant? Then, as you go through the digitizing process, information must be recorded along the way. What was the RMS error of the georeferencing step? What kinds of techniques or tests were used to ensure topological integrity? What are the definitions and content of the attribute fields added to the map? What do you know about the attribute accuracy and the completeness?

If you wait until the end, then critical information may be lost. Thus it is good practice to develop a form on which notes can be recorded during the creation of the data, organized so that the information can be easily translated to the metadata at the end. You can record these notes on paper or electronically. Even if all of the information recorded does not make it into the metadata, the notes will still provide a valuable resource for your organization. Table 15.1 shows a sample of a form that might be used. A copy of this form is stored in the mgisdata\Metadata folder for your use.

Table 15.1. Sample notes template for recording information about data layers

Project:				Feature layer:	
Compiler:		Start Date:			End Date:
___Download	___From client		___ CD/DVD	___Other:	
Original Scale:				Original format:	
Original Coordinate System:					
Originator Information (Name/Agency, contact info, and/or URL					
Publication Reference (Title, author, date, publication, version, edition, series, etc.)					
Horizontal Accuracy				Vertical Accuracy	
Field collection notes or other notes					
Processing steps and RMS when applicable				Notes on attributes/units/definitions	

About ArcGIS

The Metadata Editor

In Chapter 1 you learned how to view metadata in ArcCatalog. The Metadata toolbar (Fig. 15.2) allows the user to view the metadata in several different styles using Stylesheets. It also includes buttons for editing, creating, updating, importing, and exporting metadata.

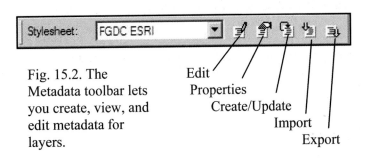

Fig. 15.2. The Metadata toolbar lets you create, view, and edit metadata for layers.

Edit
Properties
Create/Update
Import
Export

Clicking the Edit button takes you to the Metadata Editor (Fig. 15.3). It is easy to use, but its organization may take some time to learn. However, it follows the same basic layout as the metadata itself, with seven main tabs across the top corresponding to the seven main sections of metadata. Each tab is subdivided into additional tabs and panels that can be accessed and filled out. There is no set order; the tabs can be visited in any order and at any time.

Certain metadata fields are required by the FGDC in order to produce minimally compliant metadata. These sections appear with red text stating "REQUIRED:" followed by a brief description of the information to appear in that field.

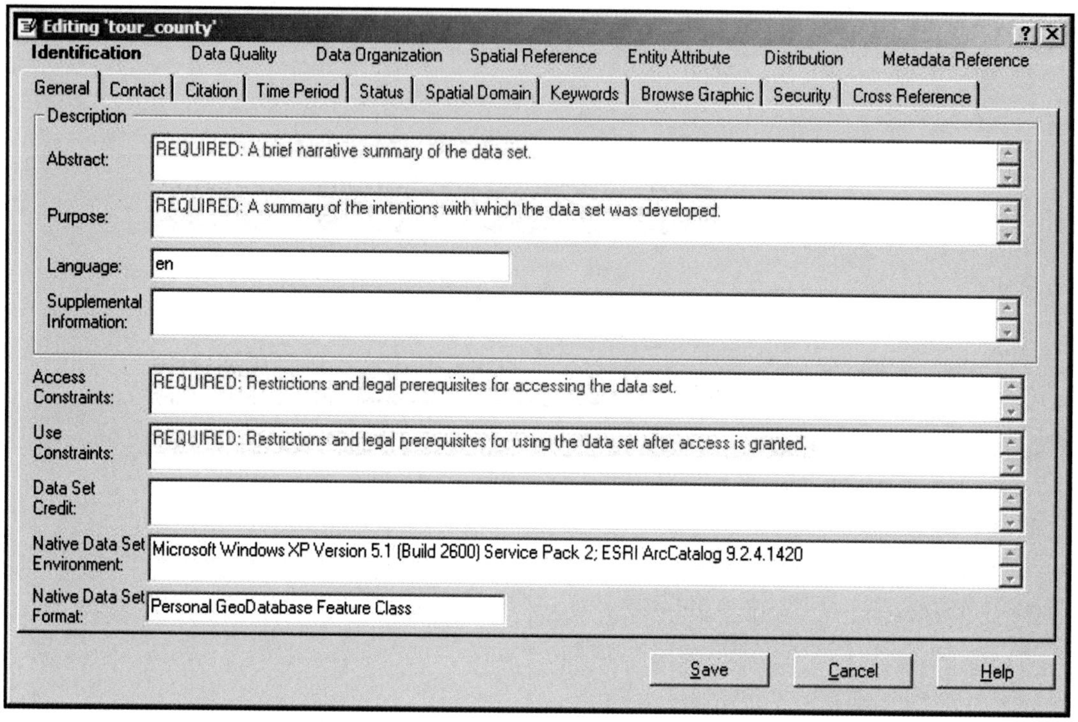

Fig. 15.3. The Metadata Editor

Some sections allow the user to add multiple instances in a certain category. For example, in developing a data set you might have used three different published sources and need to document all three. In the Source Information panel, a set of buttons allows you to add and edit sources individually (Fig. 15.4). Click the plus sign to add a record for a new source. Click the X button to remove a source. Use the arrows to move back and forth between the different source records and to enter or to edit the information for each one.

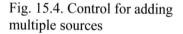

Fig. 15.4. Control for adding multiple sources

Metadata templates

Every metadata set has an impressive amount of information to be entered, but fortunately much of this information is repeated verbatim in many different data sets produced by the same organization. This fact makes the use of templates very attractive. A template contains the information that will be common to every data set produced by an organization or for a particular project. The contact information and access and use constraints, for example, are metadata fields that are likely to be consistent within an organization.

Metadata templates benefit from a hierarchical approach. Some information common to every data set for the organization can serve as the starting point. This template is further edited to include common information for a particular project, for example, the contact information for the project manager. The usual practice for a company might be to make all data available only to clients, but if a certain project was funded as an educational effort, then the usual access constraints might be edited to make the data freely redistributable. Templates might have three levels of specificity.

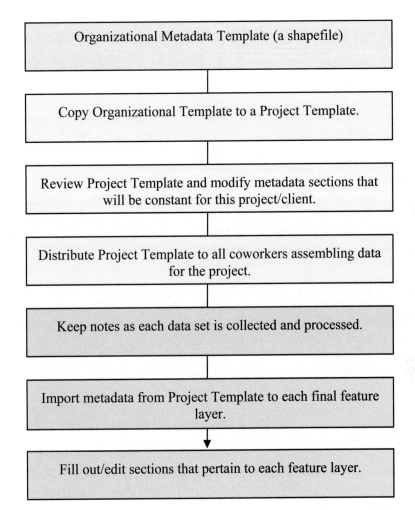

Fig. 15.5. A possible organizational process for using metadata templates

Organization level: Fields that are the same for every data set produced. Examples are the organizational contact, the access and use constraints, the metadata distribution constraints, and so on.

Project level: Fields that are the same for every data set produced for a particular project. Some of the organizational level fields might be edited for this level. Examples might be the Purpose, the Data Set Credit, the organizational contact, and some of the keywords.

Data set level: Fields that must be added for each individual data set, such as the Abstract and Purpose, the citation information, the accuracy reports, and so on.

The overall process followed might look something like Figure 15.5. The blue boxes refer to the Organizational level template, the yellow boxes to the Project level template, and the green boxes to the Individual level metadata. The user starts by creating an empty shapefile to serve as the metadata template. He then fills out all the sections common to every data set produced by the organization. This data set becomes the basic source for all subsequent metadata produced by the organization.

When a new project is initiated, the project manager copies the Organizational Template shapefile to a Project Template shapefile. She reviews the metadata and edits any portions that will be constant for the project. This might include changing some of the organizational fields or adding new information to the template. Once the Project Template is complete and has been reviewed, it is distributed to all of the workers who are producing data for the project.

As each data set is assembled, the workers take notes on the source and processing. When the data set is ready to have its metadata created, the worker imports the Project Template into the metadata for the new feature class. The data set-specific information is added to the metadata to make it complete.

The Skills Reference section has a metadata map to use as a resource. This map suggests what fields should be filled out for the organizational template, which ones might be edited or added for the project template, the fields that must be filled out for each individual data set, and the fields that are automatically generated by ArcCatalog.

Summary

➢ Metadata documents the history of a data set and provides the data quality information needed to ascertain its fitness for an intended purpose.

➢ Metadata standards and formats are established by the Federal Geographic Data Committee (FGDC).

➢ Six key data quality issues include lineage, positional accuracy, attribute accuracy, logical consistency, completeness, and temporal accuracy.

➢ Metadata are organized into seven main sections including Identification, Data Quality, Data Organization, Spatial Reference, Entity Attributes, Distribution, and Metadata Reference.

➢ ArcCatalog automatically fills out some sections of metadata.

➢ Establishing templates can reduce the work needed to prepare metadata.

Chapter Review Questions

1. What is the mission of the FGDC?

2. What is the difference between the Source and the Product when preparing metadata?

3. A raster has a cell size of 200 meters. What is the corresponding scale of the raster?

4. Why have formal guidelines for testing attribute accuracy not been established?

5. What aspects of logical consistency are most frequently tested and reported?

6. Contrast the temporal accuracy of a rivers feature class and a congressional districts feature class.

7. Which of the seven sections of metadata stores the process steps?

8. Which section of metadata would you consult to determine how to order the data and how much it costs?

9. Which section of metadata would you consult to determine if the ROADLENGTH field of a roads feature class is in kilometers or miles?

10. You download a data set from the Web and discover that it has no metadata when viewed in ArcCatalog. What is your next step?

Mastering the Skills

Teaching Tutorial

The following examples provide step-by-step instructions for doing basic tasks and solving basic problems in ArcGIS. The steps you need to do are highlighted with an arrow ➔; follow them carefully. Click on the video number in the VideoIndex to view a demonstration of the steps.

In this tutorial, we will first examine some excellent metadata examples to see what good metadata looks like. We will then create metadata for the weather map that we georeferenced in Chapter 11.

> 1➔ Start ArcCatalog and navigate to the mgisdata\Usa folder.
>
> 1➔ Expand the contents of the usdata geodatabase.

Examining metadata

Most of the feature classes in the usdata geodatabase came from the ESRI Data and Maps DVD that is distributed with the ArcGIS software. The ESRI folks are real professionals and their metadata provide excellent examples to learn from.

> 2➔ Click the states layer to highlight it, and click the Metadata tab in the display window.
>
> 2➔ Make sure the Stylesheet on the Metadata toolbar is set to FGDC ESRI.
>
> 2➔ Read the Abstract, Purpose, and Supplementary Information sections.
>
> 3➔ Click the green Status of the Data text to expand it. When you are finished reading it, click the green text again to put it away.
>
> 3➔ Expand and read the Time Period and Publication Information sections.
>
> 4➔ Expand the Data Storage and Access Information section.
>
> 4➔ Expand the Accessing the Data section and read it.
>
> 4➔ Expand the Constraints on Accessing and Using the Data and skim this long legal statement.

Next, we will look at some examples of data quality reports.

> 5➔ Click the Spatial tab at the top of the metadata.
>
> 5➔ Scroll down below the bounding coordinates information to the Lineage section.
>
> 5➔ Read the Process Step 1 description, which details the development of this data set.
>
> 5➔ Read about the data sources used to compile this data.

1. How many sources were used? _____ List them.

> 5➔ Read the Horizontal Positional Accuracy report.
>
> 6➔ Click on the quakehis feature class and go to the Spatial tab.

6➜ Read through the Process Steps, the Horizontal Positional Accuracy Report, and the Vertical Positional Accuracy Report.

7➜ Click on the spcszones feature class and go to the Spatial tab.

7➜ Skim the Process Steps and read the Horizontal Positional Accuracy Report.

Notice that if the horizontal or vertical accuracy of a data set is Unknown, it is perfectly all right to say so in the metadata.

8➜ Click the popestmt table and review the Description tab elements that are open.

8➜ Click the Spatial tab. This is a table, so the spatial section is very short.

8➜ Click the Attributes tab. Click on the first four field names and read their descriptions.

2.   What does the GQPOP90 field contain?

Creating a project template

You have been provided with a sample Organizational Template created for a fictional company called Re-GIS, Inc., that distributes weather and climate data sets by FTP download, usually at a cost. Imagine that you are the project manager for a new initiative to provide a freely distributable monthly summary of precipitation estimates based on National Weather Service radar data for the year 2008. You will create a Project Template for use with the data sets to be developed for this project.

9➜ In ArcCatalog, navigate to the mgisdata\Metadata directory and click on the OrgTemplate shapefile to highlight it.

9➜ Change the Stylesheet to FGDC for a more complete view of the metadata in the seven major sections.

9➜ Scroll through the metadata, noting the sections that are already provided and those that are REQUIRED.

10➜ Right-click the OrgTemplate shapefile and choose Copy.

10➜ Right-click the Metadata folder entry and choose Paste.

10➜ Click on the new OrgTemplateCopy shapefile twice slowly, and rename it RadProjTemplate.

11➜ Click on the RadProjTemplate shapefile to highlight it and click the Edit metadata button on the Metadata toolbar.

11➜ Click the Identification section and click the General tab.

11➜ For the Purpose, enter This data set was developed for educational use for studying the spatial distribution of rainfall events in the United States.

12➜ Update the Access Constraints to This data set is freely distributable for non-commercial purposes with inclusion of citation and full metadata.

12➔ Delete the first sentence in the Use Constraints, as it conflicts with the new Access Constraints just entered.

12➔ Click on the Contact tab and click Details. Change the Primary Contact circle to Person and enter your name as the contact person.

12➔ Click the Address tab and change the state to your state. Click OK.

Next you will enter the Citation information for this data set, showing how it should be cited by others.

13➔ Click the Citation tab.

13➔ Enter the Citation Title as **Monthly Summary of Radar Precipitation for 2008**. Click the Details button.

13➔ Enter **Re-GIS, Inc**. as the Originator.

13➔ You intend to release the data at the end of March, 2009. Enter **20090331** for the Publication Date.

13➔ Leave the other items and the Series/Publication information tab blank. Click OK.

14➔ Click the Time Period tab.

14➔ Set the Currentness Reference to **Ground Condition**.

14➔ Fill the circle for Range of Dates/Times

14➔ Enter **20080101** for the Beginning Date and **20081231** for the Ending Date.

The Status tab should reflect the condition of the data when it is published, not when it is in production.

14➔ Click the Status tab. For Progress, enter **Complete**. For Update Frequency, enter **None Planned**.

15➔ Click the Keywords tab.

15➔ Enter **precipitation** for the Keyword. Enter **None** for the Thesaurus.

15➔ Click the plus sign underneath the keyword to add another keyword.

15➔ Enter **NEXRAD**.

15➔ Click the plus again and enter another keyword, **radar**.

16➔ Click inside the Place keyword. Enter **United States**.

16➔ Click inside the Temporal keyword and enter **2008**.

16➔ Click Save to save your changes so far.

Notice that some of the items near the top have been updated with your changes. You can scroll down to see others and then continue your editing. Since a variety of layers are planned for this data set, with different sources and data quality parameters, you will leave the Data Quality section blank in the template, planning to fill it in for each feature class as it is created. The Data Organization and Spatial Reference sections will be completed automatically by ArcCatalog.

17➜ Click the Edit Metadata tool again.

17➜ Click the Distribution tab.

17➜ Click the Standard Order Process tab and change the Fees to **Free**.

17➜ Delete the statement about the valid credit card.

17➜ Click the Available Time Period tab and change the Calendar date to **20090331**.

18➜ Click the Metadata Reference section.

18➜ Enter **20090331** for the Metadata Review Date.

18➜ Click Details and enter your name as the contact person.

18➜ Click OK and Save.

Importing the template

Now you will create metadata for your first data layer for the project. Imagine that the weather map that you georeferenced in Chapter 11 will be an example of the data used to generate the final product and is to be included in the publication data set. Use the usradar image you georeferenced and put in the usdata geodatabase. (If you don't have it, use the nwsradar.img image in migsdata\Metadata.)

19➜ Click on the usradar or nwsradar.img raster to highlight it.

19➜ Click the Import metadata button on the Metadata toolbar.

19➜ Change the Format to XML.

19➜ Click the Browse button and navigate to the mgisdata\Metadata folder.

19➜ Choose the RadProjTemplate.shp.xml file and click Open.

19➜ Click OK on the Import Metadata window.

19➜ Scroll down and examine the metadata now.

The metadata for this image, which was essentially blank a moment ago, has been updated with all the information from your template. The ArcCatalog-provided fields (coordinate system, for example) are updated to reflect the current data set.

20➜ Click on the Preview tab in ArcCatalog.

20➜ Click the Thumbnail button.

20➜ Click the Metadata tab again.

The metadata has been updated to include the thumbnail you just created.

Entering metadata for a layer

Now we will complete entering the metadata for the nwsradar image. However, you will need to know some information about the source data. You had one of your coworkers investigate the Web site where the information was obtained and talk to the people at the National Weather Service who provide the radar data in the first place. She has assembled the information in Table 15.2 for you.

21➜ Click the Edit metadata button and click the Identification section.

21➜ For the Abstract, enter **This data set provides an example of the radar measurements used to develop the monthly precipitation and storm summaries.**

21➜ Under Supplemental Information, put **Largest scale when displaying the data is approximately 1:4 million.**

Table 15.2. Information about the radar image for inclusion in the metadata

Source information for NWS radar example	
Citation Title	NWS Radar Mosaic Time and date: 1718 UTC 06/20/2008
Originator	National Weather Service 1325 East-West Highway Silver Spring, MD 20910
Resolution	0.02 degrees
Scale of source	Approximately 1:4 million
Info	National mosaicked map of radar sectors for the conterminous United States. Units in DBZ from -25 to 75
URL	http://www.weather.gov/radar_tab.php
Estimated accuracy	Source resolution 0.02 degrees RMS error of transformation = 0.002 degrees.

22➜ Click the Citation tab and click the Details button.

22➜ Change the title to **Monthly Summary of United States Precipitation 2008.**

22➜ Enter the Online Linkage as **www.ReGIS.MonthlyRadar.com.**

22➜ Leave the other information as is. Remember, this tab refers to the citation for the final data product produced by your company. Click OK.

22➜ Click Save to save your changes so far.

The rest of the Identification fields should remain the same, as set up in the Project Template. We now turn to the Data Quality section.

23➜ Open the Metadata Editor and click on the Data Quality section.

23➜ You have no information on the Logical Consistency or Completeness report so leave them blank. The Cloud Cover is not applicable either, so leave it blank also.

23➜ Click the Attribute Accuracy tab.

23➜ For the Accuracy Report, enter **None Available.**

23➜ For the value, enter **DBZ (decibels).**

24➜ Click the Positional Accuracy tab.

24➜ For the Horizontal Accuracy Report, enter **Source data resolution is 0.02 degrees. RMS error associated with georeferencing is 0.002 degrees.**

25➔ Click the Source Information tab.

25➔ For Source Scale Denominator, type **4,000,000**.

25➔ For the Source Media, choose **online**.

25➔ For the Source Citation Abbreviation, choose **NWS Radar Maps**.

26➔ Click the Source Citation tab.

26➔ Click the Details button and fill out the sections as shown in Figure 15.6.

26➔ Click OK.

27➔ Click the Source Time Period of Content tab.

27➔ Enter **ground condition** for the Currentness Reference.

27➔ Fill the Single Date/Time circle and enter the Calendar Data and Time of Day as **20080629** and **17:18 UTC** (or whatever your image was).

Fig. 15.6. NWS radar citation

28➔ Click the Process Step tab.

28➔ Click the X by the Process Step near the bottom to remove any previous steps entered by the software.

28➔ Click the plus sign to enter a process step.

28➔ For Process Description enter **Georeferenced downloaded image to GCS NAD 1983 coordinate system with RMSE of 0.002 degrees.**

28➔ Enter **ArcGIS 9.3** for the software version.

28➔ Enter the Process date but leave the time blank. Click the Process Contact details and add your name and the **Re-GIS** organization. Leave the rest blank, as it was entered elsewhere. Click OK.

28➔ Enter **NWS Radar Mosaic** as the Source Used abbreviation.

28➔ Enter **Georeferenced Radar Mosaic** as the Source Produced abbreviation.

29➔ Click the plus sign to add the next Process Step.

29➔ Enter the Process Description as **Projected raster to North America Equidistant Conic coordinate system with nearest neighbor resampling and output resolution of 2000 meters.**

29➔ Fill out the rest of the window as in the previous step.

29➔ Click Save to save your edits so far.

Finally, the only section left is the Entity Attribute section.

30➔ Open the Metadata Editor and choose the Entity Attribute section.

30➔ Click the Overview Description.

30➔ For Dataset Overview, enter **Mosaicked radar measurements by the National Weather service, reported in DBZ (decibels).**

31➔ In the Entity and Attribute Overview box, enter the text shown in Figure 15.7 to document the color scheme used in the image (the values are from the scale on the original image).

31➔ Click Save.

31➔ You are finished! Examine the final metadata.

Color Scheme
Gray -25 – 5
Blue 5 – 20
Green 20 – 35
Yellow 35 – 40
Orange 40 – 50
Red 50 – 65
Pink 65 – 70
White > 70

Fig. 15.7. Color scheme for image

Importing acquired metadata

One last skill is useful. If you download a data set and the metadata is stored in an alternate file rather than as part of the data set, you can import it from a text or HTML file. Sometimes a little editing is required.

32➔ Change the Stylesheet on the Metadata toolbar to FGDC Classic.

32➔ Scroll down and examine the format of the metadata.

This format shows the typical metadata text file that might have been found before XML was used by ArcGIS and others. Often you can find files like these. If they appear in this format, it is likely they can be imported as metadata with a minimum of fuss.

Imagine that you have downloaded a 1:100,000 Digital Line Graph (DLG) file of the quadrangle boundaries in Custer County, South Dakota and imported it to a shapefile. The metadata were not included, but you found a text file elsewhere that looks as though it might help you in setting up the metadata.

33➔ Open the WordPad program on your computer.

33➔ Navigate to your mgisdata\Metadata folder and open the file 100Kdlgmeta.txt.

33➔ Examine the text file. Notice its striking similarity to the FGDC Classic metadata you were just viewing in ArcCatalog, suggesting that it might transfer.

34➔ Return to ArcCatalog.

34➔ Navigate to the mgisdata\Metadata folder and highlight the cu-quads shapefile. Notice that the metadata are essentially empty.

34➔ Click the Import metadata button.

34➔ Change the Format to FGDC CSDGM (TXT).

34➔ Click the Browse button and choose the 100Kdlgmeta.txt file. Click Open and OK.

34➔ Examine the metadata now.

The shapefile's metadata has been updated with the information from the text file, providing a starting place for further edits if necessary. However, we will not pursue that now. This is the end of the tutorial.

→ Exit ArcCatalog.

Exercises

Imagine that your class requirements include a GIS project (as it well may). Go on the Internet and find a data set that you might use for your project. Print the metadata notes form in the mgisdata\Metadata folder and fill it out for that data set.

Challenge Problem

Prepare a metadata template for your real or imagined class project.

If your instructor wants you to turn it in, set the Stylesheet in ArcCatalog to FGDC. Right-click on the metadata panel and choose Print.

Skills Reference

Editing metadata..554

Creating a metadata template 555

Importing or exporting metadata...555

A metadata reference map...556

Editing metadata

1. In ArcCatalog, navigate to the folder containing the data set to be documented.

2. Click on the data set in the folder tree to highlight it.

3. Click the Edit Metadata button.

4. Click one of the seven main sections (Fig. 15.8).

5. Click on each tab and subtab item in turn to fill out metadata. Fields with red REQUIRED text are required for FGDC-compliant metadata.

6. Click Cancel to exit without saving changes.

7. Click Save to exit and save changes.

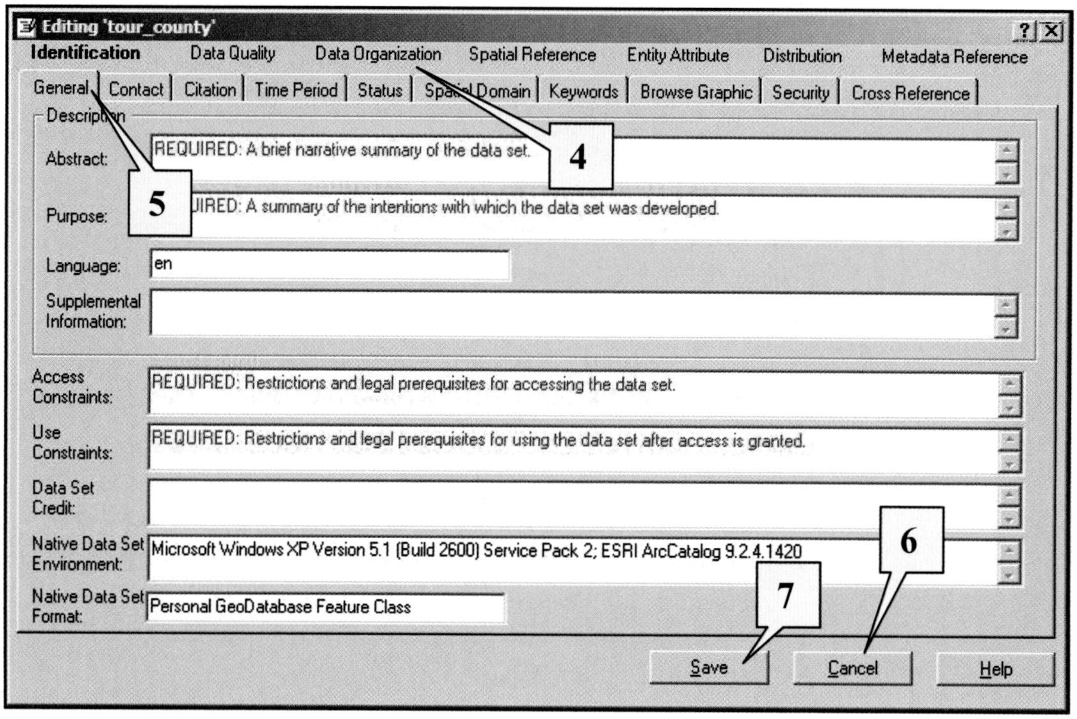

Fig. 15.8. The Metadata Editor

Creating a metadata template

1. Use the Create New Feature Class tool to create an empty shapefile and then give it a distinctive template name. The geometry type and coordinate system are not important.

2. Determine the fields that will be common to all the data sets for which the template will be used and decide what will be put into them.

3. Enter the data into the common fields using the Metadata Editor.

Importing or exporting metadata

1. Highlight the data set to be imported (exported) in ArcCatalog.

2. Click the Import metadata button (or Export metadata button).

3. Change the Format to the desired one (Fig. 15.9). XML or text is recommended. Use XML if importing from a shapefile template.

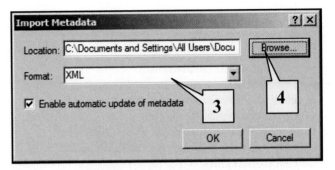

Fig. 15.9. Importing metadata

4. Click the Browse button and navigate to the folder where the import (export) file will be accessed (stored).

5. Highlight the file to be imported (or enter the name of the file to be saved).

6. Click OK.

A Metadata Reference Map

Use this map as a guide to creating metadata templates and filling in the metadata sections.

Key

To be filled in by developer for each layer
Provided by Organizational Template but might differ for Project Template
Provided by Organizational Template
Automatically generated by ArcCatalog

1. Identification

General

Abstract

Provide a brief description of the information contained in the data set.

Purpose

What is the intended use for this data set? Under what conditions is it valid?

Supplemental Information

This is a good place to list modifications made to imported data, such as subsetting it to a study area from an ESRI data set. This is also a good place to put the largest scale recommendation, Largest scale when displaying the data: 1:xxxxx. Other things may be added as appropriate.

Access Constraints

Who is allowed to access the data? The public? Only those who have licensed it?

Use Constraints

Generally, this contains a legal statement concerning those who may access the data, whether and under what conditions it may be redistributed, and a liability clause.

Data Set Credit

Credit the originator of the data set (USGS? Contractor? Map author?)

Native Data Set Environment
Native Data Set Format

Contact

Point of Contact Details

Provide the person or organization to be contacted with questions about the data set.

Citation

Citation Details

Provide information on the originator of the data set or publication as completely as possible, including publication references, URLs, agency address/phone, and so on.

Time Period

Currentness Reference

For what time period is this layer valid? Is it unlikely to change (topography), good only for a short time period (evapotranspiration), or somewhere in between (population)?

Status

Status

Is this in progress or a final version? How often are updates, if any, intended?

Spatial Domain

Bounding coordinates

Keywords

Keyword and Thesaurus

Provide at least two keywords for searching, including place name keywords

2. Data Quality
General
Logical Consistency Report
Logical consistency is largely concerned with topology. Did you test to see if the data contains dangles, gaps, or overlaps? What tests were applied, and what were the results?

Completeness Report
Provide information about omissions, selection criteria, generalization, and other processes that might impact how complete a data set is. Are all spatial entities included? For example, did you get all the wells or might some be missing? Did you subset the data from its original source? Were any criteria used in deciding which features to include (public versus private roads, for example)?

Attribute Accuracy
Accuracy Report
Summarize processes used to establish the accuracy of the attribute(s), for example, known detection limits of analyses. Evaluate detail or completeness of categorical classes. Provide known information about problems with any attributes. If no accuracy data are available, enter "Unknown."

Value/Explanation
For each attribute with a known accuracy, enter the data value and any explanation. For example, for a TMDL reading, enter "TMDL" as the value and "Detection limit xx mg/l, accuracy +/1 yy mg/l" for the Explanation.

Positional Accuracy
Horizontal Accuracy Report/Value/Explanation
If data are from a standard US federal data product, enter the national map accuracy standard of 1 in 10,000. If data are surveyed or obtained via GPS, enter the known or estimated positional accuracy of the survey or GPS unit. If data have been georeferenced, transformed, or spatially adjusted, also include the RMS error.

Source Information
General
The Source Scale Denominator is the original scale of the data set, such as 24,000 for quad data. Type of source media would be paper for a digitized map, data download/CD for a DLG, GPS unit, and so on. The Source Citation Abbreviation would be DLG, DRG, and so on.

Source Citation
Enter the original publication details of the data set: title, originator, publication date, edition, , and so on.

Source Time Period of Content
Enter Currentness reference as Ground Condition (at time of measurement) or Publication Date. Also enter whether data were collected all at one time, at multiple times, or in a range of times and enter dates. For example, if you sampled wells from January 08 to March 09, enter Range and the dates.

Process Step
Process Description
Describe one or more processing actions taken on the data before reaching its final form, for example, importing STDS quads and joining elevation attributes, merging into a single data set, projecting to current coordinate system, and clipping to study area boundary. Enter software version, process date, and the name of the person who did the processing. Include impacts on accuracy, such as an RMS error associated with georeferencing.

3. Data Organization
General
This entire section is updated automatically by ArcCatalog

4. Spatial Reference

This entire section is updated automatically by ArcCatalog.

5. Entity Attribute

Fill out either the Detailed Description tab or the Overview Description tab, whichever makes the most sense for documenting the attribute values of that particular data set.

Detailed Description

Entity Type

Attribute

General

*This section is mostly filled in for you, but, for each attribute important to the data set, you should enter a **Definition**, including units. For a field titled "TMDL," you should enter "Total Maximum Daily Load in mg/l." For an Area field, you might enter "Feature area in square km." If you wish to enter accuracies for fields, you can, but this is optional. The **Definition Source** describes who defined a particular attribute. If using Anderson land-cover categories Level I or Level II, for example, you would enter it. If it is very generic or obvious, such as acres or population, just leave this part blank. You don't necessarily have to fill in a Definition for every field, but you should do it for anything that is not obvious and/or needs interpretation, units, and so on. Think about the information YOU would want to know if you had to use this data set.*

Overview Description

This is the place to save time if you have many fields in a data set with the same type of data. For example, an agricultural table might have 20 fields of different animal counts and 30 fields of agricultural production for different crops in acres. In some cases, you might have a large table of values that are not obvious. If you are using zoning codes and the table does not include a text description, you would need to provide that information as a separate table in the database. Here you could reference the name of that table so that people know where to find out what the codes mean.

Data Set Overview

Describe the general purpose of the table, for example, "This data set lists animal counts and production of various crops by county."

Entity and Attribute Overview

Describe the main characteristics, such as animal counts in number of animals, stocking densities in cow/calf units per acre, or production in bushels. You can put them all in one entry or add multiple entries using the + button at the bottom, such as one for animal counts, one for production, and so on.

6. Distribution

This section describes whether and how the data could be ordered from a vendor/distributor, if applicable. It also is where you put your standard liability clause.

General

Provide general instructions on how to access the data, costs, if any, and what form the distribution takes.

Distributor

Enter the agency or person responsible for distributing the data and their contact information.

Standard Order Process

Enter the procedures to be followed to order or access the data.

Available Time Period

This is the time when the data will be available to the clients or public. Enter starting date of data availability and end date of availability, if known.

7. Metadata Reference

General

Provide metadata date, contact info of person or organization who created the metadata, and access and redistribution limits on metadata, if any.

Glossary

absolute pathname file pathname that starts at the drive letter

active frame the data frame that is currently in use and will respond to changes initiated by the user

address standardization conversion of a street address into specific components, such as house number, name, and type

adjacency a spatial condition that quantifies whether one feature is next to or touches another feature

affine transformation a process to translate, rotate, or skew an image to fit a new coordinate system

alias an alternate name displayed for a field in a table that does not have to follow length and character restrictions of the actual names

analysis cell size the default cell size defined for output grids during a raster analysis session

analysis extent the default extent defined for output grids during a raster analysis session

analysis mask a raster grid applied during a raster analysis session to convert all NoData cells in the mask to NoData cells in the output grid

annotation labels created from map features and stored separately as permanent objects for detailed editing

append a function that combines features from two different feature classes, often two adjacent ones

ArcToolbox a set of functions and commands for processing spatial data

ASCII international code used to store simple text letters, numbers, and symbols; synonymous with a "text file"

aspatial data data entries that are not tied, or that are only incidentally tied, to a location on the earth's surface

aspect the direction of steepest slope at a location on a surface, that is, the downhill direction

attribute fields columns in a table containing information about spatial features

attribute query an operation to extract specific records of information from a file based on values in the records

attribute tables tables containing rows of information for each feature in a spatial data file

attributes information about map features stored in columns of a table

automatic scaling fitting the contents of a data frame within a specified rectangle on the page

azimuth measurement of compass direction in 360 degrees with 0 being north and 180 being south

azimuthal projection a map projection onto a flat plane that is tangent or secant to the earth's surface

barriers objects that prevent flow or travel over a surface

bilinear a resampling method that interpolates a new cell value based on the four closest values

binary numeric data stored in base 2 as a series of ones and zeros; a computer's native data storage format

bookmarks links to a particular map area and scale for quick access

Boolean adjective describing data, operators, values, or files that contain or manipulate only true or false values

buffers delineated areas within a specified distance of a feature; buffers may be created from labels, points, lines, or polygons

byte a unit of data storage in base 2 containing eight zeros or ones and holding a number from 0 to 255

capture using the Alt-PrntScrn keys to place a copy of the active window on the clipboard for pasting into another program

cardinality the relationship between records in two tables: one-to-one, one-to-many, many-to-one, or many-to-many

categorical data data that place objects into unranked groups; examples are land-use or geology data

cell a square data element in a raster corresponding to one value representing conditions on the ground

cell size the dimensions of one square of map information in a raster data set

central meridian the central longitude of a map projection for which the x coordinate equals zero

chart a graph created in ArcMap from table data, such as bar charts or pie charts

chart maps maps showing several different attributes in chart form, with one chart for each feature

choropleth maps maps in which each feature, such as a state, is colored according to the values in a data field, such as population

class breaks break points used to classify values in a data field into a specified number of groups

classification assigning features to two or more groups based on numeric values in an attribute field

classified a raster display method that divides values into two or more groups based on their numeric values

clip to remove features and portions of features that lie outside of the features of another layer

cluster tolerance a defined distance used in topological editing, causing vertices to be made coincident if they are closer together than the tolerance distance

coded domain a rule that permits only certain specific values to be assigned to an attribute, such as land-use codes

complex edge a network entity composed of multiple linear features that behaves as a single linear feature in the network

conflict detection determining which labels will overlap each other when features are being labeled

conic projection a map projection derived by projecting latitude-longitude values on a paper cone covering a sphere

connection a link in ArcCatalog and ArcMap that points to a folder with GIS data and serves as a shortcut to frequently used folders; a link to a DBMS allowing data to be transferred

connectivity a property of linear features when they are connected to each other via junctions

containment the property of one feature including another in whole or in part

context menus computer menus that appear when certain objects on the screen are clicked, usually with the right mouse button

continuous data that take on a variety of different values and that change rapidly across a data set, such as elevation

contour a line indicating a constant value of a quantity on a surface, such as an elevation contour at 2000 ft

coordinate pair a single pair of *x* and *y* values indicating position in a planar coordinate system

coordinate space the range of *x* and *y* values onto which maps are plotted

coordinate system a specified range of units of *x-y* values onto which a map is plotted; the complete definition of the coordinate space used by a map layer, including the ellipsoid, datum, and projection

coverage the spatial data format created for, and used by, ArcInfo

cubic convolution a resampling method that interpolates a new cell value from the sixteen closest input cells

cut/fill a function that determines the different between a before and after topographic surface

cylindrical projection a map projection derived by projecting spherical latitude-longitude values onto a cylinder wrapped about a sphere

dangle a line feature that fails to connect to another line feature, leaving a gap

data frames containers in ArcMap that hold layers that are displayed and analyzed together; a map view

Data view data frame mode optimized for the display and analysis of map data

datum a combination of an earth ellipsoid and a local reference point to reduce mapping discrepancies for a particular region

dBase a database program whose file format has been adopted for the shapefile data model and tables in ArcGIS

default automatic input values assumed by a program when no values have been given

Define Projection a tool to guide the user through the task of assigning a coordinate system to a spatial data layer

defined interval a classification method in which the user specifies a specific size range for all the classes

definition query an operation to set a map layer to display only the features whose attributes meet specific criteria

degrees measurement units used in the spherical coordinate system; a circle has 360 degrees

DEM see *digital elevation model*

destination table the table that receives additional data from another table during a join operation

digital elevation model (DEM) a raster array of values representing elevations at the earth's surface

digital raster graphic (DRG) a scanned image of a USGS topographic map

digitize to convert shapes on a paper map to a digital map layer by entering or tracing vertices

discrete data that take on a relatively small number of distinct values

discrete color a display option for rasters in which every different value is assigned a random color

Display tab the tab in the Table of Contents that shows layers in the order in which they are draw from bottom to top

display units the units in which ArcMap reports the current *x-y* location of the cursor on the map

dissolve to combine features together when they share the same value for an attribute

distance join a join that combines the information from two feature tables based on the features that lie closest to each other

division units the units in which a map scale bar is measured and drawn, such as miles or kilometers

division value the length of one section of a scale bar as given in division units, such as 100 km

divisions the number of sections given to a scale bar

domains rules that determine the values that may be entered into an attribute

dot density map a map representing attribute values by a proportional number of randomly placed dots

double-precision a numeric value stored using 16 bytes of information

DRG see *digital raster graphic*

dynamic labels labels determined from an attribute and placed on a map automatically each time features are drawn and redrawn

edge a line feature that participates in a geometric network

edge snapping ensuring that new features are automatically connected to the edges of existing lines or polygons

ellipsoid a spheroidal volume with unequal axes, used to approximate the shape of the earth in map projections

end snapping ensuring that new features are automatically connected to the ends of existing line features

enterprise GIS a long-term GIS project developed by a large organization and involving many people over a long period of time

equal interval a classification method in which the user specifies a number of classes that have equal size ranges

erase an overlay function that removes features lying inside the external boundary of another polygon feature class

event layer a map layer of points created from a series of coordinate pairs in a table

Export to create a new data file from all or a subset of features in an existing one

expression a statement containing field names, values, and/or functions used to extract records in a query or calculate values in a table

extent the range of *x-y* coordinates displayed in a map or stored in a data layer

extent rectangle the range of *x-y* coordinates occupied by the features in a data layer

false easting an arbitrary *x*-coordinate translation applied to a map projection, usually to ensure that all values are positive

false northing an arbitrary *y*-coordinate translation applied to a map projection, usually to ensure that all values are positive

feature a spatial object composed of one or more *x-y* coordinate pairs and having one or more attributes in a single record of an associated table

feature class a set of similar objects with the same attributes stored together in a spatial data file

feature dataset a set of feature classes in a geodatabase that share a common coordinate system and can participate in networks and topology

feature service an Internet map layer in which the actual features may be downloaded by the user and saved locally

feature weight the priority assigned to a layer when determining which features may be drawn on top of others

FID (feature ID) a unique number assigned to every feature in a spatial data file and used for identification and tracking

field a single column of information in a data table

field definition parameters specified when creating an attribute field, including type of data, field length, precision, and scale

field length the maximum number of characters that can be stored in a text attribute field

filter a moving window applied to a raster data layer that calculates new values for the center pixel based on some function of the values in the window

fixed extent a constraint applied to a data frame to prevent changes in scale or extent in layout mode

fixed scale a constraint applied to a data frame to prevent the scale of the map from being changed in layout mode

flag a marker showing a point of interest in a network, such as a start point or endpoint for a path or a stop along a route

flat file database a database that stores data in simple files

flipping lines swapping the start and end nodes of a line feature so that it goes in the opposite direction

focal functions raster analysis functions that determine new values for a target cell based on values in a moving window around the target

GCS see *geographic coordinate system*

geocoding the matching of a location stored in a table to a spatial point feature based on a reference spatial data layer, most often applied to converting addresses to locations

geodatabases a new data model developed for ArcGIS 8 that employs recent database technology for storage and implements rules and topology

geographic coordinate system (GCS) a spherical coordinate system of degrees of latitude and longitude that is used to locate features on the earth's surface

geoid the shape of the earth as defined by mean sea level and affected by topographic and gravitational factors

geometric accuracy the accuracy with which the shape and position of features are represented

geometric interval a classification method that bases the class intervals on a geometric series in which each class is multiplied by a constant coefficient to produce the next higher class

geometric network a system of linear edges and point junctions used to model the flow of commodities, such as traffic or utilities

geoprocessing analysis of spatial data layers, such as dissolving, intersecting, and merging

georeferenced a spatial data layer that is tied to a specific location on the earth's surface for display with other data

georelational a data model that links features to attributes in a separate table using a unique feature ID code

graduated color map a map that divides numeric data from a polygon feature class into classes based on value and displays the classes with different colors

graduated symbol map a map that divides numeric data from a line or point feature class into classes based on value and displays the classes with different size or thickness of symbols

graticules grid marks of latitude and longitude placed on a map boundary

grid specific raster format native to GRID and Spatial Analyst and required for raster analysis; generic term for a raster data set

ground control points a set of points that match easily identifiable locations on two different data layers to enable georeferencing of one layer to another

hierarchical database a database that stores information in tables with permanent links between them

hillshade a raster, that displays the brightness variation of a surface as if it were illuminated by a light source at a specified azimuth and zenith angle

histogram a graph showing the number of pixels contained for each data value in a raster

image a raster data layer, usually referring to a raster that displays brightness values, as in a photograph

image service an Internet map layer that is available for viewing but that cannot be downloaded for local storage by the user

INFO an early database system upon which the Arc/Info software data model was based

inside join combining the information from two feature tables based on one feature that lies inside another

interactive labels simple graphic labels on a map that are placed by the user one at a time

interpolate to calculate values at locations between known measurements; to populate a raster with values extrapolated from a known set of point values

interactive selection choosing features by clicking them or enclosing them with the Select Features tool

intersect an operation to overlay two spatial data layers and find the areas common to both while discarding areas unique to either

intersection the property of two features touching each other in whole or in part

interval data values that follow a regular scale but have no natural zero point, such as degrees Celsius or pH

join the temporary combination of data from two tables based on a common attribute field or location

junction a point connecting two linear edges of a geometric network

kernel a moving window of values applied to a raster to calculate new values at a target in the middle of the window. See also *filter*

key an attribute field that is used to extract or match records in a table

label weights a priority rating assigned to values to determine which ones will be placed in case of an overlap

latitude a spherical unit measuring angular distance north or south from the equator

latitude of origin the reference latitude of a map projection where the y value is zero

layer a reference to a feature class and its associated properties

layer file a file that stores a pointer to spatial data along with information on how to display it

Layout view a mode of ArcMap that is used to design and create a printed map and that allows manipulation of map layers, titles, scale bars, north arrows, and more

legend a map element that displays the names and symbols used to portray layers on a map

line a spatial feature composed of a string of x-y coordinate vertices and used to represent linear features such as streets

linear regression calculation of a best-fit line showing the relationship between two variables

logical consistency a measure of how well data features represent real world features, in particular with respect to topology

logical expression a statement composed of field names, operators, and values that specifies criteria used to select records or values from a layer or table

logical network information stored in separate tables that keeps track of relationships between elements of a network

logistic regression a multilinear regression analysis in which the dependent variable takes only true or false values, or absence-presence values

longitude a spherical unit measuring angular distance east or west from the Prime Meridian

loops connected circular paths in a network in which flow is indeterminate

manual class breaks a classification scheme in which the user sets each class break to the desired value

map algebra a system that permits calculations and operations on entire raster arrays, such as adding two rasters together

map elements objects placed on a map layout, such as titles, legends, scale bars, north arrows, images, and charts

map extent the range of x-y values of the area being displayed in a map

map overlay to combine two spatial data layers, either for display or to evaluate the relationships between them

map scale the ratio of feature size on a map to its size on the ground

map topology temporary spatial relationships developed between features during editing in ArcMap to facilitate editing of features with common nodes or boundaries

map units the units of the coordinate system in which a map is stored or displayed

mask a raster layer applied during analysis to nullify unwanted cells, such as those outside a study area boundary

measurement grid a data frame grid that provides measured x-y coordinate system values around the perimeter of the frame

merge to combine two or more map features into one feature; to combine two or more features based on whether they share a common attribute value; to combine two or more data layers into a single layer

metadata information stored about data to document its source, history, management, uses, and more

minimum candidate score the lowest score at which an address may be considered as a candidate to match an address during geocoding

minimum match score the lowest score at which a candidate will be automatically matched to an address

model a sequence of steps or calculations used to convert raw data into useful information; a scheme used to understand and predict processes in the real world based on the manipulation of data

Model Builder a graphical interface used to combine existing tools to create new tools and scripts

multipart feature a single feature composed of nontouching units, such as a single state feature composed of the seven Hawaiian islands

multivariate regression an analysis used to discern predictive relationships between multiple variables

NAD see *North American Datum*

natural breaks a data classification scheme that divides values based on natural groupings or gaps between values

nearest neighbor a resampling method for discrete data that grabs the closest value to a cell center

neatline a line used to enclose one or more map elements in a rectangle

neighborhood statistics calculation of a value for a target cell or feature based on surrounding values from a defined region

network or network topology an association of linear edges and connecting points, used to model the flow of a commodity, such as traffic or utilities

network weights values associated with network elements that indicate the "cost" to traverse the element

NoData a special value used to designate that a data value is absent or unknown

nodes the beginning and end points of a line feature

nominal data values that name or identify an object, such as a street name

normalized data to divide the values of an attribute field by the total of the field or by the values in another field

North American Datum (NAD) a combination of a spheroid and reference point that is used to minimize map distortion in North America

numeric data values stored as numbers rather than as names or categories

object-oriented a programming and database approach that treats software and model elements as objects with defined properties and relationships

oblique cylindrical projection a map projection in which locations on a sphere are projected to a cylinder of paper at an arbitrary angle

ordinal a data value that indicates a rank or ordering system

origin the (0,0) point of a coordinate system

orthographic projection a map projection in which locations on a sphere are projected onto a flat piece of paper that is tangent to or intersects the sphere

orthophoto an aerial photograph that has been geometrically corrected to match a map base

overlap a spatial condition that quantifies whether one feature covers all or part of another feature

pan to move the display window to another part of the map without changing the map scale

parameters specific values associated with map projections that define how it appears

parametric arc a line feature composed of a smooth curve derived from a given radius for each segment

pathname a list of the folders that must be traversed to locate a particular file, such as c:\mgisdata\usa\states.shp

pixel a square data element in a raster corresponding to one value representing conditions on the ground

planar topology an association of feature classes in a feature dataset, established by rules regarding the spatial relationships between features, such as not overlapping each other

point a one-dimensional feature defined by a single *x-y* coordinate pair

polygon a closed two-dimensional area feature defined by three or more *x-y* coordinate pairs

precision the number of digits allotted to store a numeric value

Prime Meridian the line of zero longitude on the earth, passing through Greenwich, England

process step data assembly and geoprocessing function performed in generating a GIS data set

projection a mathematical transformation that converts spherical units of latitude and longitude to a planar *x-y* coordinate system

project-oriented GIS a GIS project with fairly limited objectives and a finite life span

proportional symbol map a map that displays attribute values with marker or line symbols that are proportional in size relative to the value of the feature

proximity the property of one feature being close to another feature

pyramid a set of rasters with different resolutions that is calculated from a raster and used to speed displays at smaller scales

quantile map a map type in which each class has approximately the same number of features

quantities data numeric attribute data

query an operation to extract records from a database according to a specified set of criteria

range domain a rule that stipulates the largest and smallest values that can be stored in a particular attribute

raster a data set composed of an array of numeric values, each of which represents a condition in a square element of ground

Raster Calculator an analysis window that applies algebraic and other operators and functions to grids to create new ones

raster model a data model that uses rasters or numeric arrays to represent real world features

ratio data data having a regular scale of measurement and a natural zero point, such as precipitation or population

reclassify to replace sets or ranges of values in a grid with different sets or ranges of values

record a row in a table containing one object

rectify to rotate, resize, or warp an image to match a map base using a selected set of ground control points

reference grid a data frame grid that provides lettered and numbered squares on a map for use with an index of the features

reference latitude the latitude of origin; see also *latitude of origin*

reference layer a layer containing special attributes needed for geocoding for which a geocoding index has been built

reference scale the display scale for which a map is designed that, when set, reduces or enlarges symbols and labels whenever the user zooms out or zooms in on a map

relate a temporary association between two tables based on a common field whereby fields may be selected based on whether they match selected records in the other table

relational database a database that stores information in tables and constructs temporary relationships between them

relative pathname path to a file that starts in the current folder

resolution the ground area represented by one cell value in a raster

RGB composite an image displayed by assigning one band of brightness information to each red, green, and blue color gun in a display monitor

RMS error (RMSE) the root mean squared differences between a set of original and translated points

rubbersheeting a process to deform a raster to fit it to a new coordinate system

scale the ratio of the size of features in a map to their size on the ground (e.g., map scale); the number of decimal places allotted to an attribute field for storing numbers

scale range the range of scales for which a data layer will be displayed, set by the user to avoid clutter or the display of layers at inappropriate scales

schema the layout plan of a geodatabase including the tables, the fields, and the relationships

script a program including ArcGIS commands and functions

secant conic projection a map projection in which spherical coordinates are projected onto a cone that intersects the sphere along two great circles

select to extract one or more features or records from a layer or table in preparation for another operation; to perform a query

Select By Attributes to choose a subset of features or table objects based on the values in one or more attribute fields

Select By Location to choose a subset of features based on their spatial relationship to other features

selected set the set of features that have been extracted prior to another operation; the result of a query

Selection tab the tab in the Table of Contents that shows the selection state and number of selected features for each layer

shapefile the spatial data model developed for, and used by, ArcView 3 and later versions

shared features features that are linked to or that share a boundary with other features

short integer an attribute field definition that uses up to five bytes to store a binary integer and that can store values up to about 62,000

simple edges linear elements in a network that always end at junctions and behave as separate entities

simple join a join that combines two layers according to common locations when a one-to-one spatial relationship exists between the layers

single symbol a map type in which every feature in a layer is displayed with the same symbol

single-precision a numeric value stored using eight bytes of information

sink a location on a network that pulls or consumes the commodity flowing through the network

sketch a provisional figure created during editing; when finished, it becomes a feature in a data layer

sketch menu a context menu accessed by right-clicking *off* a sketch during editing

slice to divide the values in a raster into a specified number of even classes

slope the drop in elevation for a specified horizontal distance, expressed as an angle or a percentage

snapping ensuring that features within a specified distance are automatically adjusted to meet at exactly the same location to avoid gaps between features

solvers programs that analyze flows or paths through a network

source a location that produces or initiates the flow of a commodity through a network; a spatial data file that provides the features for a map layer

source tab the tab in the Table of Contents that organizes the layers by pathname and allows standalone tables to be accessed

source table the table that provides the information to be appended to another table in a join

spaghetti model a model that stores spatial features as a series of *x-y* coordinates and does not store topological relationships between features

Spatial Analyst a program extension to ArcMap that is used to analyze raster data

spatial data information that is tied to a specific location on the earth's surface

spatial query an operation to extract records from a data layer based on location relative to another data layer

spatial reference the complete description of how the spatial data are stored for a feature class that includes the coordinate system, the X/Y Domain, and the precision

SPCS see *State Plane Coordinate System*

spelling sensitivity a setting used during geocoding to control the degree to which two names must resemble each other to be considered a good match

spheroid ellipsoid; see also *ellipsoid*

SQL see *Structured Query Language*

standalone table a table of information not linked to spatial data features

standard deviation a classification scheme in which the class breaks are based on the standard deviation values of the data being mapped

standard parallels parameters of conic map projections indicating the latitudes at which the cone lies tangent or secant to the sphere

State Plane Coordinate System (SPCS) a group of projections defined for different regions of the United States and designed to minimize map distortions

stereographic projection a map projection in which locations on a sphere are projected onto a flat piece of paper that is tangent to or intersects the sphere

stream digitizing method of entering vertices automatically during editing, instead of clicking to create each one

street type the part of an address that indicates the type of street, such as St, Rd, Ave, and so on

stretch to spread the values of an image to cover the entire range of symbols available, often supplemented by ignoring the tails of the distribution

Stretched a display method that spreads the data values over the entire range of symbols available

Structured Query Language (SQL) an established syntax for creating logical expressions to extract records from a database according to specified criteria

style a collection of map symbols and colors that are stored together and used together

Style Manager a window used to create symbols and manage them within collections of symbols called styles

subdivisions the number of units into which a single division of a scale bar is split, usually appearing on the left end of the scale bar

subtypes different categories assigned to features in a map layer, with each category having its own symbol and default values

suffix direction a direction appended to the end of a street name to indicate a part of the city, such as Main St *North*

summarized join a join thatcombines the records of two attribute tables with a one-to-many cardinality by assigning each output record a single value derived statistically from the many input values

surface analysis functions designed for application to rasters representing a three-dimensional surface, such as elevation

symbol a shade, line, or marker with specified shape and color used to display map features

table data stored as an array of rows and columns, with each row representing an object or feature and each column representing an attribute or property of that object

Table of Contents the ArcMap window that lists the data frames and layers in the map document

tangent conic projection a map projection in which spherical coordinates are projected upon a cone that lies tangent to the sphere along one line of latitude

target layer a layer, specified in the Editor toolbar, into which all new features are added and all edits are applied

templates a map design that can be saved and applied to many different map documents

thematic accuracy the degree to which attribute values represent the true properties in the real world

thematic mapping displaying the features of a spatial data layer based on values in its attribute table

thematic raster a raster that contains categorical or nominal data values, such as land-use codes or soil types

themes terminology in ArcView 3 used to indicate a data layer containing similar features, such as roads or states

thumbnail a small snapshot showing the appearance of a data layer and displayed in ArcCatalog to aid the user in finding data

TIN (Triangulated Irregular Network) a data model for storing surfaces as triangular facets with varying orientations

topological model a data model that stores spatial relationships between features in addition to their x-y coordinates

topology the spatial relationships between features, such as which are connected or adjacent to each other

trace to follow an existing linear feature, as when analyzing network paths or creating a new feature that follows or is offset from an existing one

trace solvers programs that analyze network flow by tracing paths in the network

transformation conversion of one coordinate system to another

transverse cylindrical projection a map projection in which spherical coordinates are converted to locations on a cylinder tangent to the sphere along a line of longitude

union an operation to create a new feature or set of features by combining all the areas from two input layers

unique values a map type in which each attribute value is assigned its own symbol

Universal Transverse Mercator (UTM) a family of map projections defined for 60 zones around the world and based on a transverse cylindrical projection

unprojected data a spatial data layer stored in a geographic coordinate system with units of degrees of latitude and longitude

UTM see *Universal Transverse Mercator*

vector model a spatial data storage method in which features are represented by one or more pairs of *x-y* coordinate values forming points, lines, or polygons

Venn diagram a diagram used in set theory to portray sets and subsets

vertex a point at which the segments of a line or polygon feature change direction

vertex menu a short context menu accessed by clicking *on* the sketch during editing

vertex snapping ensuring that new features are automatically connected to the vertices of existing line features

viewshed the area on a three-dimensional surface that is visible from a specified point or set of points

working directory a specified folder into which raster analysis output is placed by default

world file a text file containing the georeferencing information for a raster

X/Y Domain the maximum allowable range of values stored by a feature class

zenith angle an angular measure of the distance of an object above the horizon

zonal function a raster analysis function that analyzes values from one grid for zones defined by a different grid

zonal statistics raster analysis functions that calculate statistics from one grid for zones defined by a different grid

zone the combined areas of a raster grid that share the same attribute value, such as all areas with the commercial land-use code

Selected Answers

Ch. 1 Tutorial

1. 2 geodatabases, 3 coverages, 4 tables, 3 rasters, 4 layer files, 7 shapefiles, 2 feature datasets, 6 feature classes

2. Wallowa

3. 481 rows

4. Granite

5. NAD 1927 UTM Zone 13N, meters

6. ESRI, Inc. in 2000

Ch. 1 Exercises

1. Method: Use ArcCatalog to examine the geodatabase.

2. GCS_WGS_1984 or GCS_Geographic, NAD83 Oregon Statewide Lambert

3. Method: Examine the metadata.

4. Superior, 32,213 mi^2

5. Method: Preview the table and use Find.

6. Avg. 18.4, Min. 4, Max. 45

7. Method: Open folder and count rasters.

8. 894 rows, 540 columns, 30 meter cell size, six bands, NAD27 UTM Zone 13N

9. See map.

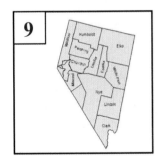

10. Upper Midwest Aerospace Consortium. Can publish a map but cannot give the data to anyone else.

Ch. 2 Tutorial

1. About 1:5 million

2. Santa Fe, about 55,859

3. About 610 miles

4. 1332 kilometers

5. North America Equidistant Conic with central meridian of -96 degrees and units of meters

6. Categorical data, unique values map

7. Ratio data, graduated symbol map

8. Points have no area

9. Ratio data, graduated color map. No normalization needed.

10. You probably chose Mt. Rainier, Mt. St. Helens, and Mt. Hood.

Ch. 2 Exercises

1. Method: Use Identify.

2. See map.

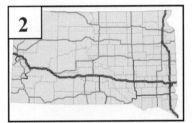

3. Use Identify to find the interstate and the Measure tool to measure it.

4. See map.

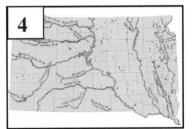

5. See map.

Ch. 2 Exercises, cont.

6. See map.

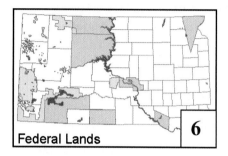

7. See map.

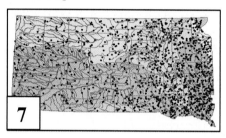

8. See map. Defined interval.

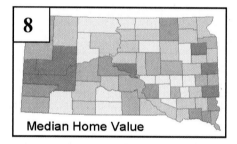

9. See map.

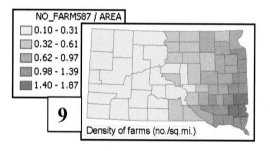

10. See map.

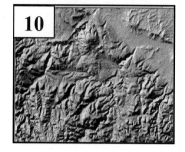

Ch. 3 Tutorial

1. Automatic scaling because the zoom/pan tools are active

2. Mercator preserves direction.

3. Annotation has a reference scale.

4. Left to right: subdivision, division, division units

Ch. 3 Exercises

Answers will vary.

Ch. 4 Tutorial

1. Wyoming

2. 19 states

3. 33,603,430 people

4. Includes District of Columbia

5. Standalone table

6. California

7. One-to-one relationship

8. One-to-many relationship

9. States is destination, districts is source.

10. 22 New England reps

11. The Pacific subregion

12. STATION_NAME and STATION

Ch. 4 Exercises

1. Method: Use Select By Attributes and Statistics and then sort selected ones.

2. New York highest number; Mississippi (or D.C.) highest percentage.

3. Method: Summarize by Sub_Region and sort results.

Ch. 4 Exercises, cont.

4. Use the FIPS code.

5. See map.

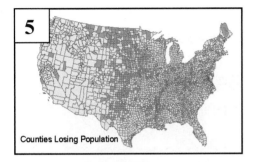

6. Texas 63, Kansas 54, Iowa 46

7. Method: Relate US States to 108th Congress and select states. Activate related records and select Democrats from those in related table.

8. 10,028,123 in state capitals; smallest 8247, largest 983,403; average 200,562.

9. Method: Add new fields and calculate percentages.

10. W N Central has 618; total population 18,910,164.

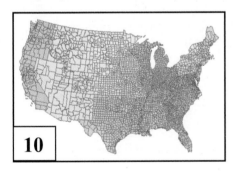

Ch. 4 Challenge

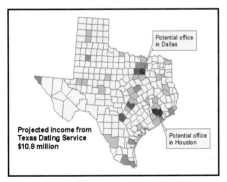

Ch. 5 Tutorial

1. All layers. This is the default.

2. 15 counties, about 310,700 people

3. 33 counties

4. 144 counties mostly in the Central Plains states

5. 104 counties

6. About 32% (1009 out of 3141)

7. Hughes County contains Pierre, SD.

8. About 89% (2798 out of 3149)

9. Brazos, Canadian, Pecos, Red, and Rio Grande

10. The Flathead and the Salt

11. 17 target counties containing 688,749 people

Ch. 5 Exercises

1. Method: Select By Attributes and use Statistics.

2. 566 counties out of 3141, or 18%

3. Method: Use Select By Attributes and do queries in steps.

4. 1010 out of 3141 counties, or 32%; 100,401,183 out of 275,247,480 people, or 36%

5. Method: Use Select By Location first, then Select By Attributes.

6. 114 cities, about 4 million people

7. Method: Select Crater Lake and then use Select By Location twice, once for volcanoes and once for interstates.

8. 91 cities. See map.

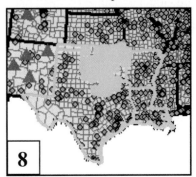

Ch. 5 Exercises, cont.

9. Method: Query for areas within 50 miles and then switch the selection.

10. Laramie, WY; Rio Rancho, NM, Sierra Vista, AZ; El Centro, CA.

Ch. 6 Tutorial

1. The counties ObjectID and FIPS fields

2. Pennington 88, Custer 35, Lawrence 23

3. St. Joseph Hospital serves most counties (12), McKennan Hospital serves most people (~145,000), and Redfield State Hospital serves greatest area (12,500 mi^2).

4. The distance units are meters.

5. 14 towns

6. Saint Anne serves the most towns (63); Sioux Valley serves the most people.

7. Meters

8. Campbell County

9. Lake Oahe received NO attributes because it did not fall completely inside a county.

10. The Lower James, nine urban areas

11. The Lower Big Sioux, about 132,000 people

Ch. 6 Exercises

1. Method: Join the gas stations to the watersheds.

2. The Medium and Low Density residential areas contain nine gas stations.

3. Method: Join gas stations to land use.

4. The Circle S Convenient, 2143 meters away from Valley Sports Bar and Grille

5. Method: Join gas stations to restaurants.

6. The Red Rock Canyon watershed, 503 units

7. The map should be similar to Figure 7.8.

8. Jordan Valley, about 268 km

9. Method: Summarized join of cities to airports.

10. Willamette NF and Deschutes NF have greatest number (12). Mt Hood NF has the greatest area (290 sq. miles).

Ch. 6 Challenge

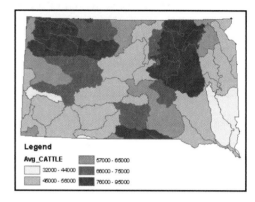

Ch. 7 Tutorial

1. ALL, NO-FID, ONLY-FID. ALL or NO-FID is best.

2. 4.86 km^2

3. 31.20 km^2

4. About 15%

Ch. 7 Exercises

1. See map.

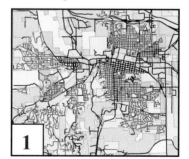

2. See map.

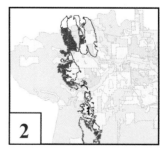

Ch. 7 Exercises, cont.

3. Method: Select the streets in your feature class from Exercise 1 and determine the total length from the Shape_Length field.

4. About 5.6 km^2

5. Method: Intersect streams with geology and determine new lengths.

6. See map.

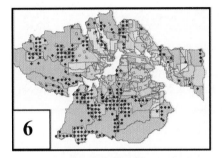

7. See map.

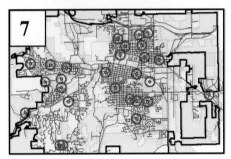

8. See map.

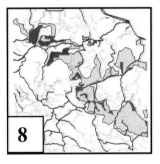

9. Method: Select forest stands and input them to the Dissolve tool, choosing SOIL_UNIT as the dissolve unit and the Mean of CROWN_COV as the statistic.

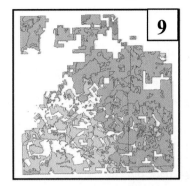

10. See map.

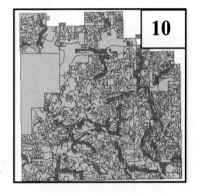

Ch. 7 Challenge

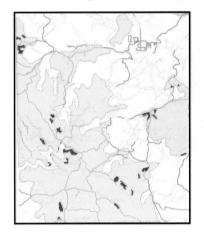

Ch. 8 Tutorial

1. About 46%

2. 2165 cells × 50 m × 50 m/1 million m^2/km^2, or about 5.4 km^2

3. About 70 degrees

577

Ch. 8 Tutorial, cont.

4. North is red; south is light blue.

5. 13 watersheds

6. 1371 meters

7. Floating-point grid; the precipitation values have decimals.

8. The Stockade Beaver watershed, 46.4 million m^3

Ch. 8 Exercises

1. Method: Reclassify the DEM.

2. See map (dark purple).

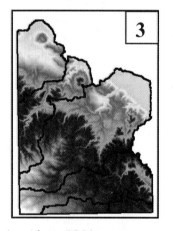

3. See map.

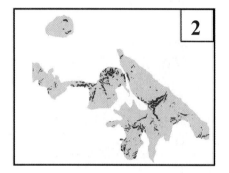

4. About 1844 meters

5. Method: Use Zonal Statistics with Summits as the zone layer and distance to streams as the value layer.

6. See map.

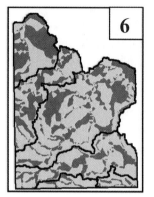

7. About 4.6 km^2

8. Madison Limestone highest average slope (14.2) ; Upper Mesozoic lowest (3.4).

9. See map.

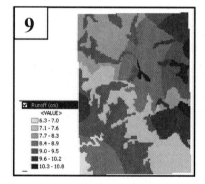

10. See map.

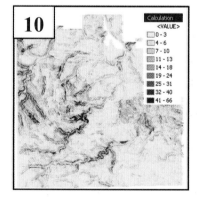

Ch. 8 Challenge

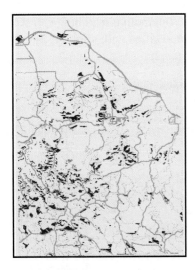

Ch. 9 Tutorial

1. The Transportation network contains a distance weight.

2. Endcaps, Galleries, Tvalves, and Water_Net_Junctions are simple junctions; Waterlines are simple edges.

3. Distance and Pressure Drop

4. About 19 edges

5. About 3900 meters

6. About 4700 meters

7. 12 differently named streets

8. Black circle

9. 17 edges

10. About 3600 meters

11. About 53 kilometers

Ch. 9 Exercises

1. See map.

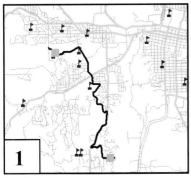

2. About 7400 meters

3. See map.

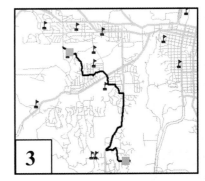

4. See map

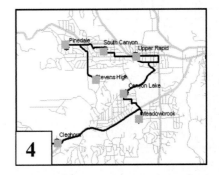

5. See map.

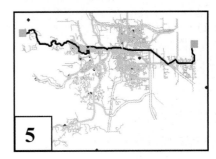

6. See map.

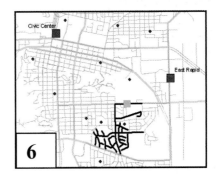

7. Method: Flag Civic Center gallery and barrier at start of north line. Trace with selection as result to get length. Move barrier to south to get north length.

Ch. 9 Exercises, cont'd.

8. Civic Center 17 km; East Rapid 10 km

9. Method: Find Upstream Accumulation from end and count edges (same as junctions).

10. Joe had set the analysis options to return only junctions; Mary was returning both edges and junctions. The cost in both cases is the same, 25.

Ch. 9 Challenge

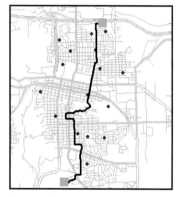

Ch. 10 Tutorial

1. Prefix direction = PREFIX, Prefix type = None, Street name = STREET, Street numbers = FRADD and TOADD, Street type = SUFFIX, Street direction = SUFFIX2

2. One Range

3. The ADDRESS field

4. Yes, there are intersections. The connector is the & character.

5. The rc_roads shapefile in mgisdata\Rapidcity

6. Shapefile

7. 64 matched and 28 tied

8. Four restaurants are unmatched.

9. The problems with the addresses are:

   1301 Omaha St; prefix W omitted.
   710 Meridian Ln; Dr, not Ln.
   2250 Haines; Ave suffix omitted.

10. US Cities with State

11. CITY_NAME and ST_ABBRV

12. 211 matched and 62 unmatched

13. First, geocoding can resolve ambiguities when two cities have the same name in two states. Interactive matching can pair up additional cities with slightly different names.

Ch. 10 Exercises

1. See map.

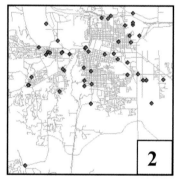

2. See map.

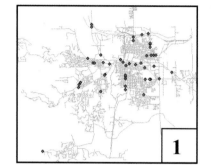

3. Method: Export the lat-long stations and distance join to geocoded stations.

4. See map.

Challenge Problem 10

Answers will vary.

Ch. 11 Tutorial

1. GCS_WGS_1984

2. About −80° and 25°

3. Map units decimal degrees; display units degrees-minutes-seconds.

Ch. 11 Tutorial, cont'd.

4. About -5550 and 1740 miles

5. About -8950000 and 2880000 meters

6. South America mostly negative; Asia mostly positive

7. Central meridian is –96° and the latitude of origin is 40°

8. Secant conic because it has two standard parallels

9. UTM Zone 11

10. UTM NAD 1927 because it defaulted to the first layer added, usfsrds

11. Define Project to create the correct label

12. Geographic Coordinate System (GCS)

Ch. 11 Exercises

1. Method: Examine properties in ArcCatalog or ArcMap.

2. The rcw_oops coordinates are in degrees but labeled as UTM Zone 13 NAD 27. Use Define Projection to assign a GCS to rcw_oops.

3. Method: Examine the data frame properties.

4. Equidistant conic, 2350 miles; Lambert, 2220 miles; Mercator, 3040 miles. Equidistant should be the closest value.

5. Method: Compare US areas to State Plane and UTM zones and choose best. For international areas, examine options available in the data frame predefined projection choices.

6. Shift central meridian of the nearest UTM zone to -89.3 to fall in the center of the state.

7. Method: Modify the central meridian and parallels of the closest State Plane zone.

8. Answers will vary.

Ch. 11 Challenge

Answers will vary.

Ch. 12 Tutorial

1. UTM Zone 13 NAD 1927

2. Angle –115 degrees

3. Map units, in this case, meters

4. The buildings don't need to be snapped.

Ch. 12 Exercises

Answers will vary, but should look something like this:

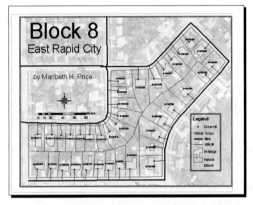

Ch. 12 Challenge

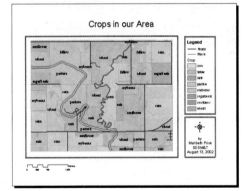

Ch. 13 Tutorial

1. Still one polygon selected

2. Multipart polygon; if you select one, both are highlighted.

Ch. 13 Exercises

1. Method: Create square, copy and paste, merge all five.

Ch. 13 Exercises, cont'd.

2. Method: Create a small and large circle with Arc tool and use smaller to clip holes in larger.

3. Method: Use constrained angle tool to create angled lines to match half of hint figure and merge them. Mirror and merge. Copy and paste, rotating each piece and snapping. Merge all lines and buffer to polygon layer.

4. See map.

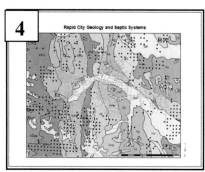

5. See map.

Ch. 13 Challenge

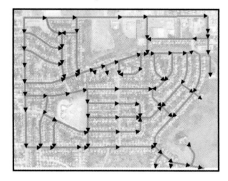

Ch. 14 Tutorial

1. NAD 1927 UTM Zone 13N

2. NAD 1927 UTM Zone 13N

3. GCS North American 1983

4. The RoadType, Direction, and StreetAbbrev would have coded domains. The NumLanes would have either a coded or a range domain.

5. The Parcel_ID field should not have a domain because it would serve little purpose for such a variable field.

6. Give Zoning field a coded value domain; give Area and ParcelValu fields range domains.

7. Give Zoning and ID fields the Duplicate policy; give Area and Parcelvalu fields the Geometry Ratio policy.

8. Give Zoning and ID fields the Default merge policy; give Area and Parcelvalu fields the Sum merge policy.

Ch. 14 Exercises

1–4. See figure.

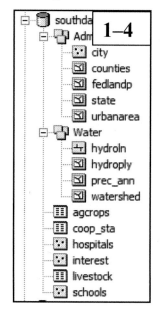

Ch. 14 Exercises, cont'd.

5. See screen captures.

Domain Name	Description
RestaurantType	Type of restaurant
NumTables	Number of Tables

Domain Properties

Field Type	Short Integer
Domain Type	Coded Values
Split policy	Default Value
Merge policy	Default Value

Coded Values:

Code	Description
1	Franchise
2	Local

5a

Domain Name	Description
NumTables	Number of Tables

Domain Properties

Field Type	Short Integer
Domain Type	Range
Minimum value	0
Maximum value	100
Split policy	Default Value
Merge policy	Default Value

Coded Values:

Code	D

5b

6–7. See layout.

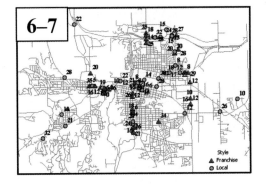

6–7

Style
▲ Franchise
◉ Local

Ch. 15 Tutorial

1. Four sources: ArcUSA 1:25M, a Census summary tape, CACI population data, and a 1987 Census of Agriculture.

2. Number of people living in non-household groups, such as nursing homes, dorms, and prisons.

Ch. 15 Exercises
Answers will vary.

Ch. 15 Challenge
Answers will vary.

A

Abbreviations, 375
Absolute pathnames, 72
Absolute X, Y option, 452, 492
Access format, 501
Accumulation (network), 332, 347, 355
Accuracy of data
    attribute accuracy, 536
    geometric, 22–23
    positional accuracy, 535–36
    temporal accuracy, 537–38
    thematic, 23
Address locators
    selecting, 371–72
    setup, 383
Address standardization, 374, 385
Add to Current Selection feature, 200
Add XY Data function, 380–81, 386
Adjacent polygons, 441, 456
Adjust Division Value (scale bar), 117
Adjust Number of Divisions (scale bar), 117
Adjust Width (scale bar), 116
Advanced tab, 61
Affine transformation, 393
Aliases, 161, 180
Alias tables, 373, 376
Analysis masks, 298
Analysis menu, Utility Network Analyst
    toolbar, 333–34
Analysis Options dialog box, 350, 351
AND operator, 193, 288
Angle box, 452
Angles, for sketching lines, 440, 453, 454–55, 488
Annotation, 117–18, 125–28
    curved, 498
    editing, 498–99
    Follow Features Options, 498
    properties, modifying, 499
Append operations, 257, 282
Applications GIS, 6
Apply, after adding individual domains, 518
ArcCatalog
    basic features, 26
    coordinate systems, 421
    demonstration of basic tasks, 35–40
    editing concurrently with ArcMap, 517
    file management, 60
    layer properties, 54
    search engine, 60–61
    setting options, 53
ArcEditor, 26, 428
ArcEdit program, 25
ArcGIS
    data files in, 26–31
    files and data sources used by, 27
    map documents, 72–73
    Metadata Editor, 541–42
    metadata templates, 542–44
        network analysis, 333–37
        overview, 25–26
        tables in, 160–61
        Utility Network Analyst toolbar, 333–34
ArcGIS Desktop, 26
ArcGIS overview, 26–31
ArcGIS tutorial
    adding connections to folders, 36–37
    contents tab, 37–38
    feature class properties, 41–42
    Internet map services, 49–50
    labeling features, 46–48
    layer files, 42–46
    metadata tab, 48–49
    preview tab, 38–40
ArcInfo, 3, 25, 26, 428
ArcMap, 428. See also Editing (ArcMap)
    adding data, 96
    adding geodatabase layers from, 516–17
    attributes, searching, 100
    data frame coordinate systems, 406
    display options, 97
    dragging and dropping files to, 95
    editing concurrently with ArcCatalog, 517
    grouping layers for display, 98
    identifying, finding, and measuring
        features, 100
    Interactive Selection, 196, 198, 203–11
    major functions, 26
    managing fields in, 186
    map document features, 72–73
    map types and features, 64–67
    measuring features, 100
    numeric data classification, 67–70
    opening a document in, 97
    options for starting, 95
    saving documents, 97
    scale bars, 116–17
    Select By Attributes, 196
    Select By Location, 196
    selecting features with, 199–200
    setting map tips, 98
    setting symbols, 57
    skills reference, 52
    Spatial Analyst tools, 294–98
    toolbar features, 96
    using Zoon/Pan tools and bookmarks, 99
    Utility Network Analyst toolbar, 333–34, 350
ArcPlot program, 25
Arc program, 25
Arc Spatial Database Engine (ArcSDE)
        software, 502
Arc tool, 464, 465, 470, 486
ArcToolbox
    basic features, 26, 31, 258
    creating practice shapefiles, 433–34
    finding tools in, 268
    map projection tools, 400
    ModelBuilder tutorial exercise, 272–78

skills reference, 52, 280–84
Spatial Analyst tools, 294–98
using, 260–61
ArcView
basic functions, 26
editing tools, 428
introduction of, 25
subtype limitations, 508, 509
Area, distortion of, 113, 391–92
AREA fields, 260, 269–70
Arguments, command-line syntax, 270
Arrows
line direction, 490
network flow, 344–45, 352
north, 131, 143
ASCII format, 159
Aspatial data, 1
Aspatial queries, 192. *See also* Attribute queries
Aspect calculations, 290, 307, 323
Attribute accuracy, 536
Attribute data. *See also* Queries; Select by Attributes
of combined features, 461, 477
default values in geodatabases, 504, 513–15, 532
domains, 504–5, 517–20, 519–20, 531 (*see also* Domains)
editing in ArcMap, 430–31, 445–46, 457
of grids, 293–94
map overlay retaining, 251–55
merge and split policies, 507, 522–25
querying, 192
raster model limitations, 20
subtypes, 508
in vector data model, 18
Attribute Editor
basic features, 430–31
skills reference, 457
tutorial exercises, 445–46
using default values with, 513–15
Attribute queries, 192. *See also* Queries; Select By Attributes
Attributes
accuracy of, 536
record selection using, 182
selecting by, 217
Attributes button, on Editor toolbar (ArcMap), 428
Attribute tables. *See also* Tables
appending, 257
basic features, 153
editing in ArcMap, 434–37
of grids, 293–94, 303
Autocomplete Polygon task, 429–30, 441, 456
Auto Layout button, 274
Automatic scaling, 115
Azimuthal projection, 391

B

Backgrounds, 148
Backup copy, 428

Badlands National Park, 8–11, 14–15
Barriers (network)
defined, 329
impact on solvers, 334, 339
tools, 353
Batch mode geocoding, 384
Bilinear resampling, 394
Binary data, 159
Binary format, 159
Binomial analysis, 10
Boldface arguments, 270
Bookmarks, 99
Boolean operators, 192–94
Boolean overlay, 288, 302–8
Boundaries
coincident, 430, 456
dissolving, 256, 265–66, 269, 281
editing shared, 460–61, 461, 473, 497
sketching, 440–42
Brackets in expressions, 199
Break values, 302
Budgets, 13, 15
Buffers, 117
basic functions, 256–57
editing features with, 462, 479–80, 494, 496
with rasters, 306
spatial queries and, 194
using from command line, 271
Buffer tool, 281
Build Geometric Network Wizard, 356–57
Building a Geodatabase, 502, 509
Buildings, creating with editing tools, 442–43
Burns Basin study area, 8
Bytes, 159

C

CAD files, 29–30
Calculate Geometry functions, 162
Callouts, 140
Canada Geographic Information System, 3
Canada thistle project case study, 8–11, 14–15
Cardinality
spatial joins, 223–24
of tables, 156
Cartesian coordinates, 228. *See also* Coordinate systems; x-y data
Case sensitivity in expressions, 194
Case studies, 4, 8–11, 14–15
Categorical data, 20, 64, 65
attribute accuracy and, 536
CD-ROM, citing data received from, 24
Cells
matching sizes for analysis, 295–96, 298
in raster model, 19
size settings, 286, 297, 318, 378
statistics, 290, 321
Cell Statistics function, 290, 321
Census Bureau, 3

Census data, 537–38
Center of rotation, 450
Central meridians, 80, 392
Chart maps, 67
Chart tools, 115
Choropleth maps, 66
Citation of data sources, 24–25
Clarke 1886, 389, 390
Classification
    numeric data, 67–70
    in quantities maps, 66
    of raster data, 292, 302–3, 320
Classified method of display, 70
Classify Values window, 320
Clear button, 199
Clip operations
    demonstration of, 268, 282
    on features, 462, 478–79, 494, 495
    overview, 251–52, 255
Clip tool, 268
Clip with Preserve, 494
Cluster tolerance, 495
Coded domains, 505, 518–19
Coincident boundaries, 430, 456
Colors
    changing, 181
    in quantities maps, 66
    raster display methods, 70–71
    in symbols, 110
    vibrating patterns of, 111
Columns, freezing/unfreezing, 56, 181
Combining features, 461–62, 477–78, 493–94
Command line (ArcToolbox), 257, 270–72
Command-line programs, 26
Command line (Raster Calculator), 295
Complete classification, 518
Completeness, of data, 537
Complex edges, 335
Computer hardware for GIS, 2
Conflict Detection options, 117
Conformal maps, 114, 392
Conic projections, 391, 392
Connections, finding in networks, 331, 345, 355
Connections to data, 36–37
Connectivity rules (network), 336, 346
Connects tab (ArcCatalog), 37–38
Containment, 195
Containment criterion, 221
Contents tab, 37–38
Content Standard for Digital Geospatial Metadata Workbook, 534
Context menus
    for editing, 429–30, 438–40, 452–54
    for map features, 446
Continent maps, 397
Continuous data
    basic features, 20, 285
    raster display options, 70
Contour function, 323

Contour lines, 290
Conversions
    raster data overview, 293
    raster to vector, 305, 309–10, 320
    vector to raster, 293, 320
COOP stations, 380
Coordinate pairs, 387
Coordinate space, 388
Coordinate systems. *See also* Geographic
        coordinate systems; Map projections
    ArcMap handling of, 397–99
    choosing, 112–14
    choosing projections, 397
    creating, 413–14
    data projection, 411–12
    data sets definitions, 421
    defined, 21
    defining for feature datasets, 503–4, 515–16
    displaying, 405–8
    distance joins and, 227–28
    editing considerations, 428
    geographic coordinate systems
        overview, 388–89
    input and output settings, 259–60
    labeling, 397–99
    managing in ArcGIS, 397–98
    map distortion, 408–11
    missing labels, 398–99
    need for, 387–88
    in rasters, 286, 295–96
    rules of precedence, 259–60
    skills reference, 54, 421
    teaching tutorial, 405–19
    troubleshooting problems in, 400–402,
        412–13
    warnings of inconsistencies, 448
Copying features, 436
Copy Parallel task, 492
Cost distance grids, 326
Cost window, 354
Count field, 293, 303
Country maps, 397
Coverages, 28–29
    editing, 428
    geodatabases *versus,* 501
    importing into geodatabases, 503, 511–13
Coverage tabs, 54
Cover# field, 29, 310
Cover-id field, 29
Create Feature Class tool, 433
Create New Feature task, 430, 441, 451
Create New Selection, 199, 341
Create New Selection method, 341
CSDGM Essential Metadata Elements, 534
Cubic convolution resampling, 394
Current workspace, setting, 270–72
Curves, creating, 464, 470, 488
Customer data, 1
Customization languages, 333

Customize function, 5
Cut/Fill function, 290, 323
Cut Polygon Features task, 441, 456, 488, 524
Cylindrical projections, 391, 392

D

Dangermond, Laura and Jack, 3
Dangle, 460
Dangles, 425
Data, tabular, 174–75
Data analysis, GIS requirements, 2
Data analysis function of GIS, 3
Data analysis state of projects, 8, 10
Database management systems (DBMS),
    1, 30, 154, 502
Databases, 1, 501–2. *See also* Geodatabases
Data collection, 9
    assessing practicalities of, 12, 13
    in Canada thistle case study, 9
    GIS requirements, 2
Data entry tools, 2
Data files. *See also* Raster data; Vector data
    raster and vector formats, 17–20
Data frame coordinate systems, 73–74, 80, 406
Data frames
    basic functions, 72
    demonstration of basic tasks, 80–82
Data frames, composing, 136
Data layers. *See* Layers
Data management tools, GIS requirements, 2
Data models, 17–20. *See also* Raster data; Vector data
Data normalization, 67
Data organization, 539
Data quality, 22–24
    attribute accuracy, 536
    completeness, 537
    lineage, 534–35
    logical consistency, 537
    metadata format, 538–39
    positional accuracy, 535–36
    temporal accuracy, 537–38
Data sets, 398. *See also* Feature datasets; Spatial data;
    Tables
Data sharing, 5
Data sources, describing in proposals, 13, 15
Data storage systems, 2
Data tables. *See* Tables
Data view mode, 81
Date tab, 61
Datums. *See also* Coordinate systems
    converting in ArcMap, 399
    introduction to, 389–90
dBase format, 27
Default merge and split policies, 507
Default values
    geodatabase attributes, 504, 513–15, 532
    grid resampling and projection, 295
Defined interval classification, 69

Define Projection tool, 399, 401–2
Definition query, 44
Deflection command, 439, 454
Degrees, 228, 388
Delete Vertex command, 438, 439, 474
Deletions
    feature classes, 527
    feature datasets, 527
    features, 451
    geodatabases, 527
    lines, 436, 437
Delta X, Y option, 452
Density functions, 289, 290
Dependent variables, 13
Descriptions for coded domains, 518, 519
Destination layers, 225–26
Destination tables, 154, 156
Determinate flow, 329, 345
Development GIS, 6
Digital elevation models (DEMs), 20, 285, 289
Digital Line Graph (DLG), 540
Digital raster graphic (DRG), 20
Digitizer tablets, 2
DIME format development, 3
Direction
    constraining for segments, 440, 443, 453
    distortion of, 113, 392
    of flow in networks, 329, 333, 344–45,
        352–53
    showing for lines, 466, 490
    weights with, 351
Direction-Distance tool, 465, 489
Direction grids, 326
Direction/Length command, 440, 453, 492
Disabled network features, 327
Disconnect button (ArcCatalog), 36
Disconnected roads, 343
Discrete color option, 70
Discrete data
    basic features, 20, 285
    raster display options, 70
Display units, setting, 80
Dissolve tool, 265
Dissolving
    with buffers, 269
    demonstration of, 265–66, 281, 283
    as spatial analysis tool, 256
Distance, distortion of, 113
Distance-Distance tool, 464, 486
Distance functions, 289
Distance joins
    coordinate systems and, 227–28
    examples, 222, 227–28
    tutorial exercises, 240–43
    when to use, 222
Distances
    cost weighting, 326
    distortion of, 392
    network weights as, 330, 340, 354, 356–57

obtaining with Measure tool, 79–80
raster calculations, 289, 326
weighting for interpolation, 312, 325
Distortion
in aerial and satellite images, 286
of area, 113, 391–92
of direction, 113, 392
of distance, 113
map distortion, 408–11
in map projections, 39–392
in planar coordinate systems, 389
by projections, 113
Distributed Database GIS, 6
Distribution section, 539
Divisions, 116, 492
Division units, 116
Division values, 116
Domains
defined, 28
editing, 509
field types for, 504–5
merge and split policies, 507, 522–25
setup, 504–5, 517–20, 531
using, 520–22
Donut polygons, creating, 479, 495
Dot density maps, 67
Double-precision field, 160
Double quotes, 168
Downstream tracing, 332, 346, 347, 355
Duplicate split policies, 507
Dynamap, 382
Dynamic labels, 46, 59, 117
converting to annotation, 142
customizing, 83–84

E

Earth-centered datums, 390
Earthquake death spatial join example, 230–32
Earthquake table, 158
Edge Flag tool, 339
Edge snapping, 426–27, 438
Edges (network)
defined, 327
flag and barrier tools, 353
simple *versus* complex, 335
weights, 329, 351
Edges (topology), 466
Editing annotation, 498–99
Editing (ArcMap)
additional sketching tools demonstrated, 469–70
additional sketching tools described, 464–65
adjacent polygons, 430, 456
attributes, 430–31, 434–37, 445–46, 457
basic tools, 425–27, 430–31
buffering, 462, 479–80, 496 (*see also* Buffers)
changing fields, 457
changing polygons, 465, 472–74, 476–83, 489–90
combining features, 461–62, 477–78, 493–94

creating polygons, 430, 440–42, 450, 456, 478–79
feature selection methods, 449–50
introduction to, 427–28, 428–29
line features, 436, 465–66, 474–76, 490–93
saving work, 458
shared features, 461–62, 473, 496–97
simple line tasks, 435–36
simple point tasks, 434–37
sketch context menus, 429–30, 438–40, 451–54
snapping features, 425–27 (*see also* Snapping)
splitting polygons, 443–44, 456
using domains, 520–22
Editing with ArcMap, 461
Editor toolbar (ArcMap), 427–28
Edit Standardization window, 374
Edit tool
basic functions, 430–31
deleting features with, 451
feature selection with, 449–50
moving features with, 450
Topology Edit tool *versus,* 466
Elk habitat study, 4
Ellipsoids, 389
Enabled network features, 327
Endpoint Arc tool, 465, 487
End snapping, 426, 438
Enterprise geodatabases, 501–2
Enterprise GIS, 6, 11
Entity attribute, 539
Environmental Systems Research Institute, Inc.
(ESRI)
GIS software, 25
origins, 3
website, 26
Environment settings
caution advised, 284
geoprocessing, 259, 296
Spatial Analyst, 294–95, 318
use when importing data, 530
viewing, 512
Equal-area projections, 114
Equal interval classification, 68
Equidistant maps, 114
Erase operations, 252, 255
Error messages, 284, 479
Errors
attribute, 536
reducing with domains, 504, 521
root mean square errors, 393
source data, 535
topology, in networks, 343–44
Establish Flow button, Utility Network Analyst
toolbar, 334
Europe Albers Equal Area Conic, 397
Evaluation stage of projects, 11
Event layer, 381, 386

Excel, 175, 190
Exclamation point, 73
Exporting data
    event layers, 381, 386
    as feature classes, 517
    selected sets, 211–12
Exporting tables, 173, 184
Expressions
    attribute queries and, 192
    in ModelBuilder, 274
    rules for writing, 192–94
Extend/Trim Features task, 493
Extensible Markup Language (XML), 538
Extensions (filename). *See* Filename extensions
Extensions to ArcGIS software, 293, 301
Extent, setting, 139
Extent rectangles, 82
Extents
    defined, 398
    grid analysis, 297, 301, 314, 318
Extract by Mask tool, 314
Extraction functions in map overlay, 251–52

F

False easting, 392
False northing, 392
Feature classes
    in CAD files, 29–30
    creating, 433, 513, 529
    deleting, 527
    function in geodatabases, 503
    importing, 530
    loading into feature datasets, 516
    loading into geodatabases, 511–13
    as networks, 327–28
    projecting, 422
    in shapefiles, 27
    in vector data model, 18
Feature classes coordinate systems, 406
Feature Class to Feature Class tool, 509, 530
Feature datasets, 18–19
    creating, 503, 515–16, 528
    creation of networks in, 333
    deleting, 527
    function in geodatabases, 503
    multiple feature classes in, 18
    relation to networks, 327–28
    spatial references, 515–16
Feature identification code (FID), 18, 28
Feature ID field, 280
Features. *See also* Line features; Point features;
        Polygons
    interactive selection of, 215
    isolating, 191
    labeling, 59, 139–41, 140
    selecting using shapes, 215
Feature services, 31
Features of interest, selecting, 191

Features to Raster dialog box, 303, 319
Feature weights, 117
Federal Geographic Data Committee, 23
Federal Geographic Data Committee (FGDC),
        48, 533
Field Calculator, 162
Field data types, 160
Field definitions, 159
Field length, 159
Field names
    repeating in expressions, 193
    restrictions, 185
    translating for geodatabases, 512
Fields. *See also* Primary Display Field
    adding and deleting in ArcCatalog, 185–86
    adding and deleting in ArcMap, 185
    adding/deleting from Preview mode, 56
    adding to feature classes, 519, 529
    adding to tables, 434
    adjusting width in tables, 180
    assignment in address locators, 371
    calculating, 187
    defined, 153
    editing and calculating, 187
    formatting, 181
    hiding, 180
    merge and split policies, 507, 518–19, 522–25
    sorting in tables, 180
    statistics, 183
    summarizing on, 184
Field types for domains, 504–5
File creation options (Spatial Analyst), 318
Filename extensions
    for shapefiles, 27
Filenames, spaces in, 297
Filtering raster data, 309
Filtering weights, 352
Find Common Ancestors solvers, 332–33
Find Connected/Disconnected solvers,
        331, 345, 355
Find Loops solver, 331, 343–44, 345–46
Find Path solver, 334, 338–42, 354
Find Path Upstream solver, 332
Find tool
    demonstration of, 79
    in network exercises, 339
Find Upstream Accumulation solvers, 332–33,
        346, 355
Finish Part command, 455
Fisher, Howard, 3
Fixed extent, 115, 139
Fixed scale, 115, 139
Fixed search radii, 325
Flags (network)
    basic functions, 329
    skills reference, 353
    toolbar tool, 334
    in tutorial exercises, 338–48
    use with generic trace solvers, 330, 331, 334

use with utility trace solvers, 332
Flat file database, 154
Flipping lines, 466, 482, 490
Floating-point grids, 293
Flow direction in networks, 329, 334, 344–45, 352–53
Flow Display Properties dialog box, 352
Flow menu, Utility Network Analyst toolbar, 334
Focal functions, 290–91
Folder connection and disconnection, 53
Folders
    pathnames, 72–73
    switching, 458
Fonts, changing, 181
Frame tab, 146
Freezing/unfreezing columns, 56
FTP server, citing data from, 24

G

Generalize lines option, 319
General topic selection, 12
General tracing options, 350
Generic trace solvers, 330–32
Geocoding. *See also* Maps
    address locator setup, 371–72
    skills reference, 382–86
    tutorial exercises, 371–81
Geocoding Options button, Review/Rematch window, 384–85
Geodatabases, 28
    adding coverages and shapefiles, 503, 511–13
    adding layers, 516–17
    advantages, 501
    creating, 503, 527
    default values, 504, 513–15
    deleting, 527
    domains, 504–5, 517–20, 520–22, 531
    feature class creation, 513–16
    feature dataset creation, 503, 515–16, 528
    file features, 28
    merge and split policies, 507, 518–19, 522–25
    network concepts, 327–28 (*see also* Networks)
    overview, 501–2
    purpose of, 333
    subtypes, 508, 509
    switching, 458
    topology in, 460
Geodatabase tabs, 54
Geographic coordinate systems. *See also* Coordinate systems
    distortions, 259–60
    raster storage in, 286
    spheroids and datums, 389–90
Geographic coordinate systems (GCS), 388–89
Geographic information systems (GIS)
    case studies, 4, 8–11, 14–15
    citing data sources, 24–25
    customized applications and, 5

history of, 3–4
    professionals in, 6
    project management, 6–8
    project planning, 11–13
    purposes and major functions, 1–3
    skills reference, 136
    trends and directions in, 4–6
    types, 11
    wireless technology, 5
Geography tab, 60
Geoids, 389
Geometric accuracy, 22–23
Geometric interval classification, 69
Geometric networks, 328, 356–57. *See also* Networks
Geometry Ratio split policies, 507
Geoprocessing
    Append and Merge tools, 257
    buffering, 256–57
    command-line method, 270–72
    defined, 251
    dissolving, 256
    environment options, 259, 296
    map overlay functions, 251–55
    map overlay tutorial, 264–78
    using ModelBuilder, 272–78
    using tools and environments, 258–60
Georeferenced rasters, 392–93
Georeferencing, 17, 19, 392–93, 414–18
Georelational data model, 18
Geospatial Metadata Fact Sheet, 533
Geospatial Metadata Quick Guide, 533
Get Unique Values button, 198
Global positioning systems (GPS), 5
Graduated color maps, 66
Graduated symbol maps, 66
Graphical user interface (GUI), 25
Graphics, adding to layouts, 143
Graphs, creating, 134–35, 151
Graticule grid, 114
Grids. *See also* Raster data
    analysis functions, 285–86, 302–4, 321
    conversion to polygons, 293
    managing, 317
    naming, 286, 293, 317
    Spatial Analyst format, 290
Ground control points, 393
GRS_1980 spheroid, 390

H

Hardware
    for GIS systems, 2
    open source, 5
Harvard Laboratory for Computer Graphics and Spatial Analysis, 3
Help system, 31–32
    ArcToolbox tools, 284
    command-line syntax information, 272

Herbicide use study, 8–11, 14–15
Hierarchical database, 154
Highlight option (Attribute Editor), 457
Hillshade maps
    creating, 92, 307, 323
    uses for, 290, 307
Histogram, 71
Hydrologic modeling, 310–11

I

Icons for file formats, 27
Identification section, of metadata, 538
Identify tool, 55
IDRISI (Clark University), 4
Image rasters display methods, 70–71, 91. *See also*
        Raster data
Image services, 30
Importing
    data files into geodatabases, 503
    feature classes, 513, 516, 530
    layers into geodatabases, 516–17
    spatial references, 433
Independent variables, 13
Indeterminate flow, 329, 345
Index tab (ArcToolbox), 268
INFO, 29
INFO databases, 25, 162
Input devices, 2
Input Features box, 511
Insert Vertex command, 439, 474
Inside joins, 222–23
Integer grids, 293, 294
Integration, 425
Interactive labeling, 117, 124–25
Interactive mode (geocoding), 372–74, 378, 385
Interactive selection, 196, 198, 203–11
Internet connections, data access via, 2
Internet Map Servers, 5
Internet map services, 49–50
Internet servers, 30–31
Internet service connection, 53
Interpolation functions, 289–90, 312–13, 325
Intersection of Inputs option, 297
Intersections
    basic exercises using, 267–68, 478–79
    basic functions, 254–55, 462
    distance-distance, 486
    finding, 485
    in ModelBuilder, 274–75
    skills reference, 280, 494
    spatial queries and, 195
    of streets, geocoding, 374
    using from command line, 274–75
Intersection tool
    skills reference, 280
    tutorial exercises, 266–67, 274–75, 478–79
Interval data, 66
INT function, 295

Introductions to proposals, 13, 14
Inverse distance weighted interpolation, 312, 325
Item names, translating, 512
Items tab, 145

J

Jenks classification method, 68
JoinAttributes option, 280
Joining features. *See* Spatial joins
JPEG files, 29, 183
Junction Flag tool, 339, 343
Junctions (network), 327, 329, 335, 353

K

Kernel density type, 324

L

Labeling
    coordinate systems, 397–99
    dynamic, 117
    features, ArcGIS, 46–48
    interactive, 117
    missing coordinate system labels, 398–99
Labels
    adding to points, 434
    annotation, 498
    dynamic, 83–84
    editing existing annotation labels, 499
    on parcels, 445–46, 524
Label weights, 117
Lambert Conformal Conic projection, 396
Landslide hazards maps, 7
Land use data, 20
Laser altimetry (LIDAR) systems, 6
Laterals (water supply), 515
Latitude-longitude values, adding to
        maps, 380, 386
Latitude of origin, 392
Latitudes, defined, 388
Layer files, 42–46
Layer geography skills reference, 55
Layers. *See also* Map overlay; Spatial joins
    adding to geodatabases, 516–17
    assessing need and source data for, 7, 9
    coordinate system changes with editing, 428
    copying features between, 436
    creating from selected features, 192, 219,
        264–66, 272–78
    purposes, 7
    selectable, 200
    selection states, 196–98
    setting selectable, 214
    shared boundaries between, 460–61
    spatial queries between, 194–96
Layouts. *See also* Maps

adding graphs and reports, 143
layout toolbar, 138
layout view mode, 81
Least-cost paths, 326, 330, 338–42, 354
Legends
adding, 128–30, 144
changing appearance of, 144–46
Length command (Sketch menu), 440, 453
LENGTH fields, 260, 269–70
LIDAR systems, 6
LIKE operator, 168, 194
Lineage, 534–35
Linear regression models, 8–9
Linear weighting, 325
Line features
coordinate pairs of, 387
deleting, 436, 437
dividing, 491, 492
joining to other lines, 226
joining to points, 226
joining to polygons, 226
modifying, 436, 465–66, 474–76
moving with Topology Edit tool, 466
as selection tools, 450
trimming and extending, 493
in vector data model, 18
Local area network, citing data received from, 24
Local maps, 397
Location, selecting by, 192, 218
Logical consistency, of features, 21, 537
Logical expressions, 154
Logical networks, 328
Logistic regression, 9, 10
Longitude, 388
Loops (network), 331–32, 343–44, 345–46, 355
Lowercase letters in expressions, 194

M

Majority filter, 309
Majority statistic, 309
Make Feature Layer tool, 273, 276
Manual class breaks, 69
Many-to-one cardinality, 157
Map addition, 288
Map algebra, 286–87, 295
Map Calculator, 304
Map distortion, 408–11
Map documents
basic features, 72–73 (*See also* Maps)
data frame coordinate systems, 73–74, 80
pathnames and, 72–73
Map extent, 115
MAPINFO, 4
Map layout functions, GIS requirements, 3
Map overlay
append operations, 282
area and length measurements, 269–70
buffers with, 268–69, 281

clip operations, 251–52, 255, 268, 282
dissolving, 256, 265–66, 283
intersections, 254–55, 266–67, 282
major functions, 251–55
preparing for, 264–66
raster-based, 287–88, 302–8
Map projections. *See also* Coordinate systems
ArcToolbox tools, 400
common systems, 395–97
principles of, 390–92
spheroids and datums for, 389–90
troubleshooting problems in,
400–402, 412–13
Maps. *See also* ArcMap; Editing (ArcMap);
Geocoding
adding legends to, 128–30
design elements, 107–10
exporting, 150
numeric data classification methods, 67–70
objectives of, 108–9
printing, 150
reviewing and printing, 133
setting page, 137
teaching tutorial, 121–35
templates, 149
types in ArcMap, 64–67
Map scale, 118
assigning, 115–16
concept overview, 63–64
defined, 63
relation to features used, 18
Map Tips, 340
Map topology
creating, 496
defined, 466
Map units, 74, 80, 387
Map window, 81
Masks, 298, 314, 379
Match Interactively button, 384, 385. *See also*
Interactive mode (geocoding)
McHarg, Ian, 3
Measurement grid, 114
Measurement units, 66
Menus, customizing, 258
Mercator, Gerhard, 3
Mercator projections, 392, 408
Merge policies, 506–7, 518–19, 522–25
Merging buffers, 480
Merging features, 257, 461, 477–78, 493
Metadata
acquiring, 539–40
data quality issues, 534–38
defined, 23
development, 540–41
entering, for a layer, 549–52
examining, 546–47
federal regulations, 48
format, 538–39
importing acquired, 552–53, 555

standards, 533–34
teaching tutorial, 546–53
templates, 542–44, 547–49, 555
viewing, 57
Metadata Editor, 541–42, 554
Metadata reference, 539
Metadata tab, 48–49
Methodologies, 13, 14–15
MGA (Intergraph), 4
Microsoft Access format, 501
Microsoft Windows commands, 297
Midpoint tool, 465, 488
Mine study proposals, 12
Minimum candidate scores, 384, 385
Minimum match scores, 384
Mirror Features task, 476–77, 491
Mirroring polygons, 476–77, 491
ModelBuilder, 258, 272–78
Model Properties window, 273
Models
    functions of, 7, 258
    methodology development and, 13
    raster and vector data, 17–20, 285
Modify Edge task, 476
Modify Feature task, 429, 465, 474, 489
Moving features, 450, 492
Multifeatures, 27
Multipart donuts, 479, 495
Multipart features, 455
Multivariate regression models, 8–9
Must Cover Each Other rule, 460
Must Not Overlap rule, 460, 462

N

NAD27, 390, 396
NAD83, 113, 390, 396
Name & location tab, 60
Naming conventions
    for field names, 512
    for grids, 293, 317
    for pathnames, 73
National Climatic Data Center, 380
National grids, 396
National Map Accuracy Standards, 535
National Park Service Geographic Information
        Systems Funding Program, 9
National Spatial Data Infrastructure (NSDI), 5
Natural breaks method of data classification, 68
Nearest neighbor resampling, 393–94
Neatlines, 132, 148
Needs assessment, 8
Negative buffers, 256, 462, 485
Neighborhood functions, 290–91, 309–10, 321
Networks
    analyzing, 333–37
    basic concepts, 327–28
    building, 335–36, 356–57
    examining properties, 347

skills reference, 349–57
tutorial exercises, 338–48
types, 328–29
Network topology, 28
New Geometric Network command, 356
NoData, 91
NoData values
    converting to, 305, 310
    displaying, 294
    for masked cells, 298, 314
    purpose of, 294
    zero values *versus*, 305
Nodes, 18, 467, 475, 496
NO-FID option, 280
Nominal data, 64, 65
Non-numeric data, 10
Nonstandard abbreviations, 375
Normalizing data in maps, 67
North American Datum 1983, 113, 390, 396
North American Datum of 1927
        (NAD27), 390, 396
North arrows, 131, 143
NOTEST schema type, 282
NOT Operator, 193
Null entries, 521, 522, 532
Numeric codes, 505
Numeric data
    basic features, 64, 65–67
    classifying, 67–70
    precision and scale, 320

O

ObjectID field, 293
Objectives, identifying, 8, 13, 14
Object-oriented data models, 18
Oblique cylindrical projections, 391
Oblique Mercator projections, 396
Observation points, 324. *See also* Viewshed
        analyses
Official projections, 400
Offsets
    due to datum discrepancies, 390
    tracing with, 487
One-to-many cardinality, 157
One-to-one cardinality, 157
Online database, citing information received
        from, 24
On-the-fly projection
    ArcMap, 409–10
    defined, 399–400
    limitations of, 400
    rules of precedence, 259
Open source software/hardware, 5
Operators (Boolean), 193–94, 287
Optional arguments, 270
Ordinal data, 65
Oregon Statewide Lambert coordinate
        system, 222, 396

Organizational databases, 502
Origins, 387
OR operator, 193
Orthographic projections, 391
Overlapping buffers, 256–57
Overlay analysis. *See* Map overlay

P

Page layouts. *See* Layouts
Parallel lines
    copying to create, 492
    sketching, 439, 443, 454
Parametric curves, 486
Parcels
    creating, 443–45
    labeling, 445–46, 524
    merge and split policies, 522–25
    modifying, 472–74
Parenthesis in expressions, 194
Pathnames, 72–73, 293, 297, 317
Paths
    finding through networks, 338–42
    between raster data points, 325
Patterns, exploring, 191
Pencil icon. *See* Sketching tools
Performance, impact of editing on, 428–29
PERIMETER fields, 260, 270
Permanent grids, 296–97
Perpendicular lines, 439, 443, 454
Perpendicular to Sketch box, 455
Personal geodatabases, 501
Personnel, GIS requirements, 2
Pesticide use study, 8–11, 14–15
Picard, Jean, 3
Pictures, adding to maps, 148
Pie chart maps, 67
Pixels, 19, 286
Placement Properties, 117
Planar topology, 28, 460–61
Planning GIS projects, 11–13
Point features
    coordinate pairs of, 387
    editing, 434–37
    joining to lines, 226, 227
    joining to other points, 226
    joining to polygons, 226
    in vector data model, 17–18
Polar projections, 391
Pollution risk spatial join example, 232–33
Polygons. *See also* Map overlay
    changing shape of, 465, 472–74, 489–90
    for clipping layers, 251–52
    coincident boundaries, 430, 456
    combining, 265, 461–62, 477–78, 493
    conversion to rasters, 293, 303, 319
    converting rasters to, 305, 310, 319
    dissolving lines between, 256, 265
    donut, 478–79, 495

    joining to lines, 226
    joining to other polygons, 226
    joining to points, 226
    line intersections with, 267–68
    logical consistency and, 537
    mirroring, 476–77, 491
    as selection tools, 450
    sketching new, 440–42, 450, 456
    splitting, 441, 443–44, 456, 488, 524–25
    in vector data model, 18
Polygon-to-polygon joins, 243–47
Population, displaying in maps, 66, 67
Positional accuracy, 535–36
Pound sign, 271
Power laws, distance weighting, 325
Precipitation, as ratio data, 65–66
Precipitation maps, 311–12
Precision of numeric fields, 23, 159, 160, 320
Presentation stage of GIS projects, 8, 11
Preview mode, 56
Primary data providers, 6
Primary Display field, 446, 457
Prime Meridian, 387
Printers, 2
Processing tolerances, 535
Process steps, 534
Program warnings, 407
Projected coordinate systems. *See also*
        Coordinate systems; Map projections
    distortions and, 259, 395–96
Projected data, ArcMap handling of, 398
Projection of raster data, 394–95
Projection parameters, 392
Projection principles, 390–92. *See also* Map
        projections
Projection systems
    State Plane Coordinate System, 396
    UTM system, 395–96
Project management, 6–8
Project-oriented GIS, 11
Project planning, 11–13
Projects (GIS)
    examples, 8–11, 14–15
    planning, 11–13
Project templates, 547–49
Project tool, 400, 401–2
Proportional symbol maps, 66
Proposals for GIS projects, 12, 14–15
Proximity, spatial queries and, 195
Proximity criterion, 221
Ptolemy's atlas, 3
Pyramids, 90
Python language, 258, 278

Q

Quantile classification, 69
Quantities maps, 66–67
Queries. *See also* Attribute data

attribute-based, 192, 206–7
Boolean operators, 192–94
common applications, 191
to create layers for overlay, 264–66
definition query, 44
as extraction functions, 251
interactive selection for, 196, 198
location-based, 207–11
in ModelBuilder, 276, 277–78
purpose of, 191–92
selection states, 196–98
skills reference, 214
spatial, 192, 194–96
on tables, 154–55
Query Builder window, 512
Query Wizard, 198
Questions for research, 12
Quote marks, 199

R

Radio transmitter collars, 4
Range domains, 505
Ranking calculators, 288
Raster Calculator
    entering expressions into, 295
    main types of functions in, 295
    tutorial exercises, 303, 305–6, 313, 315
Raster data, 319–20
    analysis functions, 285–86, 303, 321
    analysis options, 314, 318
    assigning coordinate systems to, 392–94
    converting and classifying, 293, 302–3, 305
    coordinate system information in, 286, 295–96
    display methods, 70–71
    file features, 29
    file management, 317
    georeferencing, 414–18
    grid format and attributes, 293–94
    model described, 19–20
    projecting, 419, 423
    queries, 191
    topographic functions tutorial, 306–8
    unknown values in, 91
    vector model versus, 285
Raster to Features function, 305, 310, 319
Ratio data, 65
Reclassification functions, 292, 302–3, 320
Records, 153. See also Columns
    selected sets, 154
    using attributes in selection of, 182
Rectification, 395
Reference grid, 114
Reference latitude, 392
Reference layers, 375–76
Reference scale, 118, 141
Regional maps, 397
Relate, 157
Relational database, 154

Relative pathnames, 72, 73, 273
Reliability, impact of editing on, 428–29
Rematching geocoding results, 384
Remove From Current Selection feature, 200
Renaming parameters in ModelBuilder, 276
Reprojection, on-the-fly, 399–400
Required arguments, 270
Resampling and projection, 292
Research questions, 12
Reshape Edge task, 466, 476, 497
Reshape Features task, 465, 490
Resolution, 23
    defined, 398, 504
    in raster model, 20
Review/Rematch window, 373, 378, 384–85
RGB composite method, 71, 91–92
River pollution spatial join example, 232–33
Rivers, as feature classes, 18
Roads. See also Networks; Transportation
        networks
    buffering, 479–80
    as discrete data, 20
    as feature classes, 18
    modifying, 474–76
    moving, 496–97
    as subtypes, 508
Robinson projection, 409
Rocky Mountain Research Station, 4
Root mean square error, calculating, 393
Rotate tool, 428, 450
Rubbersheeting, 393
Rule of Joining, 156, 223
Rules, topological, 460

S

Sampling, 11
Save and Load button, 199
Saving
    edits, 429, 458
    models, 276
Scale, 159. See also Map scale
    network display settings, 344–45, 353
    relation to features used, 18
    setting, 139
Scale bars, 116–17, 130–31, 147
Scale range, 45
Scanners, 2
Schema Type, 282
Scripts, 258, 278
Sea level, 389
Search radii, 324
Secant conic projections, 391
Second-order transformation, 418
Segment Deflection command, 439
Select By Attributes. See also Queries
    basic functions, 198–99
    to create layers for overlay, 264–66
    in geocoding, 375

Select By Location, 195
  basic functions, 199
  clip and erase *versus,* 251–52
Selected records, displaying, 180
Selected sets, 154
Select Elements tool, 124
Select Features tool, 517
Select Features Using a Line, 450
Select Features Using an Area, 450
Select From Current Selection feature, 200
Selection methods (editing), 449–50
Selection options, changing, 219
Selections, clearing, 216
Selections, network analysis results as,
      341, 346, 347, 351
Selection tab, 216
Server GIS, 5
SETNULL function, 295
Shades, adding, 132
Shadows
  adding, 148
  with hillshading, 323
Shape, distortion of, 113
Shapefiles
  area, length, and perimeter field
        updates, 260, 269–70
  creating, 433–34
  file features, 27–28
  geodatabases *versus,* 501
  georelational model, 161
  importing into geodatabases, 503
  introduction with ArcView, 25
  lack of topology support, 461
  multifeatures and, 27
  numeric fields in, 174
  as preferred editing format, 428
Shapefile tabs, 54
Shared features, editing, 461–62, 473, 496–97
Shortest-path tracing, 326, 354
Show Help feature (Dissolve tool), 265–66
Show/Hide Help button, 284
Simple density type, 324
Simple distance joins, 225, 229
Simple edges, 335
Simple graphs, 134–35, 151
Simple inside joins, 224
Simple joins, 224
  tutorial exercises, 237–39
  when to use, 224
Single-precision floating-point field, 160
Single quotes, 168
Single-symbol maps, 64
Sinks, 328
Sinusoidal projection, 408
Size and Position tab, 146
Sketching tools
  basic functions, 429–30
  creating features with, 451–54
  on Editor toolbar (ArcMap), 427

overview of additional tools, 465
  skills reference, 485–89
  tutorial exercises, 469–70
Sketch menu, 429, 438–40, 451
Sketch Properties button, 428
Slice, 70
Sliver polygons, 254
Slope calculations, 290, 304–5, 309–11, 323
Snapping
  network features, 339, 350, 353, 357
  perpendicular to sketch, 455
  setting options, 449
  turning on and off, 437–38
  tutorial exercises, 437–38, 469–70
  types, 425–27
  when reshaping features, 490
Snapping Environment window, 426, 437, 449
Snapping window, 469
Snap tolerance
  demonstration of, 437
  entering values, 449
  network features, 339, 350, 353, 357
Snow, John, 3
Software
  GIS requirements, 2
  open source, 5
Solve button, Utility Network Analyst toolbar,
      334
Solvers
  defined, 329
  generic, 330–32
  skills reference, 354–55
  tutorial exercises, 338–48
  utility network, 332–33
Source data, 72
Source points, 328
Source tables, 155
Spaces
  in file and folder names, 297
  in pathnames, 73, 297
Spaghetti models, 21
Spatial analysis
  append operations, 257
  buffering, 256–57
  dissolving, 256
  map overlay functions, 251–55
  map overlay tutorial, 264–78
  merge operations, 257
  overview, 251–57
Spatial Analyst. *See also* Raster data
  cell statistics functions, 321
  conversion tools, 293, 302–3, 303, 305, 319
  coordinate system management, 295–96
  density functions, 324
  displaying geocoding results, 380
  distance functions, 305
  grid format, 294
  interpolation functions, 312–13, 325
  least-cost path functions, 326

597

neighborhood functions, 309–10, 321
Options settings, 295, 313–14, 318
Raster Calculator, 295, 303, 304, 305–6
Reclassify tool, 292, 302, 320
topographic functions, 306–8, 323
turning on and off, 301, 317
zonal functions, 310–11, 322
Spatial data
    defined, 1, 17
    map document handling of, 72–73
    queries based on, 194–96
    raster and vector models, 17–20, 285
Spatial detail, assessing needs, 7
Spatial joins
    cardinality, 223–24
    choosing type of, 228–29
    coordinate systems and distance joins, 227–28
    feature geometry and, 225–27
    inside joins, 222–23
    overview, 221–23
    setup examples, 229–30
    skills reference, 249
    teaching tutorial, 237–47
    types of, 224–25
Spatially continuous data, 285
Spatially discrete data, 285
Spatial queries, 192, 194–96
Spatial references, 398, 539. *See also* Coordinate
        systems
    elements of, 398
    feature class definition, 529
    in feature datasets, 504
    importing, 433
Spatial relationships, 21, 191
Spelling sensitivity, 374, 375, 384
Spheroids, 389–90
Spline Text tool, 140
Split policies, 506–7, 518–19, 522–25
Splitting lines, 492
Splitting polygons, 441, 443–44, 456, 488, 524–25
Split tool, 428, 492
SQL Query button, 512
Square and Finish command, 443
Square corners, 455
Standalone tables, 153
Standard deviation classification, 69
Standardization of addresses, 374, 385
Standard parallels, 391
State grids, 396
State maps, 397
State Plane Coordinate System (SPCS), 21, 388, 396
Statistics
    neighborhood, 290–91, 309–10, 321
    raster cells, 291, 321
    skills reference, 56
    zonal, 291–92, 310–11, 313, 322
Stereographic projections, 391
Storage systems for GIS data, 2
Straight Line distance function, 305

Stream digitizing, 470–74
Streets, editing, 474–76. *See also* Roads
Stretched method, 70
Structured Query Language (SQL), 192, 194,
        206. *See also* Queries
Style Manager, 43
Subdivisions (scale bar), 116
Subtypes
    ArcView limitations, 508, 509
    versus changing defaults, 515
    functions of, 508
Summarize command, 170
Summarized distance joins, 225
Summarized inside joins, 225
Summarized joins, 223, 225
    examples, 228, 229
    tutorial exercises, 239
Sum merge policies, 507
Surface analysis functions, 290, 306–8, 323
Swiss cheese polygons, 495
Switch Selection feature, 200
Symbology tab, 43
Symbols. *See also* Maps
    changing color of, 57
    changing properties of, 57
    choosing, 110–12
    creation of, 58
    line direction, 490
    network flow, 344–45, 352
    psychological aspects of, 110
    in quantities maps, 66
    setting, in ArcMap, 57
Syntax (command line), 272
Systematic sampling, 11

T

Table of Contents (ArcMap), 78
Tables. *See also* Attribute data; Attribute tables
    adding and removing, 179
    adding fields to, 434
    basic functions in ArcGIS, 27, 160–61
    cardinality of, 156
    changing appearance, 165–66, 180–81
    creating new, 189
    deleting, 189
    editing and calculating fields, 162
    exporting, 173, 184
    field types, 158–60
    formats, 162
    hiding fields or creating aliases, 180
    joining and relating, 155–57, 171–73, 188
    opening, 179
    overview, 153
    queries, 154–55, 166–67
    of shapefiles, 28
    skills reference, 56, 179
    from spatial joins (*see* Spatial joins)
    statistics, 167–70

summarizing, 157–58, 170–71
  in vector model, 18
Tabular data, 174–75
Tangent conic projections, 391
Tangent Curve tool, 488
Tangent tool, 465
Target layer, 429
Task bar, 427, 429, 450
Telephone companies, 1
Temperature data, 66
Template Feature Class field, 433
Templates, 149
  creating, 547–49
  importing, 549
  sample notes template, 541
Temporal accuracy, 537–38
Temporary grids, 296–97
Tentative matches, 374–75
TEST schema type, 282
Text. *See also* Labels
  adding to maps, 132, 143
  finding, 56
Thematic accuracy, 23
Thematic mapping, 2, 18
Thematic rasters, 70, 90–92. *See also* Grids
Thistle project case study, 8–11, 14–15
3x3 averaging window, 290–91
Thumbnails, 38, 55
TIGER data, 3
Titles, adding to maps, 132, 143
Tomlinson, Roger, 3
Toolbars
  customizing, 258
  Editor, 427–28
  Spatial Analyst, 301
  Utility Network Analyst, 333–34, 350
Tools, models as, 276–77
Topic selection, 12
Topographic functions, 306–8, 323–24
Topography, 389
Topological integrity, 425
Topological models, 21
Topological rules, 537
Topologies, basic editing, 461–62
Topology
  logical consistency as editing goal, 459–61
  modeling feature behavior with, 21–22
Topology Edit tool
  basic functions, 463, 466–67
  skills reference, 497
  tutorial exercises, 475–76
Topology errors in networks, 343–44
Top Ten Metadata Errors, 533
Tordon herbicide, 8, 14
Tournachon, Gaspard Felix, 3
Traces
  defined, 333
  generic solvers, 330–32
  skills reference, 350–51

tutorial exercises, 338–48
  utility network solvers, 332–33
Trace Task menu, Utility Network Analyst
    toolbar, 334
Trace tool
  basic functions, 464
  skills reference, 487, 495
Trace Upstream/Downstream solvers, 332,
    346–47, 355
Transects, defining, 11
Transformations, 390
Transparency, 92
Transportation feature, 333
Transportation networks, 328, 340, 343–44. *See
    also* Roads
Transverse cylindrical projections, 391
Transverse Mercator projection, 396
Trim Features task, 493
Try Horizontal First option, 524
Tutorial videos, 35
Tyrwhitt, Jacqueline, 3

U

Unclassed map, 66
Underscore character, 73
Uninitialized flow, 344
Union of Inputs option, 297
Union operations
  basic functions of, 251–55
  on features, 462, 494
  skills reference, 494
Unique values classification, 70
Unique values maps, 65, 227
Universal Transverse Mercator (UTM), 388
Universal Transverse Mercator (UTM)
    system, 21
Unmatch button, 374
Unselect option, 457
Unusual characters in pathnames, 73
Upstream tracing, 332, 346–47, 355
USA Contiguous Equidistant Conic, 397
Utility Network Analyst toolbar,
    333–34, 338, 350
Utility networks
  basic features of, 328–29
  solvers for, 332–33, 344–45, 355
UTM projection system, 395–96
UTM (Universal Transverse Mercator), 21

V

Validation, 521–22
Value field, 294
Variable search radii, 324
Vector data
  conversion to grids, 292, 319
  converting rasters to, 305, 310, 319
  model described, 17–19

raster model *versus,* 285
Vector data model, benefits and drawbacks of, 19
Venn diagrams, 193
Verify button, 199
Vertex menu, 429, 438–40, 451
Vertex snapping, 426, 437–38, 438, 441, 469
Vertices
    adding and removing from features, 489–90
    editing tools, 429, 438–40, 451, 474
    inserting at midpoint, 488
    placing at intersections, 485
    for reshaping, 490, 497
    in vector data model, 18
Videos, playing, 35
Viewshed analyses, 290, 308, 324
Visible scale range, 118
Volcano proximity spatial join example, 233–34

W

Warnings (program)
    coordinate system inconsistencies, 448
    datum inconsistencies, 407
Water mains, 513–15
Watershed analysis, 309–14
Website, ESRI, 26
Weighted average merge policies, 507
Weight filters, 352
Weights (network)
    defined, 329–30
    setting options, 351–52
    setting to distance, 330, 340, 354, 356–57
    specifying in network design, 335–36, 347,
        356–57
    specifying in solvers, 330
Weights (raster analysis), 326
WGS84, 389
Wildcard characters, 168, 194
Windows commands, 297
Windows in ArcMap, 81
Windows Paint program, 183
Wireless technology, 5
Wizards Build Geometric Network, 357
Working directory for grids, 296–97, 301
Work plans, 13, 15
Workspaces, 270–72
Workstation ArcInfo, 26
World file, 393
World Geodetic survey of 1984 (WGS84), 390
World Geodetic System (WGS 1984), 113
World maps, 397
Wrapped text tools, 141
Write-protection, 72

X

XML (Extensible Markup Language), 538
XOR operator, 193
x-y coordinates. *See also* Coordinate systems
    absolute, 452, 492
    basic functions, 387
    as end product of projection, 392
    storage in feature datasets, 504
    tables containing, 161
    use in distance joins, 228
x-y data
    adding to maps, 380–81, 386
    effects of projection tools on, 400
    in raster and vector models, 17, 19–20
X/Y domain, 398, 504
XY Tolerance, 535

Z

Zonal functions, 310–11
Zonal statistics function, 291–92, 310, 322
Zone layer, 291
Zones
    grid attribute tables, 293, 310–11
    UTM system, 395–96
Zoom/Pan tools, 99
Z-scores, 10

Alphabetical Skills Index

A

Adding a legend.............................. 144
Adding a north arrow....................... 143
Adding a scale bar 147
Adding and clearing flags and barriers........ 353
Adding data 96
Adding graphics to layouts................. 143
Adding neatlines, backgrounds, and shadows148
Adding or deleting fields................... 185
Adding or removing a table 179
Adding pictures 148
Adding titles and text...................... 143
Adding *x-y* data 386
Appending layers........................... 282

B

Beginning an editing session 448
Buffering features.......................... 496
Building a simple network................. 356

C

Calculating a viewshed.................... 324
Calculating cell statistics 321
Calculating fields 187
Calculating neighborhood statistics........ 321
Calculating zonal statistics 322
Changing the appearance of a legend........ 144
Changing the results format................ 351
Changing the selection options............ 219
Classifying attributes for a map 104
Clearing a selection 216
Clipping a layer 282
Clipping features 494
Clipping to a layer 101
Composing the data frames 136
Connecting and disconnecting from folders .. 53
Connecting to an Internet service 53
Controlling display of layers 97
Converting between grids and features........ 319
Converting dynamic labels to annotation 142
Creating a density map 324
Creating a feature class.................... 529
Creating a feature dataset 528
Creating a geodatabase 527
Creating a layer from selected features 219
Creating a map based on an attribute........... 103
Creating a map from a template 149
Creating a map layer or map group layer 106
Creating a metadata template 555
Creating a new table 189
Creating a simple graph................... 151
Creating adjacent polygons 456
Creating an address locator 383
Creating and viewing thumbnails 55

Creating attribute domains........................... 531

Creating buffers 281
Creating donut polygons................... 495
Creating features with square corners.......... 455
Creating joins and relates 188
Creating multipart features 455
Creating new features 451
Creating new symbols 58

D

Defining coordinate systems of data sets 421
Deleting a geodatabase 527
Deleting a table........................... 189
Deleting feature classes 527
Deleting feature datasets.................. 527
Deleting features 451
Dissolving................................ 283
Dividing a line 491
Dragging and dropping files to ArcMap 96

E

Editing annotation........................ 498
Editing fields in a table................... 187
Editing metadata.......................... 554
Editing shared features 496
Examining network properties............. 349
Examining the coordinate system 54
Examining the coordinate system......... 101
Exporting a map as a picture file 150
Exporting a table 184
Extending lines 493

F

Finding a least-cost path 326
Finding a path 354
Finding connections and loops 355
Finding features by searching attributes 100
Finding the shortest path.................. 354
Flipping a line............................ 490

G

Geocoding addresses 384
Getting statistics for a field 183
Grouping layers for display 98

I

Identifying features 55
Importing feature classes 530
Importing or exporting metadata 555
Interactive matching 385
Interpolating between points............... 325
Intersection of features 494

L

Labeling features interactively 139
Labeling features using dynamic labels 59

M

Managing ArcCatalog files 60
Managing grids 317
Managing toolbars 96
Measuring features 100
Merging features 493
Mirroring features 491
Modifying a feature 489
Modifying a table's appearance 180
Modifying the appearance of the legend 105
Moving a feature an exact distance 492
Moving features 450

O

Opening a table 179
Opening an ArcMap document 97
Opening Excel Data 190

P

Performing a spatial join 249
Performing a trace 350
Performing an intersection 280
Previewing a table 56
Previewing layer geography 55
Printing a map 150
Projecting feature classes 422
Projecting rasters 423

R

Reclassifying a grid 320
Rematching a geocoding result 384
Reshaping a feature 490
Rotating features 450

S

Saving an ArcMap document 97
Saving changes during an edit session 458
Searching for data 60
Selecting by attributes 217
Selecting by location 218
Selecting features for editing 449
Selecting features interactively 215
Selecting features using shapes 215
Selecting records using attributes 182
Setting data frame and layer properties 101
Setting default values or domains 532
Setting general tracing options 350
Setting Map Tips 98
Setting options 53
Setting snapping options 449
Setting symbols for a layer 57
Setting the analysis environment 318
Setting the data frame coordinate system 102
Setting the reference scale 141
Setting the scale or extent 139
Setting the selectable layers 214

Setting up the map page 137
Splitting lines exactly 492
Starting ArcMap or ArcCatalog 52
Starting ArcMap 95
Starting ArcToolbox 52
Stopping an edit session 458
Summarizing on a field 184
Switching to another folder or geodatabase .. 458

T

Tracing upstream/downstream 355
Tracing with accumulation 355
Trimming lines 493
Turning on Spatial Analyst 317

U

Union of features 494
Using Copy Parallel 492
Using sketch context menus 452
Using surface functions 323
Using the Attributes window 457
Using the Layout toolbar 138
Using the Selection tab 216
Using the sketching tools 485
Using the Zoom/Pan tools and bookmarks 99
Using tools in ArcToolbox 284
Using weights when tracing 351

V

Viewing and setting layer properties 54
Viewing flow directions 352
Viewing metadata 57
Viewing the contents of a folder 54